Lecture Notes in Applied and Computational Mechanics

Volume 55

Series Editors
Prof. Dr.-Ing. Friedrich Pfeiffer
Prof. Dr.-Ing. Peter Wriggers

For other titles published in this series, go to
www.springer.com/series/4623

René de Borst • Ekkehard Ramm
Editors

Multiscale Methods in Computational Mechanics

Progress and Accomplishments

Springer

Editors
René de Borst
Department of Mechanical Engineering
Eindhoven University of Technology
Eindhoven
The Netherlands
R.d.Borst@tue.nl

Ekkehard Ramm
Institute of Structural Mechanics
University of Stuttgart
Stuttgart
Germany

ISSN 1613-7736 e-ISSN 1860-0816
ISBN 978-90-481-9808-5 e-ISBN 978-90-481-9809-2
DOI 10.1007/978-90-481-9809-2
Springer Dordrecht Heidelberg London New York

Library of Congress Control Number: 2010935928

Cover design: eStudio Calamar S.L.

Printed on acid-free paper

Springer is part of Springer Science+Business Media (www.springer.com)

Table of Contents

Part 4: Coupled Problems and Porous Media

Preface

For a long time limited computational resources restricted the scale of observation and modeling of physical systems mostly to one scale in time and space. This modus operandi was accepted although it was well known that the response at the level of practical interest is to a large extent determined by processes that occur at scales which are one to several orders of a magnitude smaller, namely the meso, micro or even nanoscales. The rapid increase in computer power and the development of efficient computational methods allow coming closer to the human dream of a continuous model through all spatial and temporal scales. However seeking solutions using numerical models simultaneously at the various length scales has proven to be beyond current computational capabilities. This perception is the starting point for the development of Multiscale Methods which is currently one of the hot research topics all over the world. Multiscale Methods either derive properties at the level of observation by repeated numerical homogenization of more fundamental physical properties several scales below (upscaling), or they devise concurrent schemes where those parts of the domain that are of interest are computed with a higher resolution than parts that are of less interest or where the solution is varying only slowly.

The present volume contains selected papers presented at the International Colloquium on Multiscale Methods in Computational Mechanics in Rolduc, the Netherlands, on 11–13 March 2009 (MMCM 2009). The contributions in the first part address multiscale methods in Computational Fluid Dynamics; in particular turbulence modeling applying the Variational Multiscale Method is described. The second part deals with materials having a distinct microstructure such as single- and poly-crystals, granular materials, aggregates embedded in a matrix, micro-laminates and nano-composites. Part three focuses on materials and structures and the interaction between their behavior on micro/meso- and macro-scales. Failure of composites and laminates, material modeling of shotcrete in tunnels, buckling of paperboard on different scales as well as material and structural optimization are investigated. The fourth part concentrates on coupled problems and porous media like concrete and charged hydrated materials.

The mentioned colloquium was organized by the German–Dutch Research Unit "Multiscale Methods in Computational Mechanics", in which a group of scientists in Germany at Universität Stuttgart and TU München and in the Netherlands at TU

Delft and TU Eindhoven have joined forces in order to address a variety of problem areas and computational solution strategies for multiscale methods in mechanics of structures, materials and flows. This joint project, which serves as a prototype for research cooperations within Europe, was financially supported by the German Research Foundation (Deutsche Forschungsgemeinschaft, DFG FOG 509), The Netherlands Technology Foundation (STW) and The Netherlands Organization for Scientific Research (NWO). This support is very much appreciated.

We also would like to thank the authors very much for their valuable contributions. Furthermore our thanks go to Springer for taking over the publication of this volume in its LNACM series.

January 2010
René de Borst, Technische Universiteit Eindhoven
Ekkehard Ramm, Universität Stuttgart

List of Authors

Acartürk, Ayhan Y.
Universität Stuttgart, Institut für Mechanik (Bauwesen), Stuttgart, Germany
acar@mechbau.uni-stuttgart.de

Akkerman, Ido
University of California, San Diego, Department of Structural Engineering, USA
IAkkerman@ucsd.edu

Allix, Olivier
L.M.T., ENS de Cachan, France
allix@lmt.ens-cachan.fr

Aslan, Ozgur
Mines ParisTech, Centre des Matériaux, Evry, France
ozgur.aslan@ensmp.fr

Badia, Santiago
Universitat Politècnica de Catalunya, International Center for Numerical Methods in Engineering (CIMNE), Barcelona, Spain
sbadia@cimne.upc.edu

Bazilevs, Yuri
University of California, San Diego, Department of Structural Engineering, USA
yuri@ucsd.edu

de Borst, René
Eindhoven University of Technology, Department of Mechanical Engineering, Eindhoven, The Netherlands
R.d.Borst@tue.nl

Bruss, Ingrid
Universität Stuttgart, Institut für Baustatik und Baudynamik, Stuttgart, Germany
bruss@ibb.uni-stuttgart.de

Calo, Victor M.
KAUST, Earth and Environmental Science and Engineering, Saudi Arabia
victor.calo@kaust.edu.sa

Cid Alfaro, Marcela V.
Corus, IJmuiden, The Netherlands

Codina, Ramon
Universitat Politècnica de Catalunya, Departament de Resistencia de Materials i Estructures a l'Enginyeria, Barcelona, Spain
ramon.codina@upc.edu

Daxner, Thomas
CAE Simulations & Solutions, Wien, Austria
daxner@ilsb.tuwien.ac.at

Dettmar, Joachim
Universität Stuttgart, Institut für Mechanik (Bauwesen), Stuttgart, Germany

Dormieux, Luc
École Nationale des Ponts et Chaussées, Laboratoire des Matériaux et Structures du Génie Civil (LMSGC), Champs-sur-Marne, France
dormieux@lmsgc.enpc.fr

Doweidar, Mohamed H.
Centro PolitĆcnico Superior de Zaragoza, Área de Mecánica de Fluidos, Zaragoza, Spain

Ehlers, Wolfgang
Universität Stuttgart, Institut für Mechanik (Bauwesen), Stuttgart, Germany
ehlers@mechbau.uni-stuttgart.de

Erhart, Andrea
DYNAmore, Stuttgart, Germany
andrea.erhart@dynamore.de

Fish, Jacob
Rensselaer Polytechnic Institute, Multiscale Science and Engineering Center, Troy, NY, USA
fishj@rpi.edu

Flatscher, Thomas
Technische Universität Wien, Institut für Leichtbau und Struktur-Biomechanik, Wien, Austria
flatscher@ilsb.tuwien.ac.at

Forest, Samuel
Mines ParisTech, Centre des Matériaux, Evry, France
samuel.forest@ensmp.fr

Fuster, Daniel
Centro Politécnico Superior de Zaragoza, Área de Mecánica de Fluidos, Zaragoza, Spain

Gamnitzer, Peter
Technische Universität München, Lehrstuhl für Numerische Mechanik, München, Germany
gamnitzer@lnm.mw.tum.de

Gawin, Dariusz
TU Lodz, Department of Building Physics and Building Materials, Lodz, Poland
gawindar@p.lodz.pl

Geurts, Bernard J.
University of Twente, Mathematical Sciences, Enschede; Eindhoven University of Technology, Department of Applied Physics, Eindhoven, The Netherlands
b.j.geurts@utwente.nl

Gosselet, Pierre
L.M.T., ENS de Cachan, France
gosselet@lmt.ens-cachan.fr

Gravemeier, Volker
Technische Universität München, Lehrstuhl für Numerische Mechanik, München, Germany
vgravem@lnm.mw.tum.de

Hackl, Klaus
Ruhr-Universität Bochum, Lehrstuhl für Allgemeine Mechanik, Bochum, Germany
klaus.hackl@rub.de

Hauke, Guillermo
Centro Politécnico Superior de Zaragoza, Área de Mecánica de Fluidos, Zaragoza, Spain
ghauke@unizar.es

Hellmich, Christian
Technische Universität Wien, Institut für Mechanik der Werkstoffe und Strukturen, Wien, Austria
christian.hellmich@tuwien.ac.at

Hettich, Thomas M.
Mahle GmbH, Stuttgart, Germany

Hilchenbach, Frédéric
Universität Stuttgart, Institut für Baustatik und Baudynamik, Stuttgart, Germany
hilchenbach@ibb.uni-stuttgart.de

Hughes, Thomas J.R.
The University of Texas at Austin, Institute for Computational Engineering and Sciences, Austin, USA
hughes@ices.utexas.edu

Hulshoff, Steven J.
Delft University of Technology, Aerospace Engineering, Delft, The Netherlands
s.j.hulshoff@tudelft.nl

Huyghe, Jacques M.R.
Eindhoven University of Technology, Department of Biomedical Engineering, Eindhoven, The Netherlands
j.m.r.huyghe@tue.nl

Jeannin, Laurent
Gaz de France (GDF), France

Kato, Junji
Universität Stuttgart, Institut für Baustatik und Baudynamik, Stuttgart, Germany
kato@ibb.uni-stuttgart.de

Kerfriden, Pierre
L.M.T., ENS de Cachan, France
kerfriden@lmt.ens-cachan.fr

Kochmann, Dennis M.
Ruhr-Universität Bochum, Lehrstuhl für Allgemeine Mechanik, Bochum, Germany
dennis.kochmann@rub.de

Kraaijeveld, Famke
Eindhoven University of Technology, Eindhoven, The Netherlands
f.kraaijeveld@tue.nl

Lee, Pilhwa
University of Colorado at Boulder, Department of Mechanical Engineering, Boulder, USA
pilhwa.lee@colorado.edu

Mang, Herbert
Technische Universität Wien, Institut für Mechanik der Werkstoffe und Strukturen, Wien, Austria
herbert.mang@tuwien.ac.at

Markert, Bernd
Universität Stuttgart, Institut für Mechanik (Bauwesen), Stuttgart, Germany
markert@mechbau.uni-stuttgart.de

Maute, Kurt
University of Colorado at Boulder, Department of Aerospace Engineering Sciences, Boulder, USA
maute@colorado.edu

Miehe, Christian
Universität Stuttgart, Institut für Mechanik (Bauwesen), Stuttgart, Germany
cm@mechbau.uni-stuttgart.de

Pahr, Dieter H.
Technische Universität Wien, Institut für Leichtbau und Struktur-Biomechanik, Wien, Austria
pahr@ilsb.tuwien.ac.at

Pesavento, Francesco
Università degli Studi di Padova, Dipartimento di Costruzioni e Trasporti, Padova, Italy
francesco.pesavento@unipd.it

Pichler, Bernhard
Technische Universität Wien, Institut für Mechanik der Werkstoffe und Strukturen, Wien, Austria
bernhard.pichler@tuwien.ac.at

Principe, Javier
Universitat Politécnica de Catalunya, International Center for Numerical Methods in Engineering (CIMNE), Barcelona, Spain
principe@cimne.upc.edu

Ramm, Ekkehard
Universität Stuttgart, Institut für Baustatik und Baudynamik, Stuttgart, Germany
ramm@ibb.uni-stuttgart.de

Rammerstorfer, Franz G.
Technische Universität Wien, Institut für Leichtbau und Struktur-Biomechanik, Wien, Austria
ra@ilsb.tuwien.ac.at

Sanahuja, Julien
Electricité de France (EDF), Department of Materials and Mechanics of Components, Moret-sur-Loing, France
julien.sanahuja@edf.fr

Schrefler, Bernhard A.
Università degli Studi di Padova, Dipartimento di Costruzioni e Trasporti, Padova, Italy
bernhard.schrefler@unipd.it

Scheiner, Stefan
The University of Western Australia, Computer Science and Software Engineering, Crawley, Perth, Western Australia
stefan.scheiner@csse.uwa.edu.au

Scovazzi, Guglielmo
Sandia National Laboratories, Albuquerque, USA
gscovaz@sandia.gov

Suiker, Akke S.J.
Delft University of Technology, Aerospace Engineering, Delft, The Netherlands
a.s.j.suiker@tudelft.nl

van der Zee, Kris
The University of Texas at Austin, Institute for Computational Engineering and Sciences, Austin, USA

van Keulen, Fred
Delft University of Technology, Department of Precision and Microsystems Engineering, Delft, The Netherlands
a.vankeulen@tudelft.nl

Wall, Wolfgang A.
Technische Universität München, Lehrstuhl für Numerische Mechanik, München, Germany
wall@lnm.mw.tum.de

de Wit, Albert J.
Dutch National Aerospace Laboratory (NLR), Amsterdam, The Netherlands
adewit@nlr.nl

Yang, Ronggui
University of Colorado at Boulder, Department of Mechanical Engineering, Boulder, USA
ronggui.yang@colorado.edu

Zäh, Dominic
Universität Stuttgart, Institut für Mechanik (Bauwesen), Stuttgart, Germany
zaeh@mechbau.uni-stuttgart.de

PART 1

Computational Fluid Dynamics

Residual-Based Variational Multiscale Theory of LES Turbulence Modeling*

Y. Bazilevs[1], V.M. Calo[2], T.J.R. Hughes[3] and G. Scovazzi[4]

[1]Department of Structural Engineering, UC San Diego, La Jolla, CA 92093-0085, USA; yuri@ucsd.edu
[2]Earth and Environmental Science and Engineering, KAUST, Saudi Arabia; victor.calo@kaust.edu.sa
[3]Institute for Computational Engineering and Sciences, UT Austin, Austin, TX 78712, USA; hughes@ices.utexas.edu
[4]1431 Computational Shock and Multi-Physics Department, Sandia National Labs, Albuquerque, NM 87185-1319, USA; gscovaz@sandia.gov

Abstract We present an aperçu of our variational multiscale theory of LES turbulence. The theory is succinctly summarized in terms of a finite-dimensional coarse-scale equation governing the resolved scales that depend parametrically upon unresolved fine scales, which in turn are defined in terms of a functional of the coarse-scale residual "lifted" to the dual of the fine-scale space, and the coarse-scale velocity field itself. We illustrate the performance of a numerical implementation of the theory with calculations of a turbulent channel flow at a friction-velocity Reynolds number of 395 and comparisons with the DNS data.

Key words: B-splines, isogeometric analysis, large eddy simulation, turbulence modeling, variational multiscale method, wall-bounded turbulence

1 Variational Multiscale Formulation of the Incompressible Navier–Stokes Equations

1.1 Incompressible Navier–Stokes Equations

We consider a space-time domain $\mathsf{Q} = \Omega \times]0, T[\subset \mathbb{R}^3 \times \mathbb{R}^+$ with lateral boundary $\mathsf{P} = \Gamma \times]0, T[$, as illustrated in the left-hand side of Figure 1. The initial/boundary-value problem consists of solving the following equations for $\boldsymbol{u} : \overline{\mathsf{Q}} \to \mathbb{R}^3$, the velocity, and $p : \mathsf{Q} \to \mathbb{R}$, the pressure (divided by the constant density),

*This is an abridged version of Bazilevs et al. [1], which may be consulted for additional implementational details and numerical results.

R. de Borst and E. Ramm (eds.), *Multiscale Methods in Computational Mechanics*, Lecture Notes in Applied and Computational Mechanics 55, DOI 10.1007/978-90-481-9809-2_1,

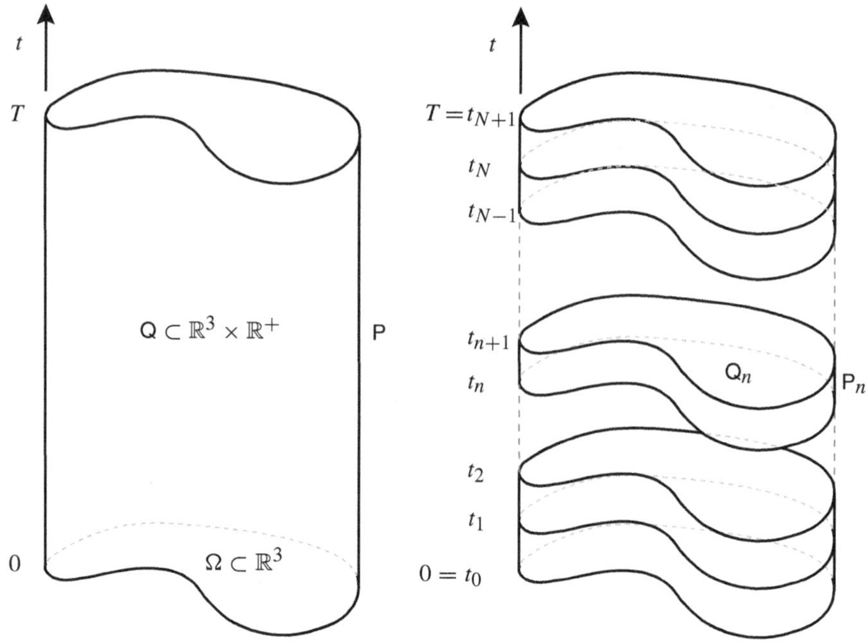

Fig. 1. Space-time domain (left) and slicing into space-time slabs (right).

$$\frac{\partial \boldsymbol{u}}{\partial t} + \nabla \cdot (\boldsymbol{u} \otimes \boldsymbol{u}) + \nabla p = \nu \Delta \boldsymbol{u} + \boldsymbol{f} \qquad \text{in Q} \tag{1}$$

$$\nabla \cdot \boldsymbol{u} = 0 \qquad \text{in Q} \tag{2}$$

$$\boldsymbol{u} = \boldsymbol{0} \qquad \text{on P} \tag{3}$$

$$\boldsymbol{u}(0^+) = \boldsymbol{u}(0^-) \qquad \text{on } \Omega \tag{4}$$

where $\boldsymbol{f} : \mathsf{Q} \to \mathbb{R}^d$ is the given body force (per unit volume); ν is the kinematic viscosity, assumed positive and constant; $\boldsymbol{u}(0^-) : \Omega \to \mathbb{R}^d$ is the given initial velocity; and $\otimes$ denotes the tensor product (e.g., in component notation, $[\boldsymbol{u} \otimes \boldsymbol{v}]_{ij} = u_i v_j$).

Equations (1)–(4) are, respectively, the linear momentum balance, the incompressibility constraint, the no-slip boundary condition and the initial condition.

Global Space-Time Variational Formulation

Let $\mathcal{V} = \mathcal{V}(\mathsf{Q})$ denote both the trial solution and weighting function spaces, which are assumed to be identical. We assume $\boldsymbol{U} = \{\boldsymbol{u}, p\} \in \mathcal{V}$ implies $\boldsymbol{u} = \boldsymbol{0}$ on P and $\int_\Omega p(t)\, d\Omega = 0$ for all $t \in \,]0, T[$. Let $(\cdot, \cdot)_\omega$ denote the L^2 inner product with respect to the domain ω. The variational formulation is stated as follows:

Find $\boldsymbol{U} \in \mathcal{V}$ such that $\forall \boldsymbol{W} = \{\boldsymbol{w}, q\} \in \mathcal{V}$:

$$B(\boldsymbol{W}, \boldsymbol{U}) = B_1(\boldsymbol{W}, \boldsymbol{U}) + B_2(\boldsymbol{W}, \boldsymbol{U}, \boldsymbol{U}) = L(\boldsymbol{W}) \tag{5}$$

with

$$B_1(W,U) = (w(T^-),u(T^-))_\Omega - \left(\frac{\partial w}{\partial t}, u\right)_{\mathsf{Q}} + (q,\nabla\cdot u)_{\mathsf{Q}} - (\nabla\cdot w, p)_{\mathsf{Q}} + \left(\nabla^s w, 2\nu\nabla^s u\right)_{\mathsf{Q}} \tag{6}$$

$$B_2(W,U,V) = -\ (\nabla w, u\otimes v)_{\mathsf{Q}} \tag{7}$$

$$L(W) = (w, f)_{\mathsf{Q}} + (w(0^+), u(0^-))_\Omega \tag{8}$$

where $V = \{v,\cdot\}$ and $\nabla^s u = \left(\nabla u + (\nabla u)^T\right)/2$. Note that $B_1(\cdot,\cdot)$ is a bilinear form and $B_2(\cdot,\cdot,\cdot)$ is a trilinear form. Assuming sufficient regularity and integrating by parts, we obtain the Euler–Lagrange form of (5)-(8):

$$0 = \left(w, \frac{\partial u}{\partial t} + \nabla\cdot(u\otimes u) + \nabla p - \nabla\cdot 2\nu\nabla^s u - f\right)_{\mathsf{Q}} + (q,\nabla\cdot u)_{\mathsf{Q}} + (w(0^+), u(0^+) - u(0^-))_\Omega \tag{9}$$

which reveals that the variational formulation implies satisfaction of the momentum equations, incompressibility constraint, and initial condition. The velocity boundary condition is built into the definition of the space $\mathcal{V}$. In summary, the variational formulation is equivalent to (1)–(4).

Sliced Space-Time Variational Formulation

Consider a *slicing* of space-time obtained by replacing $]0,T[$ by $]t_n, t_{n+1}[$, $n = 0,1,2,\ldots,N$, and summing over the space-time *slabs* Q_n (see Figure 1). The counterparts of (5)–(9) for a typical slab are:

$$B(W,U)_n = B_1(W,U)_n + B_2(W,U,U)_n = L(W)_n \tag{10}$$

$$B_1(W,U)_n = (w(t_{n+1}^-), u(t_{n+1}^-))_\Omega - \left(\frac{\partial w}{\partial t}, u\right)_{\mathsf{Q}_n} + (q,\nabla\cdot u)_{\mathsf{Q}_n} - (\nabla\cdot w, p)_{\mathsf{Q}_n} + \left(\nabla^s w, 2\nu\nabla^s u\right)_{\mathsf{Q}_n} \tag{11}$$

$$B_2(W,U,V)_n = -\ (\nabla w, u\otimes v)_{\mathsf{Q}_n} \tag{12}$$

$$L(W)_n = (w, f)_{\mathsf{Q}_n} + (w(t_n^+), u(t_n^-))_\Omega \tag{13}$$

$$0 = \left(w, \frac{\partial u}{\partial t} + \nabla\cdot(u\otimes u) + \nabla p - \nabla\cdot 2\nu\nabla^s u - f\right)_{\mathsf{Q}_n} + (q,\nabla\cdot u)_{\mathsf{Q}_n} + (w(t_n^+), u(t_n^+) - u(t_n^-))_\Omega \tag{14}$$

where, in (10)–(14), $U = \{u,p\}$ and $W = \{w,q\}$ belong to $\mathcal{V}_n = \mathcal{V}(\mathsf{Q}_n)$, the restriction of $\mathcal{V}$ to Q_n. From the Euler–Lagrange form of the equation, (14), we see that the momentum equation and incompressibility constraint are satisfied on the slab, and the solution is continuous across slab interfaces. The formulation in terms of space-time slabs exploits the causal nature of the Navier–Stokes equations and reduces the overall problem to a succession of initial/boundary-value problems on the slabs. The solution is obtained solving the variational equation on each slab successively, $n = 0, 1, 2, \ldots, N$. We emphasize that this is an *exact* formulation, entirely equivalent to (5)–(9), and (1)–(4). However, it is a more suitable starting point for the development of numerical schemes.

Remark
In order to simplify notation, we will work with the global form of the variational equation subsequently. However, all results are equally applicable to the variational equations of the individual space-time slabs.

1.2 Scale Separation

We consider a direct-sum decomposition of $\mathcal{V}$ into "coarse-scale" and "fine-scale" subspaces, $\overline{\mathcal{V}}$ and $\mathcal{V}'$, respectively,

$$\mathcal{V} = \overline{\mathcal{V}} \oplus \mathcal{V}' \tag{15}$$

$\overline{\mathcal{V}}$ is assumed to be a finite-dimensional space and it will be identified later with the space of functions with which we actually compute. In order to make the decomposition well-defined, we need to introduce a procedure for uniquely determining $\overline{\boldsymbol{U}} \in \overline{\mathcal{V}}$ and $\boldsymbol{U}' \in \mathcal{V}'$ from a given $\boldsymbol{U} \in \mathcal{V}$. This can be accomplished with the aid of a projector $\overline{\mathbb{P}} : \mathcal{V} \rightarrow \overline{\mathcal{V}}$. For example, $\overline{\mathbb{P}}$ could be the L^2-projector, H^1-projector, etc. There are infinitely many possibilities.[1] Once $\overline{\mathbb{P}}$ is selected, we know how the coarse scales approximate all scales, viz.,

$$\overline{\boldsymbol{U}} = \overline{\mathbb{P}}\boldsymbol{U} \tag{16}$$

$$\boldsymbol{U}' = \boldsymbol{U} - \overline{\mathbb{P}}\boldsymbol{U} = (\mathbb{I} - \overline{\mathbb{P}})\boldsymbol{U} \tag{17}$$

where $\mathbb{I}$ is the identity operator. Likewise, we can decompose a weighting function into its coarse- and fine-scale components:

$$\overline{\boldsymbol{W}} = \overline{\mathbb{P}}\boldsymbol{W} \tag{18}$$

$$\boldsymbol{W}' = \boldsymbol{W} - \overline{\mathbb{P}}\boldsymbol{W} = (\mathbb{I} - \overline{\mathbb{P}})\boldsymbol{W} \tag{19}$$

With these, we may decompose the original variational equation into coupled coarse-scale and fine-scale equations, viz.,

[1] The way $\overline{\boldsymbol{U}}$ is determined from $\boldsymbol{U}$ is a very important issue, and it has very significant impact on the theory to be developed. An initiatory study of typical projectors is presented in Hughes and Sangalli [8]. Not only can one envision an infinite number of possible projectors, but one can also envision an infinite number of nonlinear optimization schemes that "fit" $\overline{\boldsymbol{U}}$ to $\boldsymbol{U}$. In some applications nonlinear schemes will surely be important, an example being compressible turbulence with shocks where monotonicity is important. However, for incompressible turbulence, we feel linear projectors, such as the H^1-projector, should suffice.

$$B(\overline{\boldsymbol{W}}, \overline{\boldsymbol{U}} + \boldsymbol{U}') = L(\overline{\boldsymbol{W}}) \tag{20}$$

$$B(\boldsymbol{W}', \overline{\boldsymbol{U}} + \boldsymbol{U}') = L(\boldsymbol{W}') \tag{21}$$

where

$$\begin{aligned} B(\overline{\boldsymbol{W}}, \overline{\boldsymbol{U}} + \boldsymbol{U}') = {} & B_1(\overline{\boldsymbol{W}}, \overline{\boldsymbol{U}}) + B_1(\overline{\boldsymbol{W}}, \boldsymbol{U}') \\ & + B_2(\overline{\boldsymbol{W}}, \overline{\boldsymbol{U}}, \overline{\boldsymbol{U}}) \\ & + B_2(\overline{\boldsymbol{W}}, \overline{\boldsymbol{U}}, \boldsymbol{U}') + B_2(\overline{\boldsymbol{W}}, \boldsymbol{U}', \overline{\boldsymbol{U}}) \\ & + B_2(\overline{\boldsymbol{W}}, \boldsymbol{U}', \boldsymbol{U}') \end{aligned} \tag{22}$$

$$\begin{aligned} B(\boldsymbol{W}', \overline{\boldsymbol{U}} + \boldsymbol{U}') = {} & B_1(\boldsymbol{W}', \overline{\boldsymbol{U}}) + B_1(\boldsymbol{W}', \boldsymbol{U}') \\ & + B_2(\boldsymbol{W}', \overline{\boldsymbol{U}}, \overline{\boldsymbol{U}}) \\ & + B_2(\boldsymbol{W}', \overline{\boldsymbol{U}}, \boldsymbol{U}') + B_2(\boldsymbol{W}', \boldsymbol{U}', \overline{\boldsymbol{U}}) \\ & + B_2(\boldsymbol{W}', \boldsymbol{U}', \boldsymbol{U}') \end{aligned} \tag{23}$$

In (22), $B_2(\overline{\boldsymbol{W}}, \overline{\boldsymbol{U}}, \boldsymbol{U}')$ and $B_2(\overline{\boldsymbol{W}}, \boldsymbol{U}', \overline{\boldsymbol{U}})$ correspond to the cross-stress terms, and $B_2(\overline{\boldsymbol{W}}, \boldsymbol{U}', \boldsymbol{U}')$ corresponds to the Reynolds stress term. Equation (21) can be expressed as

$$B_{\overline{U}}(\boldsymbol{W}', \boldsymbol{U}') + B_2(\boldsymbol{W}', \boldsymbol{U}', \boldsymbol{U}') = \langle \boldsymbol{W}', \textbf{Res}(\overline{\boldsymbol{U}}) \rangle_{\mathcal{V}', \mathcal{V}'^*} \tag{24}$$

where

$$\begin{aligned} B_{\overline{U}}(\boldsymbol{W}', \boldsymbol{U}') = {} & B_1(\boldsymbol{W}', \boldsymbol{U}') \\ & + B_2(\boldsymbol{W}', \boldsymbol{U}', \overline{\boldsymbol{U}}) + B_2(\boldsymbol{W}', \overline{\boldsymbol{U}}, \boldsymbol{U}') \end{aligned} \tag{25}$$

$$\langle \boldsymbol{W}', \textbf{Res}(\overline{\boldsymbol{U}}) \rangle_{\mathcal{V}', \mathcal{V}'^*} = L(\boldsymbol{W}') - B_1(\boldsymbol{W}', \overline{\boldsymbol{U}}) - B_2(\boldsymbol{W}', \overline{\boldsymbol{U}}, \overline{\boldsymbol{U}}) \tag{26}$$

in which $\textbf{Res}(\overline{\boldsymbol{U}})$ is the coarse-scale residual "lifted" to the dual of the fine-scale space $\mathcal{V}'^*$, $\langle \cdot, \cdot \rangle_{\mathcal{V}', \mathcal{V}'^*}$ is the duality pairing, and

$$B_{\overline{U}}(\cdot, \boldsymbol{U}') = \left(\frac{d}{d\varepsilon} B(\cdot, \overline{\boldsymbol{U}} + \varepsilon \boldsymbol{U}') \right)_{\varepsilon=0} \tag{27}$$

the linearization of $B(\cdot, \overline{\boldsymbol{U}} + \boldsymbol{U}')$ about $\overline{\boldsymbol{U}}$ in the direction $\boldsymbol{U}'$. Note that the solution of (24) can be formally represented as a functional of $\overline{\boldsymbol{U}}$ and $\textbf{Res}(\overline{\boldsymbol{U}})$, namely,

$$\boxed{\boldsymbol{U}' = \boldsymbol{F}'(\overline{\boldsymbol{U}}, \textbf{Res}(\overline{\boldsymbol{U}}))} \tag{28}$$

The explicit dependence on $\overline{\boldsymbol{U}}$ in the first argument of $\boldsymbol{F}'$ emanates from the dependence of the linearized operator $B_{\overline{U}}$ on $\overline{\boldsymbol{U}}$. This expression can be inserted into (20) to "close" the finite-dimensional system for $\overline{\boldsymbol{U}}$,

$$\boxed{B\left(\overline{\boldsymbol{W}}, \overline{\boldsymbol{U}} + \boldsymbol{F}'(\overline{\boldsymbol{U}}, \textbf{Res}(\overline{\boldsymbol{U}}))\right) = L(\overline{\boldsymbol{W}})} \tag{29}$$

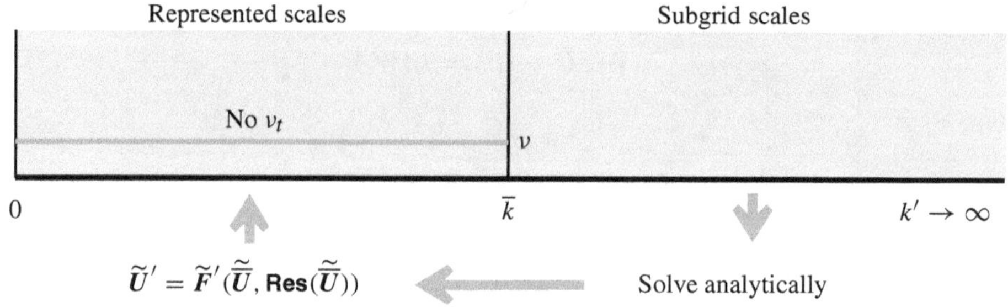

Fig. 2. The variational multiscale turbulence modeling theory is schematically illustrated. The fine, or "subgrid" scales are solved for analytically and substituted into the coarse-scale equation. The coarse scales are the represented scales in a calculation. Note that there is no *ad hoc* eddy viscosity model introduced.

(28) and (29) can be thought of in global terms or in terms of a sequence of space-time slabs. In both cases, they represent a procedure for solving the Navier–Stokes equations in terms of a scale decomposition of the solution. So far we have not discussed approximations or numerics. The solution $\boldsymbol{U} = \overline{\boldsymbol{U}} + \boldsymbol{U}'$, where $\overline{\boldsymbol{U}}$ is determined by solving (29) and $\boldsymbol{U}'$ is determined from $\overline{\boldsymbol{U}}$ through (28), is the *exact* solution of the original variational problem, (20)-(21), and (1)-(4), the Navier–Stokes initial/boundary-value problem.

Our plan for turbulence modeling is to systematically *approximate* the functional $\boldsymbol{F}'$. This will provide us with a parameterization of the fine scales in terms of the coarse scales, which can be substituted in the coarse-scale equation, "closing" it. The finite-dimensional coarse-scale equation can then be solved. In this way we obtain an approximate coarse-scale solution and an estimation of the fine scales. In summary, our variational multiscale theory of turbulence modeling is encapsulated in the following equations:

$$\boxed{\widetilde{\boldsymbol{U}}' = \widetilde{\boldsymbol{F}}'(\widetilde{\overline{\boldsymbol{U}}}, \mathbf{Res}(\widetilde{\overline{\boldsymbol{U}}}))} \tag{30}$$

$$\boxed{B(\overline{\boldsymbol{W}}, \widetilde{\overline{\boldsymbol{U}}} + \widetilde{\boldsymbol{F}}'(\widetilde{\overline{\boldsymbol{U}}}, \mathbf{Res}(\widetilde{\overline{\boldsymbol{U}}})) = L(\overline{\boldsymbol{W}})} \tag{31}$$

where $\widetilde{\boldsymbol{F}}'$ is an approximation of the exact functional $\boldsymbol{F}'$, and $\widetilde{\boldsymbol{U}}'$ and $\widetilde{\overline{\boldsymbol{U}}}$ are the approximations of $\boldsymbol{U}'$ and $\overline{\boldsymbol{U}}$, respectively. The concept underlying the model is illustrated in Figure 2. We also note that (30) constitutes an *a posteriori* estimation of the *error* in the coarse-scale solution (see [4, 9]).

Remarks

1. (31) may be thought of as playing a similar role in the variational multiscale theory as the filtered equations play in traditional turbulence modeling. Distinguishing features are (31) is finite-dimensional and closed, in contrast with the filtered equations.

2. Intuitively, the "better" the fine-scale approximation, the smaller the dimension of the coarse-scale space required, and consequently, the smaller the computational effort. It is also possible to envision a hierarchy of approximations that produce variational multiscale analogues of traditional turbulence modeling concepts, such as large eddy simulation (LES), detached eddy simulation (DES), the Reynolds averaged Navier–Stokes (RANS) approach, etc. LES represents the turbulence modeling methodology requiring the greatest computational burden, but perhaps the least complex modeling. In the following sections we will endeavor to develop a variational multiscale analogue of LES within the theoretical framework of (30) and (31).
3. It is very important to emphasize that in practice we work directly with (31), a finite-dimensional system, and we consider the solution of (31), $\tilde{\overline{\boldsymbol{U}}}$, our approximation to $\overline{\boldsymbol{U}}$, and in turn our approximation to $\boldsymbol{U}$. Recall, by design of $\overline{\mathbb{P}}$, $\overline{\boldsymbol{U}}$ is an approximation to $\boldsymbol{U}$. We do not need to solve for the fine scales and because of this (30) is completely extraneous, unless we wish to use it to estimate the error in the coarse scales. That being said, it may also be interesting to consider $\tilde{\overline{\boldsymbol{U}}}+\tilde{\boldsymbol{U}}'$ as an alternative approximation to $\boldsymbol{U}$. It will of course be necessary to assume that the coarse-scale space is sufficiently large for the philosophy of LES to be appropriate. That is, if there is a well-defined inertial sub-range, then we assume the cut-off between the coarse- and fine-scale spaces resides somewhere within it. This assumption enables us to further assume that the energy content in the fine scales is small compared with the coarse scales, an aspect of considerable importance in attempting to analytically determine the solution of the fine-scale equations.

1.3 Perturbation Series

It seems reasonable to assume that the larger the space $\overline{\mathcal{V}}$, the better the approximation of $\overline{\boldsymbol{U}}$ to $\boldsymbol{U}$, and the smaller the coarse-scale residual $\mathbf{Res}(\overline{\boldsymbol{U}}) \in \mathcal{V}'^{*}$. We further assume that if $\mathbf{Res}(\overline{\boldsymbol{U}}) = \mathbf{0}$, then $\boldsymbol{F}'(\overline{\boldsymbol{U}}, \mathbf{0}) = \mathbf{0}$, and if $\mathbf{Res}(\overline{\boldsymbol{U}})$ is "small," then $\boldsymbol{U}'$ will likewise be "small."[2] These assumptions suggest a perturbation series expansion of the form:

$$\boldsymbol{U}' = \varepsilon \boldsymbol{U}'_1 + \varepsilon^2 \boldsymbol{U}'_2 + \varepsilon^3 \boldsymbol{U}'_3 + \cdots = \sum_{k=1}^{\infty} \varepsilon^k \boldsymbol{U}'_k \tag{32}$$

where $\varepsilon = \left\| \mathbf{Res}(\overline{\boldsymbol{U}}) \right\|_{\mathcal{V}'^{*}}$. Let us rewrite (24) in terms of the proposed expansion:

$$B_{\overline{U}}\left(\boldsymbol{W}', \sum_{k=1}^{\infty} \varepsilon^k \boldsymbol{U}'_k\right) + B_2\left(\boldsymbol{W}', \sum_{k=1}^{\infty} \varepsilon^k \boldsymbol{U}'_k, \sum_{k=1}^{\infty} \varepsilon^k \boldsymbol{U}'_k\right) = \varepsilon \langle \boldsymbol{W}', \widehat{\mathbf{R}}(\overline{\boldsymbol{U}}) \rangle \tag{33}$$

---

[2]These assumptions seem physically reasonable, but rigorous mathematical justification may be difficult to obtain. The existence of nontrivial, unforced weak solutions of the Euler equations, compact in space and time, underscores the mathematical difficulties of the Navier–Stokes equations at large Reynolds numbers (see [11]).

where $\langle\cdot,\cdot\rangle = \langle\cdot,\cdot\rangle_{\mathcal{V}',\mathcal{V}'^*}$, and

$$\widehat{\mathsf{R}}(\overline{\boldsymbol{U}}) = \frac{\mathbf{Res}(\overline{\boldsymbol{U}})}{\left\|\mathbf{Res}(\overline{\boldsymbol{U}})\right\|_{\mathcal{V}'^*}} \tag{34}$$

Notice that, by linearity,

$$B_{\overline{U}}\left(\boldsymbol{W}', \sum_{k=1}^{\infty} \varepsilon^k \boldsymbol{U}'_k\right) = \sum_{k=1}^{\infty} \varepsilon^k B_{\overline{U}}\left(\boldsymbol{W}', \boldsymbol{U}'_k\right), \tag{35}$$

while the second term requires further consideration. We expand it as follows:

$$\begin{aligned}
&B_2\left(\boldsymbol{W}'\,,\, \varepsilon \boldsymbol{U}'_1 + \varepsilon^2 \boldsymbol{U}'_2 + \varepsilon^3 \boldsymbol{U}'_3 + \cdots, \varepsilon \boldsymbol{U}'_1 + \varepsilon^2 \boldsymbol{U}'_2 + \varepsilon^3 \boldsymbol{U}'_3 + \cdots\right)\\
&= \varepsilon^2 B_2\left(\boldsymbol{W}', \boldsymbol{U}'_1, \boldsymbol{U}'_1\right)\\
&+ \varepsilon^3\left[B_2\left(\boldsymbol{W}', \boldsymbol{U}'_1, \boldsymbol{U}'_2\right) + B_2\left(\boldsymbol{W}', \boldsymbol{U}'_2, \boldsymbol{U}'_1\right)\right]\\
&+ \varepsilon^4\left[B_2\left(\boldsymbol{W}', \boldsymbol{U}'_1, \boldsymbol{U}'_3\right) + B_2\left(\boldsymbol{W}', \boldsymbol{U}'_2, \boldsymbol{U}'_2\right) + B_2\left(\boldsymbol{W}', \boldsymbol{U}'_3, \boldsymbol{U}'_1\right)\right]\\
&+ \cdots
\end{aligned} \tag{36}$$

A recurrence formula can be easily deduced, by grouping coefficients of the powers of ε:

$$\begin{array}{llllllll}
\varepsilon^2 \rightarrow & B_2\left(\boldsymbol{W}', \boldsymbol{U}'_1, \boldsymbol{U}'_1\right) & & & & & & \\
\varepsilon^3 \rightarrow & B_2\left(\boldsymbol{W}', \boldsymbol{U}'_1, \boldsymbol{U}'_2\right) & + & B_2\left(\boldsymbol{W}', \boldsymbol{U}'_2, \boldsymbol{U}'_1\right) & & & & \\
\varepsilon^4 \rightarrow & B_2\left(\boldsymbol{W}', \boldsymbol{U}'_1, \boldsymbol{U}'_3\right) & + & B_2\left(\boldsymbol{W}', \boldsymbol{U}'_2, \boldsymbol{U}'_2\right) & + & B_2\left(\boldsymbol{W}', \boldsymbol{U}'_3, \boldsymbol{U}'_1\right) & & \\
\varepsilon^5 \rightarrow & \cdots & + & \cdots & + & \cdots & + & \cdots
\end{array}$$

Hence:

$$B_2\left(\boldsymbol{W}', \sum_{k=1}^{\infty} \varepsilon^k \boldsymbol{U}'_k, \sum_{k=1}^{\infty} \varepsilon^k \boldsymbol{U}'_k\right) = \sum_{k=2}^{\infty} \varepsilon^k \sum_{j=1}^{k-1} B_2\left(\boldsymbol{W}', \boldsymbol{U}'_j, \boldsymbol{U}'_{k-j}\right) \tag{37}$$

The full expansion of the equation can be compactly written as:

$$\sum_{k=1}^{\infty} \varepsilon^k B_{\overline{U}}\left(\boldsymbol{W}', \boldsymbol{U}'_k\right) + \sum_{k=2}^{\infty} \varepsilon^k \sum_{j=1}^{k-1} B_2\left(\boldsymbol{W}', \boldsymbol{U}'_j, \boldsymbol{U}'_{k-j}\right) = \varepsilon\langle \boldsymbol{W}', \widehat{\mathsf{R}}(\overline{\boldsymbol{U}})\rangle \tag{38}$$

Equating like coefficients, we obtain a sequence of linear variational problems coupled through their right-hand sides:

$$\text{For } k = 1 \quad B_{\overline{U}}\left(\boldsymbol{W}', \boldsymbol{U}'_1\right) = \langle \boldsymbol{W}', \widehat{\mathsf{R}}(\overline{\boldsymbol{U}})\rangle_{\mathcal{V}',\mathcal{V}'^*} \tag{39}$$

$$\text{For } k \geq 2 \quad B_{\overline{U}}\left(\boldsymbol{W}', \boldsymbol{U}'_k\right) = -\sum_{j=1}^{k-1} B_2\left(\boldsymbol{W}', \boldsymbol{U}'_j, \boldsymbol{U}'_{k-j}\right) \tag{40}$$

The bilinear operator $B_{\overline{U}}(\cdot,\cdot)$ is the same for all the problems in the cascade, and can be formally inverted through a Green's operator. The Green's operator concept can be introduced in an abstract sense through a *resolvent* operator:

$$G'_{\overline{U}}(\cdot) = G'(\overline{U},\cdot): \quad \mathcal{V}'^{*} \to \mathcal{V}' \tag{41}$$

$$\mathcal{F}(\cdot) \mapsto V' \tag{42}$$

such that

$$B_{\overline{U}}(W', V') = \mathcal{F}(W') \tag{43}$$

If a sequence of operators $\mathcal{F}_j : \mathcal{V}' \to \mathbb{R}$ (i.e., $\mathcal{F}_j \in \mathcal{V}'^{*}$) is defined as:

$$\text{For } k=1 \quad \mathcal{F}_1(W') = \mathcal{F}_1(W'; \widehat{\mathbf{R}}(\overline{U})) = \langle W', \widehat{\mathbf{R}}(\overline{U})\rangle_{\mathcal{V}',\mathcal{V}'^{*}} \tag{44}$$

$$\text{For } k\geq 2 \quad \mathcal{F}_k(W') = \mathcal{F}_k(W'; U'_1,\ldots,U'_{k-1}) = -\sum_{j=1}^{k-1} B_2\left(W', U'_j, U'_{k-j}\right) \tag{45}$$

then it is possible to reformulate the sequence of problems (39)–(40) as:

$$U'_k = G'_{\overline{U}}(\mathcal{F}_k) = G'(\overline{U},\mathcal{F}_k), \qquad k = 1,2,\ldots \tag{46}$$

Notice that in the cascade of problems (39)–(40) (or, equivalently, (44)–(45), or (46)) the level-k term in the expansion depends on terms on the right-hand side, which involve the coarse-scale residual and terms in the expansion from level 1 to $k-1$.

Upon substituting the U'_k's into the series (32), the powers of $\varepsilon = \left\|\mathbf{Res}(\overline{U})\right\|_{\mathcal{V}'^{*}}$ cancel out. If the series converges, it represents an exact solution to the fine-scale equation, and then (31) gives the exact solution of the coarse-scale equation. In other words, given the validity of the assumptions, the exact solution of the original Navier–Stokes system is obtained. In order to determine the exact solutions of each of the linear problems in the cascade, we need the exact Green's operator $G'_{\overline{U}}$. This is a *non-classical* Green's operator, referred to as the "fine-scale Green's operator," that in turn depends on the *classical* Green's operator and the selected projector $\overline{\mathbb{P}}$ (see Hughes and Sangalli [8]):

$$G' = G - G\overline{\mathbb{P}}^{*}(\overline{\mathbb{P}}G\overline{\mathbb{P}}^{*})^{-1}\overline{\mathbb{P}}G, \tag{47}$$

where $\overline{\mathbb{P}}^{*}$ is the adjoint of $\overline{\mathbb{P}}$. Note that the orthogonality properties

$$\overline{\mathbb{P}}G' = 0 \tag{48}$$

$$G'\overline{\mathbb{P}}^{*} = 0 \tag{49}$$

immediately follow from (47). In Hughes and Sangalli [8] it was shown, in the context of finite element approximations of the advection-diffusion equation, for the advection-dominated case, that the projector based on the H_0^1-inner product (termed the Dirichlet projector in Hughes and Oberai [5]) produced a highly *localized* fine-scale Green's operator, despite the classical Green's operator being highly *nonlocal*. In fact, for the one-dimensional case, the support of the fine-scale Green's operator was confined to individual elements, and there was no coupling between elements. It is important to realize that this is not a general feature of the fine-scale Green's operator, but one that depends crucially on the particular projector. For example, the fine-scale Green's operator produced by the L^2-projector was nonlocal in all cases.

Exact determination of the Green's function is not possible and neither is summing an infinite number of terms in the perturbation series. Consequently, two approximations are necessary in order to develop a practical solution scheme:

1. Approximation of the fine-scale Green's operator for the linearized Navier–Stokes system, $\widetilde{\boldsymbol{G}}'_{\overline{U}} \approx \boldsymbol{G}'_{\overline{U}}$.
2. Approximation of the nonlinearities by truncation of the perturbation series.

Once these approximations are made precise, we have defined a turbulence model of the form (31). This will be discussed in the next section.

Remark
It needs to be emphasized that the pathway to an approximate turbulence model identified by the above assumptions is not the only possibility, but it does seem a viable candidate for LES-type modeling within the variational multiscale method. Clearly, a more direct attack on the fully nonlinear fine-scale equation, rather than the perturbation series approach, might seem an even more propitious approach. In either case, our theoretical framework for turbulence modeling remains (30) and (31).

2 Turbulent Channel Flow

We consider an equilibrium turbulent channel flow at Reynolds number 395 based on the friction velocity and the channel half width. The computational domain is a rectangular box of size $2\pi \times 2 \times 2/3\pi$ in the stream-wise, wall-normal, and span-wise directions, respectively. A no-slip Dirichlet boundary condition is set at the wall ($y = \pm 1$), while the stream-wise and the span-wise directions are assigned periodic boundary conditions. The no-slip condition is imposed strongly, that is, velocity degrees of freedom are explicitly set to zero at the wall. Alternatively, one may enforce the no-slip conditions weakly by augmenting the discrete formulation with terms that enforce Dirichlet conditions as Euler–Lagrange conditions (see Bazilevs and Hughes [2]). Although the weak boundary condition approach was shown to be superior to the strong imposition, we did not employ it in the computations reported in this paper. The flow is driven by a constant pressure gradient, f_x, acting in the stream-wise direction. The values of the kinematic viscosity ν and the forcing f_x are set to $1.47200 \cdot 10^{-4}$ and $3.372040 \cdot 10^{-3}$, respectively.

The computations were performed on meshes of 32^3 and 64^3 elements. For both meshes we employ C^1-continuous quadratic, and C^2-continuous cubic B-Splines.

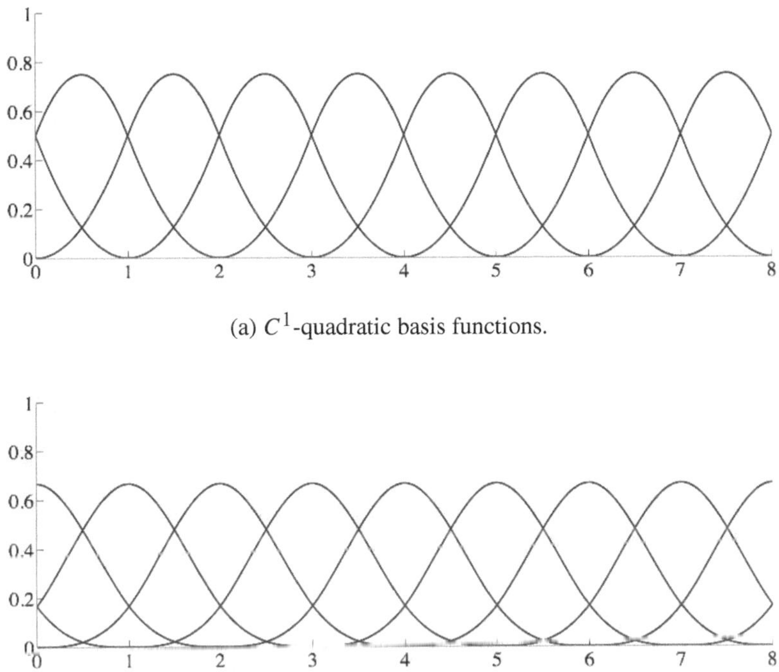

(a) C^1-quadratic basis functions.

(b) C^2-cubic basis functions.

Fig. 3. One-dimensional periodic basis functions.

For both orders, in the stream-wise and the span-wise directions the number of basis functions is equal to the number of elements in these directions (see Figure 3). On the other hand, the number of basis functions in the wall-normal direction is $n_y = n_{el} + p$, where n_{el} is the number of elements in this direction and p is the polynomial order (see Figure 4).

Numerical results for this test case are reported in the form of statistics of the mean stream-wise velocity and root-mean-square velocity fluctuations. Statistics are obtained by sampling the solution fields at the mesh knots and averaging in the stream-wise and span-wise directions as well as in time. Comparison of the statistical quantities of interest with the DNS data of Moser, Kim and Mansour [10] is made in order to assess the accuracy of the proposed turbulence modeling methodology.

Note from Figures 5 and 6 that for a mesh of 64^3 elements both quadratic and cubic solutions are in very close agreement with the DNS results.

The results for 32^3 quadratic and cubic B-splines are even better than high-fidelity spectral Galerkin LES results presented in Hughes et al. [7] and Holmen et al. [3]. We note though that the formulation utilized in [3,7] employed a fine-scale eddy viscosity model and is quite different from the one used here.

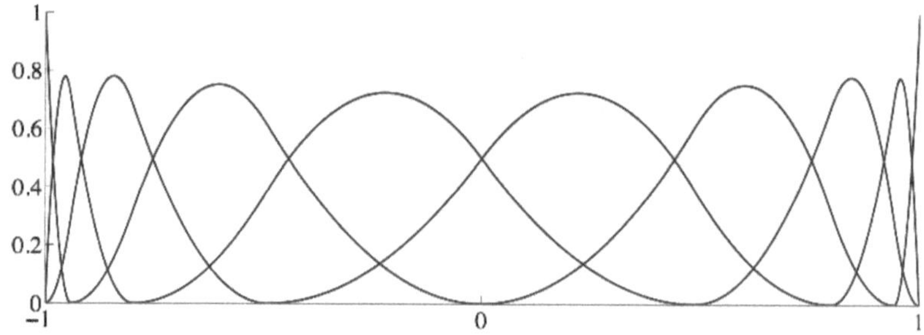

(a) Quadratic NURBS basis

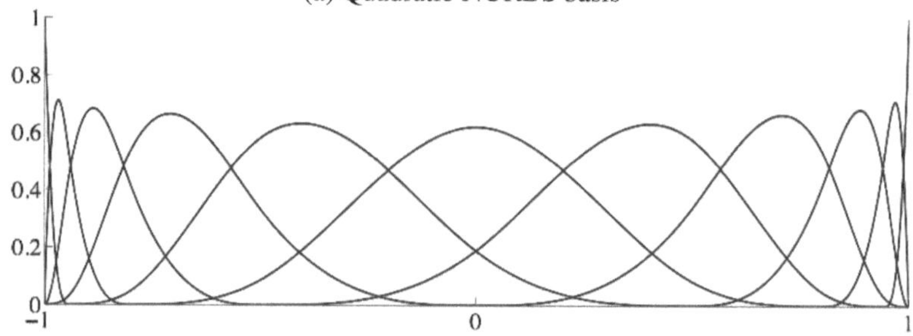

(b) Cubic NURBS basis

Fig. 4. Illustration of the wall-normal discretization for the turbulent channel flow problem. Meshes are graded towards the ends of the interval in order to better resolve boundary layers. Note that, due to the open knot vector construction (see [6] for details), the first and last basis functions are interpolatory at the endpoints of the domain, which facilitates strong imposition of no-slip Dirichlet boundary conditions.

Figure 7 shows isosurfaces of stream-wise velocity, velocity streamlines, and a series of snapshots of particles released at the channel inflow and set in motion to follow the streamlines in the boundary layer.

3 Conclusions

We presented a general variational multiscale theory suitable for LES-type turbulence modeling. The theory is derived directly from the incompressible Navier–Stokes equations and does not involve any ad hoc mechanisms. In particular, it entirely avoids use of eddy viscosities. We feel that this theory of turbulence modeling is more fundamental and logically consistent than ones derived heretofore and it has significant potential in practical engineering calculations. We might also mention that the turbulence modeling aspects remain *unaltered* when we consider laminar flows. In this sense, our methodology may be viewed as an approach for solving the incompressible Navier–Stokes equations, whether the flow under consideration is laminar or turbulent, or both. We also believe that this aspect separates our theory of turbulence modeling from predecessors.

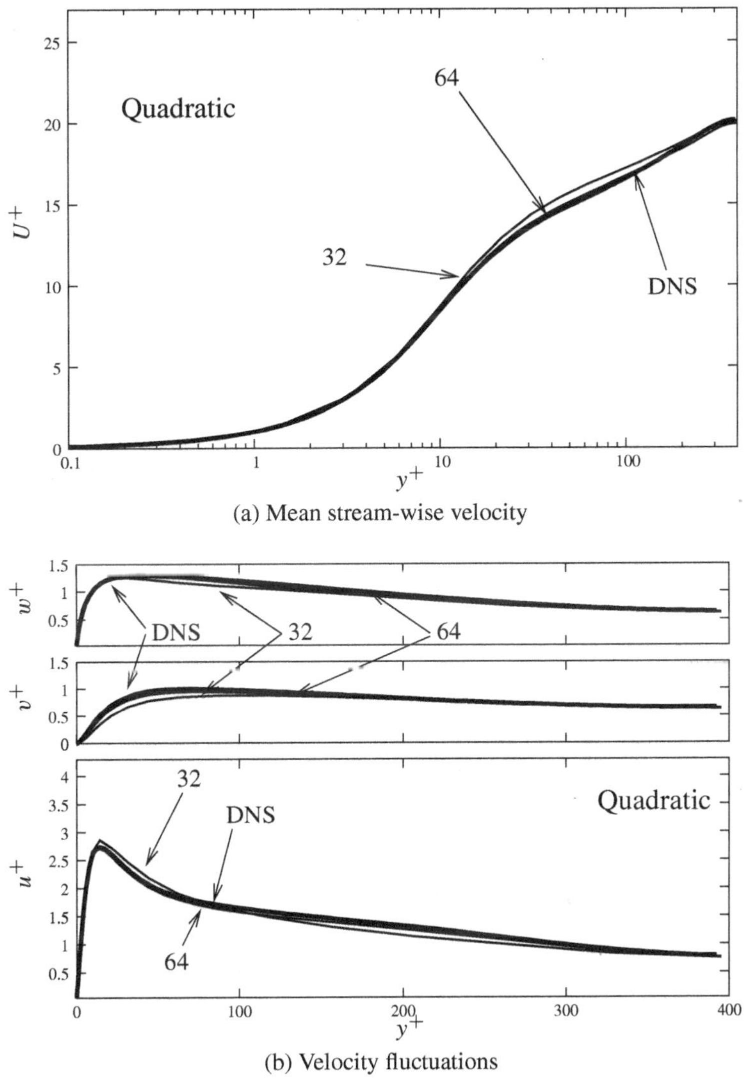

Fig. 5. Turbulent channel flow at $Re_\tau = 395$ computed using quadratic B-splines.

Given that the scale separation in the present methodology is performed with respect to the coarse-scale space actually used in the numerical computations, that is, the resolved scales, and that the fine-scale approximation is rendered well-defined by a projector used to make precise the direct sum decomposition into coarse and fine scales, it is impossible to entirely separate modeling and numerical concepts. We accept this as a fact associated with correct LES-type modeling concepts, not a shortcoming.

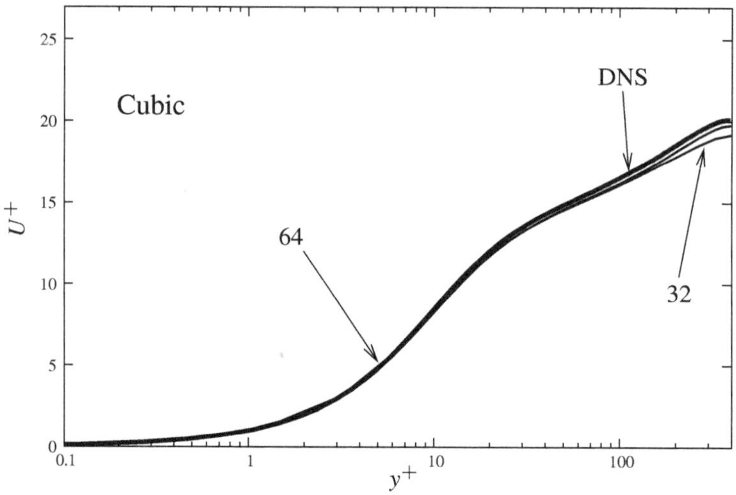

(a) Mean stream-wise velocity

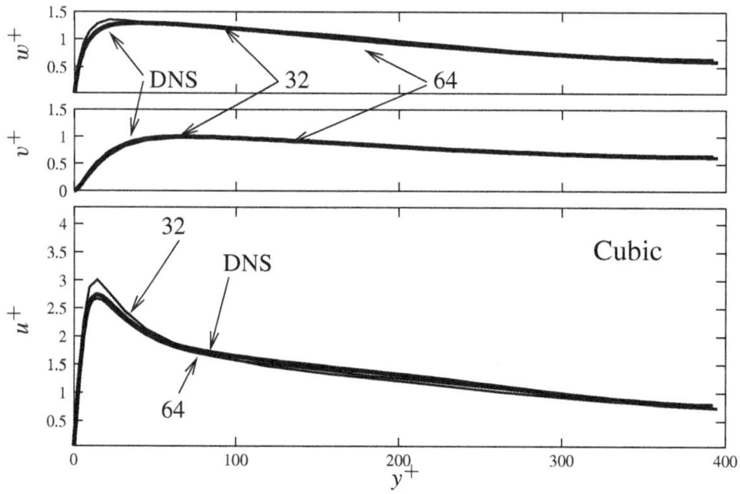

(b) Velocity fluctuations

Fig. 6. Turbulent channel flow at $Re_\tau = 395$ computed using cubic B-splines.

We also believe that our theory is more coherent mathematically than previous formulations and that it may be possible to use it as a basis of a statistical analysis of convergence and approximation. This would represent a very significant step forward for the theory of turbulence modeling, but, admittedly, a very difficult one to achieve. Nevertheless, we feel a door has been opened for the construction of a mathematical theory.

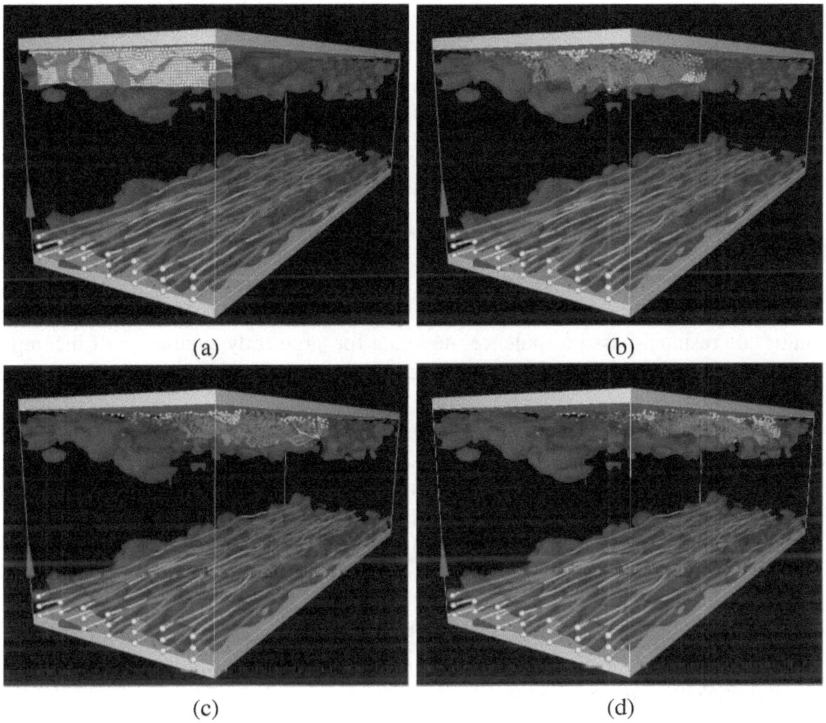

Fig. 7. Turbulent channel flow at $Re_\tau = 395$. Flow streamlines and isosurfaces of streamwise velocity. Two parallel planes of particles are released at the channel inflow close to the wall (see (a)), and are set in motion to follow the streamlines. Snapshots of the particle field are shown as particles travel down the length of the channel (see (b)-(d)). One can see the formation of the boundary layer as the particles released closer to the wall travel at slower speeds compared to the ones released in the outer layer. Also note that the particles initially released on different planes are mixed together as they approach the outflow, revealing the presence of faster and slower streaks in the boundary layer. Solution on the mesh of 32^3 quadratic NURBS was used for this visualization.

In summary, we feel a new paradigm for turbulence modeling has been established. Initial results seem to indicate its accuracy per degree of freedom is superior or, at the least, equal to any procedure proposed heretofore. Its generality and geometric flexibility also suggest it may provide a more powerful approach to turbulence calculations than previously existed.

Acknowledgements

We wish to express our appreciation for support provided by the Office of Naval Research under Contract No. N00014-03-0263, Dr. Luise Couchman, contract monitor, and Sandia National Laboratories under Contract No. 114166.

References

1. Y. Bazilevs, V.M. Calo, J.A. Cottrel, T.J.R. Hughes, A. Reali, and G. Scovazzi. Variational multiscale residual-based turbulence modeling for large eddy simulation of incompressible flows. *Computer Methods in Applied Mechanics and Engineering*, 197:173–201, 2007.
2. Y. Bazilevs and T.J.R. Hughes. Weak imposition of Dirichlet boundary conditions in fluid mechanics. *Computers and Fluids*, 36:12–26, 2007.
3. J. Holmen, T.J.R. Hughes, A.A. Oberai, and G.N. Wells. Sensitivity of the scale partition for variational multiscale LES of channel flow. *Physics of Fluids*, 16:824–827, 2004.
4. T.J.R. Hughes, G. Feijóo, L. Mazzei, and J.B. Quincy. The variational multiscale method – A paradigm for computational mechanics. *Computer Methods in Applied Mechanics and Engineering*, 166:3–24, 1998.
5. T.J.R. Hughes and A.A. Oberai. Calculation of shear stresses in Fourier–Galerkin formulations of turbulent channel flows: Projection, the Dirichlet filter and conservation. *Journal of Computational Physics*, 188(1):281–295, 2003.
6. T.J.R. Hughes, J.A. Cottrell, and Y. Bazilevs. Isogeometric analysis: CAD, finite elements, NURBS, exact geometry, and mesh refinement. *Computer Methods in Applied Mechanics and Engineering*, 194:4135–4195, 2005.
7. T.J.R. Hughes, A.A. Oberai, and L. Mazzei. Large-eddy simulation of turbulent channel flows by the variational multiscale method. *Physics of Fluids*, 13:1784–1799, 2001.
8. T.J.R. Hughes and G. Sangalli. Variational multiscale analysis: The fine-scale Green's function, projection, optimization, localization, and stabilized methods. *SIAM Journal of Numerical Analysis*, 45:539–557, 2007.
9. T.J.R. Hughes, G. Scovazzi, and L.P. Franca. Multiscale and stabilized methods. In E. Stein, R. de Borst, and T.J.R. Hughes (Eds.), *Encyclopedia of Computational Mechanics, Vol. 3, Computational Fluid Dynamics*, chapter 2. Wiley, 2004.
10. R. Moser, J. Kim, and R. Mansour. DNS of turbulent channel flow up to $Re = 590$. *Physics of Fluids*, 11:943–945, 1999.
11. V. Sheffer. An inviscid flow with compact support in space-time. *Journal of Geometric Analysis*, 3:343–401, 1993.

A Posteriori Error Estimation for Computational Fluid Dynamics: The Variational Multiscale Approach

Guillermo Hauke, Mohamed H. Doweidar and Daniel Fuster

Universidad de Zaragoza, C/María de Luna 3, 50018 Zaragoza, Spain; ghauke@unizar.es

Abstract This chapter is dedicated to using the variational multiscale method to derive an *explicit* a posteriori error estimator for computational fluid dynamics with emphasis on the scalar transport equation. The error estimator is based on a model that introduces approximations compatible with the theory of stabilized methods, yielding a very economical procedure. Yet, the strategy gives very accurate local and global error estimates, uniformly from the diffusive to the hyperbolic limit and can be easily implemented in existing computer codes.

Key words: a posteriori error estimation, explicit a posteriori estimation, residual based error estimation, stabilized methods, finite element methods, variational multiscale theory, transport equation, fluid mechanics

1 Introduction

This chapter presents a summary of the research on the development of an *explicit* a posteriori error estimator from the variational multiscale theory. The formulation is especially suited for fluid mechanics problems but it also works for elliptic problems. The initial ideas are presented in [17]. The underlaying theory is unveiled in [18], where exact error time scales are derived for one-dimensional nodally-exact solutions. Then, the theory was extended to other cases, like piecewise linear functions and residuals [14] and higher-order elements [16]. In [15, 20], the challenge of multidimensionality is addressed through the introduction of an approximate model for error estimation. Even though the model is not exact, it is shown to predict good *local* error estimates. The topic of quantities of interest is addressed in [19], but this is out of the scope of this chapter.

R. de Borst and E. Ramm (eds.), *Multiscale Methods in Computational Mechanics*, Lecture Notes in Applied and Computational Mechanics 55, DOI 10.1007/978-90-481-9809-2_2,

2 The Variational Multiscale Approach to Error Estimation

2.1 The Abstract Problem

Strong Form

Consider a spatial domain Ω with boundary Γ, which is partitioned into two non-overlapping zones Γ_g and Γ_h. Let $\boldsymbol{x}$ and $\boldsymbol{y}$ be two points in $\overline{\Omega}$. The strong form of the boundary-value problem consists of finding $u : \Omega \to \mathbb{R}$ such that for the given essential boundary condition $\mathrm{g} : \Gamma_g \to \mathbb{R}$, the natural boundary condition $h : \Gamma_h \to \mathbb{R}$, and forcing function $f : \Omega \to \mathbb{R}$, the following equations are satisfied

$$\begin{cases} \mathcal{L}u = f & \text{in } \Omega \\ u = \mathrm{g} & \text{on } \Gamma_g \\ \mathcal{B}u = h & \text{on } \Gamma_h \end{cases} \tag{1}$$

where $\mathcal{L}$ is in principle a second-order differential operator and $\mathcal{B}$, an operator defined later, acting on the boundary.

Weak Form

Given the functional solution space $\mathcal{S} \subset H^1(\Omega)$ and weighting space $\mathcal{V} \subset H^1(\Omega)$, with $H^1(\Omega)$ the Sobolev space of order one, the variational formulation for the above boundary-value problem consists of finding $u \in \mathcal{S}$ such that:

$$a(w, u) = (w, f) + (w, h)_{\Gamma_h} \qquad \forall w \in \mathcal{V} \tag{2}$$

where $(\cdot,\cdot)$ is the $L_2(\Omega)$ inner product, $(\cdot,\cdot)_\omega$ is the $L_2(\omega)$ inner product in the domain ω, and $a(\cdot,\cdot)$ is a bilinear form satisfying

$$a(w, u) = (w, \mathcal{L}u) + (w, \mathcal{B}u)_{\Gamma_h} \tag{3}$$

for all sufficiently smooth functions $w \in \mathcal{V}$ and $u \in \mathcal{S}$. Likewise, $a(\cdot,\cdot)_\omega$ is the bilinear operator with the integral extended over the domain ω.

In the case of finite element methods, the domain Ω is discretized into n_{el} non-overlapping elements with domain Ω^e and boundary Γ^e ($e = 1, 2, \ldots, n_{\mathrm{el}}$). The functions then may be discontinuous across the inter-element boundaries. Let us denote the union of element interiors as $\widetilde{\Omega}$, where $\widetilde{\Omega} = \bigcup_{e=1}^{n_{\mathrm{el}}} \Omega^e$ and the union of element boundaries minus Γ as $\widetilde{\Gamma}$, with $\widetilde{\Gamma} = \bigcup_{e=1}^{n_{\mathrm{el}}} \Gamma^e \setminus \Gamma$ also referred to as the element interfaces or inter-element boundaries.

Furthermore, let $[\![\cdot]\!]$ denote the jump operator of a function across a discontinuity, for instance, across an inter-element boundary. Following the notation of Figure 1, where the respective outward unit normals to elements Ω^+ and Ω^- are $\boldsymbol{n}^+$ and $\boldsymbol{n}^-$, the jump of $\boldsymbol{v} \cdot \boldsymbol{n}$ is defined as [23]

$$[\![\boldsymbol{v} \cdot \boldsymbol{n}]\!] = \boldsymbol{n}^+ \cdot \boldsymbol{v}^+ + \boldsymbol{n}^- \cdot \boldsymbol{v}^- \tag{4}$$

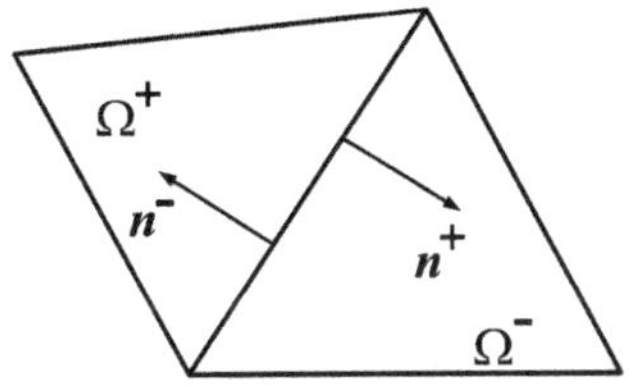

Fig. 1. Notation to define the jump across element interfaces.

2.2 The Variational Multiscale Error Estimation Paradigm

The variational multiscale theory [22] revisits the concept of scale splitting, where the solution and weighting functions are decomposed according to

$$\begin{aligned} u &= \bar{u} + u' \\ w &= \bar{w} + w' \end{aligned} \tag{5}$$

with the corresponding resolved $\bar{u} \in \bar{\mathcal{S}}$ and unresolved scales $u' \in \mathcal{S}'$ finite element spaces. Here, u represents the *exact solution*; $\bar{u}$, the *resolved scales* or finite element solution; and u', the *unresolved scales, subgrid scales* or *error*.

The above decomposition can be substituted in the variational formulation (2). Because of its bilinearity, the weak form is equivalent to these two subproblems,

$$a(\bar{w}, \bar{u}) = -a(\bar{w}, u') + (\bar{w}, f) + (\bar{w}, h)_{\Gamma_h} \tag{6}$$

$$a(w', u') = -a(w', \bar{u}) + (w', f) + (w', h)_{\Gamma_h} \tag{7}$$

the first one for the resolved scales $\bar{u}$ and the second one for the unresolved scales u'.

For smooth functions within the element interiors, but rough across the inter-element boundaries, such as finite element functions, integration by parts of $a(w', \bar{u})$ leads to

$$\begin{aligned} a(w', \bar{u}) &= \sum_{e=1}^{n_{\mathrm{el}}} a(w', \bar{u})_{\Omega^e} \\ &= \sum_{e=1}^{n_{\mathrm{el}}} \{(w', \mathcal{L}\bar{u})_{\Omega^e} + (w', \mathcal{B}\bar{u})_{\Gamma^e}\} \\ &= (w', \mathcal{L}\bar{u})_{\tilde{\Omega}} + (w', [\![\mathcal{B}\bar{u}]\!])_{\tilde{\Gamma}} + (w', \mathcal{B}\bar{u})_{\Gamma_h} \end{aligned} \tag{8}$$

Consequently, the fine-scale subproblem (7) can be written as

$$\begin{aligned} a(w', u') &= -(w', \mathcal{L}\bar{u})_{\tilde{\Omega}} - (w', [\![\mathcal{B}\bar{u}]\!])_{\tilde{\Gamma}} - (w, \mathcal{B}\bar{u})_{\Gamma_h} \\ &\quad + (w', f) + (w', h)_{\Gamma_h} \\ &= -(w', \mathcal{L}\bar{u} - f)_{\tilde{\Omega}} - (w', [\![\mathcal{B}\bar{u}]\!])_{\tilde{\Gamma}} - (w', \mathcal{B}\bar{u} - h)_{\Gamma_h} \end{aligned} \tag{9}$$

Furthermore, the fine-scale problem (9) can be solved analytically using the Green's function of the fine-scale problem $g'(\boldsymbol{x}, \boldsymbol{y})$ [24, 25]. The result, extended here for Neumann problems, is that the fine scales are given by

$$
\boxed{\begin{aligned} u'(\boldsymbol{x}) = &-\int_{\widetilde{\Omega}_y} g'(\boldsymbol{x},\boldsymbol{y})\,(\mathcal{L}\bar{u}-f)(\boldsymbol{y})\,\mathrm{d}\Omega_y - \int_{\widetilde{\Gamma}_y} g'(\boldsymbol{x},\boldsymbol{y})\,([\![\mathcal{B}\bar{u}]\!])(\boldsymbol{y})\,\mathrm{d}\Gamma_y \\ &-\int_{\Gamma_{h\,y}} g'(\boldsymbol{x},\boldsymbol{y})\,(\mathcal{B}\bar{u}-h)(\boldsymbol{y})\,\mathrm{d}\Gamma_y \end{aligned}} \tag{10}
$$

which can be used to estimate u', the error of the resolved scales $\bar{u}$.

Remarks.

1. The same result can be expressed via the Green's function of the adjoint fine-scale problem g'^{*} through the identity $g'(\boldsymbol{x},\boldsymbol{y}) = g'^{*}(\boldsymbol{y},\boldsymbol{x})$.
2. Note that the integral is non-local and, therefore, the effect of the error is, in principle, non-local. For instance, the error at a singularity can influence the precision at distant places. This is called the *pollution* effect [2]. For certain problems, the pollution effect is locally restrained.
3. The error estimator is residual based. The contemplated residuals include:
   i. Element interior residuals, $\mathcal{L}\bar{u} - f$ in $\widetilde{\Omega}$.
   ii. Inter-element residuals, $[\![\mathcal{B}\bar{u}]\!]$ on $\widetilde{\Gamma}$.
   iii. Natural boundary condition residual, $\mathcal{B}\bar{u} - h$ on Γ_h.
4. Other sources of errors, such as Dirichlet boundary condition approximations or boundary approximation, are not contemplated here.

3 The Smooth Paradigm for Error Estimation

The smooth a posterior error estimator assumes that the finite element solution $\bar{u}$ is exact on Γ^e, that is, the error $u' = 0$ on Γ^e.

In this situation, the boundary integrals of (10) vanish and the errors decouple from element to element. Furthermore, the *fine-scale* Green's function $g'(\boldsymbol{x},\boldsymbol{y})$ is exactly represented by the *element* Green's function $g_e(\boldsymbol{x},\boldsymbol{y})$, which for *linear* finite element spaces satisfies within each element

$$
\begin{cases} \mathcal{L}g_e = \delta_y & \text{in } \Omega^e \\ g_e = 0 & \text{on } \Gamma^e \end{cases} \tag{11}
$$

In the above equation, $\delta_y(\boldsymbol{x}) = \delta(\boldsymbol{x}-\boldsymbol{y})$ represents the Dirac delta distribution. Indeed, g_e vanishes on Γ^e, and the a posteriori error estimator (10) simplifies to

$$
\boxed{u'(\boldsymbol{x}) = -\int_{\Omega^e_y} g_e(\boldsymbol{x},\boldsymbol{y})\,(\mathcal{L}\bar{u}-f)(\boldsymbol{y})\,\mathrm{d}\Omega_y \qquad \text{on } \Omega^e} \tag{12}
$$

That is, the error in one element Ω^e is determined by information confined inside the element. The preceding paradigm (12) is exact, for instance, in the case of one-dimensional linear problems solved with stabilized methods or in the case of one-dimensional Poisson problems solved with the Galerkin method.

3.1 Intrinsic Error Time Scales

Throughout this section, it will be assumed that the residual $\mathcal{L}\bar{u} - f \in \mathcal{P}_0$ is *piecewise constant* and the functional spaces are formed by piecewise linear functions. Then, the smooth multiscale error estimator (12) yields,

$$\begin{aligned} u'(\boldsymbol{x})|_{\Omega^e} &= -\int_{\Omega^e} g_e(\boldsymbol{x}, \boldsymbol{y})(\mathcal{L}\bar{u} - f)(\boldsymbol{y})\, \mathrm{d}\Omega_y \\ &= -(\mathcal{L}\bar{u} - f) \int_{\Omega^e} g_e(\boldsymbol{x}, \boldsymbol{y})\, \mathrm{d}\Omega_y \\ &= -(\mathcal{L}\bar{u} - f)\, b_0^e(\boldsymbol{x}) \end{aligned} \tag{13}$$

that is,

$$\boxed{u'(\boldsymbol{x})|_{\Omega^e} = -(\mathcal{L}\bar{u} - f)\, b_0^e(\boldsymbol{x})} \tag{14}$$

where the function $b_0^e(\boldsymbol{x})$ is defined as

$$b_0^e(\boldsymbol{x}) = \int_{\Omega^e} g_e(\boldsymbol{x}, \boldsymbol{y})\, \mathrm{d}\Omega_y \tag{15}$$

This new function is easier to compute than the element Green's function since (for linear elements) it is also the solution of the problem

$$\begin{cases} \mathcal{L}b_0^e = 1 & \text{in } \Omega^e \\ b_0^e = 0 & \text{on } \Gamma^e \end{cases} \tag{16}$$

which is precisely a *residual-free bubble* [4,5,7,11]. Thus, residual-free bubbles are estimators of u'.

Remarks.

1. Note that for multi-dimensional elliptic problems, bubble functions are not good error estimators. However, they are for problems obeying the underlying assumptions of this section. See also [6] where the computation of the bubble reproduces the unresolved scales and, therefore, the error.
2. Bubble functions were exploited in [28] for estimating the H^1-seminorm error in the diffusive limit. However, bubble functions alone do not reproduce the error due to the inter-element jumps, whereas (10) includes a mechanism to take them into account.
3. Once the bubble is available, the error can be predicted in any norm.
4. For a one-dimensional elliptic problem with piecewise constant f, the Galerkin finite element solution $\bar{u}$ is nodally exact. Then, expression (14) is exact.

Estimates in the L_2 Norm

As a consequence, the L_2 norm of (14) is

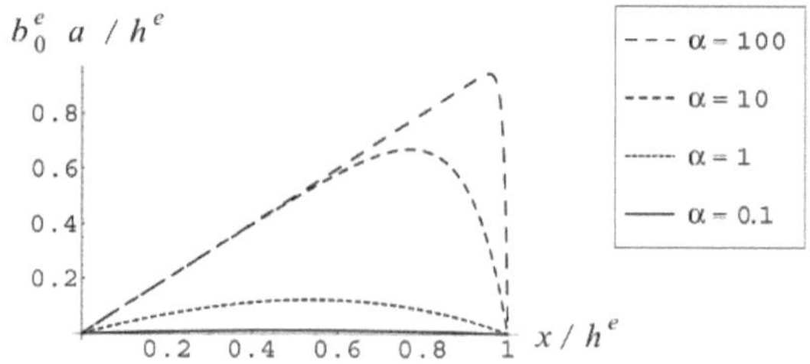

Fig. 2. Bubble function $b_0^e(x)$ for one-dimensional advection-diffusion problem.

$$||u'(\boldsymbol{x})||_{\Omega^e} = ||(\mathcal{L}\bar{u} - f)\ b_0^e(\boldsymbol{x})||_{\Omega^e} \tag{17}$$

yielding for piecewise constant residuals

$$||u'(\boldsymbol{x})||_{\Omega^e} = |\mathcal{L}\bar{u} - f|\ ||b_0^e(\boldsymbol{x})||_{\Omega^e} \tag{18}$$

Equation (18) reflects the correct time scale for error estimation, $\tau_{L_2}^e$, which is defined as

$$\boxed{\tau_{L_2}^e = \frac{1}{\sqrt{\Omega^e}}\ ||b_0^e(\boldsymbol{x})||_{\Omega^e}} \tag{19}$$

Consequently, (18) becomes

$$\boxed{||u'(\boldsymbol{x})||_{\Omega^e} = \tau_{L_2}^e\ ||\mathcal{L}\bar{u} - f||_{\Omega^e}} \tag{20}$$

Remarks.

1. The above expression is constant free.
2. The error time scale $\tau_{L_2}^e$ can be precomputed like the flow intrinsic time scale τ_{flow}^e.
3. The variational multiscale approach can be used to estimate the error in other norms, such as the H^1 seminorm [18].

Example: One-Dimensional Advection-Diffusion

Let us apply the above ideas to the one-dimensional advection-diffusion equation,

$$\mathcal{L}u = au_{,x} - \kappa u_{,xx} \tag{21}$$

with constant coefficients, linear elements of length h_e and piecewise constant f. In this case, clearly, the required assumptions are satisfied since the residual is piecewise constant.

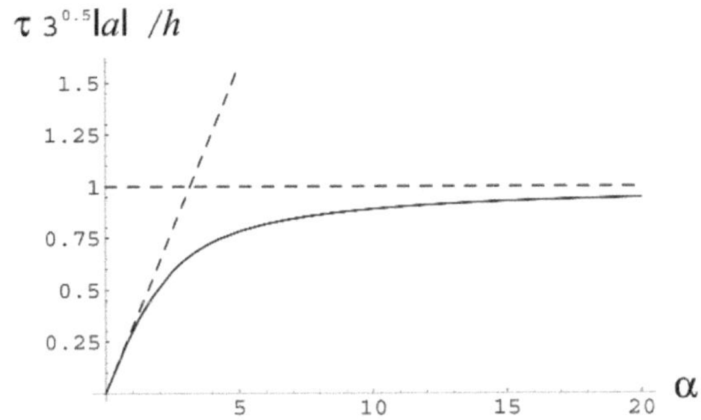

Fig. 3. Exact and asymptotic error time scales $\tau^e_{L_2}$ for one-dimensional advection-diffusion.

Exact Time Scale

In this case, the bubble function is (see Figure 2)

$$b^e_0(x) = \frac{x}{a} + \frac{h_e}{a} \frac{e^{ax/\kappa} - 1}{1 - e^{ah_e/\kappa}} \tag{22}$$

which produces the following error time scale (see Figure 3)

$$\boxed{\tau^e_{L_2} = \frac{h_e}{\sqrt{3}|a|} \frac{1}{4\alpha} \left[(24 + 16\alpha^2 + 6\alpha / \sinh(\alpha)^2 (2\alpha - 3\sinh(2\alpha)) \right]^{1/2}} \tag{23}$$

Asymptotic Result

The asymptotic proper time scale for error estimation can be obtained as

$$\boxed{\tau^e_{L_2} = \min\left(\frac{h_e}{\sqrt{3}\,|a|}, \frac{h_e^2}{2\sqrt{30}\,\kappa} \right) = \frac{h_e}{\sqrt{3}\,|a|} \min\left(1, \frac{\alpha}{\sqrt{10}}\right)} \tag{24}$$

with

$$\alpha = \frac{h_e|a|}{2\kappa} \qquad \text{element Peclet number} \tag{25}$$

Figure 4 shows the global efficiency I_{eff} of one-dimensional advection-diffusion problems for various parameter choices. Recall that the efficiency is defined as the ratio of a norm of the estimated error to the norm of the true error. Thus, it can be observed that the method is exact.

Remarks.

1. Note that the error time scale is a little larger than the stabilizing intrinsic time scale used in SUPG and other stabilized methods [8, 10], which is given by

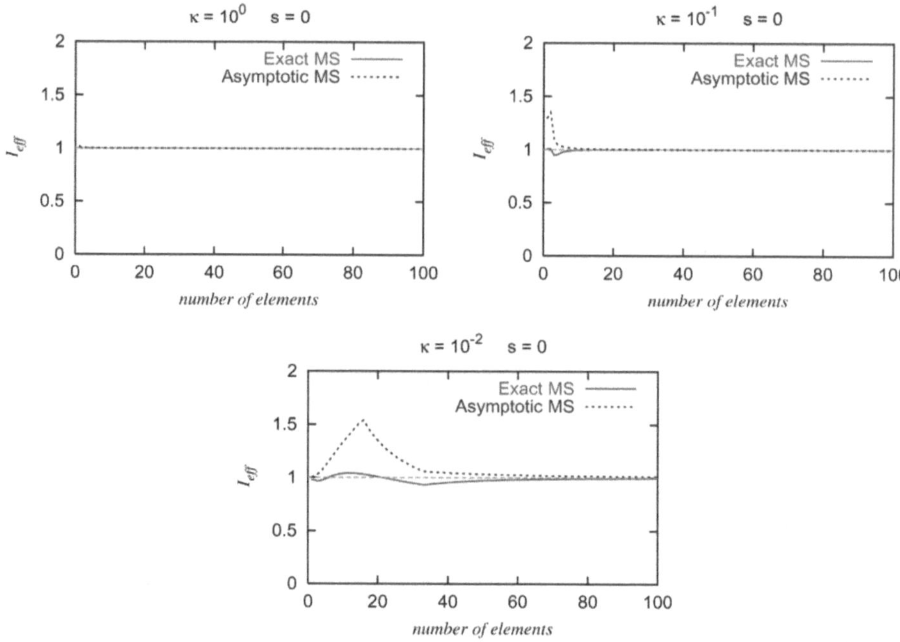

Fig. 4. L_2 global efficiency index for one-dimensional advection-diffusion.

$$\tau^e_{\text{flow}} = \coth\alpha - \frac{1}{\alpha} \tag{26}$$

and asymptotically

$$\tau^e_{\text{flow}} = \min\left(\frac{h_e}{2\,|a|}, \frac{h_e^2}{12\,\kappa}\right) = \frac{h_e}{2\,|a|}\min\left(1, \frac{\alpha}{3}\right) \tag{27}$$

2. For further examples, the reader is referred to [18].

3.2 Error Upper Bounds

Departing from (12) and taking the absolute value

$$\begin{aligned} |u'(\boldsymbol{x})| &= \left| \int_{\tilde{\Omega}_y} g_e(\boldsymbol{x}, \boldsymbol{y})\,(\mathcal{L}\bar{u} - f)(\boldsymbol{y})\,\mathrm{d}\Omega_y \right| \\ &\leq ||g_e(\boldsymbol{x}, \boldsymbol{y})\,(\mathcal{L}\bar{u} - f)(\boldsymbol{y})||_{L_1(\Omega^e_y)} \end{aligned} \tag{28}$$

By Hölders inequality (see, for instance, Brenner and Scott [3]),

$$|u'(\boldsymbol{x})| \leq ||g_e(\boldsymbol{x}, \boldsymbol{y})||_{L_p(\Omega^e_y)}\; ||\mathcal{L}\bar{u} - f||_{L_q(\Omega^e)} \tag{29}$$

with $1 \leq p, q \leq \infty$, $1/p + 1/q = 1$ and always that the above integrals and norms are defined. Note also that nothing has been assumed about the polynomial representation of the residual.

Taking now the L_r norm,

$$||u'(\boldsymbol{x})||_{L_r(\Omega^e)} \leq \left\| \, ||g_e(\boldsymbol{x}, \boldsymbol{y})||_{L_p(\Omega^e_y)} \right\|_{L_r(\Omega^e_x)} ||\mathcal{L}\bar{u} - f||_{L_q(\Omega^e)} \tag{30}$$

where $|| \, ||g_e(\boldsymbol{x}, \boldsymbol{y})||_{L_p(\Omega^e_y)}||_{L_r(\Omega^e_x)}$ is an error scale which can be precomputed.

3.3 Relation to the Flow Time Scale Parameter

Selecting above $p = 1, q = \infty, r = 1$, then,

$$||u'(\boldsymbol{x})||_{L_1(\Omega^e)} \leq ||g_e(\boldsymbol{x}, \boldsymbol{y})||_{L_1(\Omega^e_x \times \Omega^e_y)} \, ||\mathcal{L}\bar{u} - f||_{L_\infty(\Omega^e)} \tag{31}$$

But for $g_e \geq 0$, the definition of τ^e_{flow} [22] is precisely

$$\tau^e_{\text{flow}} = \frac{1}{\text{meas}(\Omega^e)} ||g_e(\boldsymbol{x}, \boldsymbol{y})||_{L_1(\Omega^e_x \times \Omega^e_y)} \tag{32}$$

Therefore, the flow time scale parameter τ^e_{flow} gives an upper bound in the L_1 norm of the solution error as a function of the L_∞ norm of the residual. Thus,

$$\boxed{||u'(\boldsymbol{x})||_{L_1(\Omega^e)} \leq \tau^e_{\text{flow}} \, \text{meas}(\Omega^e) \, ||\mathcal{L}\bar{u} - f||_{L_\infty(\Omega^e)}} \tag{33}$$

Remark. For piecewise constant residuals and nodally exact solutions, the above expression holds with the equal sign.

3.4 Extensions

All the above results can be extended to more general cases, such as higher-order elements [15], piecewise linear residuals [14] and other norms [18].

4 Multidimensional Model

The fine-scale Green's function is the distribution that characterizes the behavior of the numerical error, and emanates from the proper projection of the global Green's function. Therefore, it depends on the operator (with the corresponding geometry and boundary conditions), on the finite element space and on the method (or projector). Furthermore, $\mathcal{S}'$ is the kernel of the projector that defines the method (see [21] for further details).

Even though (10) is exact for a linear problem, the calculation of the fine-scale Green's function is far more complex than that of the solution of the original problem. In order to avoid this complexity, we aim here at developing a simple strategy based on approximations compatible with the theory of stabilized methods.

4.1 A Model for the Error Distribution

The error representation (10) can be split into errors stemming from element interior residuals and boundary element residuals, namely

$$u'(\boldsymbol{x}) = u'_{\text{int}}(\boldsymbol{x}) + u'_{\text{bnd}}(\boldsymbol{x}) \tag{34}$$

Therefore, using the triangle inequality, we can write [15]

$$||u'(\boldsymbol{x})|| \leq ||u'_{\text{int}}(\boldsymbol{x})|| + ||u'_{\text{bnd}}(\boldsymbol{x})|| \tag{35}$$

Element Interior Error

In [21] it was shown that for certain types of variational methods, such as stabilized methods, the error distribution is practically local. For these methods, the *fine-scale* Green's function can be approximated by the *element* Green's function $g_e(\boldsymbol{x}, \boldsymbol{y})$. Following [14, 17, 18] the error due to element interiors is modeled as the smooth paradigm (12),

$$\boxed{u'_{\text{int}}(\boldsymbol{x}) \approx -\int_{\Omega_y^e} g_e(\boldsymbol{x}, \boldsymbol{y})\,(\mathcal{L}\bar{u} - f)(\boldsymbol{y})\, \mathrm{d}\Omega_y \qquad \text{on } \Omega^e} \tag{36}$$

As stated earlier, (36) is exact for element-edge-exact solutions.

By Hölders inequality (see Brenner and Scott [3]) we can develop an (approximate) upper bound,

$$|u'_{\text{int}}(\boldsymbol{x})| \leq ||g_e(\boldsymbol{x}, \boldsymbol{y})||_{L_p(\Omega_y^e)}\, ||\mathcal{L}\bar{u} - f||_{L_q(\Omega^e)} \tag{37}$$

with $1 \leq p, q \leq \infty$, $1/p + 1/q = 1$. Taking the L_r norm,

$$\boxed{||u'_{\text{int}}(\boldsymbol{x})||_{L_r(\Omega^e)} \leq \Big|\Big|\, ||g_e(\boldsymbol{x}, \boldsymbol{y})||_{L_p(\Omega_y^e)} \Big|\Big|_{L_r(\Omega_x^e)} ||\mathcal{L}\bar{u} - f||_{L_q(\Omega^e)}} \tag{38}$$

Remark. This contribution is vital for error estimation in the hyperbolic limit.

Element Boundary Error

The inter-element boundary errors are approximated within each element as

$$\boxed{u'_{\text{bnd}}(\boldsymbol{x}) \approx -\int_{\Gamma_y^e} g'(\boldsymbol{x}, \boldsymbol{y})\,([\![\mathcal{B}\bar{u}]\!])(\boldsymbol{y})\, \mathrm{d}\Gamma_y \qquad \text{on } \Omega^e} \tag{39}$$

where the jump definition has been formally extended to encompass the natural boundary condition residual,

$$[\![\mathcal{B}\bar{u}]\!] = \begin{cases} \mathcal{B}\bar{u} - h & \text{on } \Gamma^e \cap \Gamma_h \\ 0 & \text{on } \Gamma^e \cap \Gamma_g \end{cases} \tag{40}$$

Note that the inter-element boundary integral on $\widetilde{\Gamma}$ has been replaced by the boundary integral along Γ^e. Since this integral contains the jump of the flux, $\mathcal{B}\bar{u}$, it increases the non-locality by one layer of elements around Ω^e.

Again, by Hölders inequality

$$|u'_{\text{bnd}}(\boldsymbol{x})| \leq ||g'(\boldsymbol{x}, \boldsymbol{y})||_{L_p(\Gamma^e_y)} \, ||[\![\mathcal{B}\bar{u}]\!]||_{L_q(\Gamma^e)} \tag{41}$$

with $1 \leq p, q \leq \infty$, $1/p + 1/q = 1$. Taking the L_r norm,

$$\boxed{||u'_{\text{bnd}}(\boldsymbol{x})||_{L_r(\Omega^e)} \leq \Big|\Big| \, ||g'(\boldsymbol{x}, \boldsymbol{y})||_{L_p(\Gamma^e_y)} \Big|\Big|_{L_r(\Omega^e_x)} ||[\![\mathcal{B}\bar{u}]\!]||_{L_q(\Gamma^e)}} \tag{42}$$

Remarks.

1. This contribution is vital for error estimation in the diffusive limit.
2. Larson and Målqvist [27] state that most of the error lays within one layer of elements around Ω^e.

4.2 Norms Based on the L_∞ Norm of the Residual

Because the Green's function may not be very smooth in multidimensional applications, the choice $p = 1$ and $q = \infty$ stands out. Then, typical choices for r are $r = 1$ or $r = 2$. In this case, if $g_e(\boldsymbol{x}, \boldsymbol{y})$ does not change sign in Ω^e,

$$\Big|\Big| \, ||g_e(\boldsymbol{x}, \boldsymbol{y})||_{L_1(\Omega^e_y)} \Big|\Big|_{L_r(\Omega^e_x)} = ||b^e_0(\boldsymbol{x})||_{L_r(\Omega^e_x)} \tag{43}$$

where the function $b^e_0(\boldsymbol{x})$ is a *residual-free bubble* (see Eq. (16)). Then, the error time scales can be defined as [18]

$$\tau^e_{L_1} = \frac{1}{\text{meas}(\Omega^e)} ||b^e_0(\boldsymbol{x})||_{L_1(\Omega^e)}, \qquad \tau^e_{L_2} = \frac{1}{\text{meas}(\Omega^e)^{1/2}} ||b^e_0(\boldsymbol{x})||_{L_2(\Omega^e)} \tag{44}$$

Remark. Note that whereas the norms of the element Green's function depend on the dimensionality of the problem, the error time scales do not. The time scale has the advantage that allows comparison between the various norm choices and the flow time scale.

Now, an approximate analysis [20] shows that the norm of the fine-scale Green's function on Γ^e can be approximately linked to the norm of the element Green's function in the domain Ω^e by

$$\Big|\Big| \, ||g'(\boldsymbol{x}, \boldsymbol{y})||_{L_1(\Gamma^e_y)} \Big|\Big|_{L_r(\Omega^e_x)} \approx \frac{1}{2} \frac{\text{meas}(\Gamma^e)}{\text{meas}(\Omega^e)} \Big|\Big| \, ||g_e(\boldsymbol{x}, \boldsymbol{y})||_{L_1(\Omega^e_y)} \Big|\Big|_{L_r(\Omega^e_x)} \tag{45}$$

4.3 Summary of the Model

$$\boxed{\begin{aligned} ||u'(x)||_{L_r(\Omega^e)} \leq\ & \text{meas}(\Omega^e)^{1/r}\, \tau_{L_r} \\ & \times \left(||\mathcal{L}\bar{u} - f||_{L_\infty(\Omega^e)} + \tfrac{1}{2} \frac{\text{meas}(\Gamma^e)}{\text{meas}(\Omega^e)} ||[\![\mathcal{B}\bar{u}]\!]||_{L_\infty(\Gamma^e)} \right) \\ & \text{on } \Omega^e \end{aligned}} \quad (46)$$

Remarks.

1. This is an explicit a posteriori error estimator.
2. It is very similar to the estimator proposed by [29] and its extensions, with the advantage that it is constant-free, that is, the method gives the error constant.
3. The error constants are dimensionally correct.
4. In principle, the theory allows the error estimate in the norm of choice.
5. In [1], the stabilizing parameter is tuned to predict the error norm exactly in the one-dimensional case. Here, this parameter is predicted by the theory and there is no need to solve local problems. The present method also is exact for one-dimensional solutions which are nodally exact.

5 Multidimensional Error Scales for the Bilinear Quad

In this section, we calculate the 2D error time scales for the bilinear quad. For the linear triangle, see [20].

5.1 Hyperbolic Limit

In the hyperbolic limit, the residual-free bubble is the solution of the problem

$$\begin{cases} |\boldsymbol{a}|\nabla_a b_0^e = 1 & \text{in } \Omega^e \\ b_0^e = 0 & \text{on } \Gamma_{\text{in}}^e \end{cases} \quad (47)$$

where $\boldsymbol{a}$ is the velocity; ∇_a, the derivative in the direction of $\boldsymbol{a}$ and Γ_{in}^e denotes the inflow portion of the boundary, where $\boldsymbol{a} \cdot \boldsymbol{n} < 0$. The solution can be expressed as (see Figure 5)

$$b_0^e(\boldsymbol{x}) = \begin{cases} \dfrac{y}{|\boldsymbol{a}| \sin\theta} & y < \dfrac{a_y}{a_x} x \\ \dfrac{x}{|\boldsymbol{a}| \cos\theta} & y > \dfrac{a_y}{a_x} x \end{cases} \quad (48)$$

For a rectangular element of sides h_x and h_y, the norms of $b_0^e(\boldsymbol{x})$ yield the corresponding error time scales,

$$\tau_{L_1}^e = \frac{||b_0||_{L_1(\Omega^e)}}{\text{meas}(\Omega^e)} = \frac{h_{\text{flow}}}{2|\boldsymbol{a}|} \left(1 - \frac{1}{3} \frac{a_y}{a_x} \frac{h_x}{h_y}\right) \leq \frac{h_{\text{flow}}}{2|\boldsymbol{a}|}$$

$$\tau_{L_2}^e = \frac{||b_0||_{L_2(\Omega^e)}}{\text{meas}(\Omega^e)^{1/2}} = \frac{h_{\text{flow}}}{\sqrt{3}|\boldsymbol{a}|} \sqrt{1 - \frac{1}{2} \frac{a_y}{a_x} \frac{h_x}{h_y}} \leq \frac{h_{\text{flow}}}{\sqrt{3}|\boldsymbol{a}|}$$

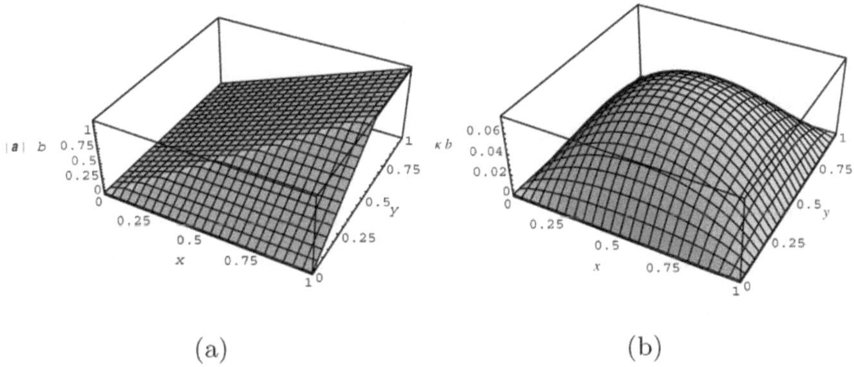

Fig. 5. Bubble function $b_0^e(x, y)$ for the rectangular element in (a) the hyperbolic limit for a velocity of $\boldsymbol{a} = (\sin 30, \cos 30)$; (b) the diffusive limit.

where h_{flow} is the longest length of the element along the streamwise direction.

Remarks.

1. The above upper bound for the L_1 error scale, calculated here for the reactangular bilinear quad and any velocity direction, was suggested by [12]. An attempt to include flow directionality into the flow intrinsic time scale was given in [13].
2. For $\theta = 0$, the one-dimensional error time scales are recovered.
3. Let us recall that in 1D $\tau_{\text{flow}} = \tau_{L_1}$ [18].

5.2 Elliptic Limit

In the elliptic limit, the residual-free bubble is the solution of the problem

$$\begin{cases} \kappa \Delta b_0^e = 1 & \text{in } \Omega^e \\ b_0^e = 0 & \text{on } \Gamma^e \end{cases} \tag{49}$$

where κ is the diffusion coefficient.

The solution can be expressed as the series [9] (see Figure 5)

$$b_0^e(\boldsymbol{x}) = \frac{16}{\pi^4 \kappa} \sum_{m=1(\text{odd})}^{\infty} \sum_{n=1(\text{odd})}^{\infty} \frac{1}{\frac{n^2}{h_x^2} + \frac{m^2}{h_y^2}} \frac{1}{nm} \sin\left(\frac{n\pi}{h_x} x\right) \sin\left(\frac{m\pi}{h_y} y\right) \tag{50}$$

which leads to the following error time scales:

$$\tau_{L_1} = \frac{64}{\pi^6 \kappa} \sum_{m=1(\text{odd})}^{\infty} \sum_{n=1(\text{odd})}^{\infty} \frac{1}{\frac{n^2}{h_x^2} + \frac{m^2}{h_y^2}} \frac{1}{(nm)^2} \tag{51}$$

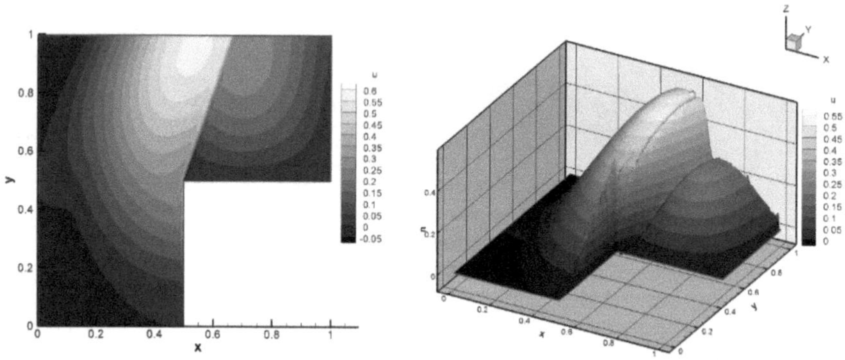

Fig. 6. L-shaped domain problem. Exact solution for $\kappa = 10^{-6}$.

$$\tau_{L_2} = \frac{8}{\pi^4 \kappa} \sqrt{\sum_{m=1(\text{odd})}^{\infty} \sum_{n=1(\text{odd})}^{\infty} \frac{1}{\left(\frac{n^2}{h_x^2} + \frac{m^2}{h_y^2}\right)^2} \frac{1}{(nm)^2}} \tag{52}$$

Particular Case. The error scales for $h_x = h_y$ are calculated as

$$\tau_{L_1} = \frac{h_x^2}{28.45\kappa}, \qquad \tau_{L_2} = \frac{h_x^2}{24.24\kappa} \tag{53}$$

Remarks.

1. Note that the error time scales are approximately half of the one-dimensional diffusive flow time scale.
2. The error time scales for the L_1 and L_2 norms are similar.
3. For the reaction limit, see [20].

6 Numerical Example: L-shaped Domain Problem

This problem, described in [26], has zero essential boundary conditions along the domain boundary, which has the shape of an L. The parameters of the problem are $\boldsymbol{a} = (1, 3)$, two values of $\kappa = 10^{-6}, 1$, $s = -1$ and the independent force term $f(x)$,

$$f(x, y) = 100r(r - 0.5)(r - 1/\sqrt{2}) \tag{54}$$

with $r^2 = (x - 0.5)^2 + (y - 0.5)^2$. Figure 6 shows the exact solution for the low viscosity case.

Figures 7 and 8 show the global and local efficiencies for the advection dominated problem. Again, results are reasonably good, with global efficiencies close to one and correct local efficiencies in areas where the error is large.

For the diffusion dominated problem (Figures 9 and 10), the global efficiencies are about unity with accurate predictions of the local error (with local efficiencies varying between 0.5 and 10).

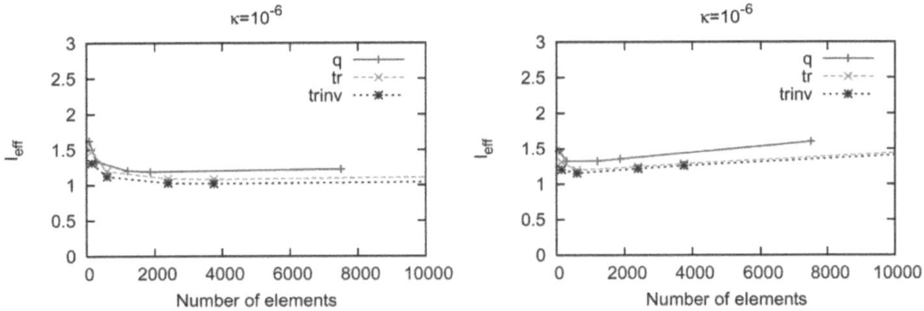

Fig. 7. L-shaped domain problem. Efficiencies for the L_1 (left) and L_2 (right) norms. Advection dominated flow. Legend: q stands for quads; tr, for triangles with the diagonal aligned with the flow direction; trinv, triangles with the diagonal opposite to the flow direction.

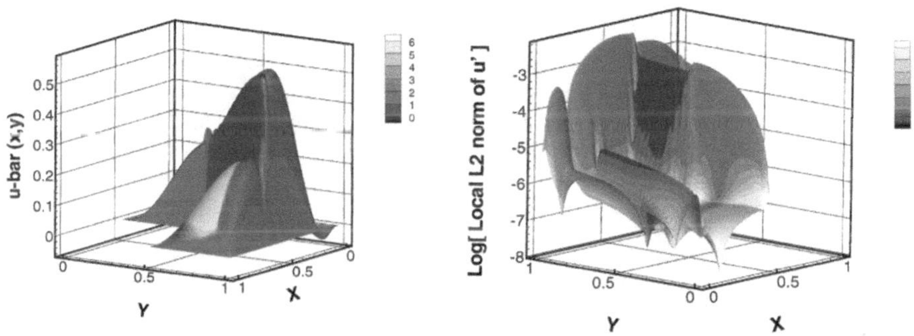

Mesh $h_x = 0.01$.

Fig. 8. L-shaped domain problem. Local efficiencies for the L_2 norm plotted over the finite element solution (left) and over the L_2 norm of local exact error (right). Advection dominated flow.

7 Adaptivity

In the adaptive process, given a finite element solution on a mesh, the next step is to design an ideal mesh that would result in a solution with the desired accuracy. In this work, that is achieved by prescribing the ideal element size as a function of space, leading to coarsening or refinement.

Given an error tolerance in the L_2 norm, $e_{\text{tol}} = ||u - \bar{u}||_{\Omega^e} = ||u'_{\text{tol}}||_{\Omega^e}$, the norm of the estimated L_2 error $\eta_e^{(i)}$ and the mesh size distribution $h_e^{(i)}$ at iteration i, standard arguments lead to the mesh size distribution at iteration $(i+1)$,

$$h_e^{(i+1)} = \left[\frac{e_{\text{tol}}}{\eta_e}\right]^{1/p} h_e^{(i)} \tag{55}$$

with p the order of convergence of the solution.

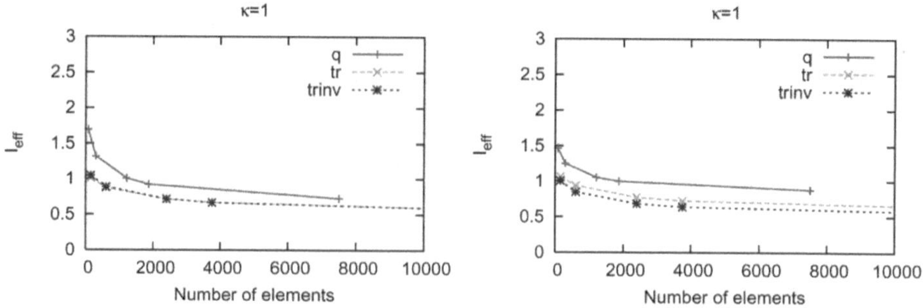

Fig. 9. L-shaped domain problem. Efficiencies for the L_1 (left) and L_2 (right) norms. Diffusion dominated flow.

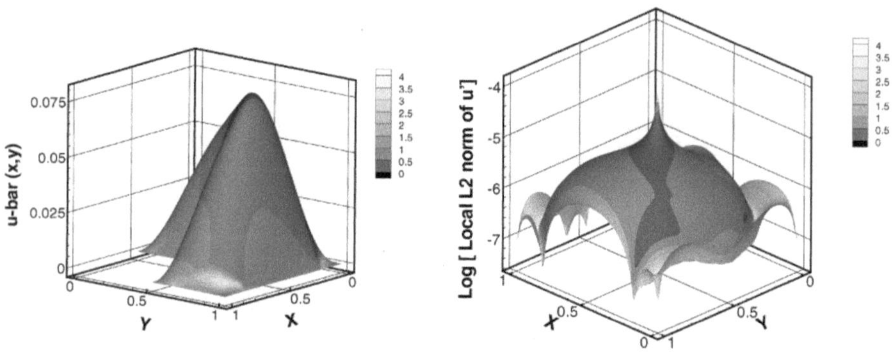

Mesh $h_x = 0.01$.

Fig. 10. L-shaped domain problem. Local efficiencies for the L_2 norm plotted over the finite element solution (left) and over the L_2 norm of local exact error (right). Diffusion dominated flow.

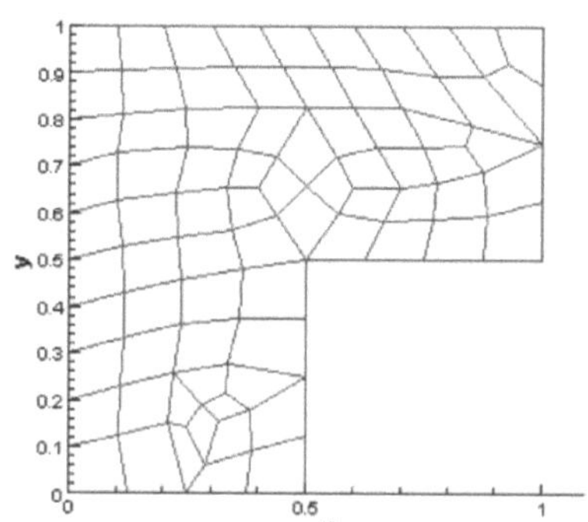

Fig. 11. L-shaped domain problem. Initial mesh.

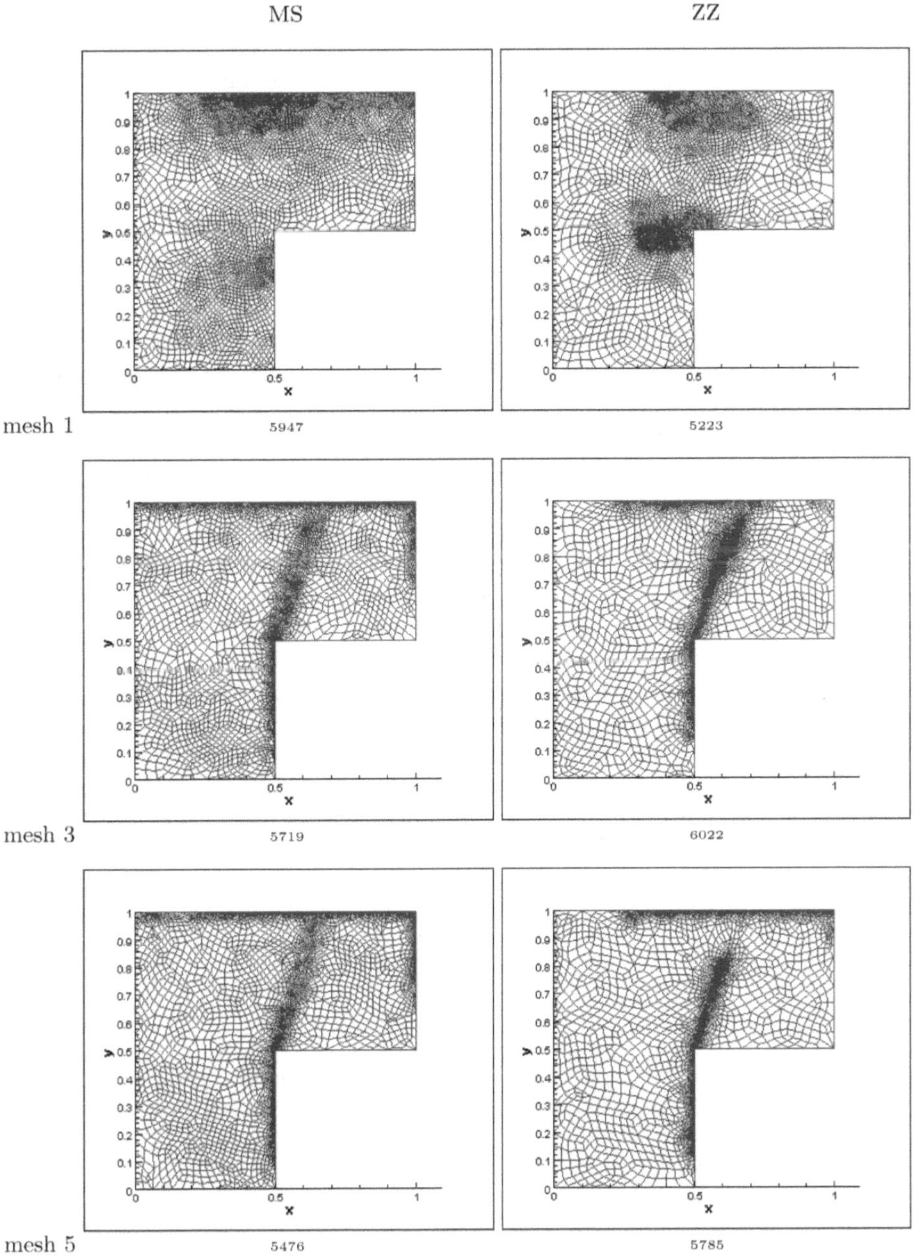

Fig. 12. L-shaped domain problem. Adaptive process for the MS and ZZ methods.

Figure 11 shows the initial mesh and Figure 12, the adaptation process based on the L_2 norm. The mesh produced by the ZZ [30] method concentrates all the elements near the boundary and interior layers, however, missing the boundary layer at the right boundary and the outflow portion of the interior boundary layer. The variational multiscale method (MS) produces a graded refinement, concentrating elements along all the length of the layers.

8 Conclusions

An *explicit* a posteriori error estimator has been derived from the variational multiscale theory. The error estimator, which is based on a model that introduces approximations compatible with the theory of stabilized methods, includes both, element interior and inter-element residuals. The element interior residual is key for predicting the error in the hyperbolic regime, whereas the inter-element residual, for the diffusive regime.

It has been shown that the error constants can be written in the form of *error* time scales, which have been calculated explicitly from element Green's functions.

Numerical examples confirm that in all cases the global efficiencies are very close to one and that the local efficiencies are a good approximation of the true error, mainly in areas where the errors are large.

The good behavior of the present technique can be explained by the fact that it solves *a-priori* the local dual problems at the element level, therefore, attaining similar accurancy to the more sophisticated implicit methods. Secondly, the error distribution of methods stemming from H_0^1 projection is practically local, and, therefore, for these methods most of the error information remains at the element level.

Thus, the proposed a-posteriori error estimator leads to a very economical and robust technique for transport problems computed with stabilized methods and can be readily implemented in existing computer codes.

References

1. Agarwal, A.N. and Pinsky, P., Stabilized element residual method (SERM): A posteriori error estimation for the advection-diffusion equation. *Journal of Computational and Applied Mathematics*, 74:3–17, 1996.
2. Ainsworth, M. and Oden, J.T., *A Posterior Error Estimation in Finite Element Analysis*. John Wiley & Sons, 2000.
3. Brenner, S.C. and Scott, L.R., *The Mathematical Theory of Finite Element Methods*. Springer-Verlag, 2002.
4. Brezzi, F., Bristeau, M.O., Franca, L.P., Mallet, M., and Rogé, G., A relationship between stabilized finite element methods and the Galerkin method with bubble functions. *Comput. Meth. Appl. Mech. Engrng.*, 96:117–129, 1992.
5. Brezzi, F., Franca, L.P., Hughes, T.J.R., and Russo, A., $b = \int g$. *Comput. Meth. Appl. Mech. Engrng.*, 145:329–339, 1997.

6. Brezzi, F., Hauke, G., Marini, L.D., and Sangalli, G., Link-cutting bubbles for the stabilization of convection-diffusion-reaction problems. *Math. Models Methods Appl. Sci.*, 13:445–461, 2003.
7. Brezzi, F. and Russo, A.: Choosing bubbles for advection-diffusion problems. *Math. Models Methods Appl. Sci.*, 4:571–587, 1994.
8. Brooks, A.N. and Hughes, T.J.R., Streamline upwind/petrov-galerkin formulations for convection dominated flows with particular emphasis on the incompressible navier-stokes equations. *Comput. Meth. Appl. Mech. Engrng.*, 32:199–259, 1982.
9. Courant, R. and Hilbert, D., *Methods of Mathematical Physics*, Vol. I. John Wiley & Sons, 1989.
10. Franca, L.P., Frey, S.L., and Hughes, T.J.R., Stabilized finite element methods: I. Application to the advective-diffusive model. *Comput. Meth. Appl. Mech. Engrng.*, 95:253–276, 1992.
11. Franca, L.P. and Russo, A.: Deriving upwinding, mass lumping and selective reduced integration by residual-free bubbles. *Appl. Math. Lett.*, 9:83–88, 1996.
12. Franca, L.P. and Valentin, F.: On an improved unusual stabilized finite element method for the advective-reactive-diffusive equation. *Comput. Meth. Appl. Mech. Engrng.*, 190:1785–1800, 2000.
13. Harari, I., Franca, L.P., and Oliveira, S.P., Streamline design of stability parameters for advection-diffusion problems. *J. Comput. Phys.*, 171:115–131, 2001.
14. Hauke, G. and Doweidar, M.H., Intrinsic scales and a posteriori multiscale error estimation for piecewise-linear functions and residuals. *Int. J. Comput. Fluid Dynamics*, 20:211–222, 2006.
15. Hauke, G., Doweidar, M.H., and Fuster, D., Multiscale multi-dimensional explicit a-posteriori error estimation for fluid dynamics. In: P. Wesseling, E. Oñate, J. Périaux (Eds.), *European Conference on Computational Fluid Dynamics*, pp. 1–13, 2006.
16. Hauke, G., Doweidar, M.H., Fuster, D., Gomez, A., and Sayas, J.: Application of variational a-posteriori multiscale error estimation to higher-order elements. *Comput. Mech.*, 38:382–389, 2006.
17. Hauke, G., Doweidar, M.H., and Miana, M.: The multiscale approach to error estimation and adaptivity. *Comput. Meth. Appl. Mech. Engrng.*, 195:1573–1593, 2006.
18. Hauke, G., Doweidar, M.H., and Miana, M.: Proper intrinsic scales for a-posteriori multiscale error estimation. *Comput. Meth. Appl. Mech. Engrng.*, 195:3983–4001, 2006.
19. Hauke, G. and Fuster, D., Variational multiscale a-posteriori error estimation for quantities of interest. *J. Appl. Mech.*, 76:021,201(1–6), 2009.
20. Hauke, G., Fuster, D., and Doweidar, M.H., Variational multiscale a-posteriori error estimation for the multi-dimensional transport equation. *Comput. Meth. Appl. Mech. Engrg.*, 197:2701–2718, 2008.
21. Hughes, T. and Sangalli, G., Variational multiscale analysis: the fine-scale green's function, projection, optimization, localization and stabilized methods. *SIAM J. Numer. Anal.*, 45(2):539–557, 2007.
22. Hughes, T.J.R., Multiscale phenomena: Green's functions, the Dirichlet-to-Neumann formulation, subgrid scale models, bubbles and the origins of stabilized methods. *Comput. Meth. Appl. Mech. Engrng.*, 127:387–401, 1995.
23. Hughes, T.J.R., *The Finite Element Method: Linear Static and Dynamic Finite Element Analysis*. Dover Publications, 2000.
24. Hughes, T.J.R., Feijoo, G.R., Mazzei, L., and Quincy, J.B., The variational multiscale method: A paradigm for computational mechanics. *Comput. Meth. Appl. Mech. Engrng.*, 166:3–24, 1998.

25. Hughes, T.J.R. and Sangalli, G.: Variational multiscale analysis: The fine-scale Green's function, projection, optimization, localization and stabilized methods. *SIAM J. Numer. Anal.*, 45(2):539–557, 2007.
26. John, V.: A numerical study of a posteriori error estimators for convection-diffusion equations. *Comput. Methods Appl. Mech. Engrg.*, 190:757–781, 2000.
27. Larson, M.G. and Målqvist, A.: Adaptive variational multiscale methods based on a posteriori error estimation: Energy norm estimates for elliptic problems. *Comput. Methods Appl. Mech. Engrg.*, 196:2313–2324, 2007.
28. Russo, A.: A posteriori error estimators via bubble functions. *Math. Models Methods Appl. Sci.*, 1:33–41, 1996.
29. Verfürth, R.: A posteriori error estimators for convection-diffusion equations. *Numer. Math.*, 80:641–663, 1998.
30. Zienkiewicz, O.C. and Zhu, J.Z.: A simple error estimator in the finite element method. *Int. J. Numer. Methods Engrg.*, 24:337–357, 1987.

Advances in Variational Multiscale Methods for Turbulent Flows

Peter Gamnitzer[1], Volker Gravemeier[1,2] and Wolfgang A. Wall[1]

[1]Institute for Computational Mechanics, Technische Universität München, Boltzmannstr. 15, 85747 Garching, Germany; gamnitzer,wall@lnm.mw.tum.de
[2]Emmy Noether Research Group "Computational Multiscale Methods for Turbulent Combustion", Technische Universität München, Boltzmannstr. 15, 85747 Garching, Germany; vgravem@lnm.mw.tum.de

Abstract This article presents some of our recent approaches to variational multiscale methods for turbulent flows. The residual-based variational multiscale method using a dynamic subgrid-scale approximation is presented as well as the algebraic variational multiscale-multigrid method for large eddy simulation of turbulent flows. The influence of isogeometric representations is briefly reviewed and current research directions are described in the context of turbulent fluid-structure interaction. Preliminary results for turbulent channel flow on moving meshes are provided.

Key words: Turbulent flows, large eddy simulation, variational multiscale method, isogeometric concepts, ALE

1 Introduction

Let $\boldsymbol{f}$ be a body force, $\boldsymbol{\varepsilon}(\boldsymbol{u}) = \frac{1}{2}(\nabla\boldsymbol{u}+\nabla\boldsymbol{u}^T)$ the rate of deformation tensor and ν the kinematic fluid viscosity. Then, for incompressible flow, a velocity-pressure solution $(\boldsymbol{u}, p)$ of the Navier–Stokes equations given as

$$\frac{\partial \boldsymbol{u}}{\partial t} + \nabla \circ \big(\boldsymbol{u} \otimes \boldsymbol{u}\big) + \nabla p - 2\nu\nabla \circ \boldsymbol{\varepsilon}(\boldsymbol{u}) - \boldsymbol{f} = \boldsymbol{0} \quad \text{and} \quad \nabla \circ \boldsymbol{u} = 0 \qquad (1)$$

on domain Ω, subject to initial and boundary conditions, is assumed to represent all scales involved in laminar and turbulent flows.

Turbulent flows represent a multiscale problem due to the wide range of scales they contain. This reaches from the characteristic length of the geometry ℓ down to the Kolmogorov length scale η. The ratio of these scales is proportional to $\ell/\eta = \mathrm{Re}^{3/4}$ based on the Reynolds number $\mathrm{Re} = u\ell/\nu$, where u is a characteristic velocity, see e.g. [1]. A numerical resolution of all scales in a high-Reynolds-number turbulent flow is still computationally unaffordable for flows of engineering interest.

In this article, a survey will be given of several recent developments in the area of variational multiscale methods for turbulent flows (VMM). In contrast to filtering in

R. de Borst and E. Ramm (eds.), *Multiscale Methods in Computational Mechanics*, Lecture Notes in Applied and Computational Mechanics 55, DOI 10.1007/978-90-481-9809-2_3,

traditional large eddy simulation (LES), the variational multiscale approach to large eddy simulation as described in [2] introduces scale separation in the weak form of (1),

$$B(\boldsymbol{v}, p, \boldsymbol{u}; \boldsymbol{u}) + (q, \nabla \circ \boldsymbol{u})_\Omega = (\boldsymbol{v}, \boldsymbol{f})_\Omega, \tag{2}$$

where

$$\begin{aligned} B(\boldsymbol{v}, p, \boldsymbol{u}; \boldsymbol{u}) := \left(\boldsymbol{v}, \frac{\partial \boldsymbol{u}}{\partial t}\right)_\Omega &+ (\boldsymbol{v}, \nabla \circ (\boldsymbol{u} \otimes \boldsymbol{u}))_\Omega - \\ -(\nabla \circ \boldsymbol{v}, p)_\Omega &+ 2\nu(\boldsymbol{\varepsilon}(v), \boldsymbol{\varepsilon}(\boldsymbol{u}))_\Omega, \end{aligned}$$

by variational projection. In equation (2), $\boldsymbol{v}$ and q denote velocity and pressure test functions, elements of the infinite-dimensional test function space W. The symbol $(.,.)_\Omega$ denotes the inner product of $L^2(\Omega)$. For the ease of notation, we assume zero Dirichlet boundary conditions on the whole boundary $\partial\Omega$ of domain Ω, implying that the space of solution functions V is equivalent to W. Without loss of generality, we will stick to this simplification below. For a review addressing VMM methods in laminar and turbulent flows, the reader is referred to [3], and to [4] for a comparison of VMM-based LES and classical LES.

The first method reviewed in this paper, the residual-based variational multiscale method (resVMM), is based on a two-scale separation. In this approach, unresolved scales are taken into account by an approximate analytical solution of the subgrid-scale problem as described in [5]. This residual-based approximation is usually based on the assumption that the current residual instantaneously defines the subgrid-scale velocities. In [6], an ordinary differential equation was introduced to account for the dynamic character of the subgrid scales, and the properties of that method were studied for laminar flow problems. In [7], the impact of this dynamic subgrid approximation was investigated in detail for turbulent flow in a channel. Section 2 will summarize the theory and key results.

The resVMM is compared to a method based on a three-scale separation, the algebraic variational multiscale-multigrid method. This method uses level-transfer operators based on plain aggregation algebraic multigrid methods to separate scales. In general, three-scale methods allow modeling to be confined to the smaller of the resolved scales, and hence to preserve consistency with respect to the larger of the resolved scales. See, e.g., [8–13] for alternative developments on the basis of a three-scale method. The algebraic variational multiscale-multigrid method for large eddy simulation was introduced in [14]. Section 3 will be dedicated to this new approach.

One of the keys for obtaining good results is a proper choice of trial and test function space. Results for example in [5, 15] indicate that non-uniform rational B-spline (NURBS) shape functions as introduced in [16] are superior to finite elements using Lagrangian polynomials. Section 4 was included to look into this recent achievement.

Before finally concluding, an outlook to further research directions in VMM for turbulent flows will be provided in Section 5 by addressing applications on moving meshes as they arise, for instance, in fluid-structure interaction.

2 Residual-Based Variational Multiscale Method with Dynamic Subgrid Scales

The residual-based variational multiscale method using dynamic subgrid scales as introduced and used in [7] is based on the residual-based variational multiscale method described in [5] and the time-dependent subgrid-scale approach introduced in [6]. To separate large and small scales, a direct sum decomposition $V = V^h \oplus \hat{V}$ of trial and test function spaces is introduced. This implies a decomposition of solution and test functions into a resolved-scale part denoted by $(.)^h$ and an unresolved-scale part $\hat{(.)}$:

$$\boldsymbol{u} = \boldsymbol{u}^h + \hat{\boldsymbol{u}}, \quad p = p^h + \hat{p}; \qquad \boldsymbol{v} = \boldsymbol{v}^h + \hat{\boldsymbol{v}}, \quad q = q^h + \hat{q}. \tag{3}$$

Using this decomposition, the weak form of the Navier–Stokes equations is decoupled into a resolved-scale equation

$$\begin{aligned} &B(\boldsymbol{v}^h, p^h, \boldsymbol{u}^h; \boldsymbol{u}^h) - (\boldsymbol{v}^h, \boldsymbol{f})_\Omega + (q^h, \nabla \circ \boldsymbol{u}^h)_\Omega + \\ &+ \left(\boldsymbol{v}^h, \frac{\partial \hat{\boldsymbol{u}}}{\partial t}\right) - (\nabla \circ \boldsymbol{v}^h, \hat{p})_\Omega - 2\nu(\nabla \circ \boldsymbol{\varepsilon}(\boldsymbol{v}^h), \hat{\boldsymbol{u}})_\Omega - \\ &-(\nabla q^h, \hat{\boldsymbol{u}})_\Omega - ((\boldsymbol{u}^h \circ \nabla)\boldsymbol{v}^h, \hat{\boldsymbol{u}})_\Omega - \\ &-((\hat{\boldsymbol{u}} \circ \nabla)\boldsymbol{v}^h, \boldsymbol{u}^h)_\Omega - ((\hat{\boldsymbol{u}} \circ \nabla)\boldsymbol{v}^h, \hat{\boldsymbol{u}})_\Omega = 0 \end{aligned} \tag{4}$$

and an unresolved-scale equation. In the framework of the resVMM in general, a closure for the resolved-scale equation is provided by an approximation to the solution of the unresolved-scale momentum equation. This approximation is made based on $\boldsymbol{r}_M^h(\boldsymbol{u}^h, p^h)$, the momentum's residual in convective form. In contrast to the resVMM [5], the time-dependent subgrid-scale approach introduced in [6] employs a different closure equation for the subgrid problem:

$$\left[\begin{array}{c} \dfrac{1}{\tau_M^{\mathrm{rVM}}}\hat{\boldsymbol{u}} = -\boldsymbol{r}_M^h(\boldsymbol{u}^h, p^h) \\ \left(\dfrac{\partial \hat{\boldsymbol{u}}}{\partial t}, \boldsymbol{v}^h\right)_\Omega \approx \boldsymbol{0} \end{array}\right] \longleftrightarrow \quad \frac{\partial \hat{\boldsymbol{u}}}{\partial t} + \frac{1}{\tau_M^{\mathrm{td}}}\hat{\boldsymbol{u}} = -\boldsymbol{r}_M^h(\boldsymbol{u}^h, p^h), \tag{5}$$

where τ_M^{td} is an algebraic parameter depending on viscosity, velocity and grid size. In addition, τ_M^{rVM} depends on the time-step size of the applied time integration to account for the emerging reaction-type term in the time-discretized weak form. Using another algebraic parameter, τ_C, and the residual of the continuity equation, $r_C^h(\boldsymbol{u}^h)$, the subgrid-scale pressure is defined as is usual in stabilized methods:

$$\frac{1}{\tau_C}\hat{p} = -\nabla \circ \boldsymbol{u}^h = -r_C^h(\boldsymbol{u}^h) \tag{6}$$

Using the subgrid closure equations above, a streamline upwind and pressure stabilized Petrov–Galerkin (SUPG/PSPG/grad-div) formulation according to [17–19] plus cross- and Reynolds-stress terms according to [5] plus time-dependent terms according to [6] is achieved. It is completed by initial conditions for both resolved- and subgrid-scale velocities.

One option for discretizing the resulting system of equations in time is the generalized-α method, see [20, 21]. For details on that particular time integration method and its application, the reader is referred to [7]. Any other time-integration scheme may alternatively be used.

In [7], computations of turbulent channel flows at various Reynolds numbers Re_τ were performed, comparing the results of resVMM with and without time-dependent subgrid scales. Those results and results for flow around a square cylinder at Re $= 22000$, which will be shown in Section 4, clearly indicate a limited impact of the dynamic representation of the unresolved scales on the flow statistics of resolved-scale quantities. Even though it is known from [22] that the resVMM approach described in [5] looses its accuracy for small time-step sizes, only a limited impact is observable over a surprisingly wide range of time-step sizes as investigated in [7] for the channel flow problem. In contrast to the resolved-scale behavior, it was that in the range of time-step sizes usually employed for turbulent channel flow computations, the subgrid quantities in the resVMM approach depend heavily on the time-step size. Taking into account the dynamic character of the subgrid scales was proven to be a remedy for that problem, leading to a more robust representation of unresolved scales independent of the time-step size used.

3 The Algebraic Variational Multiscale-Multigrid Method

In the algebraic variational multiscale-multigrid method described in [14], the resolved scales in equation (3) are separated further into large and small resolved scales, resulting in a three-scale separation

$$\boldsymbol{u} = \underbrace{\boldsymbol{u}^{3h} + \delta\boldsymbol{u}^{h}}_{\boldsymbol{u}^h} + \hat{\boldsymbol{u}} \qquad \boldsymbol{v} = \underbrace{\boldsymbol{v}^{3h} + \delta\boldsymbol{v}^{h}}_{\boldsymbol{v}^h} + \hat{\boldsymbol{v}}. \tag{7}$$

The superscript $3h$ indicates the usual aggregation factor 3 in aggregation-based algebraic multigrid methods. The scale separation is introduced in a purely algebraic manner using a matrix representation of the scale separation operator,

$$\boldsymbol{S}_h^{3h} := \boldsymbol{P}_{3h}^{h}\boldsymbol{R}_h^{3h}, \tag{8}$$

operating on the velocity degrees of freedom $\boldsymbol{U}^h$. In equation (8), $\boldsymbol{P}_{3h}^{h}$ is the prolongation operator matrix generated using plain aggregation algebraic multigrid (PA-AMG). The restriction operator $\boldsymbol{R}_h^{3h}$ is the transpose of $\boldsymbol{P}_{3h}^{h}$. Using this scale-separating operator matrix, the degrees-of-freedom representations $\boldsymbol{U}^{3h}$, $\delta\boldsymbol{U}^h$ of $\boldsymbol{u}^{3h}$, $\delta\boldsymbol{u}^h$ are given as

$$\boldsymbol{U}^{3h} = \boldsymbol{S}_h^{3h}\boldsymbol{U}^h \qquad \text{and} \qquad \delta\boldsymbol{U}^h = \left[\boldsymbol{I} - \boldsymbol{S}_h^{3h}\right]\boldsymbol{U}^h. \tag{9}$$

According to scale separation (7), the resolved-scale equation (4) is split into three equations:

- the momentum equation for the large resolved scales

$$B(\boldsymbol{v}^{3h}, p^h, \boldsymbol{u}^h; \boldsymbol{u}^h) - (\boldsymbol{v}^{3h}, \boldsymbol{f})_\Omega + (\boldsymbol{v}^{3h}, \dot{\hat{\boldsymbol{u}}})_\Omega - (\nabla \circ \boldsymbol{v}^{3h}, \hat{p})_\Omega - \\ -2\nu(\nabla \circ \boldsymbol{\varepsilon}(\boldsymbol{v}^{3h}), \hat{\boldsymbol{u}})_\Omega - ((\boldsymbol{u}^h \circ \nabla)\boldsymbol{v}^{3h}, \hat{\boldsymbol{u}})_\Omega - \\ -((\hat{\boldsymbol{u}} \circ \nabla)\boldsymbol{v}^{3h}, \boldsymbol{u}^h)_\Omega - ((\hat{\boldsymbol{u}} \circ \nabla)\boldsymbol{v}^{3h}, \hat{\boldsymbol{u}})_\Omega = 0 \tag{10}$$

- the momentum equation for the small resolved scales

$$B(\delta\boldsymbol{v}^{h}, p^h, \boldsymbol{u}^h; \boldsymbol{u}^h) - (\delta\boldsymbol{v}^{h}, \boldsymbol{f})_\Omega + (\delta\boldsymbol{v}^{h}, \dot{\hat{\boldsymbol{u}}})_\Omega - (\nabla \circ \delta\boldsymbol{v}^{h}, \hat{p})_\Omega - \\ -2\nu(\nabla \circ \boldsymbol{\varepsilon}(\delta\boldsymbol{v}^{h}), \hat{\boldsymbol{u}})_\Omega - ((\boldsymbol{u}^h \circ \nabla)\delta\boldsymbol{v}^{h}, \hat{\boldsymbol{u}})_\Omega - \\ -((\hat{\boldsymbol{u}} \circ \nabla)\delta\boldsymbol{v}^{h}, \boldsymbol{u}^h)_\Omega - ((\hat{\boldsymbol{u}} \circ \nabla)\delta\boldsymbol{v}^{h}, \hat{\boldsymbol{u}})_\Omega = 0 \tag{11}$$

- the continuity equation

$$(q^h, \nabla \circ \boldsymbol{u}^h)_\Omega = (\nabla q^h, \hat{\boldsymbol{u}})_\Omega. \tag{12}$$

Since equal-order interpolation for velocity and pressure is used, stability of the numerical scheme requires a pressure-stabilizing Petrov–Galerkin term in the last equation (12) as in Section 2:

$$(q^h, \nabla \circ \boldsymbol{u}^h)_\Omega = -(\nabla q^h, \tau_M^{\text{rVM}} \boldsymbol{r}_M^h)_\Omega \tag{13}$$

In equation (10), the impact of unresolved scales on large resolved scales is neglected based on the assumption that their respective influence is relatively weak.

$$B(\boldsymbol{v}^{3h}, p^h, \boldsymbol{u}^h; \boldsymbol{u}^h) = (\boldsymbol{v}^{3h}, \boldsymbol{f})_\Omega \tag{14}$$

Finally, in equation (11), the impact of the unresolved scales on the small resolved scales is modeled by a Smagorinsky-type model

$$B(\delta\boldsymbol{v}^h, p^h, \boldsymbol{u}^h; \boldsymbol{u}^h) + 2\nu_T^{\delta h}(\boldsymbol{\varepsilon}(\delta\boldsymbol{v}^h), \boldsymbol{\varepsilon}(\delta\boldsymbol{u}^h))_\Omega = (\delta\boldsymbol{v}^h, \boldsymbol{f})_\Omega, \tag{15}$$

with the subgrid viscosity defined as

$$\nu_T^{\delta h} = (C_S h)^2 \cdot \sqrt{\boldsymbol{\varepsilon}(\delta\boldsymbol{u}^h) : \boldsymbol{\varepsilon}(\delta\boldsymbol{u}^h)} \tag{16}$$

Summing up, equations (13), (14) and (15) can be reassembled to

$$B(\boldsymbol{v}^h, p^h, \boldsymbol{u}^h; \boldsymbol{u}^h) + (q^h, \nabla \circ \boldsymbol{u}^h)_\Omega + 2\nu_T^{\delta h}(\boldsymbol{\varepsilon}(\delta\boldsymbol{v}^h), \boldsymbol{\varepsilon}(\delta\boldsymbol{u}^h))_\Omega + \\ +(\nabla q^h, \tau_M^{\text{rVM}} \boldsymbol{r}_M^h)_\Omega = (\boldsymbol{v}^h, \boldsymbol{f})_\Omega \tag{17}$$

For some applications it turned out beneficial to add other stabilization terms such as the grad-div term in equation (17). Note that those terms would have a direct impact on all resolved scales, in contrast to the small-scale Smagorinsky model introduced above which affects only the small resolved scales directly. Referring once more to the algebraic character of the scale separation, the result of equation (17) stated as a matrix system reads

$$\begin{bmatrix} \boldsymbol{K}^{hh} + \boldsymbol{M}^{hh}_{\mathrm{FSSV}} \left(\boldsymbol{I} - \boldsymbol{S}^{3h}_{h} \right) & \boldsymbol{G}^{hh} \\ \overline{\boldsymbol{C}}^{hh} & \overline{\boldsymbol{D}}^{hh} \end{bmatrix} \cdot \begin{bmatrix} \boldsymbol{U}^{h} \\ \boldsymbol{P}^{h} \end{bmatrix} = \begin{bmatrix} \boldsymbol{F}^{h} \\ \overline{\boldsymbol{F}}^{h} \end{bmatrix}. \tag{18}$$

$\boldsymbol{K}^{hh}$ contains the transient, (nonlinear) convective and viscous term, $\boldsymbol{G}^{hh}$ the pressure term, and $\overline{\boldsymbol{C}}^{hh}$ the continuity equation. All entries with an overbar are generated or modified by the PSPG stabilization. $\boldsymbol{M}^{hh}_{\mathrm{FSSV}}$ results from the fine-scale subgrid viscosity term. The scale separating operator matrix $\boldsymbol{S}^{3h}_{h}$ allows the application of $\boldsymbol{M}^{hh}_{\mathrm{FSSV}}$ exclusively to the smaller of the resolved scales.

Numerical results obtained with this method can be found in [14, 23], showing excellent performance for applications to turbulent channel flow, turbulent flow in a driven cavity and flow around a square-section cylinder.

4 Using NURBS in Residual-Based Variational Multiscale Methods

In resVMM, the unresolved scales are driven by the residual of the momentum equation evaluated at integration points. And since the shape function type has a considerable impact on the residual it also has a direct impact on the modeled unresolved scales. Shape functions which seem to be better suited than standard polynomial functions are non-uniform rational B-splines

$$N_i^p(u) = \frac{w_i \cdot B_i^p(u)}{\sum_{j=1}^{p+1} w_j \cdot B_j^p(u)}, \tag{19}$$

where w_i is the i-th weight and $B_i^p(u)$ is the i-th B-spline polynomial of degree p defined by recursion on a series of $2p + 2$ knot values u_k, see [24] for an introduction. One feature making NURBS-based FEM especially attractive for multiscale simulations is its high solution quality over the complete scale spectrum. Hence, the subgrid-scale terms can really focus on accounting for physical effects and do not have to account for the shortcomings of the coarse-scale method.

The use of NURBS shape-functions in finite elements was initiated in [16], where an isogeometric FEM is defined, and important properties such as positiveness of shape functions and their variation-diminishing behavior are presented. In isogeometric FEM, the domain is decomposed into several patches on which the solution functions and the geometry are described using NURBS shape functions:

$$\begin{bmatrix} \boldsymbol{x}(\xi,\eta,\zeta) \\ \boldsymbol{u}(\xi,\eta,\zeta) \\ p(\xi,\eta,\zeta) \end{bmatrix} = \sum_{i,j,k} \begin{bmatrix} \boldsymbol{x}_{i,j,k} \\ \boldsymbol{u}_{i,j,k} \\ p_{i,j,k} \end{bmatrix} \cdot \underbrace{N_i^p(u(\xi)) \cdot N_j^p(v(\eta)) \cdot N_k^p(w(\zeta))}_{=:N_I^p(u(\xi),v(\eta),w(\zeta))} \tag{20}$$

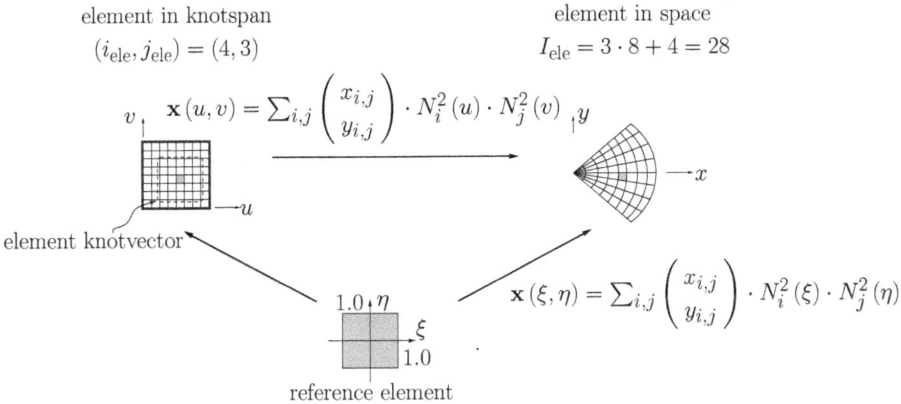

Fig. 1. In FEM, shape functions are integrated on a reference element. For NURBS shape functions, the direct mapping between reference element and actual element in space has to be done via the 'element in the knotspan', which has access to the element knot vector required for the shape function recursion (all knot values inside the dashed box). The position of the element in the knotspan is defined by the global ID of the element in space and can be used to access the specific element knot vector and the associated control points carrying the degrees of freedom.

Table 1. Number of nonzeros on matrix (nnz) and ratio between nnz and number of unknowns (ndof). Note that in comparison to serendipity elements, full quadratic elements would have an even larger nnz as well as a larger bandwidth.

Element type	64^3 trilinear	32^3 quad. serendipity	32^3 quad. NURBS
nnz	1.138e+08	1.235e+08	6.718e+07
ndof	1,064,960	536,576	139,264
$\frac{\text{nnz}}{\text{ndof}}$	106.8	230.1	482.4

Due to the cartesian structure of the knotspan of each patch, the triple index i, j, k can be converted into the unique ID I of the associated control point $(\boldsymbol{x}_I, w_I)$ using the position $i_{ele}, j_{ele}, k_{ele}$ of the element in the knotspan and the unique ID I_{ele} of the actual element, see Figure 1 for a two-dimensional example. Isogeometric finite elements were successfully applied to turbulent flows, for example in [5, 15] to turbulent channel flow.

A question often raised in the context of NURBS approximations is how the choice of the shape function influences the sparsity pattern of the system matrix involved in the computation. Table 1 contains numbers for these quantities for various discretizations of a turbulent channel flow configuration. Although the NURBS approximation leads to the least sparse representation of the three approaches, the

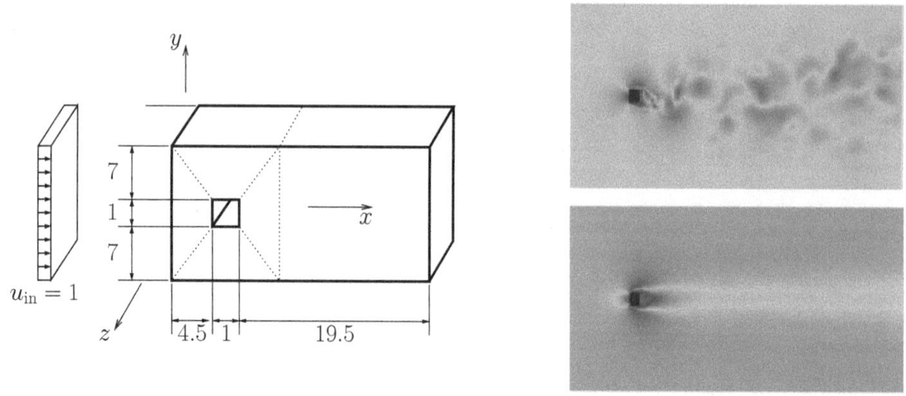

Fig. 2. Turbulent flow around a square cylinder. Setup (left), instantaneous velocity field (top right) and velocity field averaged in time and space along the z-axis (bottom right). Dotted lines in the setup sketch on the left indicate patch boundaries.

number of nonzeros on the system matrix, which is of interest for matrix-vector multiplications required in an iterative solver, is much smaller than in the other two approaches. Hence, the NURBS shape functions do not only provide the best results of these three approximations, see [5, 15], but are also the approximation with the smallest number of nonzeros on the system matrix.

Another point of interest is the introduction of initial conditions so that their spatial distribution fits to the method. Since NURBS shape functions are not interpolating in general, a direct setting of values at control points is not feasible. Instead, we propose the solution of the following least-squares problem for the initial control-point values $\boldsymbol{u}^h_{0,I}$:

$$\min_{\boldsymbol{u}^h_0} \left\| \sum_I N_I \cdot \boldsymbol{u}^h_{0,I} - \boldsymbol{u}_0 \right\|^2 \tag{21}$$

Note that this minimization problem is equivalent to the solution of a linear system using a "mass" matrix as the system matrix.

Finally, we want to present the first results of a resVMM with time-dependent subgrid scales on NURBS elements. The problem under consideration is the flow around a square cylinder at $\mathrm{Re} = 22\,000$, a benchmark presented in [25], see Figure 2 for the flow configuration. The example contains an unsteady separation of the flow over the cylinder, resulting in a three dimensional vortex shedding in the cylinder wake, see Figure 3. The computations were done on 33 000 quadratic NURBS elements distributed in 5 patches. In order to better approximate the boundary layer on the cylinder surface, a hyperbolic mesh stretching (tanh-type with stretching constant $C = 2.5$) was applied to the respective knot-vectors. A time-step length $dt = 0.05$ is chosen. Periodic boundary conditions in z-direction and no-slip boundary conditions on the cylinder surface are assumed.

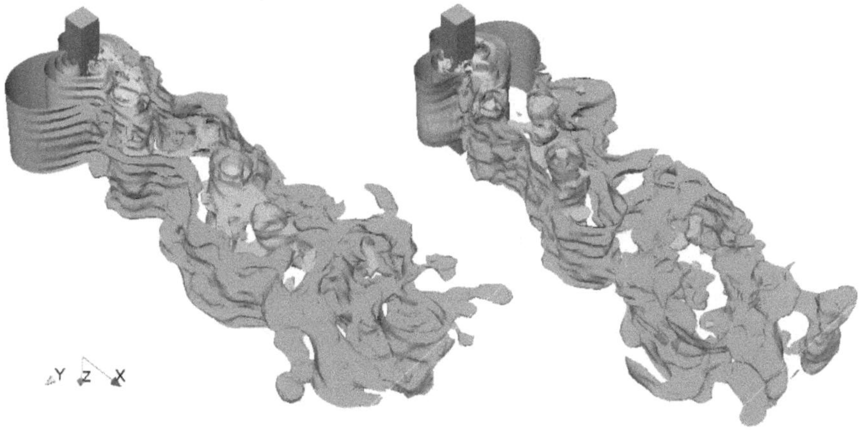

Fig. 3. Pressure isosurfaces in the cylinder wake for two time-step sizes. The three-dimensionality of turbulent flows can be clearly observed behind the cylinder.

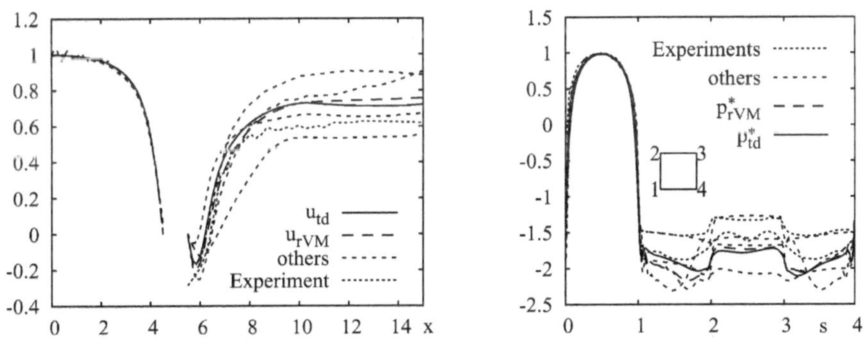

Fig. 4. Mean x-velocity along the x-axis (left) and dimensionless pressure along cylinder surface (right). Experimental data and results from other computations based on finite volume or finite difference methods are taken from [25]. The reader is referred to the respective paper for an exact assignment of the results and further information.

Results for the space (z-direction) and time averaged x-velocity along the x-axis and for the averaged dimensionless pressure

$$p^* := \frac{p - p_{\text{in}}}{\frac{1}{2}\rho \cdot u_{\text{in}}^2}$$

along the cylinder surface are given in Figure 4. In this formula, p_{in} and u_{in} denote the inlet velocity and pressure and ρ the fluid density. Although the mesh is coarse and the boundary layer is not adequately resolved, the results achieved are in quite good agreement with the experimental data presented in [25]. Note that a significant difference between the resVMM with (td) and without (rVM) a time-dependent subgrid-scale approximation is not observed also for this example. Even the oscilla-

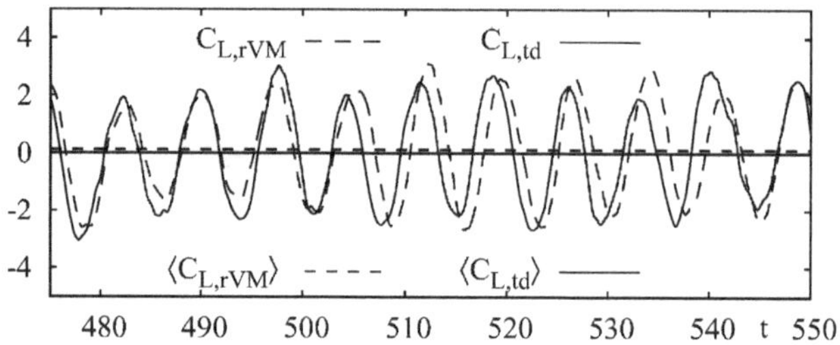

Fig. 5. Oscillation of lift coefficients for standard resVMM ($C_{\mathrm{L,rVM}}$) and resVMM using a time-dependent subgrid-scale approximation ($C_{\mathrm{L,td}}$). The dashed lines, $\langle C_{\mathrm{L,rVM}}\rangle$ and $\langle C_{\mathrm{L,td}}\rangle$, denote the respective time-averaged mean values.

tion of the lift coefficient, as displayed in Figure 5, is similarly reproduced by both approaches.

5 Towards a Residual-Based Variational Multiscale Method for Turbulent Fluid-Structure Interaction

For fluid-structure interaction (FSI) problems, the solution of the Navier–Stokes equation has to be computed on a deforming domain. This can be done, for example, using an Arbitrary Lagrangian–Eulerian (ALE) formulation of the Navier–Stokes equation

$$\begin{aligned}\frac{\partial \boldsymbol{u}(\boldsymbol{\phi}(\boldsymbol{\xi},t),t)}{\partial t}\big(\boldsymbol{\phi}^{-1}(\boldsymbol{x},t),t\big) + (\boldsymbol{c}\,(\boldsymbol{x},t)\circ\nabla)\,\boldsymbol{u}\,(\boldsymbol{x},t) + \nabla\, p(\boldsymbol{x},t) - \\ - 2\nu\nabla\circ\boldsymbol{\varepsilon}\big(\boldsymbol{u}(\boldsymbol{x},t)\big) = \boldsymbol{f}\,(\boldsymbol{x},t)\end{aligned}$$

$$\nabla\circ\boldsymbol{u}(\boldsymbol{x},t) = 0$$

In this equation, $\boldsymbol{\phi}$ is the mesh motion mapping and $\boldsymbol{c}$ the ALE convective velocity:

$$\boldsymbol{c}(\boldsymbol{x},t) = \boldsymbol{u}(\boldsymbol{x},t) - \boldsymbol{u}^G(\boldsymbol{x},t) = \boldsymbol{u}(\boldsymbol{x},t) - \left(\frac{\partial}{\partial t}\boldsymbol{\phi}\right)\big(\boldsymbol{\phi}^{-1}(\boldsymbol{x},t),t\big) \tag{22}$$

See, e.g., [26] for a derivation of these equations.

The application of the resVMM to FSI problems appears to be straightforward at first sight. Successful applications of the three-scale VMM to turbulent fluid flow computations on moving meshes are reported in [27].

To estimate the potential of the resVMM for turbulent FSI computations, preliminary test simulations of turbulent channel flow at $\mathrm{Re}_\tau = 395$ on a deformable mesh were performed. The mesh motion, defined as

$$\boldsymbol{\phi}(\boldsymbol{x}_0,t) = \boldsymbol{x}_0 + \sin(t)\cdot(1-(\boldsymbol{x}_{0,y})^2)\cdot 1.2, \tag{23}$$

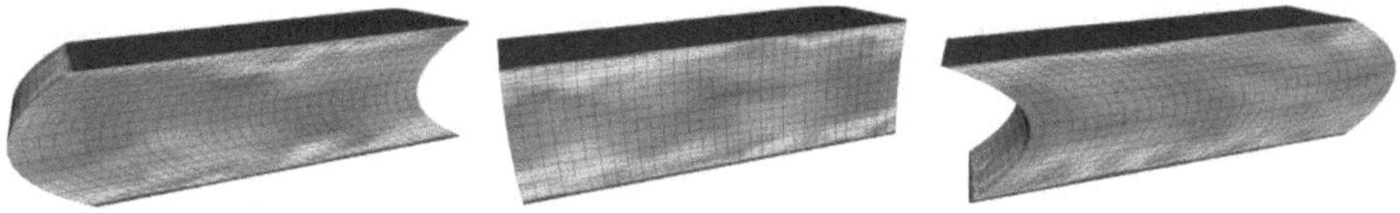

Fig. 6. Test case for turbulent channel flow on moving grids.

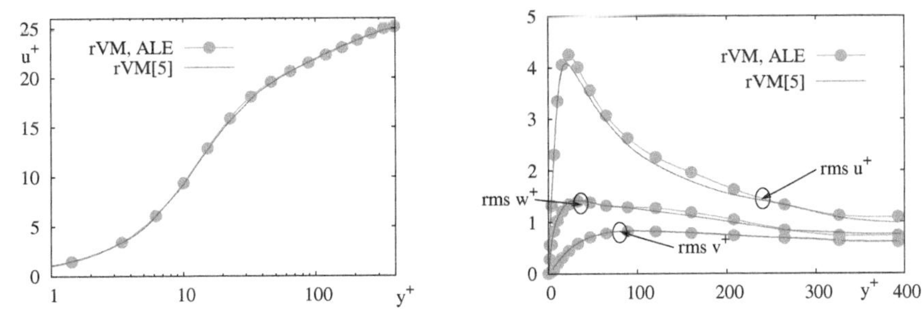

Fig. 7. Results for turbulent channel flow computations on 32^3 trilinear elements at $Re_\tau = 395$. The solution on the deforming domain (rVM, ALE) is in excellent agreement with the solution on the fixed mesh taken from [5] (rVM [5]).

leads to deformations as depicted in Figure 7. For simplicity, this mesh-motion is chosen such that it preserves the distance between nodes and walls. This allows for a nodal-based spatial averaging in wall-parallel planes. The maximum of the grid velocity induced by ϕ is approximately equal to the mean streamwise velocity. For details on mesh, geometry, material parameters and sampling, the reader is referred to the setting presented in [7]. In Figure 7, results from a computation on a mesh deforming according to equation (23) are compared to a result on a fixed mesh taken from [5]. The results for moving and fixed grid turned out to be almost identical. The good performance for this simple test case motivates further investigation of this approach in more complex configurations. This will be part of our future work.

6 Conclusion

In this article, several recent developments in the context of the variational multiscale method for the numerical solution of turbulent flows have been briefly reviewed. First, two selected variants of variational multiscale methods for large eddy simulation were discussed, one based on a two-scale and one based on a three-scale separation. The section on a residual-based variational multiscale method based on a two-scale separation focused on the time-dependency of the unresolved scales. Although a dynamic subgrid-scale approximation improves the representation of the unresolved scales, the influence on the resolved-scale results is relatively small. The fact that the stabilization parameter for the dynamic subgrid-scale approximation does not depend on the time-step size may be exploited for performing a parameter tuning for particular problems such as boundary-layer flows. For example, better choices for the characteristic element lengths in boundary-layer meshes might improve the quality of the resolved-scale solution, and this improvement would not be restricted to a particular choice of the time-step size. The algebraic variational multiscale-multigrid method for large eddy simulation was furthermore introduced, representing a three-scale approach. Within this method, scales are separated using operators taken from plain aggregation algebraic multigrid methods.

From the subsequent discussion of isogeometric analysis, two results shall be highlighted once more. First, the sparsity pattern resulting from a NURBS-based method is comparable to the corresponding pattern resulting from a traditional approach using Lagrangean polynomials. However, the total number of degrees of freedom for a quadratic NURBS approximation is significantly smaller than the respective number for a traditional approximation of comparable accuracy, even if serendipity elements are used. Second, for turbulent flow around a square cylinder, results of a resVMM using a dynamic subgrid-scale approximation on a NURBS discretization have been presented. They indicate that, even for coarse discretizations, NURBS-based elements allow for a high-quality subgrid-scale approximation resulting in quite accurate predictions of mean flow quantities.

Acknowledgement

The present study is supported by a grant of the "Deutsche Forschungsgemeinschaft" (DFG) through project WA 1521/2, i.e. project II.D. within the DFG Research Unit 509 "Multiscale Methods in Computational Mechanics". This support is gratefully acknowledged. Furthermore, the second author is grateful to the DFG for the support via the Emmy Noether program.

References

1. S.B. Pope, *Turbulent Flows*, Cambridge University Press, 2000.
2. T.J.R. Hughes, L. Mazzei, and K.E. Jansen, Large eddy simulation and the variational multiscale method, *Computing and Visualization in Science*, 3:47–59, 2000.
3. V. Gravemeier, The variational multiscale method for laminar and turbulent flow, *Archives of Computational Methods in Engineering – State of the Art Reviews*, 13:249–324, 2006.
4. V. John, On large eddy simulation and variational multiscale methods in the numerical simulation of turbulent incompressible flows, *Applications of Mathematics*, 51(4):321–353, 2006.
5. Y. Bazilevs, V.M. Calo, J.A. Cottrell, T.J.R. Hughes, A. Reali, and G. Scovazzi, Variational multiscale residual-based turbulence modeling for large eddy simulation of incompressible flows, *Computer Methods in Applied Mechanics and Engineering*, 197:173–201, 2007.
6. R. Codina, J. Principe, O. Guasch, and S. Badia, Time dependent subscales in the stabilized finite element approximation of incompressible flow problems, *Computer Methods in Applied Mechanics and Engineering*, 196:2413–2430, 2007.

7. P. Gamnitzer, V. Gravemeier, and W.A. Wall, Time-dependent subgrid scales in residual-based large eddy simulation of turbulent channel flow, *Computer Methods in Applied Mechanics and Engineering*, submitted.
8. S.S. Collis, Monitoring unresolved scales in multiscale turbulence modeling, *Physics of Fluids*, 13(6):1800–1806, 2001.
9. V. Gravemeier, W.A. Wall, and E. Ramm, A three-level finite element method for the instationary, incompressible Navier–Stokes equations, *Computer Methods in Applied Mechanics and Engineering*, 193:1323–1366, 2004.
10. T.J.R. Hughes, A.A. Oberai, and L. Mazzei, Large eddy simulation of turbulent channel flows by the variational multiscale method, *Physics of Fluids*, 13(6):1784–1799, 2001.
11. B. Koobus and C. Farhat, A variational multiscale method for the large eddy simulation of compressible turbulent flows on unstructured meshes – Application to vortex shedding, *Computer Methods in Applied Mechanics and Engineering*, 193(15/16):1367–1383, 2004.
12. V. John and S. Kaya, A finite element variational multiscale method for the Navier–Stokes equations, *SIAM Journal on Scientific Computing*, 26(5):1485–1503, 2005.
13. V. Gravemeier, Scale-separating operators for variational multiscale large eddy simulation of turbulent flows, *Journal of Computational Physics*, 212: 400–435, 2006.
14. V. Gravemeier, M.W. Gee, M. Kronbichler, and W.A. Wall, An algebraic variational multiscale-multigrid method for large eddy simulation of turbulent flow, *Computer Methods in Applied Mechanics and Engineering*, doi:10.1016/j.cma.2009.05.017.
15. I. Akkerman, Y. Bazilevs, V.M. Calo, T.J.R. Hughes, and S. Hulshoff, The role of continuity in residual-based variational multiscale modeling of turbulence, *Computational Mechanics*, 41:371–378, 2008.
16. T.J.R. Hughes, J.A. Cottrell, and Y. Bazilevs, Isogeometric analysis: Cad, finite elements, nurbs, exact geometry and mesh refinement, *Computer Methods in Applied Mechanics and Engineering*, 194(39/41):4135–4195, 2005.
17. A.N. Brooks and T.J.R. Hughes, Streamline upwind/Petrov–Galerkin formulations for convection dominated flows with particular emphasis on the incompressible Navier–Stokes equation, *Computer Methods in Applied Mechanics and Engineering*, 32:199–259, 1982.
18. T.J.R. Hughes, L.P. Franca, and M. Balestra, A new finite element formulation for computational fluid dynamics: V. Circumventing the Babuska–Brezzi condition: A stable Petrov–Galerkin formulation of the Stokes problem accommodating equal-order interpolation, *Computer Methods in Applied Mechanics and Engineering*, 59(1):85–99, 1986.
19. P. Hansbo and A. Szepessy, A velocity-pressure streamline diffusion finite element method for the incompressible Navier–Stokes equations, *Computer Methods in Applied Mechanics and Engineering*, 84:175–192, 1990.
20. J. Chung and G.M. Hulbert, Time integration algorithm for structural dynamics with improved numerical dissipation: The generalized-alpha method, *Journal of Applied Mechanics*, 60(2):371–375, 1993.
21. K.E. Jansen, C.H. Whiting, and G.M. Hulbert, A generalized-α method for integrating the filtered Navier–Stokes equations with a stabilized finite element method, *Computer Methods in Applied Mechanics and Engineering*, 190(3/4):305–319, 2000.
22. M.C. Hsu, Y. Bazilevs, V.M. Calo, T.E. Tezduyar, and T.J.R. Hughes, Improving stability of stabilized and multiscale formulations in flow simulations at small time steps, *Computer Methods in Applied Mechanics and Engineering*, doi:10.1016/j.cma.2009.06.019.
23. V. Gravemeier, M. Kronbichler, M.W. Gee, and W.A. Wall, An algebraic variational multiscale-multigrid method for large eddy simulation: Generalized-α time integration,

Fourier analysis and application to turbulent flow past a square-section cylinder, to be submitted.
24. L. Piegl and W. Tiller, *The NURBS Book*, Springer, 2000.
25. W. Rodi, J.H. Ferziger, M. Breuer, and M. Pourquié, Status of large eddy simulation: Results of a workshop, *Journal of Fluids Engineering*, 119:248–262, 1997.
26. J. Donea and A. Huerta, *Finite Element Methods for Flow Problems*, Wiley, 2003.
27. A. Rajesekharan and C. Farhat, Application of a variational multiscale method for large eddy simulation of turbulent flows on moving/deforming unstructured grids, *Finite Elements in Analysis and Design*, 45:272–279, 2009 .

Variational Germano Approach for Multiscale Formulations

Ido Akkerman[1], Steven J. Hulshoff[2], Kris G. van der Zee[3] and René de Borst[4]

[1]Structural Engineering, University of California, San Diego, 9500 Gilman Dr., La Jolla, CA 92093, USA; IAkkerman@ucsd.edu
[2]Aerospace Engineering, Delft University of Technology, Kluyverweg 1, 2629 HS Delft, The Netherlands; S.J.Hulshoff@tudelft.nl
[3]Institute for Computational Engineering and Sciences, The University of Texas at Austin, 201 East 24th Street, Austin, TX 78712, USA
[4]Mechanical Engineering, Eindhoven University of Technology, P.O. Box 513, WH 1.125, 5600 MB Eindhoven, The Netherlands

Abstract In this chapter the recently introduced Variational Germano procedure is revisited. The procedure is explained using commutativity diagrams. A general Germano identity for all types of discretizations is derived. This relation is similar to the Variational Germano identity, but is not restricted to variational numerical methods. Based on the general Germano identity an alternative algorithm, in the context of stabilized methods, is proposed. This partitioned algorithm consists of distinct building blocks. Several options for these building blocks are presented and analyzed and their performance is tested using a stabilized finite element formulation for the convectionŰdiffusion equation. Non-homogenous boundary conditions are shown to pose a serious problem for the dissipation method. This is not the case for the least-squares method although here the issue of basis dependence occurs. The latter can be circumvented by minimizing a dual-norm of the weak relation instead of the Euclidean norm of the discrete residual.

Key words: Variational Germano, stabilized finite elements, variational multiscale

1 Introduction

Many numerical methods, Large-eddy simulation (LES) methods in particular, employ algorithmic parameters for enhancing stability and accuracy of the method. As such, these parameters are of paramount importance. However, optimal values for these parameters are often not exactly known. Usually these parameters depend on the discretization characteristics such as mesh size and order of approximation. In these cases a dynamic procedure similar to the one introduced by Germano [1] can be used to automatically choose parameters in an objective fashion.

The original approach by Germano was proposed for tuning the Smagorinsky constant, the key parameter in the popular Smagorinsky model used for closing the

R. de Borst and E. Ramm (eds.), *Multiscale Methods in Computational Mechanics*, Lecture Notes in Applied and Computational Mechanics 55, DOI 10.1007/978-90-481-9809-2_4,

filtered LES equations. Therefore the original Germano approach is cast in a filtering framework. The approach works as follows. First a solution is computed using a initial guess of the constant. This solution is filtered using a filter width larger than the original filter width. Combining these two solutions results in a relation for the Smagorinsky constant.

In the recent work of Oberai and Wanderer [2] the Germano approach is cast in a variational form by replacing the filter by a variational projector. For a variety of problems this approach has proven to give satisfying results [3–5]. Similar approaches have been presented by Hoffman [6] and Onate et al. [7]. Hoffman casts the whole procedure in wavelet form, but the method is essentially identical to that of Oberai. Onate et al. generate a superior solution by means of stress recovery, instead of an inferior solution by an additional coarse projection, and compare the two solutions in a fashion similar to the procedure of Oberai.

In this work we continue in the direction established by Oberai. The numerical method is interpreted as a discrete projector, which is graphically expressed using commutativity diagrams. By combining multiple diagrams, a general graphical explanation for the Germano approach is found. This clarifies the process of tuning a numerical method, for example via the parameters it incorporates. After these generic developments a specific algorithm for tuning stabilization parameters in stabilized finite element formulations is proposed.

In the proposed algorithm the coarse-scale equations are used directly, without subtracting the large-scale equations. This results in a simplified algorithm with clearly identifiable components, including reconstruction and interpolation of the stabilization parameter, and relaxation.

Two main types of reconstruction will be discussed, the dissipation method and the Least-Squares method. The standard dissipation method is shown to break down in case of non-homogenous boundary conditions. Alternatives based on a Lagrange-multiplier approach are therefore investigated. For the Least-Squares method, non-homogenous boundary conditions do not pose a problem. However, the straightforward Least-Squares approach based on the Euclidean norm of the residual turns out to be basis dependent. This is shown both theoretically and numerically. A revised method incorporating a dual norm is thus proposed.

For clarity, the presented concepts are demonstrated using a stabilized method for the 1D convection-diffusion equation. This allows straightforward assessment of competing approaches, as the exact expression for the stabilization parameter is known.

2 General Discrete Germano Identity

2.1 Numerical Method as a Discrete Projector

Assume one is interested in the solution u of the following abstract problem:

$$\mathcal{L}u = f \tag{1}$$

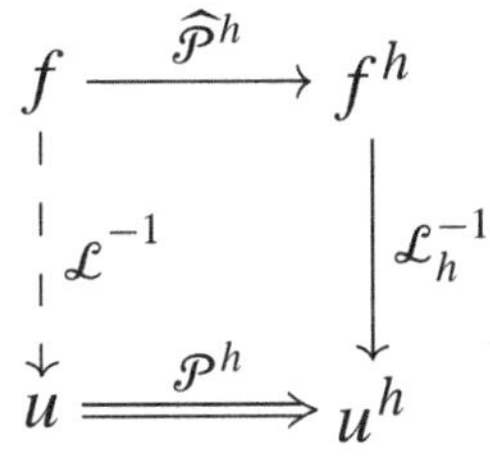

Fig. 1. Numerical method implicitly defines a projector.

Usually the problem is too difficult to solve analytically and a closed form expression for the solution is not available. An approximate solution can be obtained by resorting to a numerical method. The associated discretization limits the space of approximate solutions and as such the numerical method implicitly defines a mapping,

$$\mathcal{P}^h : \mathcal{V} \longrightarrow \mathcal{V}^h \qquad\qquad u^h = \mathcal{P}^h u \tag{2}$$

which maps the exact solution onto a discrete subspace. Consistency of the numerical method requires this mapping to be a projector,

$$u^h = \mathcal{P}^h u^h \qquad \longrightarrow \qquad \mathcal{P}^h = \mathcal{I} \text{ on } \mathcal{V}^h \tag{3}$$

This projection is schematically presented in the commutativity diagram of Figure 1. $\widehat{\mathcal{P}}^h$ indicates the discretization projector of the force term.

In the diagram the dashed line indicates the unavailable inversion of the operator in the continuous setting, which would have resulted in the exact solution. The numerical method is indicated by the two solid lines, resulting in the approximate solution. The projector implied by the numerical method is represented by the double solid line.

2.2 Inverse Implication: Projector Implies Numerical Method

The idea of the Germano approach is to reverse this implication. In other words, to define a projector and construct a numerical method such that its solution is the desired projection of the exact solution. Or, in a weaker form, the solution approximates the desired projection reasonably well. This reversed implication is presented schematically in Figure 2.

The dashed line again indicates the unavailable inversion of the continuous operator. The solid line indicates the desired projector, while the double lines indicate the implied numerical method.

The variational multiscale method [8–10] is constructed with this implication in mind. How well the numerical solution approximates the projection of the exact solution depends on the small scale reconstruction. When the small scales, or at least the terms in the large scale equation involving the small scales, are reconstructed exactly, then the numerical method results in a solution which is the desired projection.

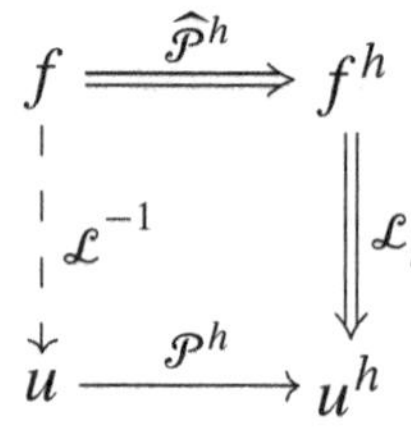

Fig. 2. Projector defines the numerical method.

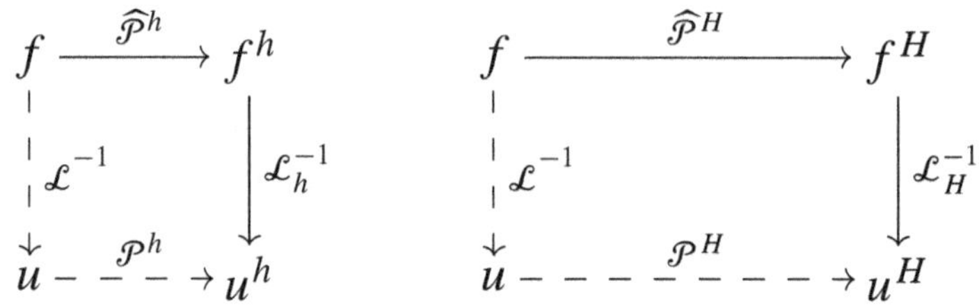

Fig. 3. Commutativity diagram for a fine and coarse discretization.

The difference between numerical solution and the projected solution is an indication of optimality. A natural choice for quantitative measurement of this difference is the norm associated with desired projection. Desirable features of the numerical solution need to be phrased in terms of the desired projector and the associated discrete norm. For example, the demand for the absence of spurious wiggles needs to be rephrased in an objective sense like nodal exactness or optimality in H_0^1-norm.

2.3 Multilevel Commutativity

When a numerical method is capable of reproducing a desired projector, then the commutativity diagram as given in Figure 2 holds. One can expect this commutativity to hold for all meshes, including the coarser mesh indicated by H in Figure 3. Combining the two relationships shown in Figure 3 results in the diagram as given in Figure 4.

This diagram is not very useful by itself. However, depending on the scale-separation projector and the meshes involved, the projector onto the coarse mesh, $\mathcal{P}^H$, can be decomposed in a projector onto the fine scale, $\mathcal{P}^h$, and a projector from the fine scale onto the coarse scales,$\mathcal{P}_h^H$. In other words: there exists a projector P_h^H such that,

$$\mathcal{P}_h^H : \mathcal{V}^h \rightarrow \mathcal{V}^H \qquad\qquad u^H = \mathcal{P}^H u = \mathcal{P}_h^H \mathcal{P}^h u \tag{8}$$

This projector can be defined when, for example, the desired projector is defined using an inner product and the coarse space is a subset of the fine space, $\mathcal{V}^H \subset \mathcal{V}^h$. Using the projector P_h^H and the combined commutativity results in a relationship between fine and coarse meshes as represented in Figure 5.

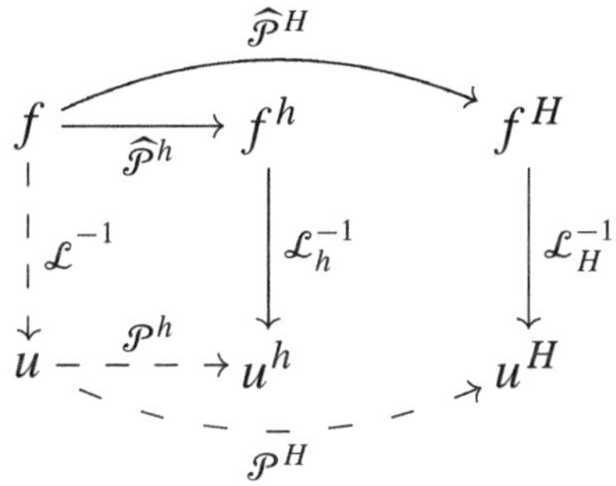

Fig. 4. Merged commutativity diagram for a fine and coarse discretization.

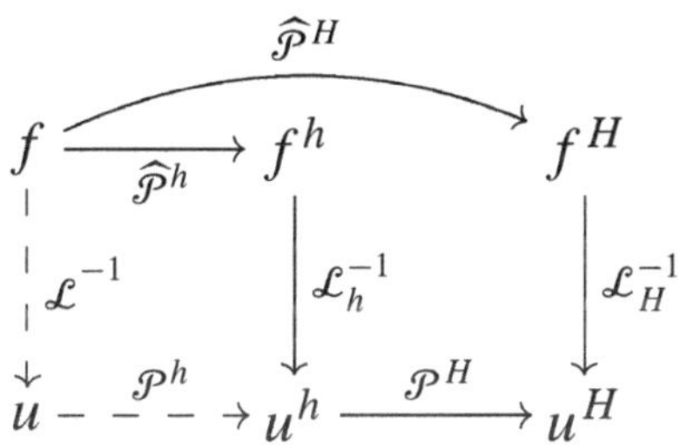

Fig. 5. Merged commutativity diagram for a fine and coarse discretization

In general the source projector can not be decomposed in a projector onto the fine scales and a projector from these fine scales onto the coarse scales. For instance, in stabilized methods the mesh-dependent stabilization parameter is part of the projection which prohibits the decomposition. Fortunately a decomposition of the source projector is not required. A decomposition of the scale separation projector is sufficient.

2.4 Discrete Germano Identity

Due to the decomposition of the projector there are two computable routes to the coarse solution u^H. In Figure 5 these two routes consist of solely solid lines. These two different ways of computing the coarse solution are:

$$\begin{aligned} u^H &= \mathcal{L}_H^{-1} f^H = \mathcal{L}_H^{-1} \widehat{\mathcal{P}}^H f \\ u^H &= \mathcal{P}_h^H u^h = \mathcal{P}_h^H \mathcal{L}_h^{-1} f^h = \mathcal{P}_h^H \mathcal{L}_h^{-1} \widehat{\mathcal{P}}^h f \end{aligned} \tag{10}$$

The first option is to compute the coarse solution directly, the second option is to compute the fine solution first and project this solution onto the coarse mesh. Using these two equations for the coarse-scale solution results in the vector relation

$$\mathcal{L}_H^{-1} \widehat{\mathcal{P}}^H f = \mathcal{P}_h^H \mathcal{L}_h^{-1} \widehat{\mathcal{P}}^h f \tag{11}$$

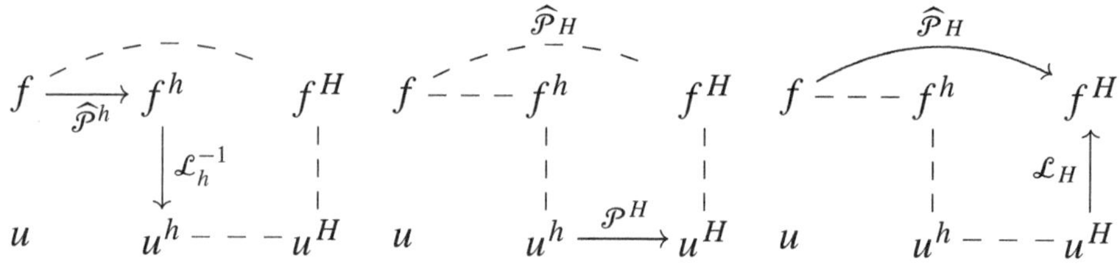

Fig. 6. Graphical representation of fixed-point iteration.

which is a necessary, but not necessarily sufficient, condition for the numerical method to satisfy the commutativity relation shown in f Figure 3. Assuming this relation to be independent of the source term leads to an operator equation,

$$\mathcal{L}_H^{-1}\widehat{\mathcal{P}}^H = \mathcal{P}_h^H \mathcal{L}_h^{-1}\widehat{\mathcal{P}}^h \tag{12}$$

However, due to the inverse operators this relation and the former vector relation are not very useful. This problem can be partially circumvented by decomposing the commutativity relation into the following three relations

$$\begin{aligned} \mathcal{L}_h u^h &= \widehat{\mathcal{P}}^h f \\ u^H &= \mathcal{P}_h^H u^h \\ \mathcal{L}_H u^H &= \widehat{\mathcal{P}}^H f \end{aligned} \tag{13}$$

where f is the known source and $\mathcal{P}_h^H$ is the known projector, derived from the desired projector. The unknowns are the fine and coarse solution u^h and u^H *and* the numerical method, that is $\mathcal{L}_h$,$\widehat{\mathcal{P}^h}$ and $\mathcal{L}_H$,$\widehat{\mathcal{P}^H}$.

2.5 Partitioned Approach

Although monolithic approaches are possible, the easiest way to proceed is to use a partitioned solution strategy. This is graphically depicted in Figure 6. The first step is a straightforward computation of the fine solution. The second step is a projection step giving a coarse solution. This coarse solution is then finally used in the third step, which is basically a relation for determining the optimal numerical method.

In the figure it is easy to see the three steps together are equivalent to the complete commutativity relation as given in Figure 5. To avoid scaling ambiguities with respect to the mesh-size, multiple coarse meshes might be required when determining the optimal numerical method (see Section 3.3 for further details).

3 Discrete Germano Approach for Stabilized Methods

For the case of stabilized finite element methods the commutativity relations of equation (13) become

$$\begin{aligned}\langle w^h, \mathcal{L}u^h - f\rangle + (\mathcal{L}^s w^h, \tau_h(\mathcal{L}u^h - f))_{\cup\Omega_e} &= 0 \qquad \forall w^h \in \mathcal{V}^h \\ u^H - \mathcal{P}_h^H u^h &= 0 \\ \langle w^H, \mathcal{L}u^H - f\rangle + (\mathcal{L}^s w^H, \tau_H(\mathcal{L}u^H - f))_{\cup\Omega_e} &= 0 \qquad \forall w^H \in \mathcal{V}^H \end{aligned} \tag{14}$$

where τ_H and τ_h are the coarse and fine stabilization parameters for residual-based stabilization of the coarse-mesh and fine-mesh problems. The partitioned solution strategy as presented in the previous section leads to the procedure given by Algorithm 1. The algorithm starts with a given fine stabilization parameter. Each iteration this parameter is updated until a fixed point is reached. The first two steps, solving for the fine and coarse solutions, are self-explanatory. However, the final three steps need some clarification.

choose initial τ_h
repeat
  step 1) solve for u^h using τ_h
  step 2) obtain u^H by projecting u^h on coarse mesh(es)
  step 3) compute τ_H using u^H
  step 4) construct τ_h using τ_H
  step 5) relax τ_h using previous τ_h's
until fixed-point is reached

Algorithm 1: The fixed-point procedure for stabilized methods.

In general the stabilization parameters are expected to vary in space. This is the case when mesh size h or physical parameters vary in space or when the problem is nonlinear. A representation of the spatial variation of τ requires a separate discretization for τ. Although an approach for finding spatially varying stabilization parameters will be presented, the examples considered have globally constant stabilization parameters.

The equations used for finding the coarse stabilization parameter are over constrained: a scalar stabilization parameter is to be determined while a vector of equations is available. Therefore these equations need to be manipulated such that one equation is obtained which unambiguously defines the coarse stabilization parameter. In the following section two approaches are discussed.

3.1 Computation of Coarse Stabilization Parameter: Dissipation Method

In this approach the weight function w^H in the last equation of (14) is chosen to be the coarse solution u^H. This results in an equation for the energy balance

$$\langle u^H, \mathcal{L}u^H - f\rangle + \tau_H(\mathcal{L}^s u^H, (\mathcal{L}u^H - f))_{\cup\Omega_e} = 0 \tag{15}$$

From (15) the coarse stabilization parameter which gives the anticipated dissipation can be extracted,

$$\tau_H = -\frac{\langle u^H, \mathcal{L}u^H - f\rangle}{(\mathcal{L}^s u^H, \mathcal{L}u^H - f))_{\cup\Omega_e}} \tag{16}$$

hence the name of this approach.

Non-Homogenous Boundary Conditions

To allow u^H to be chosen for the w^H requires the weak formulation to be cast in a Lagrange multiplier form. Where the boundary conditions are enforced via the weak form:

$$\begin{aligned}\langle w^H, \mathcal{L}u^H - f\rangle + \tau_H(\mathcal{L}^s w^H, \mathcal{L}u^H - f))_{\cup\Omega_e}& \\ +\langle w^H, \lambda^H\rangle_\Gamma + \langle \eta^H, u_\Gamma - u\rangle_\Gamma &=0\end{aligned} \tag{17}$$

where λ^H and η^H are the Lagrange multiplier and its corresponding test function. Setting the test functions equal to the coarse solution, $w^H = u^H$ and $\eta^H = \lambda^H$ respectively, and rearranging yields:

$$\tau_H = -\frac{\langle u^H, \mathcal{L}u^H - f\rangle + \langle \lambda^H, u_\Gamma\rangle_\Gamma}{(\mathcal{L}^s u^H, \mathcal{L}u^H - f))_{\cup\Omega_e}} \tag{18}$$

This relation still requires the determination of the coarse Lagrange multiplier. This can be done by either reconstruction on the fine level followed by a projection onto the coarse level or a direct reconstruction on the coarse level. In either case equation (17) can be used to obtain the following expression,

$$\langle w^h, \lambda^h\rangle_\Gamma = -\langle w^h, \mathcal{L}u^h - f\rangle - \tau_h(\mathcal{L}^s w^h, \mathcal{L}u^h - f))_{\cup\Omega_e} \tag{19}$$

where $\eta^h = 0$ and w^h should be chosen appropriately to obtain a relation for the Lagrange multiplier λ^h.

3.2 Computation of Coarse Stabilization Parameter: Least-Squares Method

A naive least-squares approach would minimize the Euclidean norm of the vector relation. However, this will result in a basis-dependent formulation, as shown in the following paragraph. In the subsequent sections a basis independent least-squares formulation for scalar and finite element τ_H will be presented.

Starting from the last equation of (14)

$$\langle w^H, \mathcal{L}u^H - f\rangle + \tau_H(\mathcal{L}^s w^H, \mathcal{L}u^H - f))_{\cup\Omega_e} =0 \qquad \forall w^H \in \mathcal{V}^H \tag{20}$$

using the following finite element basis

$$\text{span}\{\phi_1, \phi_2, .., \phi_n\} = \mathcal{V}^H \tag{21}$$

results in the following vector relation:

$$G_i + \tau_H M_i = 0 \qquad \forall i \in [1, 2, .., n] \tag{22}$$

with the shorthand notation

$$\begin{aligned} G_i &= \langle \phi_i, \mathcal{L}u^H - f \rangle \\ M_i &= (\mathcal{L}^s \phi_i, \mathcal{L}u^H - f)_{\cup \Omega_e} \end{aligned} \tag{23}$$

Minimizing the Euclidean norm of this vector relation with an additional penalty term

$$\tau_H = \arg\inf_{\tau \in \mathbb{R}} (G_i + \tau_H M_i)(G_i + \tau M_i) + \lambda(\tau - \tau_{ref})^2 \tag{24}$$

can be done by setting the derivative with respect to τ equal to zero

$$0 = M_i(G_i + \tau_H M_i) + \lambda(\tau_H - \tau_{ref}) \tag{25}$$

which gives

$$\tau_H = \frac{\lambda \tau_{ref} - M_i G_i}{\lambda + M_i M_i} \tag{26}$$

However, this is not a basis-independent formulation. The dependence on the basis is easily shown by taking a scaled basis,

$$\text{span}\{\alpha_1 \phi_1, \alpha_2 \phi_2, \ldots, \alpha_n \phi_n\} = \mathcal{V}^H \tag{27}$$

resulting in

$$\tau_H = \frac{\lambda \tau_{ref} - \sum \alpha_i M_i G_i}{\lambda + \sum \alpha_i M_i M_i} \tag{28}$$

which is clearly dependent on the scaling α_i. As such this approach is not appealing.

Basis Independent Least-Squares Method

A basis-independent least-squares method can be obtained by solving the following minimization relation:

$$\tau_H = \arg\inf_{\tau \in \mathbb{R}} \left(\| \mathcal{L}u^H - f \|_\tau^2 + \lambda(\tau - \tau_H^{ref})^2 \right) \tag{29}$$

The first term is the dual norm of the formulation, while the second term penalizes deviations from a specified reference parameter. It will be shown that this term guarantees coercivity. This helps to determine the stability parameter in case the residual term vanishes, which would otherwise make the stability parameter undetermined. The dual norm is defined as

$$\|\mathcal{L}u^H - f\|_\tau^2 = \sup_{w^H \in \mathcal{V}^H} \left(\frac{\langle w^H, \mathcal{L}u^H - f \rangle + (\mathcal{L}^s w^H, \tau(\mathcal{L}u^H - f))_{\cup\Omega_e}}{\|w^H\|_X} \right)^2$$

where X indicates an appropriate norm, such as the L_2-norm or H_0^1-norm. Using the Riesz representations,

$$\begin{aligned} \langle w^H, \mathcal{L}u^H - f \rangle &= (w^H, G^H)_X & \forall w^H \in \mathcal{V}^H \\ (\mathcal{L}^s w^H, \mathcal{L}u^H - f))_{\cup\Omega_e} &= (w^H, M^H)_X & \forall w^H \in \mathcal{V}^H \end{aligned} \tag{30}$$

defined using the inner product related to the X-norm, the dual norm can be written as:

$$\|u^H\|_{\mathcal{L}_H}^2 = \sup_{w^H \in \mathcal{V}^h} \left(\frac{(w^H, G^H + \tau_H M^H)_X}{\|w^H\|_X} \right)^2 \tag{31}$$

The supremum is attained when $w^H = G^H + \tau_H M^H$ resulting in

$$\tau_H = \arg\inf_{\tau \in \mathbb{R}} \left(\|G^H + \tau M^H\|_X^2 + \lambda(\tau - \tau_H^{ref})^2 \right) \tag{32}$$

The infimum can be found by setting the derivative with respect to τ equal to zero

$$(M^H, G^H + \tau_H M^H)_X + \lambda(\tau_H - \tau_H^{ref}) = 0 \tag{33}$$

resulting in

$$\tau_H = \frac{\lambda \tau_H^{ref} - (M^H, G^H)_X}{\lambda + (M^H, M^H)_X} \tag{34}$$

which is basis independent. However, this comes at the price of two extra projection problems, which are needed for obtaining the Riesz representations.

Basis Independent Least-Squares Method for a Spatial Varying Stability Parameter

A spatially varying τ can be approximated using finite elements. It is convenient to chose the space as that of the primal variables. In this case the following relation needs to be solved:

$$\tau_H = \arg\inf_{\tau \in \mathcal{V}^H} \left(\|\mathcal{L}u^H - f\|_\tau^2 + \lambda \|\tau - \tau_H^{ref}\|_Y^2 \right) \tag{35}$$

where Y indicates a norm, which should guarantee coercivity when appropriately chosen.

Using the τ dependent Riesz representations,

$$(\mathcal{L}^s w^H, \tau(\mathcal{L}u^H - f)))_{\cup\Omega_e} = (w^H, M^H(\tau))_X \qquad \forall w^H \in \mathcal{V}^H \tag{36}$$

and proceeding with similar steps as in the previous section we arrive at:

$$(M^H(w^H), M^H(\tau_H))_X + \lambda(w^H, \tau_H)_Y$$
$$= \lambda(w^H, \tau_H^{ref})_Y - (M^H(w^H), G^H)_X \qquad \forall w^H \in \mathcal{V}^H \quad (37)$$

From this the necessity for introducing the λ-term is clear. The term ensures the problem is coercive. Hence it guarantees the existence of a unique solution.

The approach is potentially very computationally expensive. The Riesz representant $M^H(w^H)$ for each test function needs to be determined by a projection. This results in a number of solves equal to the dimension of the finite element space used for approximating τ_H. This issue can be partially circumvented by either lumping the resulting matrix or choosing nodal inner products, resulting in nodal projections. In both cases an appropriate basis for spanning the finite element space needs to be selected, partially destroying the basis-independent property of the method.

3.3 Computation of Fine Stabilization Parameter

The stabilization parameters on the different meshes need to be related to each other. This can be achieved by relating all stabilization parameters to their respective mesh size,

$$\tau_h = \tau(\boldsymbol{c}, h) \tag{38}$$

where the constants $\boldsymbol{c}$, from here on called stabilization constants, are unknown and need to be determined. In order to avoid ambiguities the number of coarse stabilization parameters should be equal to the number of unknown constants. In this case an array of coarse stabilization parameters can be introduced

$$\boldsymbol{\tau} = \left[\, \tau_{H_1}\ \tau_{H_2}\ \ldots\ \tau_{H_n} \,\right]^T \tag{39}$$

assuming they are linear with respect to the stabilization constants gives:

$$\boldsymbol{\tau} = \frac{\partial \boldsymbol{\tau}}{\partial \boldsymbol{c}} \boldsymbol{c}, \qquad \boldsymbol{c} = \left(\frac{\partial \boldsymbol{\tau}}{\partial \boldsymbol{c}}\right)^{-1} \boldsymbol{\tau} \tag{40}$$

which results in the following equation for the fine stabilization parameter:

$$\tau = \frac{\partial \tau}{\partial \boldsymbol{c}} \cdot \boldsymbol{c} = \frac{\partial \tau}{\partial \boldsymbol{c}} \cdot \left(\frac{\partial \boldsymbol{\tau}}{\partial \boldsymbol{c}}\right)^{-1} \boldsymbol{\tau} \tag{41}$$

from the appearance of the inverse matrix it becomes clear why the number of coarse levels should equal the number of unknown coefficients in the chosen structure for the stabilization parameter. It needs to be stressed that the proposed method is not limited to linear dependence of the stabilization parameter with respect to the constants. However, assuming linearity shows the clear need for multiple coarse levels in certain cases. This requirement remains in nonlinear cases when the solution can be obtained by either Newton iterations or a direct nonlinear inverse.

The convergence of the fixed-point iteration algorithm can be enhanced by extrapolation using previous fine stabilization parameter iterants [14, 15].

4 1D Convection-Diffusion

In this section the stabilized formulation for 1D convection-diffusion and its related Germano equations are presented. This formulation is selected due to its simplicity and the availability of a nodally-exact stabilization parameter.

4.1 Stabilized Formulation

This results in the following variational-multiscale stabilized formulation:

Find $u^h \in V^{h,p}_{u_\Gamma}$ *such that*

$$(w^h, au^h_x) + (w^h_x, \nu u^h_x) + (aw^h_x + \nu w^h_{xx}, \tau(au^h_x - \nu u^h_{xx}))_{\cup\Omega_e}$$
$$= (w^h, f) + (aw^h_x + \nu w^h_{xx}, \tau f)_{\cup\Omega_e} \qquad \forall w^h \in V^{h,p}_0 \tag{42}$$

where $V^{h,p}_{u_\Gamma}$ and $V^{h,p}_0$ are the usual conforming finite-element spaces of piecewise polynomials, satisfying non-homogenous and homogenous boundary conditions, respectively. The superscripts h and p indicate mesh size and interpolation order.

4.2 Stabilization Parameter Structure

We define the stabilization parameter as follows:

$$\tau = \frac{h}{2|a|}\xi(Pe_h), \qquad Pe_h = \frac{h|a|}{2\nu} \tag{43}$$

where ξ is the upwind function, which is a function of the element Peclet number Pe_h. Choosing a stabilization parameter τ now reduces to specifying an upwind function ξ in terms of the element Peclet number Pe_h.

For linear elements the nodally exact upwind function is

$$\xi = \coth(Pe_h) - \frac{1}{Pe_h} \tag{44}$$

see for instance [11]. This results in a stable formulation which converges optimally [12]. This upwind function (44) has linear behavior in the diffusive limit and constant behavior in the convective limit, corresponding to a quadratic and linear behavior of the stabilization parameter with respect to the mesh size.

For higher-order elements the best choice for the stabilization parameter is less clear. Optimal order of convergence can be obtained [12], but it is not clear whether these choices of the stabilization parameter are also optimal in the absolute sense. Furthermore, the definition of the stabilization parameter involves an inverse estimate which is not always easy to determine [13]. One thing is clear, however, in the diffusive limit the stabilization parameter should have quadratic behavior and linear behavior in the convective limit, with respect to the mesh size.

For the Germano procedure the stabilization parameter is specified as

$$\tau = c_1 h^{c_2} \tag{45}$$

which is able to represent both limits of the exact stabilization parameter. This is the structure of the stabilization parameter proposed by Oberai [2]. Note that alternatives are available [14].

When an appropriate structure for τ is not known, (45) is a safe bet, as it is able to represent every monomial. In [14], it is also demonstrated that even when the specified structure is inappropriate, the discrete Germano method will do its best to reproduce the behaviour of the exact structure.

4.3 Dissipation Method for Homogeneous Boundary Conditions

We first consider homogeneous boundary conditions for which the solution is an element of the weight space. Hence the relations for τ can be found by setting $w^h = u^h$ in the variational form. This gives:

$$\tau_H = -\frac{(u^H, au_x^H - f) + (u_x^H, \nu u_x^H)}{(au_x^H + \nu u_{xx}^H, au_x^H - \nu u_{xx}^H - f)_{\cup\Omega_e}} \tag{46}$$

or

$$\tau_H = -\frac{(u^H, au_x^H - f) + (u_x^H, \nu u_x^H)}{(au_x^H, au_x^H - f)_{\cup\Omega_e}} \tag{47}$$

for the case of vanishing second-order derivatives.

4.4 Dissipation Method for Non-Homogeneous Boundary Conditions

For non-homogeneous boundary conditions the solution is not an element of the weight space. However, the relation for τ can still be derived by considering an equivalent formulation in which the boundary conditions are applied using Lagrange multipliers. For this formulation the boundary conditions on both the test and trial function space are removed, allowing the solution to be used as test function. The boundary conditions are then enforced by augmenting the weak form. The Lagrange multiplier formulation satisfies the boundary conditions exactly, and in that sense it is equivalent to the original formulation. The Lagrange multiplier formulation reads,

Find $(u^h, \lambda_l, \lambda_r) \in V^h \times \mathbb{R} \times \mathbb{R}$ *such that*

$$\begin{aligned}(w^h, au_x^h) + (w_x^h, \nu u_x^h) + (aw_x^h + \nu w_{xx}^h, \tau(au_x^h - \nu u_{xx}^h))_{\cup\Omega_e} \\ - w(0)\lambda_l + w(1)\lambda_r - \zeta_l u(0) + \zeta_r u(1) \\ = (w^h, f) + (aw_x^h + \nu w_{xx}^h, \tau f)_{\cup\Omega_e} - \zeta_l u_l + \zeta_r u_r \\ \forall (w^h, \zeta_l, \zeta_r) \in V^h \times \mathbb{R} \times \mathbb{R}\end{aligned} \tag{48}$$

To obtain a useful relation for τ the values of the Lagrange multipliers need to be determined in terms of the approximate solution. This can be done by choosing appropriate test functions. The left Lagrange multiplier is found by choosing $l \in V^h$ such that, $l(0) = 1$ and $l(1) = 0$ which gives

$$\lambda_l = (l, au_x^h - f) + (l_x, \nu u_x^h) + (al_x + \nu l_{xx}, \tau(au_x^h - \nu u_{xx}^h - f))_{\cup\Omega_e} \tag{49}$$

The right Lagrange multiplier is found by choosing $r \in V^h$ such that, $r(0) = 0$ and $r(1) = 1$ which gives

$$-\lambda_r = (r, au_x^h - f) + (r_x, \nu u_x^h) + (ar_x + \nu r_{xx}, \tau(au_x^h - \nu u_{xx}^h - f))_{\cup\Omega_e} \tag{50}$$

From here there are two ways to proceed, both of which will be discussed in the following sections.

Reconstruction of the Lagrange Multipliers on Coarse Mesh

In this case the values of the Lagrange multipliers, λ_l^H and λ_r^H are reconstructed on the coarse mesh.

$$\begin{aligned}\lambda_l^H &= (l, au_x^H - f) + (l_x, \nu u_x^H) + (al_x + \nu l_{xx}, \tau(au_x^H - \nu u_{xx}^H - f))_{\cup\Omega_e}\\ \lambda_r^H &= -(r, au_x^H - f) - (r_x, \nu u_x^H) - (ar_x + \nu r_{xx}, \tau(au_x^H - \nu u_{xx}^H - f))_{\cup\Omega_e}\end{aligned}$$

where $l, r \in V^H$ are coarse test functions. Substituting these Lagrange multiplier values and setting $w^H = u^H$, gives after some rearranging,

$$\begin{aligned}(\widehat{u}^H, au_x^H) + (\widehat{u}_x^H, \nu u_x^H) + (a\widehat{u}_x^H + \nu\widehat{u}_{xx}^H, \tau_H(au_x^H - \nu u_{xx}^H))_{\cup\Omega_e}&\\ = (\widehat{u}^H, f) + (a\widehat{u}_x^H + \nu\widehat{u}_{xx}^H, \tau f)_{\cup\Omega_e}&\end{aligned} \tag{51}$$

with $\widehat{u}^H = u^H - u_l l - u_r r$. In this expression it is assumed that the boundary conditions are satisfied exactly, $u(0) = u_l$ and $u(1) = u_r$. The relation for the coarse stabilization parameter now becomes,

$$\tau_H = -\frac{(\widehat{u}^H, au_x^H - f) + (\widehat{u}_x^H, \nu u_x^H)}{(a\widehat{u}_x^H + \nu\widehat{u}_{xx}^H, au_x^H - \nu u_{xx}^H - f)_{\cup\Omega_e}} \tag{52}$$

or

$$\tau_H = -\frac{(\widehat{u}^H, au_x^H - f) + (\widehat{u}_x^H, \nu u_x^H)}{(a\widehat{u}_x^H, au_x^H - f)_{\cup\Omega_e}} \tag{53}$$

when the second-order derivatives vanish.

These relations are similar to equations (46) and (47) obtained for the non-homogeneous case, the only difference being the modified test function $\widehat{u}^h$. However, this test function $\widehat{u}^h \in V_0^h$ is completely arbitrary as both test functions l and r are arbitrary except for their boundary conditions. Whether or not the value of τ_H is influenced by this arbitrary test function, $\widehat{u}^h$, is not clear. Therefore this method does not seem to be acceptable.

Injection of the Lagrange Multipliers from Fine Mesh

In this case the values of the Lagrange multipliers, λ_l^H and λ_r^H are projected from the fine mesh, in a manner similar to the solution. As the Lagrange multipliers are only point values straightforward injection will do. Setting the test function equal to the solution $w^H = u^H$ and assuming that the boundary conditions are satisfied exactly results in

$$(u^H, au_x^H - f) + (u_x^H, \nu u_x^H) - u_l\lambda_l + u_r\lambda_r \\ + (au_x^H + \nu u_{xx}^H, \tau_H(au_x^H - \nu u_{xx}^H - f))_{\cup\Omega_e} = 0 \quad (54)$$

using

$$(u^H, au_x^H) = \tfrac{1}{2}a(u_r^2 - u_l^2) \quad (55)$$

this can be rewritten to yield

$$\tau_H = -\frac{\frac{1}{2}a(u_r^2 - u_l^2) - (u^H, f) + \nu\|u_x^H\|^2 - u_l\lambda_l + u_r\lambda_r}{(au_x^H + \nu u_{xx}^H, au_x^H - \nu u_{xx}^H - f)_{\cup\Omega_e}} \quad (56)$$

or

$$\tau_H = -\frac{\frac{1}{2}a(u_r^2 - u_l^2) - (u^H, f) + \nu\|u_x^H\|^2 - u_l\lambda_l + u_r\lambda_r}{a^2\|u_x^H\|^2 - (au_x^H, f)_{\cup\Omega_e}} \quad (57)$$

when the second-order derivatives vanish.

This method does not introduce any arbitrariness, in contrast to the method presented in the previous section. Therefore this does seem to be an acceptable method.

4.5 Naive Least-Squares Method

For both homogeneous and non-homogeneous boundary conditions the least-squares method is the same. For the naive basis-dependent version the following vectors need to be constructed:

Find $(g^H, m^H) \in V_0^{H,p} \times V_0^{H,p}$ *such that*

$$g_i = (\phi_i, au_x^H - f) + (\phi_{i_x}^H, \nu u_x^H) \qquad \forall i$$
$$m_i = (a\phi_{i_x}^H + \nu\phi_{i_{xx}}^H, au_x^H - \nu u_{xx}^H - f)_{\cup\Omega_e} \qquad \forall i \quad (58)$$

These can then be used to obtain a relation for τ

$$\tau_H = -\frac{m_i g_i}{m_j m_j} \quad (59)$$

which can be undetermined when m_i becomes zero. This occurs when the exact solution happens to lie exactly in the finite element space, and hence the residual vanishes completely.

4.6 Basis-Independent Least-Squares Method

For the basis-independent version the Riesz representations need to be constructed using the following projections:

$$\begin{array}{|l|}\hline \textit{Find } (g^H, m^H) \in V_0^{H,p} \times V_0^{H,p} \textit{ such that} \\ (w^H, g^H)_X = (w^H, au_x^H - f) + (w_x^H, \nu u_x^H) \qquad \forall w^H \in V_0^H \\ (w^H, m^H)_X = (aw_x^H + \nu w_{xx}^H, au_x^H - \nu u_{xx}^H - f)_{\cup\Omega_e} \qquad \forall w^H \in V_0^H \qquad (60) \\ \hline \end{array}$$

This can then be used to obtain a relation for τ

$$\tau_H = \frac{\lambda \tau_H^{ref} - (m^H, g^H)_X}{\lambda + (m^H, m^H)_X} \tag{61}$$

From this it can be clearly seen that the penalty parameter λ prevents division by zero and forces τ to the reference value in cases for which it would otherwise be undetermined.

5 Numerical Results

In this section some numerical results for the convection-diffusion problem are given. These results show the viability of the Germano approach. The issues described for the dissipation and least-squares method, as presented in the previous sections, are also encountered in the experiments.

A mesh consisting of 24 linear elements is used for the finest level. The chosen structure for the stabilization parameter involves two unknown coefficients, hence two coarser levels are necessary. These are obtained by coarsening the fine mesh with a factor 2 and 4, respectively. The H_0^1-projector is used for obtaining the coarse scale solutions. In a 1D setting with linear shape functions this is equivalent to nodal interpolation.

5.1 Homogenous Boundary Conditions

As benchmark a problem with homogenous boundary conditions and unit forcing is used. The obtained upwind function as function of the element Peclet number is given in Figure 7.

The dissipation method and both least-squares methods find exactly the same upwind function. Furthermore, the obtained upwind function seems to be similar to the one presented by Oberai [2].

In Figure 7, the right most part of a result obtained by the Germano procedure is plotted together with a result obtained with the Galerkin method and one obtained with a stabilized formulation with the nodally exact stabilization parameter. Form this it can be seen that the Germano approach approximates the exact projection quite well.

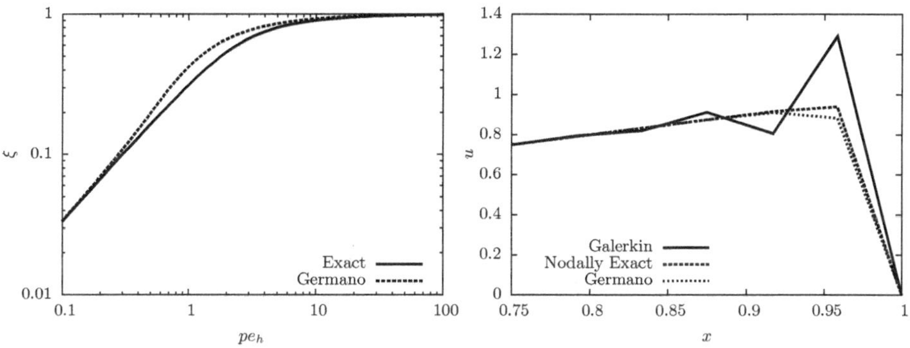

Fig. 7. Upwind functions and example solutions for problem with homogenous boundary conditions.

5.2 Non-Homogenous Boundary Conditions

In this section the same approaches as in the previous section are used. However, they are applied to a problem with non-homogenous boundary conditions ($u(0) = 0$ and $u(1) = 1$) and no forcing. Both least-square methods give results identical to the results obtained with homogenous boundary conditions, while the dissipation method breaks down as presented in the left plot of Figure 8.

In fact, when using the identity equation (47) the procedure brakes down completely, leading to no results at all. The upwind function presented in Figure 8 is obtained with an identity similar to (47),

$$\tau_H = \frac{(u^H, f) + (u_x^H, au^H) - (u_x^H, \nu u_x^H)}{(au_x^H, au_x^H - f)_{\cup\Omega_e}} \tag{62}$$

which is based on a weak formulation in conservative form. These two formulations, as given in equations (47) and (62), should be identical. Hence, this result reconfirms the non-applicability of the dissipation method in the case of non-homogenous boundary conditions.

In the previous section two different Lagrange multiplier approaches were presented which could potentially solve the non-homogenous boundary condition issue for the dissipation method. Results obtained with these approaches are presented in Figure 8.

From the left plot it can be seen that injection of the Lagrange multipliers results in an erroneous upwind function in the diffusive limit, while the right plot seems to validate the reconstruction approach. However, it should be noted that these results are obtained by reconstructing the Lagrange multipliers using global test functions while the same approach using local test functions for reconstruction of the Lagrange multipliers breaks down. This clearly indicates the arbitrariness involved.

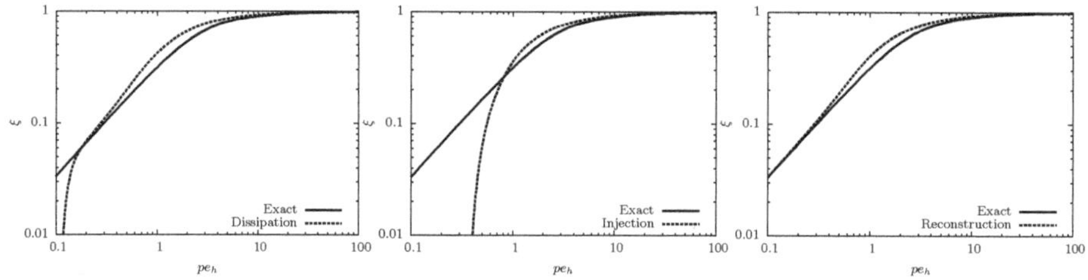

Fig. 8. Upwind functions for problem with non-homogenous boundary conditions. Obtained without, with injected and with reconstructed Lagrange multipliers respectively.

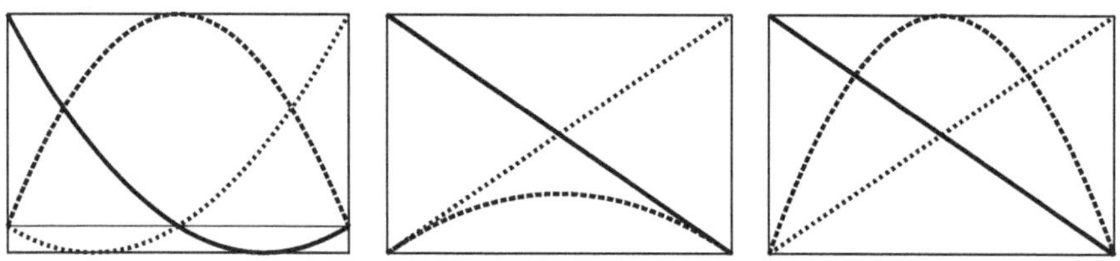

Fig. 9. Different basis functions for spanning $\mathcal{V}^{h,2}$.

5.3 Basis Dependence of the Least-Squares Method

In the previous numerical experiments both least-squares approaches yielded identical upwind functions. This is due to the linear basis functions and uniform mesh used. Therefore, quadratic elements are used in order to show the basis-dependence of the naive least-squares approach and validate the basis independence of the dual-norm approach. All other parameters are the same as in the previous section. Three different bases, plotted in Figure 9, are used for spanning this finite element space.

The left plot shows the nodal basis, while the other plots show two modal bases. For one basis the interior bubble is defined by the multiplication of the two linear shape functions resulting in a mid-element value of 1/4. For the other basis the bubble function has a mid-element value of unity.

The upwind functions obtained with these bases are plotted in Figure 10.

The upwind functions obtained by the basis-independent approach are confirmed to be independent of the basis used for spanning the finite element space. While the basis-dependent approach clearly exhibits dependence of the basis in the convective limit.

Values for the upwind function in the convective limit obtained by the least-squares methods are given in Table 1.

The resulting difference in the solutions for $Pe_h = 50$ obtained by the different methods are given in Figure 11. The basis-dependent results deviate only moderately from the basis-independent result. This is because the different bases considered are reasonable choices. However, the results indicate there is an arbitrary component,

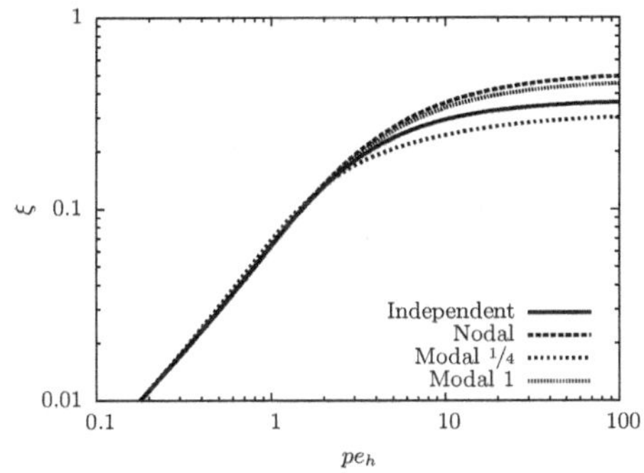

Fig. 10. Upwind functions obtained with different least-squares methods.

Table 1. Deviation of upwind function at $pe_h = 10^3$ due to different bases.

Basis	ξ	%
Independent	0.369	–
Nodal	0.513	39%
Modal 1/4	0.311	–16%
Modal 1	0.468	27%

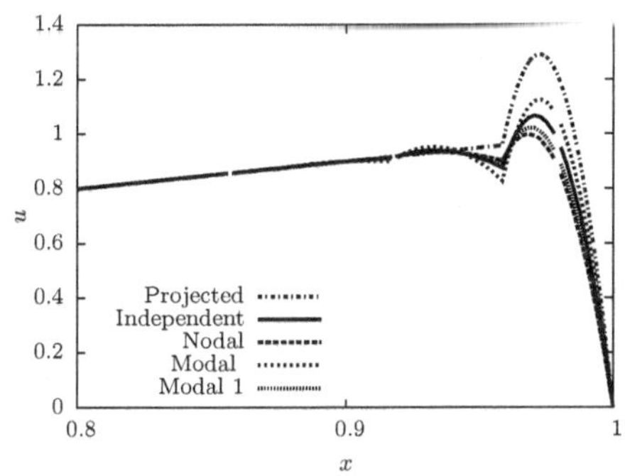

Fig. 11. Solution for $Pe_h = 50$ obtained with different least-squares methods.

namely the basis, in the formulation when the Euclidean norm of the residual is minimized. While minimizing the dual-norm eliminates this issue.

5.4 Computational Cost

As presented in [15] on the order of 5 to 10 iterations need to be taken to arrive at a converged solution. This is relatively expensive when the underlying problem is linear. For such cases the rate of convergence might be improved by solving the coupled problem with a monolithic Newton method.

Most problems of interest are either nonlinear or unsteady, or both (e.g. turbulence). For nonlinear problems the iterations for obtaining the parameters can be combined with the nonlinear iterations. For unsteady problems, convergence should be rapid as a good initial guess is available. In certain cases, even explicit updates might be considered if the stabilisation terms are anticipated to have only higher-order effects. In summary, it is anticipated that for most problems of interest, the additional costs associated with the proposed method will not be prohibitive.

6 Conclusion

By analyzing the variational Germano identity via commutativity diagrams, a general Germano identity for all types of discretizations has been derived, which can justifiably be named the discrete Germano identity. This relation is similar to the variational Germano identity, but is not restricted to variational numerical methods. Based on the discrete Germano identity a new algorithm, in the context of stabilized methods, is proposed. This partitioned algorithm consists of distinct building blocks. Besides the straightforward steps of solving on the fine mesh and projecting onto the coarse mesh, this procedure includes a reconstruction of the coarse mesh stabilization parameters and an extrapolation of these parameters to the fine mesh. The partitioning of the algorithm is possible as the reconstruction uses the coarse scale equations directly, without combining them with the large scale equations.

It was demonstrated using a number of examples that the reconstruction of coarse-mesh stabilization parameters requires special care, as naive approaches can lead to invalid or basis-dependent results.

First, dissipation-based approaches were considered. It was shown that for non-homogeneous boundary conditions the standard dissipation method is invalid. This problem can be circumvented using Lagrange multipliers, however computing these multipliers on the fine mesh and injecting them into the coarse mesh gives poor results, while computing them directly on the coarse mesh results in a formulation with an arbitrary component.

Next, two approaches based on least-squares were considered. Using least-squares, non-homogeneous boundary conditions can be handled without difficulty. However, it was shown both theoretically and by numerical experiment that least-square methods based on straightforward minimization of the Euclidean norm of the discrete residual lead to basis-dependent results. Therefore an alternate least-squares method was proposed, based on minimization of the dual norm. With the addition of a penalty parameter to prevent division by zero, it was demonstrated that this method is both robust and basis independent.

Acknowledgments

This research is supported by the Dutch Technology Foundation STW, applied science division of NWO and the Technology Program of the Ministry of Economic Affairs.

References

1. M. Germano, U. Piomelli, P. Moin, and W.H. Cabot. A dynamic subgrid-scale eddy viscosity model. *Phys. Fluid*, 3:1760–1765, 1991.
2. A.A. Oberai and J. Wanderer. A dynamic approach for evaluating parameters in a numerical method. *Int. J. Numer. Meth. Fluids*, 62:50–71, 2005.
3. A.A. Oberai and J. Wanderer. A dynamic multiscale viscosity method for the spectral approximation of conservation laws. *Comput. Methods Appl. Mech. Engrg.*, 195:1778–1792, 2006.
4. A.A. Oberai and J. Wanderer. Variational formulation of the Germano identity for the Navier-Stokes equations. *Journal of Turbulence*, 6(7): 2005.
5. A.A. Oberai and J. Wanderer. Optimal numerical solution of PDEs using the variational Germano identity. *Comput. Methods Appl. Mech. Engrg.*, 197:2948-2962, 2008.
6. J. Hoffman. Dynamic subgrid modeling for time dependent convection-diffusion-reaction equations with fractal solutions. *Int. J. Numer. Meth. Fluids*, 40:583–592, 2002.
7. E. Onate, J. Garcia, and S. Idelsohn. Computation of the stabilization parameter for the finite element solution of advection-diffusion problems. *Int. J. Numer. Meth. Fluids*, 25:1385–1407, 1997.
8. T.J.R. Hughes. Multiscale phenomena: Green's functions, the Dirichlet-to-Neumann formulation, subgrid scale models, bubbles and the origins of stabilized methods. *Comput. Methods Appl. Mech. Engrg.*, 127:387–401, 1995.
9. T.J.R. Hughes, G.R. Feijoo, L. Mazzei, and J.B. Quincy. The variational multiscale method – A paradigm for computational mechanics. *Comput. Methods Appl. Mech. Engrg.*, 166:3–24, 1998.
10. T.J.R. Hughes and G. Sangalli. Variational multiscale analysis: the fine-scale green's function, projection, optimization, localization, and stabilized methods. *SIAM J. Numer. Anal*, 45:539–557, 2007.
11. A.N. Brooks and T.J.R. Hughes. Streamline Upwind/Petrov–Galerkin formulations for convection dominated flows with particular emphasis on the incompressible Navier-Stokes equations. *Comput. Methods Appl. Mech. Engrg.*, 32:199–259, 1982.
12. L.P. Franca, S.L. Frey, and T.J.R. Hughes. Stabilized finite element methods: I. Application to the advective-diffusive model. *Comput. Methods Appl. Mech. Engrg.*, 95:253–276, 1992.
13. I. Harari and T.J.R. Hughes. What are C and h?: Inequalities for the analysis and design of finite element method. *Comput. Methods Appl. Mech. Engrg.*, 97:157–192, 1992.
14. I. Akkerman. Adaptive variational multiscale formulations using the discrete Germano approach. *Dissertation, University of Technology Delft*, 2009.
15. I. Akkerman, K.G. van der Zee, and S.J. Hulshoff. A Variational Germano approach for Stabilized Finite element methods. *Comput. Methods Appl. Mech. Engrg.* Accepted.

Dissipative Structure and Long Term Behavior of a Finite Element Approximation of Incompressible Flows with Numerical Subgrid Scale Modeling

Ramon Codina, Javier Principe and Santiago Badia

Universitat Politècnica de Catalunya, 08034 Barcelona, Spain;
ramon.codina@upc.edu, principe@cimne.upc.edu, sbadia@cimne.upc.edu

Abstract In this chapter we summarize a finite element formulation for incompressible flows based on a two-scale decomposition of the velocity field, where the sub-grid scales are modeled numerically. The main features of the formulation are the choice for the space of the sub-grid scales, their time dependency and the fact that they are accounted for in all the terms where they appear. We present the main results obtained for this formulation, with emphasis on its dissipative structure and stability behavior in the long term, which give arguments to support the claim that it is able to model turbulent flows without any additional turbulence model.

Key words: Incompressible flows, scale splitting, dynamic sub-grid scales, long term behavior, large eddy simulation, energy budget

1 Introduction

The objective of this paper is to summarize the finite element formulation for incompressible flows that has been developed in our group during the last years, starting with [7], and also to present some of the latest results [2, 10, 12, 15, 23]. In particular, issues related to stability, the dissipative structure of the formulation, the energy transfer between unresolved and resolved scales (including a discussion about the possibility to produce backscatter) and the long term behavior will be addressed. Our purpose is to explain the main ideas without technicalities, even at the expense of some occasional lack of precision. Likewise, we do not pretend to be exhaustive in the bibliographical references. The reader is addressed to the bibliography included in the papers listed above for more details.

The formulation we will present can be framed within the variational multiscale concept introduced in [17, 18]. In fact, the starting point is a *two scale* decomposition of the unknowns, the velocity $\boldsymbol{u}$ and the pressure p, defined in a spatial domain Ω and in a time interval $[0, T]$, including the possibility to let $T \rightarrow \infty$. The original motivation of this type of formulations was to justify the so called *stabilized finite element methods*. In the case of the incompressible Navier-Stokes equations,

R. de Borst and E. Ramm (eds.), *Multiscale Methods in Computational Mechanics*, Lecture Notes in Applied and Computational Mechanics 55, DOI 10.1007/978-90-481-9809-2_5,

stabilization is required to avoid the need for using velocity-pressure interpolations satisfying the inf-sup condition and to deal with convection-dominated flows (see, for example, [5, 14] for an overview of different methods and background).

Let V and Q be the spaces where velocities and pressures must belong for $t \in [0, T]$. In all what follows, the time variable will be left continuous. Regularity in time of the unknowns will be specified when required. Let also $V_h \subset V$ and $Q_h \subset Q$ be finite element spaces to approximate the velocity and the pressure, respectively, *in space*.

The starting idea of the formulation to be presented is to split the velocity and the pressure as

$$\boldsymbol{u} = \boldsymbol{u}_h + \tilde{\boldsymbol{u}}, \quad p = p_h + \tilde{p}, \tag{1}$$

where $\boldsymbol{u}_h$, p_h belong to the finite element spaces and $\tilde{\boldsymbol{u}}$ and $\tilde{p}$ are what we will call the subgrid scales or, simply, *the subscales*. The way these are modeled defines the particular numerical approximation. We can identify the finite element components of the solution as the *resolved* scales, whereas the subscales are the *unresolved* scales.

Splitting (1) is usually used together with several approximations when stabilized finite element methods need to be justified. However, our approach is to accept the multiscale decomposition with *all its consequences* and, in particular:

- To consider the subscales to be time dependent.
- To keep the nonlinear terms involving $\tilde{\boldsymbol{u}}$ in the Navier-Stokes equations.

Both items will be described in what follows. Some of the rewards are

- The dependence of the subscales with the time step size of the time integration scheme becomes clear.
- Global conservation of momentum can be obtained.
- Modeling of the subcales defines automatically a model for the extra stresses appearing in a LES-like approach .

The first two issues are discussed in detail in [12] (see also references therein). The third item will be also discussed here.

Let us analyze the implications of keeping the nonlinear terms involving $\tilde{\boldsymbol{u}}$ in the Navier–Stokes equations. Using (1), the convective term will lead to

$$\begin{aligned} \nabla \cdot (\boldsymbol{u} \otimes \boldsymbol{u}) &= \nabla \cdot (\boldsymbol{u}_h \otimes \boldsymbol{u}_h) + \nabla \cdot (\boldsymbol{u}_h \otimes \tilde{\boldsymbol{u}}) + \nabla \cdot (\tilde{\boldsymbol{u}} \otimes \boldsymbol{u}_h) + \nabla \cdot (\tilde{\boldsymbol{u}} \otimes \tilde{\boldsymbol{u}}) \\ &\equiv (\mathrm{I}) + (\mathrm{II}) + (\mathrm{III}) + (\mathrm{IV}). \end{aligned} \tag{2}$$

Obviously, (I) would be the only term appearing in a Galerkin approximation, whereas the rest are the contributions from the velocity subscale. It can be shown that when this subscale is modeled, the term that provides numerical stability is (II), in the sense that it allows control of the convective derivative and the pressure gradient, in order to deal with convection dominated flows. One can also show that (III) leads to global momentum conservation [12].

Because of its resemblance with similar terms in turbulence models, (IV) in (2) raises the question of whether keeping the contribution from the subscales in the convective term could be viewed as a turbulence model or not. This possibility was mentioned in [7], which contrasts with the option in [19], where a large-eddy-simulation (LES) model is used to represent the subgrid scales (see Remark 6 in [7] and, for background on LES models, [22]). In [3] the possibility to model turbulence using only numerical ingredients within the variational multiscale context is fully and successfully exploited. The role of numerical stabilization terms as turbulence models was also envisaged in [13, 16], for example.

By analogy with LES models, the different terms appearing in (2) could be termed as follows:

$$
\begin{aligned}
\text{(II)+(III)} &= \boldsymbol{u}_h \otimes \tilde{\boldsymbol{u}} + \tilde{\boldsymbol{u}} \otimes \boldsymbol{u}_h : && \text{Cross stress} \\
\text{(IV)} &= \tilde{\boldsymbol{u}} \otimes \tilde{\boldsymbol{u}} \quad : && \text{Reynolds stress} \\
\text{(II)+(III)+(IV)} &= \boldsymbol{u}_h \otimes \boldsymbol{u}_h - \boldsymbol{u} \otimes \boldsymbol{u} : && \text{Subgrid scale tensor}
\end{aligned}
$$

Some comments in the line of viewing an approximation to the subscales in (2) as an effective LES model will be provided in the following.

The other crucial point of our approach is to keep the time dependency of the velocity subscale. Thus, the velocity time derivative can be split as

$$\partial_t \boldsymbol{u} = \partial_t \boldsymbol{u}_h + \partial_t \tilde{\boldsymbol{u}}. \tag{3}$$

The first term would be the only one kept if the time derivative of the subscales is neglected. In this situation, the subscales were termed *quasi-static* in [7], in contrast to *dynamic subscales* if they are considered to be time-dependent. As shown in [12], the second term leads to a correct behavior of time integration schemes and better accuracy. In particular, in [1] stability and convergence for the Stokes problem is proved without any restriction on the time step size and the stabilization parameters on which the formulation depends.

The third crucial ingredient does not follow from assuming the consequences of the two-scale decomposition, but from the choice of the space for the subscales. As explained later, our analysis relies heavily on the fact that we consider this space orthogonal to the finite element space.

As indicated earlier, apart from presenting the numerical formulation we propose, which is described in Section 2, our purpose is to discuss issues related to its dissipative structure and its long term behavior. Let us summarize the main results to be explained.

Assuming that

- orthogonal subgrid scales are used,
- the finite element mesh is capable of capturing (part of) the inertial range,
- the assumptions of classical statistical fluid mechanics apply,

one can show that *the dissipation introduced by (II) + (III) in (2) is proportional to the molecular dissipation of the physical subgrid scales*. The meaning of the assumptions listed and the above conclusion is explained in Section 3.

Also in Section 3 we will state the energy budget presented in [23]. We will show that *the finite element component and the subgrid component have a proper scale separation only if they are orthogonal*, in the sense that the total kinetic energy is the sum of the kinetic energy of $\boldsymbol{u}_h$ plus the kinetic energy of $\tilde{\boldsymbol{u}}$. We will also explain why *dynamic subscales ($\partial_t \tilde{\boldsymbol{u}} \neq \mathbf{0}$) are needed to model backscatter.*

In Section 4 we will list the results of the long term stability analysis conducted in [2]. We will see that *dynamic subscales allow us to obtain the correct long term stability of the space-discrete scheme.*

Some numerical simulations, which try to show that the numerical model we propose can be used in real applications and has the correct behavior, are presented in Section 5. A summary and some concluding remarks close the paper in Section 6.

2 Formulation

In this section we summarize the formulation we propose. For more details, including bibliographical references, see [7, 12].

2.1 Continuous problem

The Navier–Stokes problem for an incompressible fluid consists of finding a velocity $\boldsymbol{u}$ and a pressure p solution of the initial and boundary value problem

$$
\begin{aligned}
\partial_t \boldsymbol{u} + \boldsymbol{u} \cdot \nabla \boldsymbol{u} - \nu \Delta \boldsymbol{u} + \nabla p = \boldsymbol{f} \quad & \text{in } \Omega,\ t > 0, \\
\nabla \cdot \boldsymbol{u} = 0 \quad & \text{in } \Omega,\ t > 0, \\
\boldsymbol{u} = \mathbf{0} \quad & \text{on } \Gamma,\ t > 0, \\
\boldsymbol{u} = \boldsymbol{u}^0 \quad & \text{in } \Omega,\ t = 0.
\end{aligned}
$$

In these equations, ν is the kinematic viscosity, $\boldsymbol{f}$ is the vector of body forces, $\boldsymbol{u}^0$ is the initial condition and $\Gamma = \partial\Omega$.

Let $V = H_0^1(\Omega)^d$ and $Q = L_0^2(\Omega)$ (L^2 functions with zero mean), d being the number of space dimensions. Let us denote by $(\cdot, \cdot)$ the L^2 inner product in Ω and by $\langle f, g \rangle$ the integral of two functions f and g over Ω, whenever this integral makes sense. When the integral is computed over a region ω, we will denote it as $\langle f, g \rangle_\omega$. Let also $L^2(0, T; V)$ be the set of functions whose V-norm in space is L^2 in time and $\mathcal{D}'(0, T; Q)$ the set of "functions" whose Q-norm in space is a distribution in time. The weak form of the Navier–Stokes equations can be written as follows: find $[\boldsymbol{u}, p] \in L^2(0, T; V) \times \mathcal{D}'(0, T; Q)$ such that

$$
(\partial_t \boldsymbol{u}, \boldsymbol{v}) + \langle \boldsymbol{u} \cdot \nabla \boldsymbol{u}, \boldsymbol{v} \rangle + \nu(\nabla \boldsymbol{u}, \nabla \boldsymbol{v}) - (p, \nabla \cdot \boldsymbol{v}) = \langle \boldsymbol{f}, \boldsymbol{v} \rangle, \quad \forall \boldsymbol{v} \in V, \tag{4}
$$

$$
(q, \nabla \cdot \boldsymbol{u}) = 0, \quad \forall q \in Q. \tag{5}
$$

2.2 Subgrid Scale Decomposition

Let us consider a finite element partition $\{K\}$ of the computational domain Ω. As explained earlier, the starting point of the formulation to be presented is the splitting (1). For simplicity, we will not consider pressure subscales (see [6, 7] for an analysis of their inclusion). Thus, if $\tilde{\boldsymbol{u}}$ is a certain approximation to the exact velocity subscale, the splitting we consider is $\boldsymbol{u} \approx \boldsymbol{u}_* := \boldsymbol{u}_h + \tilde{\boldsymbol{u}},\ p \approx p_h$. When inserted into (4)-(5) this yields:

$$
\begin{aligned}
&(\partial_t \boldsymbol{u}_h, \boldsymbol{v}_h) + \langle \boldsymbol{u}_* \cdot \nabla \boldsymbol{u}_h, \boldsymbol{v}_h \rangle + \nu(\nabla \boldsymbol{u}_h, \nabla \boldsymbol{v}_h) - (p_h, \nabla \cdot \boldsymbol{v}_h) + (q_h, \nabla \cdot \boldsymbol{u}_h) \\
&\quad + (\partial_t \tilde{\boldsymbol{u}}, \boldsymbol{v}_h) - \sum_K \langle \tilde{\boldsymbol{u}}, \boldsymbol{u}_* \cdot \nabla \boldsymbol{v}_h + \nu \Delta \boldsymbol{v}_h + \nabla q_h \rangle_K \\
&\quad + \sum_K \langle \tilde{\boldsymbol{u}}, \nu \boldsymbol{n} \cdot \nabla \boldsymbol{v}_h + q_h \boldsymbol{n} \rangle_{\partial K} = \langle \boldsymbol{f}, \boldsymbol{v}_h \rangle, \qquad (6)
\end{aligned}
$$

$$
\begin{aligned}
&(\partial_t \tilde{\boldsymbol{u}}, \tilde{\boldsymbol{v}}) + \sum_K \langle \boldsymbol{u}_* \cdot \nabla \tilde{\boldsymbol{u}} - \nu \Delta \tilde{\boldsymbol{u}}, \tilde{\boldsymbol{v}} \rangle_K + \sum_K \langle \nu \boldsymbol{n} \cdot \nabla \tilde{\boldsymbol{u}}, \tilde{\boldsymbol{v}} \rangle_{\partial K} \\
&\quad + \sum_K \langle \partial_t \boldsymbol{u}_h + \boldsymbol{u}_* \cdot \nabla \boldsymbol{u}_h - \nu \Delta \boldsymbol{u}_h + \nabla p_h, \tilde{\boldsymbol{v}} \rangle_K \\
&\quad + \sum_K \langle \nu \boldsymbol{n} \cdot \nabla \boldsymbol{u}_h - p_h \boldsymbol{n}, \tilde{\boldsymbol{v}} \rangle_{\partial K} = \langle \boldsymbol{f}, \tilde{\boldsymbol{v}} \rangle. \qquad (7)
\end{aligned}
$$

These discrete variational equations must hold for all test functions $[\boldsymbol{v}_h, q_h] \in V_h \times Q_h$ and $\tilde{\boldsymbol{v}} \in \tilde{V}$, where $\tilde{V}$ is the space of subscales to be defined. It is observed that some terms have been integrated by parts within each element.

2.3 Simplifying Assumptions

Apart from taking the pressure subscale to be zero, no approximations have been made to arrive at (6)-(7). Different approximations will lead precisely to different formulations within the same framework. As mentioned in Section 1, one usually considers $\partial_t \tilde{\boldsymbol{u}} \approx \boldsymbol{0}$ and takes $\boldsymbol{u}_* \approx \boldsymbol{u}_h$ as advection velocity in (6)-(7). However, these are precisely approximations that *we do not necessarily assume*.

Firstly, let us describe the *space of subscales* $\tilde{V}$, that is, the space where $\tilde{\boldsymbol{u}}$ belongs for t fixed. A particular feature of our approach is to take it L^2 *orthogonal to the finite element space*, that is to say, $\tilde{V}$ is taken as a subspace of $V_h^{\perp}$. As it will be mentioned later, this choice has important theoretical consequences.

Next, we will use the approximation $\tilde{\boldsymbol{u}} \approx \boldsymbol{0}$ on ∂K for each element domain K of the finite element partition. That could be understood as approximating the velocity subscale by a space of bubble functions. However, the heuristic Fourier argument proposed in [7] also allows us to explain why the effect of the subscales on the element boundaries can be neglected compared to the effect in the element interiors. Nevertheless, this approximation can be relaxed following the ideas suggested in [11].

Finally, the essential approximation is

$$\sum_K \langle \boldsymbol{u}_* \cdot \nabla \tilde{\boldsymbol{u}} - \nu \Delta \tilde{\boldsymbol{u}}, \tilde{\boldsymbol{v}} \rangle_K \approx \sum_K \tau_K^{-1} \langle \tilde{\boldsymbol{u}}, \tilde{\boldsymbol{v}} \rangle_K, \tag{8}$$

where τ_K is a set of algorithmic parameters computed within each element K as

$$\tau_K^{-1} = \frac{c_1 \nu}{h_K^2} + \frac{c_2 \|\boldsymbol{u}_*\|_{L^\infty(K)}}{h_K}. \tag{9}$$

Here, h_K is a characteristic length of K and c_1 and c_2 are algorithmic constants that depend only on the degree of the finite element approximation being used.

Expression (9) can be motivated also by a heuristic Fourier analysis [7]. In fact, what is important is its asymptotic behavior in terms of h_K, ν and $\|\boldsymbol{u}_*\|_{L^\infty(K)}$. We also would like to stress that the introduction of τ_K comes from the approximation of a *spatial* operator, as it is clearly seen from (8). Therefore, we never include a dependence of τ_K on the temporal discretization.

2.4 Final Formulation

The approximations described allow us to formulate a method that can be effectively implemented and that is the formulation we propose. It consists of finding $\boldsymbol{u}_h \in L^2(0,T;V_h)$ and $p_h \in \mathcal{D}'(0,T;Q_h)$ such that

$$\begin{aligned}
&(\partial_t \boldsymbol{u}_h, \boldsymbol{v}_h) + \langle \boldsymbol{u}_* \cdot \nabla \boldsymbol{u}_h, \boldsymbol{v}_h \rangle + \nu(\nabla \boldsymbol{u}_h, \nabla \boldsymbol{v}_h) - (p_h, \nabla \cdot \boldsymbol{v}_h) + (q_h, \nabla \cdot \boldsymbol{u}_h) \\
&\quad - \sum_K \langle \tilde{\boldsymbol{u}}, \boldsymbol{u}_* \cdot \nabla \boldsymbol{v}_h + \nu \Delta \boldsymbol{v}_h + \nabla q_h \rangle_K = \langle \boldsymbol{f}, \boldsymbol{v}_h \rangle, \qquad (10) \\
&(\partial_t \tilde{\boldsymbol{u}}, \tilde{\boldsymbol{v}}) + \sum_K \tau_K^{-1} \langle \tilde{\boldsymbol{u}}, \tilde{\boldsymbol{v}} \rangle_K \\
&\quad + \sum_K \langle \boldsymbol{u}_* \cdot \nabla \boldsymbol{u}_h - \nu \Delta \boldsymbol{u}_h + \nabla p_h, \tilde{\boldsymbol{v}} \rangle_K = \langle \boldsymbol{f}, \tilde{\boldsymbol{v}} \rangle. \qquad (11)
\end{aligned}$$

These equations must hold for all $[\boldsymbol{v}_h, q_h] \in V_h \times Q_h$ and $\tilde{\boldsymbol{v}} \in \tilde{V}$.

A complete numerical analysis of (10)-(11) would include stability and convergence estimates as well as a qualitative analysis of the associated dynamical system. Moreover, in the context of stabilized finite element methods this analysis should be conducted in *norms that do not explode as* $\nu \to 0$ *and allow for any velocity-pressure interpolation.* Whereas the second requirement could be considered not essential by those that favor the use of inf-sup stable velocity-pressure interpolations, the first is a must. From the numerical point of view, estimates that explode with ν are *completely useless* if the formulation is intended to be applied to large Reynolds number flows and, obviously, to model turbulence.

We are still far from the objective described and, in fact, we are not aware of any numerical formulation even close to the target. Certainly, we have some partial (and minor) results. If convection is not an issue and the only concern is the pressure interpolation, we have stability and convergence estimates for the stationary counterpart of (10)–(11) in [9] and for a first order time discretization in [4]. The linearized stationary problem, accounting also for convection, is analyzed in full detail in [8], whereas the transient Stokes problem is analyzed in [1].

The following two sections describe aspects related to the dissipative structure of problem (10)–(11) and its long term stability.

3 Dissipative Structure and Backscatter

In this section we describe the dissipative structure of the proposed formulation. First, a local balance of energy will allow us to investigate the flow of energy between the finite element component and the subscales (that is to say, between the resolved and the unresolved scales). We will then present a global balance of energy which shows that it is not necessary to use a LES model apart from the one inherent to the numerical approximation described. Finally, we will discuss the possibility of modeling backscatter and show in a numerical example that it can effectively take place.

In order to highlight the importance of taking $\tilde{V}$ orthogonal to V_h, we will also consider the possibility of using (8) in (6)–(7) without this orthogonality enforcement. The L^2 projection from V to $\tilde{V}$ will be denoted by $\tilde{P}$.

3.1 Local Kinetic Energy Balance Equations

Let R be a region formed by a patch of elements, and let $\boldsymbol{t}_R$ be the flux on ∂R, which may include both the flux of stresses (tractions) and convective fluxes. For simplicity, suppose that τ is constant (computed with a characteristic velocity and element length, see Section 4). If $\mathcal{L}_u \boldsymbol{v} := \boldsymbol{u}_* \cdot \nabla \boldsymbol{v} - \nu \Delta \boldsymbol{v}$, taking $\boldsymbol{v}_h = \boldsymbol{u}_h$, $q_h = p_h$ and $\tilde{\boldsymbol{v}} = \tilde{\boldsymbol{u}}$ in (6)–(7), using (8) and neglecting the subscales on the interelement boundaries, we get:

$$\begin{aligned}
&\frac{1}{2}\frac{\mathrm{d}}{\mathrm{d}t}\|\boldsymbol{u}_h\|_R^2 + \nu\|\nabla \boldsymbol{u}_h\|_R^2 + (\partial_t \tilde{\boldsymbol{u}}, \tilde{P}(\boldsymbol{u}_h))_R \\
&\quad + \sum_{K\subset R} \langle \tilde{\boldsymbol{u}}, \tilde{P}(\mathcal{L}_u^* \boldsymbol{u}_h - \nabla p_h)\rangle_K = W_R(\boldsymbol{u}_h), \qquad (12)\\
&\frac{1}{2}\frac{\mathrm{d}}{\mathrm{d}t}\|\tilde{\boldsymbol{u}}\|_R^2 + \tau^{-1}\|\tilde{\boldsymbol{u}}\|_R^2 + (\tilde{P}(\partial_t \boldsymbol{u}_h), \tilde{\boldsymbol{u}})_R \\
&\quad + \sum_{K\subset R} \langle \tilde{\boldsymbol{u}}, \tilde{P}(\mathcal{L}_u \boldsymbol{u}_h + \nabla p_h)\rangle_K = \langle \boldsymbol{f}, \tilde{\boldsymbol{u}}\rangle_R, \qquad (13)
\end{aligned}$$

where $W_R(\boldsymbol{u}_h) = \langle \boldsymbol{f}, \boldsymbol{u}_h\rangle_R + \langle \boldsymbol{t}_R, \boldsymbol{u}_h\rangle_{\partial R}$ and $\mathcal{L}_u^* \boldsymbol{v} := -\boldsymbol{u}_* \cdot \nabla \boldsymbol{v} - \nu \Delta \boldsymbol{v}$. In (12) we have neglected the term $\langle \boldsymbol{u}_* \cdot \nabla \boldsymbol{u}_h, \boldsymbol{u}_h\rangle$. If $\boldsymbol{u}_*$ is not divergence free (and it is not in the approximated problem), the term $\langle \boldsymbol{u}_* \cdot \nabla \boldsymbol{u}_h, \boldsymbol{v}_h\rangle$ in (6) can be replaced by the skew-symmetric form

$$\begin{aligned}
&\frac{1}{2}\langle \boldsymbol{u}_* \cdot \nabla \boldsymbol{u}_h, \boldsymbol{v}_h\rangle - \frac{1}{2}\langle \boldsymbol{u}_* \otimes \boldsymbol{u}_h, \nabla \boldsymbol{v}_h\rangle \\
&\quad = \langle \boldsymbol{u}_* \cdot \nabla \boldsymbol{u}_h, \boldsymbol{v}_h\rangle + \frac{1}{2}\langle \nabla \cdot \boldsymbol{u}_*, \boldsymbol{u}_h \cdot \boldsymbol{v}_h\rangle - \frac{1}{2}\langle \boldsymbol{n} \cdot \boldsymbol{u}_*, \boldsymbol{u}_h \cdot \boldsymbol{v}_h\rangle_\Gamma
\end{aligned}$$

without altering the stability and consistency of the formulation. The last term in this expression has been kept to show that only the definition of the flux $\boldsymbol{t}_R$ will change if the integral is performed in a region R interior to Ω.

From (12)–(13) we may draw the first important conclusion. Suppose that $\nu\Delta\boldsymbol{u}_h$ is negligible (because $\boldsymbol{u}_h$ is linear within each element or because ν is very small or because $\tilde{P}(\Delta\boldsymbol{u}_h) \approx \boldsymbol{0}$, as we will see below). Let us define

$$
\begin{aligned}
\mathcal{K}_h &:= \frac{1}{2}\|\boldsymbol{u}_h\|_R^2, & \tilde{\mathcal{K}} &:= \frac{1}{2}\|\tilde{\boldsymbol{u}}\|_R^2 && \text{Kinetic energy of } \boldsymbol{u}_h \text{ and of } \tilde{\boldsymbol{u}}\\
\mathcal{M}_h &:= \nu\|\nabla\boldsymbol{u}_h\|_R^2, & \tilde{\mathcal{M}} &:= \tau^{-1}\|\tilde{\boldsymbol{u}}\|_R^2 && \text{Dissipation of } \boldsymbol{u}_h \text{ and of } \tilde{\boldsymbol{u}}\\
\mathcal{P}_h &:= W_R(\boldsymbol{u}_h), & \tilde{\mathcal{P}} &:= \langle \boldsymbol{f}, \tilde{\boldsymbol{u}}\rangle_R && \text{External power on } \boldsymbol{u}_h \text{ and on } \tilde{\boldsymbol{u}}\\
\mathcal{T} &:= \sum_{K\subset R} \langle \tilde{\boldsymbol{u}}, \tilde{P}(\mathcal{L}_u^*\boldsymbol{u}_h - \nabla p_h)\rangle_K & & && \text{Energy transfer}
\end{aligned}
$$

Note that the energy transfer term $\mathcal{T}$, when considered in the equation for the finite element component, can be thought as the *numerical dissipation of the formulation*.

From (12)–(13) we see that only if $\tilde{V}$ is a subspace of $V_h^{\perp}$ the energy balance in region R can be written as

$$
\boxed{
\begin{aligned}
\frac{\mathrm{d}}{\mathrm{d}t}\mathcal{K}_h + \mathcal{M}_h + \mathcal{T} &= \mathcal{P}_h\\
\frac{\mathrm{d}}{\mathrm{d}t}\tilde{\mathcal{K}} + \tilde{\mathcal{M}} - \mathcal{T} &= \tilde{\mathcal{P}}
\end{aligned}
}
\tag{14}
$$

Therefore, *there is a scale separation in the kinetic energy balance only if the subscales are orthogonal to the finite element space.*

3.2 Global Kinetic Energy Balance Equations

Our thesis that the numerical approximation of the velocity subscales defines a LES model is sustained not only by numerical experiments, but also by some physical reasoning. We present one of these physical arguments next.

For the following discussion we may assume quasi-static subscales (an important implication of considering dynamics subscales is presented in the next subsection) and no body forces applied to the fluid. Neglecting $\nu\Delta\boldsymbol{u}_h$ and calling $\tilde{P} = P_h^{\perp}$ the projection orthogonal to the finite element space V_h, from (11) it follows that

$$
\tilde{\boldsymbol{u}} = -\tau P_h^{\perp}(\boldsymbol{u}_* \cdot \nabla\boldsymbol{u}_h + \nabla p_h),
$$

so that the energy balance equation (12) when $R = \Omega$ yields

$$
\frac{1}{2}\frac{\mathrm{d}}{\mathrm{d}t}\|\boldsymbol{u}_h\|^2 + \nu\|\nabla\boldsymbol{u}_h\|^2 + \int_{\Omega} \varepsilon_{\text{num}} = 0, \tag{15}
$$

where

$$\varepsilon_{\text{num}} := \tau |P_h^{\perp}(\boldsymbol{u}_* \cdot \nabla \boldsymbol{u}_h + \nabla p_h)|^2 \geq 0 \tag{16}$$

is the global numerical dissipation at each point. We may now compare the energy budget (15) with what would be obtained for the continuous problem, using a LES model and approximating numerically a LES model. Let ε_{LES} be the pointwise dissipation associated to a certain LES model (see [22]) and $\bar{\boldsymbol{u}}$ the filtered velocity field resulting from this model. If the LES model is approximated numerically, let $\bar{\boldsymbol{u}}_h$ be the approximation to $\bar{\boldsymbol{u}}$, $\varepsilon_{\text{LES}}^h$ the approximation to the LES dissipation and $\bar{\varepsilon}_{\text{num}}$ the numerical dissipation inherent to the scheme, for example (16) replacing $\boldsymbol{u}_h$ by $\bar{\boldsymbol{u}}_h$ if the formulation we propose is used.

The counterpart of (15) for the continuous problem, a LES model and a numerical approximation of a LES model would respectively be:

$$\frac{1}{2}\frac{\mathrm{d}}{\mathrm{d}t}\|\boldsymbol{u}\|^2 + \int_{\Omega} \varepsilon_{\text{mol}} = 0, \qquad \varepsilon_{\text{mol}} = \nu |\nabla \boldsymbol{u}|^2, \tag{17}$$

$$\frac{1}{2}\frac{\mathrm{d}}{\mathrm{d}t}\|\bar{\boldsymbol{u}}\|^2 + \nu \|\nabla \bar{\boldsymbol{u}}\|^2 + \int_{\Omega} \varepsilon_{\text{LES}} = 0, \tag{18}$$

$$\frac{1}{2}\frac{\mathrm{d}}{\mathrm{d}t}\|\bar{\boldsymbol{u}}_h\|^2 + \nu \|\nabla \bar{\boldsymbol{u}}_h\|^2 + \int_{\Omega} \bar{\varepsilon}_{\text{num}} + \int_{\Omega} \varepsilon_{\text{LES}}^h = 0. \tag{19}$$

Let us discuss Lilly's argument [20]. Suppose that the flow is turbulent, with fully developed and isotropic turbulence, and that a LES model is used to capture the main flow features. In the inertial range of the Kolmogorov spectrum one may assume that all the kinetic energy of the flow is contained in the large scales, and that the molecular dissipation of these scales is negligible, that is,

$$\frac{\mathrm{d}}{\mathrm{d}t}\|\boldsymbol{u}\|^2 \sim \frac{\mathrm{d}}{\mathrm{d}t}\|\bar{\boldsymbol{u}}\|^2, \qquad \nu \|\nabla \bar{\boldsymbol{u}}\|^2 \sim 0.$$

Comparing the localized versions of (17) and (18) it turns out that these assumptions imply

$$\varepsilon_{\text{LES}}(\bar{\boldsymbol{u}}) \sim \varepsilon_{\text{mol}}(\boldsymbol{u}). \tag{20}$$

This is the basic requirement of a LES model: the dissipation it introduces must be proportional (equal, in the best case) to the molecular dissipation.

The question now is whether the numerical dissipation (16) satisfies this requirement. It is shown in [15] that *if the mesh size h belongs to the inertial range and the classical assumptions of statistical fluid mechanics apply (as described for example in [21]), then*

$$\varepsilon_{\text{num}}(\boldsymbol{u}_h) \sim \varepsilon_{\text{mol}}(\boldsymbol{u}). \tag{21}$$

Furthermore, like in LES models we may consider that $\boldsymbol{u}_h$ carries all (or most of) the kinetic energy of the flow. In this case, the energy balance equations (15) and (18) are formally identical. Moreover, if (20) and (21) both hold, from (19) it follows that *approximating a LES model with a numerical scheme with a dissipation satisfying (21) is clearly redundant.* Of course the numerical approximation is unavoidable. Hence, what is unnecessary is the use of a LES model.

3.3 Backscatter

In this subsection we use again a heuristic reasoning to analyze the possibility to model backscatter by our numerical formulation.

Using the notation introduced heretofore, backscatter can be defined by condition $\mathcal{T} < 0$, where $\mathcal{T}$ is the energy transfer term appearing in (14). This means that energy is supplied from the unresolved (small) scales to the resolved (finite element) scales. Physically, it is known that this can happen only at isolated spatial points and time instants. Numerically, the model should be such that $\mathcal{T}$ can be negative only if region R is small enough (possibly a single element) and at a few time steps of the time discretization.

As before, we will consider the subscales orthogonal to the finite element space and that $\boldsymbol{f} \in V_h$. From (11) we have that

$$\tilde{\boldsymbol{u}} = -\tau[\partial_t \tilde{\boldsymbol{u}} + \tilde{P}(\mathcal{L}_u \boldsymbol{u}_h + \nabla p_h)],$$

which upon substitution in the expression of $\mathcal{T}$ yields

$$\mathcal{T} = \underbrace{\sum_{K \subset R} \tau \langle \tilde{P}(\mathcal{L}_u \boldsymbol{u}_h + \nabla p_h), \tilde{P}(-\mathcal{L}_u^* \boldsymbol{u}_h + \nabla p_h) \rangle_K}_{(\int_R \varepsilon_{\text{num}})} + \underbrace{\sum_{K \subset R} \tau \langle \partial_t \tilde{\boldsymbol{u}}, \tilde{P}(-\mathcal{L}_u^* \boldsymbol{u}_h + \nabla p_h) \rangle_K}_{(\int_R \beta_{\text{num}})}.$$

If viscous terms are negligible (or $\Delta \boldsymbol{v}_h$ is approximated by the discrete Laplacian $\Delta_h \boldsymbol{v}_h$) and $\boldsymbol{r} = -P_h^{\perp}(\boldsymbol{u}_* \cdot \nabla \boldsymbol{u}_h + \nabla p_h)$ we have

$$\mathcal{T} = \int_R \varepsilon_{\text{num}} + \int_R \beta_{\text{num}} = \sum_K \tau \left(\langle \boldsymbol{r}, \boldsymbol{r} \rangle_K - \langle \partial_t \tilde{\boldsymbol{u}}, \boldsymbol{r} \rangle_K \right). \tag{22}$$

From this expression it immediately follows that *when orthogonal subscales are used, backscatter is possible only if the subscales are dynamic*, since otherwise the second term in (22) vanishes and the first one is obviously non-negative. If the time derivative of the subscales is included in (22), *nothing can be said about the sign of* $\mathcal{T}$, and therefore the possibility to model backscatter is open. However, the numerical model will be physically admissible if $\mathcal{T}$ is, in a certain average sense, non-negative. This is what we will justify now heuristically.

Assume $\boldsymbol{r}$ is periodic in t, of period T_0, and so is $\tilde{\boldsymbol{u}}$. If

$$\boldsymbol{r}(\boldsymbol{x}, t) = \sum_{n=0}^{\infty} (A_n(\boldsymbol{x}) \cos(\omega_n t) + B_n(\boldsymbol{x}) \sin(\omega_n t)), \quad \omega_n = \frac{2\pi n}{T_0},$$

$$\tilde{\boldsymbol{u}}(\boldsymbol{x}, t) = \sum_{n=0}^{\infty} (\tilde{A}_n(\boldsymbol{x}) \cos(\omega_n t) + \tilde{B}_n(\boldsymbol{x}) \sin(\omega_n t)),$$

Fig. 1. Averaged velocity field (left) and instantaneous velocity field (right).

imposing

$$\partial_t \tilde{\boldsymbol{u}} + \tau^{-1}\tilde{\boldsymbol{u}} = \boldsymbol{r}(t),$$

it is found that

$$\tilde{A}_n(\boldsymbol{x}) = \frac{A_n\tau^{-1} - B_n\omega_n}{\tau^{-2} + \omega_n^2}, \qquad \tilde{B}_n(\boldsymbol{x}) = \frac{A_n\omega_n + B_n\tau^{-1}}{\tau^{-2} + \omega_n^2}.$$

Averaging over a period and assuming τ constant in region R we get

$$\begin{aligned}\frac{1}{T_0}\int_0^{T_0} \mathcal{T}\,\mathrm{d}t &= \frac{1}{T_0}\int_0^{T_0}\int_R (\varepsilon_{\mathrm{num}} + \beta_{\mathrm{num}})\mathrm{d}t \\ &= \frac{\tau}{T}\int_0^{T}\sum_K (\langle \boldsymbol{r}, \boldsymbol{r}\rangle_K - \langle \partial_t \tilde{\boldsymbol{u}}, \boldsymbol{r}\rangle_K)\,\mathrm{d}t \\ &= \tau \int_R \sum_{n=0}^{\infty} \frac{1}{1 + (\omega_n\tau)^{-2}}(A_n^2 + B_n^2).\end{aligned}$$

If $\beta_{\mathrm{num}} = 0$ (quasi-static subscales) the result we would have obtained is the same without the term $(\omega_n\tau)^{-2}$, from where we may conclude that $\int_R \beta_{\mathrm{num}}$ can be negative. However, even though the average energy transfer is smaller with dynamic subscales, *it is, on average, positive*. Thus, this physical restriction holds in the case considered, regardless of the period T_0.

3.4 Flow over a Surface Mounted Obstacle

We close this section with a numerical experiment showing that *backscatter can certainly be found in numerical experiments*. This example is taken from [23], to which the reader is referred for details of the calculation.

The problem consists in modeling the flow over a surface mounted obstacle, consisting in a cylinder of square cross section. Just to have a feeling of the flow, the time-averaged iso-velocity contours and the instantaneous velocity contours in the mid section of the channel where the flow takes place are plotted in Figure 1. The numerical dissipation $\mathcal{T}$ averaged over time in the same section is shown in Figure 2. It is observed that it is more localized using dynamic subscales that quasi-static subscales.

The interesting fact of this numerical simulation is that backscatter is found, obviously using dynamic subscales (no backscatter can be found with quasi-static subscales). If we compute the numerical dissipation $\mathcal{T}$ for each element and at each time

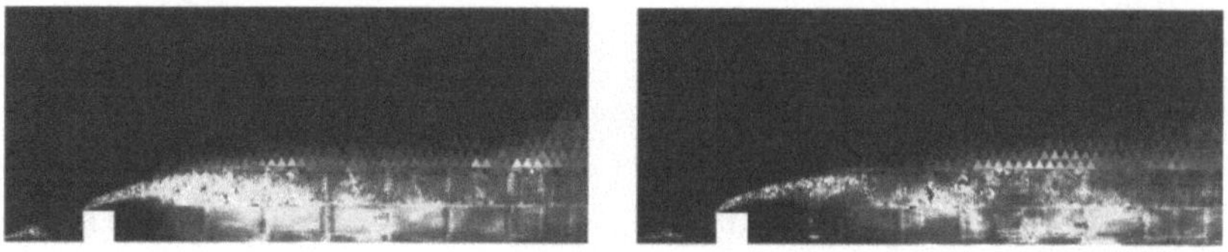

Fig. 2. Averaged numerical dissipation obtained with quasi-static subscales (left) and with dynamic subscales (right).

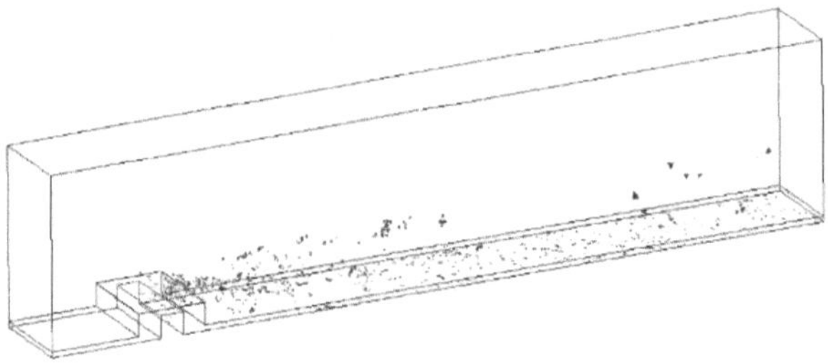

Fig. 3. Elements with instantaneous negative numerical dissipation using dynamic subscales.

step of the time discretization, it turns out that it is negative at some elements and some time steps. In Figure 3 the elements where this happens at a certain time step are marked in dark gray.

4 Long Term Stability

In this section we present the results of a complete stability analysis we have performed for formulation (10)–(11), which can be found in [2]. We will state the stability result for both $T < \infty$ and $T = \infty$, which is our main interest.

Our analysis is based in fact on a simplified version of (10)–(11). First, we assume orthogonal subscales (we envisage a lot of difficulties in the analysis if this is not assumed). This allows us to delete the viscous operator in the terms involving the subscales in (10)–(11). In the case of orthogonal subscales, this does not deteriorate the consistency and stability of the formulation [8]. We may think for example that $\Delta \boldsymbol{v}_h|_K$ is approximated by the discrete Laplacian $\Delta_h \boldsymbol{v}_h|_K$, and therefore $P_h^\perp(\Delta \boldsymbol{v}_h|_K) \approx P_h^\perp(\Delta_h \boldsymbol{v}_h|_K) = \boldsymbol{0}$.

The two real additional assumptions that we make are the following:

- $\boldsymbol{u}_* \approx \boldsymbol{u}_h$ as advection velocity.
- $\forall K \; \tau_K^{-1} = \tau^{-1} = \frac{c_1 \nu}{h^2} + \frac{c_2 U}{h}$, where $U = \frac{\|\boldsymbol{u}_h\|_{L^p(\Omega)}}{|\Omega|^{1/p}}$, with $2 < p < \infty$.

The first assumption simplifies greatly the nonlinearity of the problem. The second basically states that a single τ is used for the whole computational domain. This will weaken the strength of the results to be presented, but nevertheless they are important enough, as we will see.

With the assumptions described, the method reduces to

$$(\partial_t \boldsymbol{u}_h, \boldsymbol{v}_h) + \langle \boldsymbol{u}_h \cdot \nabla \boldsymbol{u}_h, \boldsymbol{v}_h \rangle + \nu(\nabla \boldsymbol{u}_h, \nabla \boldsymbol{v}_h) - (p_h, \nabla \cdot \boldsymbol{v}_h) + (q_h, \nabla \cdot \boldsymbol{u}_h) - \langle \tilde{\boldsymbol{u}}, \boldsymbol{u}_h \cdot \nabla \boldsymbol{v}_h + \nabla q_h \rangle = \langle \boldsymbol{f}, \boldsymbol{v}_h \rangle, \tag{23}$$

$$(\partial_t \tilde{\boldsymbol{u}}, \tilde{\boldsymbol{v}}) + \tau^{-1} \langle \tilde{\boldsymbol{u}}, \tilde{\boldsymbol{v}} \rangle + \langle \boldsymbol{u}_h \cdot \nabla \boldsymbol{u}_h + \nabla p_h, \tilde{\boldsymbol{v}} \rangle = \langle \boldsymbol{f}, \tilde{\boldsymbol{v}} \rangle, \tag{24}$$

which must hold for all test functions $[\boldsymbol{v}_h, q_h]$, $\tilde{\boldsymbol{v}}$, and in the time interval $[0, T]$.

Estimates for $T < \infty$.

Formally (see [2] for technical details) we may take $\boldsymbol{v}_h = \boldsymbol{u}_h$, $q_h = p_h$, $\tilde{\boldsymbol{v}} = \tilde{\boldsymbol{u}}$ in (23)–(24). Integrating then on $[0, t']$, $t' \leq T$, one gets:

$$\|\boldsymbol{u}_h(t')\|^2 + \|\tilde{\boldsymbol{u}}(t')\|^2 + \int_0^{t'} \nu \|\nabla \boldsymbol{u}_h\|^2 \mathrm{d}t + \int_0^{t'} \tau^{-1} \|\tilde{\boldsymbol{u}}\|^2 \mathrm{d}t \leq \int_0^{t'} \frac{1}{\nu} \|\boldsymbol{f}\|_{-1}^2 \mathrm{d}t + \|\boldsymbol{u}^0\|^2,$$

from where

$$\boxed{\begin{array}{ll} \|\boldsymbol{u}_h\| \in L^\infty(0, T), & \nu^{1/2} \|\nabla \boldsymbol{u}_h\| \in L^2(0, T), \\ \|\tilde{\boldsymbol{u}}\| \in L^\infty(0, T), & \tau^{-1/2} \|\tilde{\boldsymbol{u}}\| \in L^2(0, T). \end{array}} \tag{25}$$

The interest of these estimates is that we obtain *the same bounds for the finite element component as for the Galerkin method plus additional stability on the subscales.* This is also true for the estimates presented next. In [2] it is shown how to translate the stability obtained for the subscales to additional stability for the finite element component.

Estimates in $L^\infty(0, \infty; L^2(\Omega)^d)$.

In this case our analysis strategy is similar to that used for continuous problems (see for example [24]). Taking $\boldsymbol{v}_h = \boldsymbol{u}_h$, $q_h = p_h$, $\tilde{\boldsymbol{v}} = \tilde{\boldsymbol{u}}$ in (23)–(24) and using the classical Gronwall lemma it is found that

$$\boxed{\|\boldsymbol{u}_h\| \in L^\infty(0, \infty), \qquad \|\tilde{\boldsymbol{u}}\| \in L^\infty(0, \infty).} \tag{26}$$

Similar remarks to those made for (25) apply to (26). In this case we can also prove that

$$\lim_{t \to \infty} \sup (\|\boldsymbol{u}_h\| + \|\tilde{\boldsymbol{u}}\|) \leq C \frac{|\Omega|^{2/d}}{\nu} \|\boldsymbol{f}\|_{L^\infty(0,\infty; L^2(\Omega)^d)},$$

from which we conclude that:

- There is a $L^2(\Omega) \oplus L^2(\Omega)$-absorbing set in $V_h \oplus \tilde{V}$ (not only in V_h).
- The diameter of this absorbing set is bounded by the Reynolds number, Re (or, equivalently, by $1/\nu$).

Estimates in $L^\infty(0,\infty;H^1(\Omega)^2)$ *(case* $d=2$*).*

Finally, we can prove estimates in the strong norm of $L^\infty(0,\infty;H^1(\Omega)^2)$. As for the continuous problem, this is only possible in the two-dimensional case.

Let $\tau_0^{-1} = \frac{\nu}{h^2} + \frac{U_0}{h}$, where $U_0 \le U$ is a characteristic velocity. Using the uniform Gronwall lemma it is found that

$$\boxed{\nu^{1/2}\|\nabla \boldsymbol{u}_h\| \in L^\infty(0,\infty), \quad \tau_0^{-1/2}\|\tilde{\boldsymbol{u}}\| \in L^\infty(0,\infty).} \tag{27}$$

Once again, we obtain stability not only for the finite element component, but also for the subscales.

Define now the quantities

$$\begin{aligned}
a_1 &:= \tilde{t}\left(\|\boldsymbol{f}\|^2_{L^\infty(0,\infty;L^2(\Omega)^d)} + U_0^4\right),\\
a_2 &:= (\nu^{-1}|\Omega| + \tilde{t})\nu^{-1}|\Omega|\|\boldsymbol{f}\|^2_{L^\infty(0,\infty;L^2(\Omega)^d)},\\
a_3 &:= \nu^{-2}(\nu^{-4}|\Omega|^2\|\boldsymbol{f}\|^2_{L^\infty(0,\infty;L^2(\Omega)^d)} + 1)a_2.
\end{aligned}$$

Then, it can be shown that

$$\lim_{t\to\infty}\sup\left(\nu\|\nabla \boldsymbol{u}_h\|^2 + \tau_0^{-1}\|\tilde{\boldsymbol{u}}\|^2\right) \le C\left(a_1 + \frac{a_2}{\tilde{t}}\right)\exp(a_3),$$

from where conclude that

- There is a $H^1(\Omega)\oplus\tilde{H}(\Omega)$-absorbing set in $V_h \oplus \tilde{V}$, where $\tilde{H}(\Omega) = \tau_0^{-1/2}L^2(\Omega)$.
- The diameter of this absorbing set is bounded by $\exp(\mathrm{Re}^4)$. This is useless from the numerical point of view, but
- Since $H^1(\Omega)$ is compact in $L^2(\Omega)$, the $L^2(\Omega)$-absorbing set in V_h is in fact an attractor.

The general conclusion of all these results is that the long term dynamics engendered by the numerical formulation (23)–(24) are what one could expect. To obtain all these results, particularly estimates in $L^\infty(0,T)$ (including the case $T=\infty$) it is crucial to have the time derivative of the subscales, that is to say, the subscales need to be dynamic.

5 Long Term Simulations

In this section we present without details the results of two numerical simulations with the objective of comparing the performance of (10)–(11) for fully developed turbulent flows with LES models. Since our claim is that these models are not necessary, we will compare the results obtained with and without one of them, in particular Smagorinsky's model. In this case, a turbulent viscosity $\nu_{\mathrm{LES}} = 0.01h^2(\nabla^S \boldsymbol{u}_h : \nabla^S \boldsymbol{u}_h)^{1/2}$ is added to the molecular viscosity ν.

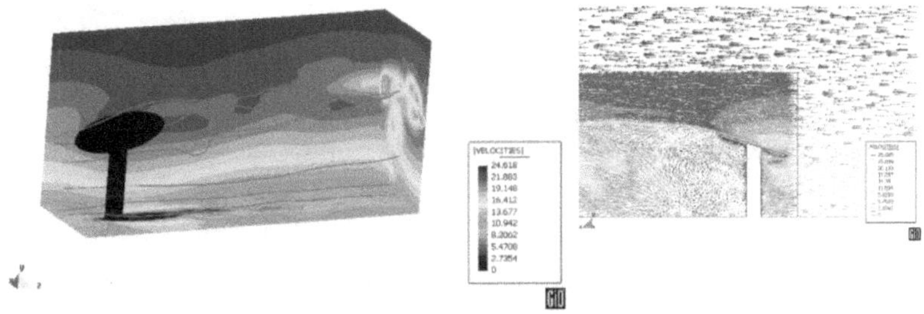

Fig. 4. Left: pressure contours and some particle trajectories. Right: velocity field in the mid section.

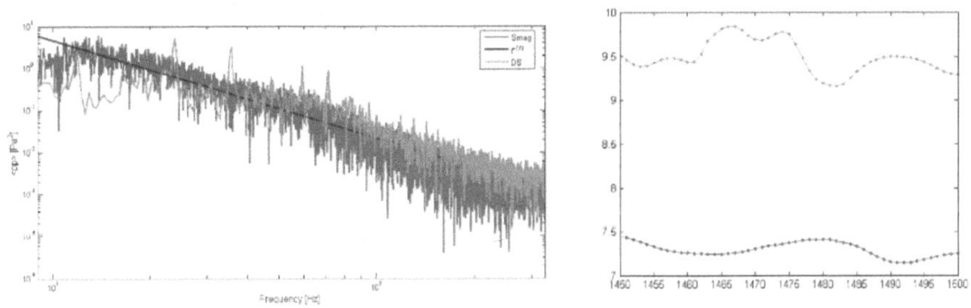

Fig. 5. Comparison of dynamic subscales (light gray in both pictures) and quasi-static subscales with Smagorinsky's model. Left: numerical pressure spectra (theoretical slope = $-7/3$), right: detail of the pressure evolution (at the center of the plate).

5.1 Flow over a Plate

The first example is the flow over a circular plate supported on a column and inclined. This example has been taken from [15], which may be consulted for details. Again, Figure 4 shows some results of the numerical simulation to provide a qualitative feeling of the flow.

What we wish to compare in this example is the Kolmogorov spectrum obtained with and without Smagorinky's model. The pressure spectra at the center of the plate obtained in both cases are shown in Figure 5. It is observed that in both cases the $-7/3$ slope is well approximated. In fact, these spectra agree well with experimental results [15]. The interesting fact is that, even though both spectra are similar, the dynamical behavior obtained *without* Smagorinky's model is richer that with it, as it could be expected. This is a general trend that we have observed and demonstrated in Figure 5 (right), where a window of the temporal evolution of the pressure at the center of plate is depicted.

5.2 Flow around a Telescope

The second example we present is intended to demonstrate that (10)–(11) is indeed applicable to real flow problems. It consists in the aerodynamic analysis around a

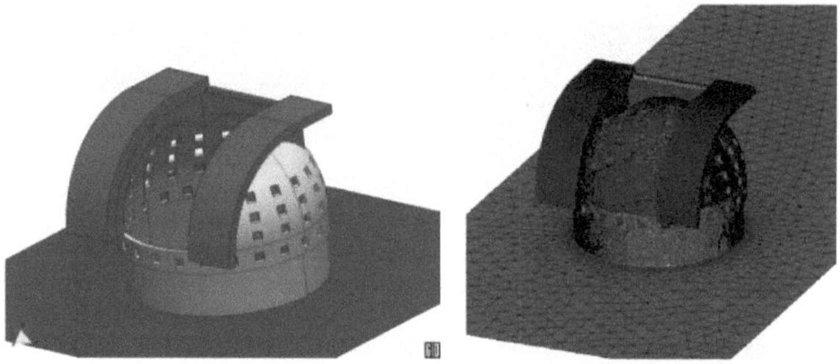

Fig. 6. Geometry of telescope enclosure (left) and the surface mesh for the numerical approximation (right).

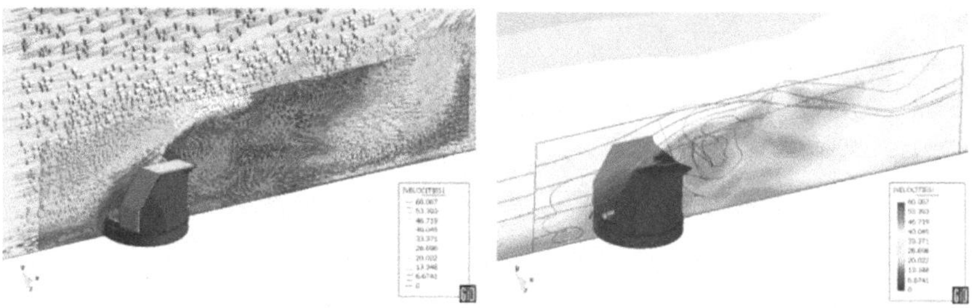

Fig. 7. Contours of velocity norm, together with velocity vectors (left) and pressure contours, together with some particle trajectories (right) for the flow over a telescope enclosure.

building enclosing a large telescope. Modeling turbulence is crucial to determine the optical quality of the site where the telescope is placed.

The geometry of the building and the finite element mesh used to discretize it are shown in Figure 6. To understand the behavior of the flow, Figure 7 shows the velocity vectors at a certain time of the calculation and the pressure contours, as well as the trajectories of some particles.

The scientifically relevant issue is whether the model is able to capture the inertial range of the Kolmogorov spectrum. The pressure spectra at a point inside the building for different flow conditions are shown in Figure 8. It is observed that these spectra display the correct $-7/3$ slope.

In order to compare the results obtained with Smagorinsky's model (SMA) with quasi-static subscales, quasi-static subscales without Smagorinsky's model (QSS) and dynamic subscales without Smagorinsky's model (DS), we have listed in Table 1 some parameters of the turbulence along a vertical line in the middle of the telescope enclosure. In particular, values for the mean velocity U and turbulence intensity I, defined by

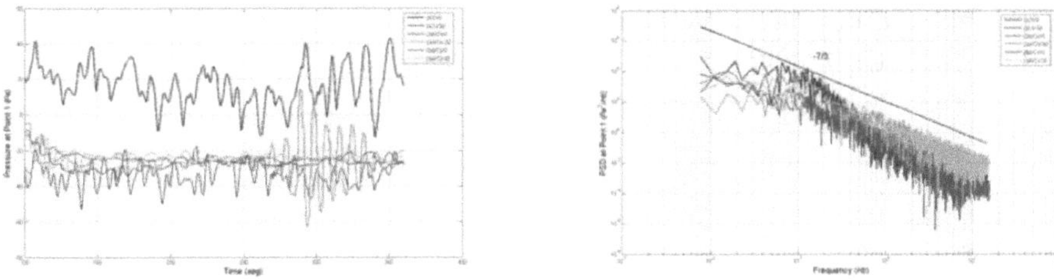

Fig. 8. Pressure history and pressure spectrum at a point inside the enclosure for different flow conditions.

Table 1. Mean velocity and turbulence intensity along a vertical line inside the telescope enclosure.

H	U SMA	U QSS	U DS	I SMA	I QSS	I DS
6.93	1.06	1.08	1.21	0.18	0.21	0.20
13.08	1.00	1.02	1.01	0.23	0.25	0.25
19.70	1.61	1.64	1.65	0.46	0.55	0.54
26.65	5.19	5.25	5.33	0.08	0.11	0.11
32.86	5.27	5.26	5.25	0.03	0.06	0.06
39.44	3.21	3.22	3.44	0.09	0.11	0.12
46.13	1.89	1.99	2.05	0.17	0.35	0.33
52.51	1.58	1.67	1.98	0.25	0.44	0.44
59.41	1.37	1.44	1.40	0.22	0.43	0.42
66.12	1.08	1.07	1.04	0.35	0.67	0.65
72.13	0.87	1.02	1.01	0.30	0.41	0.42

$$U(\boldsymbol{x}) := \frac{1}{T}\int_0^T |\boldsymbol{u}_h(\boldsymbol{x},t)|\,\mathrm{d}t, \quad I^2(\boldsymbol{x}) := \frac{1}{U^2(\boldsymbol{x})T}\int_0^T (|\boldsymbol{u}_h(\boldsymbol{x},t)| - U(\boldsymbol{x}))^2\,\mathrm{d}t \tag{28}$$

are listed in Table 1 against the height (H) of the point to the ground. In (28), T is a time window, large enough to consider the results (almost) stationary.

As expected, results are much more dissipative using Smagorinsky's model than without it. However, there is no clear trend in whether QSS is more or less dissipative that DS. In fact, results in this case are very similar.

6 Conclusions

In this work we have summarized the formulation we have been developing to approximate the incompressible Navier–Stokes equations. We have also explained some of its features related to the simulation of fully developed turbulent flows and long term stability.

The key ingredients of our formulation are

- time dependent subgrid scales,

- contribution from the subscales in the advection velocity,
- subgrid scales orthogonal to the finite element space,

together with a certain definition of the stabilization parameter. These ingredients are essential to obtain analytically:

- the correct dissipative structure (balance of energy between resolvable and subgrid scales),
- a numerical dissipation proportional to the molecular one (which poses the question on the need to use LES models),
- the possibility to model backscatter,
- correct weak and strong stability estimates in the long term.

Numerical experiments show that:

- the numerical solution is stable in the long term,
- backscatter can certainly be found,
- the $-7/3$ pressure spectrum slope in the inertial range can be observed *without* using any LES model.

Acknowledgments

Several people have participated in the research presented here. In particular, Oriol Guasch and Juan V. Gutiérrez-Santacreu have collaborated closely in the results presented in [12, 15] and [2], respectively, and summarized here. The participation of Florian Henke, Daniel Pérez-Sánchez and Christian Muñoz in the elaboration of the numerical results is also acknowledged. The numerical results presented in Section 5 have been conducted in the context of project *ELT-DS Technology development programme towards a European Extremely Large Telescope*, within the FP6-2003 of the European Comission.

References

1. S. Badia and R. Codina. On a multiscale approach to the transient Stokes problem. Transient subscales and anisotropic space-time discretization. *Applied Mathematics and Computation*, 207:415–433, 2009.
2. S. Badia, R. Codina, and J.V. Gutiérrez-Santacreu. Long term stability estimates and existence of global attractors in a finite element approximation of the Navier–Stokes equations with numerical sub-grid scale modeling. Submitted.
3. Y. Bazilevs, V.M. Calo, J.A. Cottrell, T.J.R. Hughes, A. Reali, and G. Scovazzi. Variational multiscale residual-based turbulence modeling for large eddy simulation of incompressible flows. *Computer Methods in Applied Mechanics and Engineering*, 197:173–201, 2007.
4. J. Blasco and R. Codina. Space and time error estimates for a first order, pressure stabilized finite element method for the incompressible Navier–Stokes equations. *Applied Numerical Mathematics*, 38:475–497, 2001.

5. R. Codina. Comparison of some finite element methods for solving the diffusion-convection-reaction equation. *Computer Methods in Applied Mechanics and Engineering*, 156:185–210, 1998.
6. R. Codina. A stabilized finite element method for generalized stationary incompressible flows. *Computer Methods in Applied Mechanics and Engineering*, 190:2681–2706, 2001.
7. R. Codina. Stabilized finite element approximation of transient incompressible flows using orthogonal subscales. *Computer Methods in Applied Mechanics and Engineering*, 191:4295–4321, 2002.
8. R. Codina. Analysis of a stabilized finite element approximation of the Oseen equations using orthogonal subscales. *Applied Numerical Mathematics*, 58:264–283, 2008.
9. R. Codina and J. Blasco. Analysis of a pressure-stabilized finite element approximation of the stationary Navier–Stokes equations. *Numerische Mathematik*, 87:59–81, 2000.
10. R. Codina and J. Principe. Dynamic subscales in the finite element approximation of thermally coupled incompressible flows. *International Journal for Numerical Methods in Fluids*, 54:707–730, 2007.
11. R. Codina, J. Principe, and J. Baiges. Subscales on the element boundaries in the variational two-scale finite element method. *Computer Methods in Applied Mechanics and Engineering*, 198:838–852, 2009.
12. R. Codina, J. Principe, O. Guasch, and S. Badia. Time dependent subscales in the stabilized finite element approximation of incompressible flow problems. *Computer Methods in Applied Mechanics and Engineering*, 196:2413–2430, 2007.
13. P.A.B. de Sampaio, P.H. Hallak, A.L.G.A. Coutinho, and M.S. Pfeil. A stabilized finite element procedure for turbulent fluid-structure interaction using adaptive time-space refinement. *International Journal for Numerical Methods in Fluids*, 44:673–693, 2004.
14. J. Donéa and A. Huerta. *Finite Element Methods for Flow Problems*. John Wiley & Sons, 2003.
15. O. Guasch and R. Codina. A heuristic argument for the sole use of numerical stabilization with no physical LES modelling in the simulation of incompressible turbulent flows. UPCommons, `http://hdl.handle.net/2117/3021`, Submitted.
16. J. Hoffman and C. Johnson. A new approach to computational turbulence modeling. *Computer Methods in Applied Mechanics and Engineering*, 195:2865–2880, 2006.
17. T.J.R. Hughes. Multiscale phenomena: Green's function, the Dirichlet-to-Neumann formulation, subgrid scale models, bubbles and the origins of stabilized formulations. *Computer Methods in Applied Mechanics and Engineering*, 127:387–401, 1995.
18. T.J.R. Hughes, G.R. Feijóo, L. Mazzei, and J.B. Quincy. The variational multiscale method–a paradigm for computational mechanics. *Computer Methods in Applied Mechanics and Engineering*, 166:3–24, 1998.
19. T.J.R. Hughes, L. Mazzei, and K.E. Jansen. Large eddy simulation and the variational multiscale method. *Computing and Visualization in Science*, 3:47–59, 2000.
20. D.K. Lilly. The representation of small-scale turbulence theory in numerical simulation experiments. In H.H. Goldstine (Ed.), *Proc. IBM Scientific Computing Symp. on Environmental Sciences*, 1967.
21. A.S. Monin and A.M. Yaglom. *Statistical Fluid Mechanics: Mechanics of Turbulence. Volume I*. MIT Press, Cambridge, MA, 1971.
22. S.B. Pope. *Turbulent Flows*. Cambridge University Press, 2000.
23. J. Principe, R. Codina, and F. Henke. The dissipative structure of variational multiscale methods for incompressible flows. *Computer Methods in Applied Mechanics and Engineering*, 199:791–801, 2010.
24. R. Temam. *Infinite-Dimensional Dynamical Systems in Mechanics and Physics*. Springer, 1988.

Large-Eddy Simulation of Multiscale Particle Dynamics at High Volume Concentration in Turbulent Channel Flow

Bernard J. Geurts

Multiscale Modeling and Simulation, J.M. Burgers Center, Faculty EEMCS, University of Twente, P.O. Box 217, 7500 AE Enschede, The Netherlands; b.j.geurts@utwente.nl
and
Anisotropic Turbulence, Fluid Dynamics Laboratory, Department of Applied Physics, Eindhoven University of Technology, P.O. Box 513, 5600 MB Eindhoven, The Netherlands

Abstract Large-eddy simulations (LES) of a vertical turbulent channel flow laden with a large number of solid point particles are discussed. This flow is relevant for a better understanding of co-current turbulent gas-particle flows, as encountered in riser reactors. The particle volume fraction is chosen to be high, it equals about 1.3%. This necessitates the use of a four-way coupling model for the discrete particle dynamics. The Euler–Lagrangian method is adopted, which means that the fluid dynamics is governed by the Navier–Stokes equations and that the motion of individual particles is tracked by solving their equation of motion, taking both the particle-fluid and particle-particle interactions into account. LES results are compared to single-phase channel flow to investigate the effect of the particles on turbulent statistics. Due to particle-fluid interactions the mean fluid profile is flattened and the boundary layer is thinner. Compared to single-phase turbulent flow, the streamwise turbulence intensity of the gas phase is increased, while the normal and spanwise turbulence intensities are reduced, as is also observed in experimental data. The four-way coupled simulations are also compared with two-way coupled simulations, in which the inelastic collisions between particles are neglected. The latter comparison demonstrates that the collisions have a large influence on the main statistics of both phases, expressed for example by the stronger coherent particle structures observed in four-way coupling.

Key words: Large-eddy simulation, channel flow, particle-laden turbulence, four-way coupling

1 Introduction

Many flows of relevance to large-scale chemical processing involve solid catalyst particles at significant concentrations embedded in a carrying gas-flow. Control over the spatial distribution of these particles, especially its homogeneity, is essential in order to provide a chemical processing that is as complete and uniform as possible,

R. de Borst and E. Ramm (eds.), *Multiscale Methods in Computational Mechanics*, Lecture Notes in Applied and Computational Mechanics 55, DOI 10.1007/978-90-481-9809-2_6,

that is consistent with modern environmental requirements and that does not constitute a strong safety hazard. This provides the main context for this study which is directed toward understanding the fundamental aspects of the dynamics of the embedded, interacting particles, and to develop a simulation strategy with which the central up-scaling of laboratory-scale experiments to realistic industrial settings can be supported.

The dynamics of the embedded particle-ensemble is quite complex and interacts nonlinearly with the carrying gas-flow. The particles are dragged along by this carrying gas-flow and exchange momentum with it. Moreover, the solid particles interact among each other, e.g., through inelastic particle-particle collisions. In case only particle-fluid interactions are incorporated the description is referred to as 'two-way coupled' while a 'four-way coupled' formulation arises when also the particle-particle interactions are included [5]. At sufficiently low particle volume fraction ψ two-way coupling is adequate. However, with increasing ψ the collisions will become dynamically significant and the computationally more involved four-way coupling will become required. We will show that these collisions strongly influence the main statistical fluid properties and amplify the 'self-organization' of the embedded particles in coherent swarms.

The two-phase gas-solid flow is governed by an interplay between the convective gas-flow nonlinearity, the particle-fluid and the particle-particle interactions. These effects may accumulate and significantly change basic turbulence properties such as mean flow and turbulence intensities. A large-scale dynamic flow-structuring may arise affecting the flow-statistics compared to the case with no or only weak interactions. These flow-alterations constitute the so-called modulation of turbulence [12] which, e.g., seriously complicates the prediction of the up-scaling of flow-phenomena from laboratory-scale experiments to industrial-scale settings.

Turbulent gas-solid flows have been studied experimentally (e.g., [20,27,30,39]) and with simulations. Simulations can be performed using a two-fluid model in which the solid phase is modeled as a fluid using continuous variables (e.g., [11,24, 30]). This approach is quite well established and may be used to investigate statistical properties of multi-phase flows [7]. A promising, more recent, direction to solve two-phase flows is to enforce the no-slip condition on the boundary of each particle using front tracking methods (see e.g., [6,38]). No additional modeling assumptions are required, but the amount of particles that can be calculated is currently around 1000. In this paper we consider a third approach, the discrete particle method in which the Navier–Stokes equations which govern the fluid in a Eulerian framework are combined with a Lagrangian tracking of the motion of each individual particle. The forces between the fluid and each particle are modeled with a drag law and all collisions between particles are treated with a deterministic approach [15,22]. The modeling is more refined than in two-fluid models, while it presently allows 1,000–10,000 times more particles than the direct front tracking methods.

The discrete particle technique will be combined with Large-Eddy Simulation (LES) of the fluid flow. Large-eddy simulation solves the large flow scales like Direct Numerical Simulation (DNS), but models the effect of the small scales with a subgrid-model (see the reviews by Pope [32], Sagaut [36] and Geurts [9]). It is considerably more efficient than Direct numerical simulation (DNS), which resolves all turbulent scales in the flow. These techniques are able to give proper detailed descriptions of the turbulence in a channel flow. LES/DNS of channel flows supplemented with a discrete particle model have been reported a number of times (e.g., [1,2,23,37,44,45]). However, the total solids volume fraction in these studies remains rather small (0.01%) and most of these works employ one- or two-way coupling. An exception is Yamamoto et al. [44] who started to investigate the influence of particle-particle interactions in LES of channel flow with particle volume fractions up to 0.014%. They found that even in such dilute regimes the effects of collisions are significant.

The purpose of this paper is to present LES of a channel flow in which the particle volume concentration is an order of magnitude higher than existing Euler–Lagrangian studies in literature and hence closer to industrial applications. The discrete particle module developed by Hoomans et al. [15] will be used, in which the spherical particles have a finite size and all (inelastic) collisions are taken into account. A subgrid closure needs to be adopted for the LES-equations of the gas phase for which we will mainly adopt the recently developed model by Vreman [43].

The organization of this paper is as follows. In Section 2 we extensively present the simulation method. Results of a large number of channel flow simulations are presented in Section 3, focusing on turbulence modulation, the differences between two- and four-way coupling and coherent particle structures. Finally, concluding remarks are collected in Section 4.

2 Mathematical Formulation

In this section we specify the mathematical formulation of the simulation model for the turbulent gas-solids flow. In Section 2.1 the equations governing the gas-phase are described. The treatment of the solids-phase is specified in Section 2.2. The subgrid modeling for the turbulent stresses that arise in the large-eddy simulation is introduced in Section 2.3 and, finally, the numerical method is discussed in Section 2.4.

2.1 The Gas Phase

The computational model distinguishes a gas phase and a solids phase. The embedded solid particles are considered to be small compared to convective turbulent length-scales. This allows to effectively approximate the equations for the gas phase in terms of flow through a (time- and position-dependent) porous medium [15,21,33,48]. The local, instantaneous particle concentration determines the fluid-volume-fraction that is accessible to the gas phase. At particle-volume-fractions around 1–2% we may restrict to the standard Navier–Stokes equations that govern a compressible flow with appropriate forcing terms:

$$\partial_t \rho + \partial_j(\rho u_j) = 0, \tag{1}$$

$$\partial_t(\rho u_i) + \partial_j(\rho u_i u_j) = -\partial_i p + \partial_j \sigma_{ij} + \rho a_{ext} \delta_{i3} + f_i, \tag{2}$$

$$\partial_t e + \partial_j((e+p)u_j) = \partial_j(\sigma_{ij} u_i) + \rho a_{ext} u_3 + f_i u_i - \partial_j q_j. \tag{3}$$

where the symbols ∂_t and ∂_j denote the partial differential operators $\partial/\partial t$ and $\partial/\partial x_j$ respectively. Furthermore, ρ is the density, $\mathbf{u}$ the velocity, p the pressure and $e = p/(\gamma - 1) + \frac{1}{2}\rho u_k u_k$ the total energy per volume unit. The constant γ denotes the ratio of specific heats $C_P/C_V = 1.4$. The coordinate x_3 denotes the streamwise direction of the channel flow, x_2 is the normal and x_1 is the spanwise direction. Throughout, we will frequently interchange the symbols x_1, x_2 and x_3 by x, y and z and u_1, u_2 and u_3 by u, v and w respectively. The domain is rectangular and the channel width, height and depth equal $L_2 = 0.05m$, $L_3 = 0.30m$ and $L_1 = 0.075m$ respectively. Periodic boundary conditions are assumed for the stream- and spanwise directions.

The viscous stress σ_{ij} equals $2\rho\nu S_{ij}$ where ν is the fluid viscosity and the strain-rate is defined by

$$S_{ij}(u) = \frac{1}{2}\partial_i u_j + \frac{1}{2}\partial_j u_i - \tfrac{1}{3}\delta_{ij}\partial_k u_k. \tag{4}$$

The heat-flux q_j is defined as $-\kappa\partial_j T$ where T is the temperature and κ the heat-conductivity coefficient. Pressure, density and temperature are related to each other by the equation of state for an ideal gas $\rho R_{gas} T = M_{gas} p$, where $R_{gas} = 8.314$ J/(mol K) is the universal gas constant and $M_{gas} = 0.0288$ kg/mol is the molar mass of the gas.

The symbol a_{ext} represents the acceleration caused by external forces on the gas phase. It corresponds to a mean pressure gradient over the channel that is needed to maintain a desired mass flux. The forcing term f_i denotes the contributions due to the momentum-exchange of the particles to the flow. The external acceleration a_{ext} is a function of time only and its level is such that the total fluid mass flow is constant. In all cases the average mean gas velocity is identical, $U_m = 3.92$ m/s. The sink-terms f_i are induced by an effective relative motion of the particles with respect to the gas which gives rise to drag forces on the fluid (see next subsection).

We are interested in a section of a riser flow with a vertical centerline velocity U_c of about 4 m/s. The parameters of the fluid in the riser are close to those for air. The initial fluid density is uniform and equals $\rho_g = 1.0kg/m^3$. The viscosity equals $\nu = 3.47 \cdot 10^{-5}$ m^2/s and the heat-conductivity is obtained from the assumption that the Prandtl number equals one. The value of the viscosity is chosen such that $Re_\tau = 180$ for channel flow without particles and $U_c = 4.5$ m/s. The Kolmogorov length-scale in channel flow equals about $\eta^+ \approx 1.5$ in wall-units [32], which implies $\eta \approx 0.2$ mm.

2.2 The Solids Phase

The number of solid particles in the channel flow equals $N_p \approx 4 \cdot 10^5$. During the simulations the motion of all these particles was tracked, starting from an initially uniform distribution of particles throughout the flow-domain. The initial velocity of each particle was taken equal to the local initial velocity of the gas-phase. The particle diameter and density are $d_p = 0.4$ mm and $\rho_p = 1500$ kg/m^3, respectively. With the parameters above the average volume fraction of the particles equals 0.013. The Stokes response-time, defined as

$$\tau_p = \frac{\rho_p d_p^2}{18\mu}, \tag{5}$$

equals 0.4 s. The Stokes number equals 10, based on the Kolmogorov time derived from the average dissipation of the unladen flow. In this paper only a single, rather high, value of the Stokes response-time will be adopted in order to emphasize the dynamic effects of the embedded particles. This provides a characteristic, demanding case of turbulent gas-solid flow which is used to assess the feasibility and accuracy of Euler–Lagrangian LES.

The motion of every individual particle i in the system is calculated from Newton's second law:

$$m_i \frac{d\mathbf{v}_i}{dt} = V_i \beta (\mathbf{u} - \mathbf{v}_i) + m_i g \mathbf{e}_z + \mathbf{f}_i^{pp} + \mathbf{f}_i^{pw}, \tag{6}$$

where m_i denotes the mass, $\mathbf{v}_i$ the velocity, V_i the volume of the i-th particle and $\mathbf{e}_z$ is the unit vector in the z-direction. The gravitational acceleration equals $g = -9.81$ m/s^2, which is opposite to the mean flow direction. The forces on the right hand side of the equation represent standard drag, gravity, particle-particle interaction ($\mathbf{f}_i^{pp}$) and particle-wall interaction ($\mathbf{f}_i^{pw}$), respectively. We return to this momentarily. The general equation of motion for a single particle derived by Maxey and Riley [25] contains additional forces, such as added mass and history terms. However, the comparison with DNS results performed by Bagchi and Balachandar [3] did not show improvements when these forces were included. In the present case, the particle density is much larger than the fluid density. Correspondingly, these additional forces, including buoyancy effects are relatively small and can be neglected [2].

The symbol β in the drag term is the inter-phase momentum transfer coefficient. The flow is sufficiently dilute to employ the correlation of Wen and Yu [46]:

$$\frac{\beta d_p^2}{\mu} = \frac{3}{4} C_D \mathrm{Re}; \quad C_D = \begin{cases} 24(1 + 0.15\mathrm{Re}^{0.687})/\mathrm{Re}; & \mathrm{Re} < 10^3 \\ 0.44 & \mathrm{Re} > 10^3, \end{cases} \tag{7}$$

$\mathrm{Re} = \rho |\mathbf{u} - \mathbf{v}_\mathrm{p}| \mathrm{d}_\mathrm{p} / \mu$ is the particle Reynolds number, which is evaluated at the particle position.

Bagchi and Balachandar [3] investigated the validity of the standard drag law for particles with a diameter $1.5\eta < d < 10\eta$, where η is the Kolmogorov length-scale. They found that the time-averaged drag is accurately predicted and insensitive to whether the fluid velocity is measured at the particle center, or obtained by averaging over a fluid volume of the order of the particle size. Instantaneous drag is reasonably well predicted for moderate particle sizes, e.g., $d < 4\eta$. The diameter of the particles in the present study equals 2η and is within this region. Hence, we will assume the drag law to be adequately representative of the dominant particle-motion physics, as far as the single particle dynamics is concerned. Since we consider volume fractions in the 1% range we treat collisional effects separate from the standard drag dynamics and ignore corrections from the average proximity of neighboring particles.

The collision model used in this work is based on the hard-sphere model developed by Hoomans and coworkers [15, 16]. The original formulation of this model was developed for two-dimensional flows. Here, we extend it to be applicable to turbulent flow in three spatial dimensions. As the particles are assumed small compared to the convective turbulent length-scales, this extension involves no principal changes but only some technical complications associated with the transition from two to three spatial dimensions. These are specified in detail below.

In the collision model that characterizes $\mathbf{f}_i^{pp}$ (and $\mathbf{f}_i^{pw}$) in the equation of motion it is assumed that the interaction forces are impulsive and therefore all other finite forces are negligible during collision. We closely follow Hoomans et al. [15]. Consider two colliding spheres a and b with position vectors $\mathbf{x}_a$ and $\mathbf{x}_b$. The particle velocities prior-to-collision are indicated by the subscript 0 and the relative velocity at the contact point c (i.e., just after the collision) is defined as $v_{ab} = v_{a,c} - v_{b,c}$.

For a binary collision of these spheres the following equations can be derived by applying Newton's second and third laws:

$$m_a(\mathbf{v}_a - \mathbf{v}_{a,0}) = -m_b(\mathbf{v}_b - \mathbf{v}_{b,0}) = \mathbf{J}, \tag{8}$$

$$\frac{I_a}{R_a}(\omega_a - \omega_{a,0}) = \frac{I_b}{R_b}(\omega_b - \omega_{b,0}) = -\mathbf{n} \times \mathbf{J}, \tag{9}$$

where $\mathbf{J}$ is the momentum vector, R is the particle radius and $\mathbf{n}$ is the unit vector directed along $\mathbf{x}_a - \mathbf{x}_b$. The moment of inertia I is given by

$$I = \frac{2}{5} m R^2. \tag{10}$$

Equations 8 and 9 can be rearranged to obtain

$$\mathbf{v}_{ab} - \mathbf{v}_{ab,0} = \frac{7\mathbf{J} - 5\mathbf{n}(\mathbf{J} \cdot \mathbf{n})}{2m_{ab}}, \tag{11}$$

where m_{ab} is the reduced mass given by

$$m_{ab} = \left(\frac{1}{m_a} + \frac{1}{m_b}\right)^{-1}. \tag{12}$$

In order to calculate the post-collision velocities, a closure model consisting of three parameters is used to describe the momentum vector $\mathbf{J}$. The first parameter is the coefficient of normal restitution, $(0 \le e \le 1)$:

$$\mathbf{v}_{ab} \cdot \mathbf{n} = -e(\mathbf{v}_{ab,0} \cdot \mathbf{n}). \tag{13}$$

The second parameter is the coefficient of dynamic friction $(\mu_f \ge 0)$:

$$|\mathbf{n} \times \mathbf{J}| = -\mu_f(\mathbf{n} \cdot \mathbf{J}). \tag{14}$$

The third parameter is the coefficient of tangential restitution $(0 \le \beta_0 \le 1)$:

$$\mathbf{v}_{ab} \cdot \mathbf{t} = -\beta_0(\mathbf{v}_{ab,0} \cdot \mathbf{t}). \tag{15}$$

Combining Eqs. (11) and (13) yields the following expression for the normal component of the momentum vector:

$$J_n = -(1+e)m_{ab}\mathbf{v}_{ab,0} \cdot \mathbf{n}. \tag{16}$$

For the tangential component, two types of collisions can be distinguished, i.e. sticking or sliding collisions. If the tangential component of the relative velocity is sufficiently high in comparison to the coefficients of friction and tangential restitution, gross sliding occurs throughout the whole duration of the contact and the collision is of the sliding type. The non-sliding collisions are of the sticking type. When β_0 is equal to zero, the tangential component of the relative velocity becomes zero during a sticking collision. When β_0 is greater than zero in such a collision, reversal of the tangential component of the relative velocity will occur. The criterion to determine the type of collision on basis of pre-collision information is as follows:

$$J_t = \begin{cases} -\frac{2}{7}(1+\beta_0)m_{ab}\mathbf{v}_{ab,0} \cdot \mathbf{t} \text{ if } \mu_f J_n \geq \frac{2}{7}(1+\beta_0)m_{ab}\mathbf{v}_{ab,0} \cdot \mathbf{t} \\ -\mu_f J_n \qquad\qquad \text{otherwise,} \end{cases} \tag{17}$$

where the two equations respectively describe collisions of the sticking and sliding type. Given the definition of $\mathbf{J}$ in Eqs. (16) and (17), the post-collision velocities can now be calculated from Eqs. (8) and (9). In particle-wall col the momentum equations of the gas phase, which is computed from:

$$\mathbf{f}(\mathbf{r}) = \frac{1}{V_{cell}} \int_{V_{cell}} \sum_{i=0}^{N_p} V_i \beta(\mathbf{u} - \mathbf{v}_i) D(\mathbf{r}_i - \mathbf{r}) dV \tag{18}$$

where V_{cell} represents the nonuniform volume of the local fluid grid cell which contains the location vector $\mathbf{r}$ at which the drag term needs to be computed. The distribution function D locally distributes the reaction force acting on the gas phase to the Eulerian grid via volume weighting (see [15] for more details). The integral over a computational cell is performed in order to include the contributions to the momentum transfer arising from all particles that are contained in the respective grid-cells at any instant in time. The finite volume discretization is formulated in terms of the volume averaged momentum transfer per grid cell.

The collision model is based on the hard-sphere model developed by Hoomans and coworkers [15, 16], who applied this model to two-dimensional flow first. This model has also been validated and frequently applied to three-dimensional flows (e.g., [4, 14], and references therein). In the collision model it is assumed that the interaction forces are impulsive and therefore all other finite forces are negligible during collision.

The particle collision characteristics play an important role in the overall system behavior as was shown by Hoomans et al. [15] and Goldschmidt et al. [11]. For this reason realistic collision properties of the particles are supplied to the model. The parameters used in the treatment of the discrete particles are summarized in

Table 1. Parameters used in the treatment of the discrete particles. Each particle is spherical, corresponds to a Stokes response-time $\tau_p = 0.4$ s, associated with a viscosity of the gas phase of $\mu = 3.47 \cdot 10^{-5}$ kg m^{-1}s^{-1}. The solids phase has an average volume fraction $\langle\psi\rangle = 0.013$ and the mass-load is given by $\langle\psi\rangle\rho_p/\rho_g = 19.5$.

Symbol	Value	Description
d_p	$4 \cdot 10^{-4}$ m	particle diameter
ρ_p	$1.5 \cdot 10^3$ kg m^{-3}	particle density
N_p	419904	number of particles
e	0.97	normal restitution coefficient
β_0	0.33	tangential restitution coefficient
μ_f	0.1	friction coefficient

Table 1. Here ψ is the local solid volume fraction, which does not occur in the present model equations, but is computed for evaluation purposes (see Section 3.3). Collisions between particles are monitored as follows. For each particle, say A, a neighbor-list is kept. This includes all particles that are located within a certain radius of particle 'A'. Particles nearer the top of the list are closer to 'A', while particles that are further separated are stored lower on the list. After each collision among the entire particle swarm, the neighbor lists are updated when necessary.

2.3 Subgrid-Modeling

In order to make large-scale turbulent flow simulations at high particle volume fractions feasible, the gas phase is described using large-eddy simulation. This is obtained by applying spatial filtering to the flow equations in order to reduce their dynamical complexity. The filter is defined by

$$\overline{a} = \int G(x,\xi)a(\xi)d\xi \tag{19}$$

where $\overline{a}$ denotes a filtered flow variable and G the filter-kernel. For the compressible equations we use a density-weighted filter,

$$\widetilde{a} = \overline{\rho a}/\overline{\rho}, \tag{20}$$

the so-called Favre filter and originally proposed by Reynolds [34]. If the convolution filter is applied to the governing equations the result may be expressed in terms of the LES-template [9]: $\mathrm{NS}(\overline{U}) = R(U,\overline{U})$ where the original and filtered variables are defined by $U = [\rho, u_j, p, T]$; $\overline{U} = [\overline{\rho}, \widetilde{u}_j, \overline{p}, \widetilde{T}]$. The spatial filtering yields a 'closure-residual' $R(U,\overline{U})$ which contains, e.g., the filtered forcing term $\overline{f}_i$ and the divergence of the turbulent stress tensor

$$\tau_{ij} = \overline{\rho u_i u_j} - \overline{\rho u_i}\,\overline{\rho u_j}/\overline{\rho} = \overline{\rho}\{\widetilde{u_i u_j} - \widetilde{u}_i\widetilde{u}_j\} \tag{21}$$

The only closure term modeled in this paper is this turbulent stress tensor. The subgrid terms that result from the filtering of the diffusive viscous fluxes are neglected. Likewise, subgrid contributions arising from filtering the momentum exchange between the discrete particles and the fluid as represented by the drag law are also neglected. This is a reasonable assumption for the present application, as we consider relatively coarse particles. The particles are slightly larger than the Kolmogorov length-scale and the Stokes response time is an order of magnitude larger than the Kolmogorov time. Therefore the motion of the particles is mainly influenced by the large-scale eddies in the flow [1, 18, 19].

Simulations were performed using different subgrid-models for the turbulent stress tensor τ_{ij}. As the results were found to be quite insensitive of the adopted subgrid model, compared to the dynamic effects of the particles, we only include the eddy-viscosity model as described in [10,41,43]. This model employs the following positive invariant of the gradient model m_{ij},

$$B = m_{11}m_{22} - m_{12}^2 + m_{11}m_{33} - m_{13}^2 + m_{22}m_{33} - m_{23}^2, \tag{22}$$

and defines the eddy-viscosity

$$\nu_e = c\sqrt{\frac{B}{(\partial_j \tilde{u}_i)(\partial_j \tilde{u}_i)}} \tag{23}$$

where $c = 0.07$. The dissipation of this eddy-viscosity and the exact subgrid dissipation were shown to vanish for precisely the same class of flows [43]. This model was proposed independently of the work of Nicoud and Ducros [29], who constructed a similar eddy-viscosity. That eddy-viscosity is not based on the gradient model, but on the square of the velocity gradient matrix.

2.4 The Numerical Method

The equations for the fluid phase are solved with a second-order finite volume method, based on central differencing on a collocated grid. For details we refer to [8, 42]. In the latter work, the second-order numerical method was found to be sufficiently accurate in LES, also with respect to the dissipation of kinetic energy.

The channel flow is solved on the domain $3H \times 2H \times 12H$. The length of the domain in the spanwise (x_1) direction is 1.5 times smaller than for most DNS/LES of single-phase channel flow in order to limit the amount of carried particles and their collisions to a manageable number. As a point of reference, we also performed a single-phase DNS for this computational domain. Specifically, we used an incompressible Fourier–Chebyshev method with $64 \times 128 \times 128$ modes. The mean and rms profiles (shown in the next subsection) were verified to be identical to those of standard DNS-databases for $Re_\tau = 180$ (e.g., [17,26,28,40]).

The large-eddy simulations presented in this paper involve $32 \times 64 \times 64$ grid cells. The grid is only nonuniform in the normal direction and symmetric with respect to the plane $x_2 = 0$. The grid-points in the left-half of the channel are defined by

$$x_{2,j}/H = -1 + \frac{\sinh(aj/N_2)}{\sinh(a/2)} \quad \text{with } j = 0, \ldots, N_2/2 \text{ and } a = 6.5 \qquad (24)$$

(see [40]). The first grid point of the wall is at $x_{2,1} = 0.2$ mm, corresponding to $y^+ = 1.5$. The grid is sufficiently fine to have a well-resolved LES of channel flow at $Re_\tau = 180$, according to common criteria (see [31]).

The discretization in time is explicit: forward Euler for the particles, and a four-stage Runge–Kutta scheme for the fluid convective, viscous and pressure terms, using coefficients $\frac{1}{4}$, $\frac{1}{3}$, $\frac{1}{2}$ and 1. The most costly part of the simulation method is the discrete particle model, which requires about 80% of the computation time in case of four-way coupled simulations. The Euler forward method was adopted in combination with first-order accurate tri-linear interpolation to obtain the Eulerian velocity field at the particle location. It is combined with sufficiently small time-steps to yield adequate accuracy for the particle trajectories. The particle time-step is about 1/5 of that of the fluid time-step. The simulations run until at least $t = 5$ s, while statistics are accumulated between $t = 3$ s and $t = 5$ s. The averaging time of 2 s corresponds to $20H/u_\tau$ in terms of the wall shear-velocity u_τ, twice as large as a typical averaging time in single-phase channel flow. With a Stokes response-time of $\tau_p = 0.4$ s, the particles are evolved for 12.5 τ_p and the accumulation of the statistics is over $5\tau_p$.

3 Results

In this section we will compare results obtained in 'clean' riser-flow with the particle-laden case, using four different subgrid models. First we will consider the clean and four-way coupled cases and quantify the turbulence modulation of the gas phase arising from the presence of the particles (Section 3.1). In Section 3.2 we will isolate the effects of the particle collisions and compare the four-way coupled results with two-way coupled simulations. Finally we will show the occurrence of coherent particle structures in our simulations (Section 3.3).

3.1 Turbulence Modulation

The particle-phase strongly alters the fluid mean flow. Figure 1 shows the fluid mean flow profile normalized with the maximum U_m (a) and the wall shear velocity u_τ (b), based on the four-way coupled formulation. Relative to the clean channel we observe that the particles give rise to a strongly reduced boundary layer thickness and a flatter velocity profile. It corresponds to a larger skin-friction coefficient and, consequently, a larger Re_τ based on the fluid velocity, which increases from 180 to 300. The effects of the embedded particles on the developing flow are also reflected by the profile in the logarithmic region. Compared to the clean case an approximately logarithmic velocity profile develops for $10^{-3} < x_2 < 10^{-2}$, i.e., corresponding to $20 < y^+ < 200$, but at a much larger Von Kármán 'constant' [13]. In the following all velocity statistics are normalized with U_m, which is the same for all calculations.

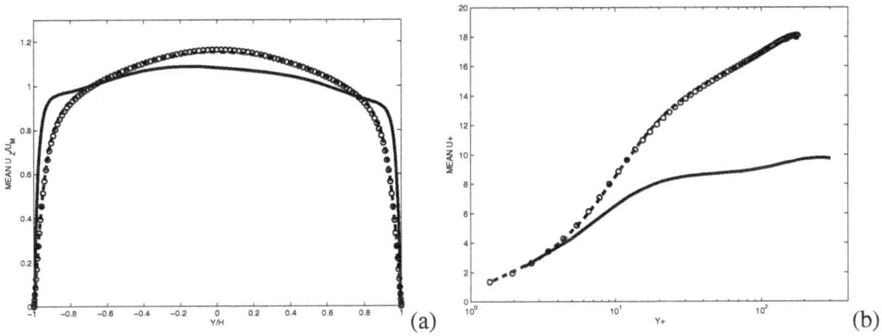

Fig. 1. Mean streamwise fluid velocity $\langle u_z \rangle$: linear (a) and logarithmic (b) for a particle-laden flow, comparing clean flow (dashed) with four-way coupling (solid). DNS results of clean flow are indicated by circles.

Figure 2 shows the turbulence intensities of the fluid phase. The turbulence modulation by coarse particles leads to an increased streamwise turbulence intensity and decreased transverse and spanwise intensities. This observation appears generally in line with existing experimental data. Although a direct comparison with physical experiments faces important difficulties in view of differences in flow-conditions, volume fractions and particle properties, an interesting analogy with the simulation findings may be drawn. The experiments by Kulick et al. [20] adopt smaller particles at lower volume fraction. In this regime a *decrease* in all components of the turbulence intensity was noted. This appears to contrast the present simulation findings. However, in a study by Tsuji et al. [39] larger particles at higher volume fraction were studied including particles of 0.5 mm in diameter, with Stokes response-time $\tau_p \approx 0.6$ s. This situation is close to our simulation setting in a number of respects. In such cases an *increase* in streamwise intensities in the core region of the channel was reported. For still coarser particles Tsuji et al. [39] found an even stronger increase of turbulence intensity across the entire diameter of the pipe.

3.2 Effects of Collisions

In this subsection we quantify the effects of the collisions between particles. For this purpose we compare predictions of the four-way with the two-way coupled simulation.

In Figure 3 we collected the mean streamwise fluid and solids velocity profiles. Considering the mean flow, both the two- and four-way coupling cause a higher skin-friction coefficient and results in a near-wall profile quite similar to the four-way coupled case. The correspondence of the near-wall fluid velocity profiles in the two- and four-way coupled descriptions reflects the interaction of the particles with the solid channel walls which were treated identically in both the two-way coupled and the four-way coupled cases. The inelastic collisions with the walls create a low-velocity-layer directly adjacent, which effectively acts analogous to an increased

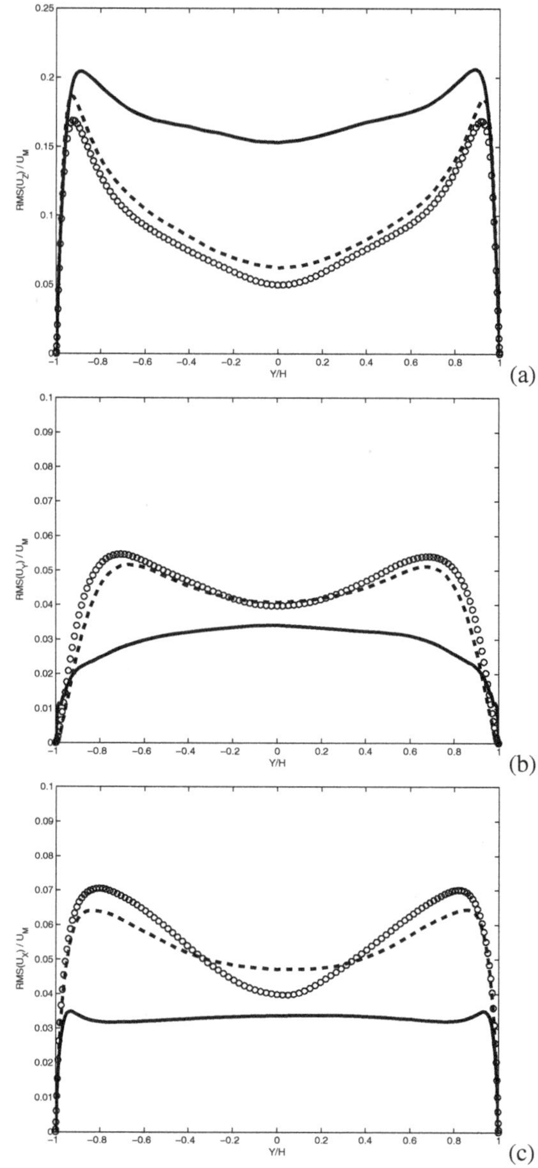

Fig. 2. Turbulence intensities of the fluid phase; streamwise (a), normal (b) and spanwise direction (c), comparing clean flow (dashed) with four-way coupling (solid). DNS results of clean flow are indicated by circles.

wall-roughness and hence yields an increased skin-friction coefficient. The absence of inelastic particle-particle collisions in the bulk of the flow is responsible for the absence of a flattening of the fluid velocity profile. The prediction of the bulk flow

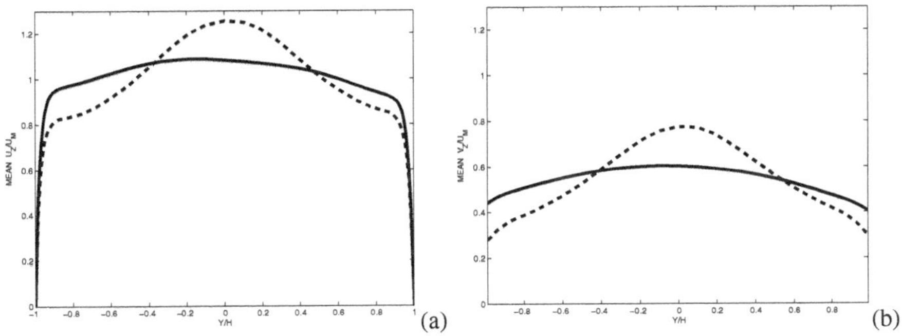

Fig. 3. Mean fluid velocity (a) and mean solids velocity (b), comparing two-way (dashed) with four-way coupling (solid) using LES.

away from the boundary layers is quite different when comparing the two-way and the four-way approaches. The two-way description is seen to give rise to a somewhat localized 'center-jet' in which the fluid velocity is up to about 60% larger than the velocity at the edge of the boundary layer. In contrast, the four-way coupling gives rise to a slightly flatter velocity profile compared to the clean channel; the particle-particle collisions evidently allow to avoid the 'center-jet' as discussed above.

The consequences of collisions for the mean solids profile is shown in Figure 3(b). The strong center-jet in $\langle u_z \rangle$ observed in the two-way coupling model, also arises in $\langle v_z \rangle$. The shape of the four-way coupled solids velocity profile is much flatter and quite similar to experimental data [39]. We also observe that the near-wall solids velocity is positive and does not drop to zero (except when the distance between the measuring point and the channel wall is smaller than the particle radius). This is consistent with the observation that the mean fluid velocity profile in the boundary layer is enhanced by the forces of the particles on the fluid. It confirms experimental data with measurements below $y^+ = 40$ in a horizontal channel [35]: close to the wall particles move faster than the fluid, while in the outer layer the opposite occurs.

The striking differences between the resulting dynamics in the two-way and the four-way coupled descriptions of the solids-phase are directly related to the inter-particle collisions. These collisions diffuse kinetic energy of the solids in the normal direction. As a consequence the mean solids velocity profile flattens. Due to the coupling between phases through the drag force, this collisional diffusion also flattens the gas mean velocity profile.

The particle volume fraction distribution is shown in Figure 4. A characteristic turbophoresis effect is visible in terms of an approximately 15% higher concentration near the solid walls. Turbophoresis has been observed in many experiments [47] and simulations. However the factor by which the particle concentration near the wall is increased, relative to the average bulk-value, depends strongly on the precise flow-regime that is considered. The relatively small turbophoresis effect observed in the present simulations may be attributed to the fact that the particles are coarse and the mass load is high. To verify this, we performed a four-way coupled simulation

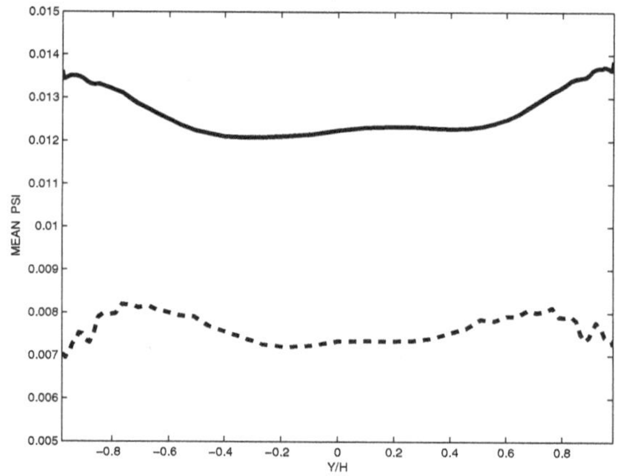

Fig. 4. Solid volume fraction $\langle\psi\rangle$ obtained with LES: two-way (dashed) compared with four-way coupling (solid). Notice that $\langle\psi\rangle$ is shifted downward by 0.005 for the two-way results for clarity.

with particles with a much smaller diameter ($d_p = 0.04$ mm) and, consequently, a lower particle concentration ($\psi = 1.3 \cdot 10^{-5}$). In that case strong turbophoresis was observed; the particle concentration near the walls increased with a factor of 30 relative to the mean bulk-concentration. When two-way coupling is used in combination with coarse, slowly responsive particles, no appreciable turbophoresis remains as is seen in Figure 4.

3.3 Coherent Particle Structures

In this subsection we will consider the dynamic self-organization that arises due to the 'competition' between the structuring associated with the inelastic particle collisions and the bursting of particle-clusters due to the underlying tendency of the clean flow to develop strong turbulence. The observed flow-structuring displays an interesting dynamic behavior which will be illustrated in terms of characteristic instantaneous snapshots of the particle concentration. These display qualitatively the sequence of formation and destruction of quite large coherent regions of considerably increased particle densities. We will also show that the four-way coupling model gives rise to large-scale coherent particle swarms which are much weaker when the two-way coupling model is used.

In order to characterize the flow-structuring we concentrate on visualizing the instantaneous particle volume fraction ψ at different times in Figure 5. A contour value of $\psi = 0.03$ is selected, while the solid volume fraction attains a maximum of approximately 0.1 and an average of 0.013. From these snapshots one may infer the formation of large-scale coherent structures in the particle concentration.

At the particle volume fractions considered here, the use of the full four-way coupling is essential. This is illustrated in Figure 6 in which we compare a structured particle field associated with four-way coupling, with a structure-less field arising in

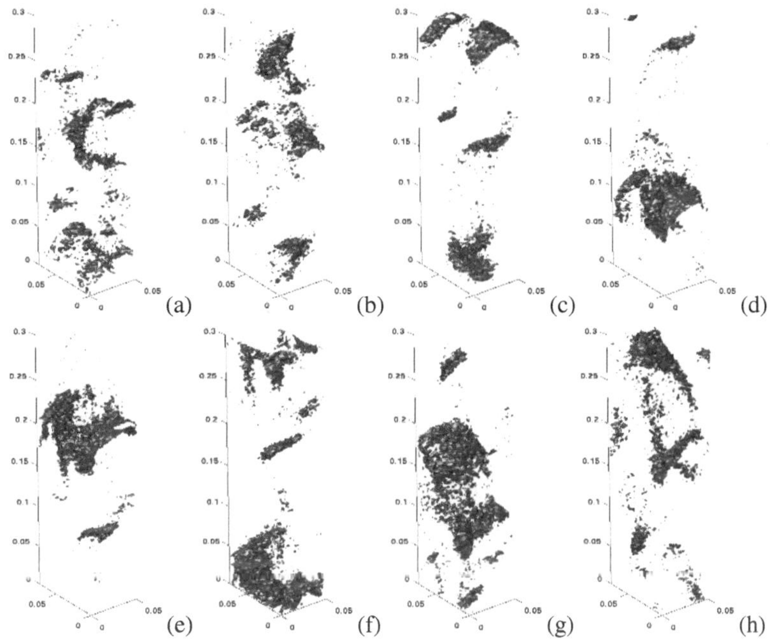

Fig. 5. Snapshots of the particle volume fraction showing iso-surfaces at $\psi = 0.03$ for four-way coupling; $t = 3.1$ s (a) with steps of 0.05 s until $t = 3.45$ s (h).

the two-way coupling model. The particle-particle interactions play a crucial role in the formation of coherent particle structures. These qualitative impressions further establish that four-way coupling cannot be replaced by the computationally more appealing two-way coupling.

4 Concluding Remarks

In this paper we presented large-eddy simulation results of particle-laden turbulent flow in a vertical riser. This flow is relevant, e.g., to chemical processing and an understanding of the fundamental dynamics of this flow is essential in order to properly predict up-scaling of processes from a laboratory scale to settings which are of industrial importance. We showed that already at a modest particle volume fraction of about 1.5% the particle-particle interactions play an important role in the development of the flow. The computationally more accessible two-way coupling model proved to give rise to predictions, which for slowly responsive particles and the present particle volume fraction, lack a turbophoresis effect and show the occurrence of a fairly strong 'center-jet' which was not recorded in experimental studies. The present particle volume fraction is much larger than in previous studies of plane channel flow, and the effects of inter-particle collisions found in the present work are

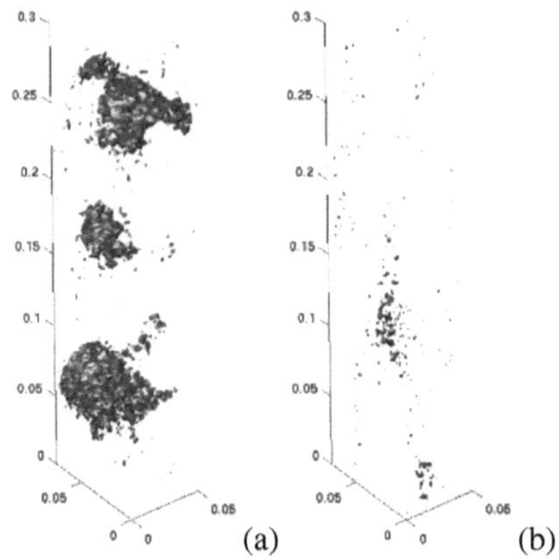

Fig. 6. Granular clustering in coherent particle-swarms is strongly associated with the four-way coupling description. Snapshot of the particle volume fraction at $t = 4$ s compare the four-way coupling (a) with the two-way coupling (b). The iso-surfaces shown correspond to $\psi = 0.03$.

much larger than the collisional effects reported by Yamamoto et al. [44], who found a mild effect of collisions in their case.

The presence of a large number of interacting particles leads to a strong modulation of the turbulence in the channel. Relative to a clean channel the coupling between particles and fluid and between particles with other particles and walls through inelastic collisions is mainly responsible for the reduction in the thickness of the boundary layer and the corresponding strong increase in the skin-friction. Moreover, the log-layer that is characteristic of wall-bounded flows was seen to be retained in the particle-laden case but with a much larger Von Kármán 'constant'. Turbulent intensities in the normal and spanwise directions were reduced whereas the streamwise turbulent intensity was found to be amplified by the presence of coarse particles. The conclusions regarding turbulence modulation in the plane channel flow are generally consistent with the experimental findings for coarse particles embedded in turbulent pipe flow.

The inelastic collisions are mainly responsible for retaining turbophoresis in case particles with high Stokes response-time are used, and a flattening of the mean particle velocity distribution. These interactions also gave rise to the occurrence of dynamic self-organization of the embedded particles in coherent swarms.

Acknowledgments

The author is grateful to A.W. Vreman (AKZO-Nobel), J.G.M. Kuerten (Eindhoven University of Technology) and J.A.M. Kuipers, N.D. Deen (University of Twente) for fruitful collaborations. This work was sponsored by the Stichting Nationale Computerfaciliteiten (National Computing Facilities Foundation, NCF), project SH-061, for the use of supercomputer facilities, with financial support from the Nederlandse Organisatie voor Wetenschappelijk Onderzoek (Netherlands Organization for Scientific Research, NWO).

References

1. Armenio, V., Piomelli, U., and Fiorotto, V., Effect of the subgrid scales on particle motion. *Phys. Fluids*, 11:3030, 1999.
2. Armenio, V. and Fiorotto, V., The importance of the forces acting on particles in turbulent flows. *Phys. Fluids*, 13:2437, 2001.
3. Bagchi, P. and Balachandar, S., Effect of turbulence on the drag and lift of a particle. *Phys. Fluids*, 15:3496–3513 (2003).
4. Delnoij, E., Kuipers, J.A.M., and van Swaaij, W.P.M., A three-dimensional CFD model for gas-liquid bubble columns *Chem. Eng. Sci.*, 54:2217–2226, 1999.
5. Elghobashi, S. and Truesdell, G.C., On the two-way interaction between homogeneous turbulence and dispersed solid particles. I: Turbulence modification. *Phys. Fluids A*, 5:1790–1801, 1993.
6. Feng, Z.-G. and Michaelides, E.E., Proteus: A direct forcing method in the simulation of particulate flows. *J. Comp. Phys.*, 202:20–51, 2005.
7. Ferry, J. and Balachandar, S., A fast Eulerian method for disperse two-phase flow, *Int. J. of Multiphase Flow*, 27:1199–1226, 2001.
8. Geurts, B.J. and Kuerten, J.G.M., Numerical aspects of a block-structured flow solver. *J. Eng. Math.*, 27:293, 1993.
9. Geurts, B.J., *Elements of Direct and Large-Eddy Simulation*. Edwards Publishing, 2003.
10. Geurts, B.J. and Vreman, A.W., Dynamic self-organization in particle-laden turbulent channel flow. *Int. J. Heat Fluid Flow*, 27:945–954, 2006.
11. Goldschmidt, M.J.V., Beetstra, R., and Kuipers, J.A.M., Comparison of the kinetic theory of granular flow with 3D hard-sphere discrete particle simulations. *Chem. Eng. Sci.*, 57:2059-2075, 2002.
12. Gore, R.A. and Crowe, C.T., Effect of particle size on modulating turbulence intensity. *Int. J. Multiphase Flow*, 15:279–285, 1989.
13. Hinze, J.O., *Turbulence: An Introduction to Its Mechanism and Theory*. McGraw-Hill, New York, 1975.
14. Van der Hoef, M.A., van Sint Annaland M., Deen N.G., and Kuipers, J.A.M., Numerical simulation of gas-solid fluidized beds: A multiscale modeling strategy. *Ann. Rev. Fluid Mech.*, 40:47–70, 2008.
15. Hoomans, B.P.B., Kuipers, J.A.M., Briels, W.J., and van Swaaij, W.P.M., Discrete particle simulation of bubble and slug formation in a two-dimensional gas-fluidized bed: A hard-sphere approach. *Chem. Eng. Sci.*, 51:99–118, 1996.
16. Hoomans, B.P.B., Granular dynamics of gas-solid two-phase flows. PhD-Thesis, University of Twente, 1999.
17. Howard, R.J.A. and Sandham, N.D., Simulation and modeling of a skewed turbulent channel flow. *Flow, Turbulence and Combustion*, 65:83–109, 2000.
18. Kuerten, J.G.M., Subgrid modeling in particle-laden channel flow. *Phys. Fluids*, 18:025108, 2006.
19. Kuerten, J.G.M. and Vreman, A.W., Can turbophoresis be predicted by large-eddy simulation? *Phys. Fluids*, 17:011701, 2005.

20. Kulick, J.D., Fessler, J.R., and Eaton, J.K., Particle response and turbulence modification in fully developed channel flow. *J. Fluid Mech.* 277:109–134, 1994.
21. Lakehal, D., Smith, B.L., and Milelli, M., Large-eddy simulation of bubbly turbulent shear flows. *J. Turbulence*, 3:25, 2002.
22. Li, Y., McLaughlin, J.B., Kontomaris, K., and Portela, L., Numerical simulation of particle-laden turbulent channel flow. *Phys. Fluids* 13:2957–2967, 2001.
23. Marchioli, C., Giusti, A., Salvetti, M.V., and Soldati, A., Direct numerical simulation of particle wall transfer and deposition in upward turbulent pipe flow. *Int. J. of Multiphase Flow*, 29:1017–1038, 2003.
24. Mathiesen, V., Solberg, T., and Hjertager, B.H., An experimental and computational study of multi-phase flow behavior in a circulating fluidized bed. *Int. J. Multiphase Flow*, 26:387–419, 2000.
25. Maxey, M.R. and Riley, J., Equation of motion for a small rigid sphere in a turbulent fluid flow. *Phys. Fluids*, 26:883, 1983.
26. Moin, P. and Kim, J., Numerical investigation of turbulent channel flow. *J. Fluid Mech.*, 118:341, 1982.
27. Moran, J.C. and Glicksman, L.R., Mean and fluctuating gas phase velocities inside a circulating fluidized bed. *Chem. Eng. Science*, 58:1867–1878, 2003.
28. Moser, R.D., Kim, J. and Mansour, N.N., Direct numerical simulation of turbulent channel flow up to Ret = 590. *Phys. Fluids*, 11:943–945, 1999.
29. Nicoud, F. and Ducros, F., Subgrid-scale stress modeling based on the square of the velocity gradient tensor. *Flow, Turbulence and Combustion*, 62:183–200, 1999.
30. Nieuwland, J.J., Hydrodynamic modeling of gas-solid two-phase flows. PhD-Thesis, University of Twente, 1995
31. Piomelli, U. and Balaras, E., Wall-layer models for large-eddy simulations. *Ann. Rev. Fluid Mech.*, 34:349, 2002.
32. Pope, S.B., *Turbulent Flows*. Cambridge University Press, 2000.
33. Powers, J.M., Two-phase viscous modeling of compaction of granular materials. *Phys. Fluids*, 16:2975–2990, 2004.
34. Reynolds, O., On the dynamical theory of incompressible viscous fluids and the determination of the criterion. *Phil. Trans.*, 186:123, 1895.
35. Righetti, M. and Romano, G.P., Particle-fluid interactions in a plane near-wall turbulent flow. *J. Fluid Mech.*, 505:93–121, 2004.
36. Sagaut, P., *Large Eddy Simulation for Incompressible Flows*. Springer Verlag, 2001.
37. Squires, K.D. and Simonin, O., Recent advances and perspective of DNS and LES for dispersed two-phase flow. In *Proceedings of the 10th workshop on two-phase flow predictions*, Merseburg, pp. 152–163, 2002.
38. Tryggvason, G., Bunner, B., Esmaeeli, A., Juric, D., Al-Rawahi, N., Tauber, W., Han, J., Nas, S., and Jan, Y.-J., A front-tracking method for the computations of multiphase flow. *J. Comp. Phys.*, 169:708–759, 2001.
39. Tsuji, Y., Morikawa, Y., and Shiomi, H., LDV measurements of an air-solid two-phase flow in a vertical pipe. *J. Fluid Mech.*, 139:417–434, 1984.
40. Verstappen, R.W.C.P., and Veldman, A.E.P., Symmetry-preserving discretization of turbulent flow. *J. Comp. Phys.*, 187:343–368, 2003.
41. Vreman, A.W., Geurts, B.J., Deen, N.G., Kuipers, J.A.M., and Kuerten, J.G.M., Two- and four-way coupled Euler–Lagrangian large-eddy simulation of turbulent particle-laden channel flow. *Flow, Turbulence and Combustion*, 82:47–71, 2009.
42. Vreman, B., Geurts, B., and Kuerten, H., Comparison of numerical schemes in large-eddy simulation of the temporal mixing layer. *Int. J. Num. Meth. Fluids*, 22:297, 1996.

43. Vreman, A.W., An eddy-viscosity model for turbulent shear-flow: Algebraic theory and applications. *Phys. Fluids*, 16:3670–3681, 2004.
44. Yamamoto, Y., Potthoff, M., Tanaka, T. Kajishima, T., and Tsuji, Y., Large-eddy simulation of turbulent gas-particle flow in a vertical channel: Effect of considering inter-particle collisions. *J. Fluid Mech.*, 442:303–334, 2001.
45. Wang, Q. and Squires, K.D., Large eddy simulation of particle deposition in a vertical turbulent channel flow. *Int. J. Multiphase Flow*, 22:667, 1996.
46. Wen, Y.C. and Yu, Y.H., Mechanics of fluidization. *Chem. Eng. Prog. Symp. Ser.*, 62:100–111, 1966.
47. Young, J. and Leeming, A., A theory of particle deposition in turbulent pipe flow. *J. Fluid Mech.*, 340:129–159, 1997.
48. Zhang, D.Z. and Prosperetti, A., Momentum and energy equations for disperse two-phase flows and their closure for dilute suspensions. *Int. J. Multiphase Flow*, 23:425–453, 1997.

PART 2

Materials with Microstructure

An Incremental Strategy for Modeling Laminate Microstructures in Finite Plasticity – Energy Reduction, Laminate Orientation and Cyclic Behavior

Klaus Hackl and Dennis M. Kochmann

Lehrstuhl für Allgemeine Mechanik, Ruhr-Universität Bochum,
D-44780 Bochum, Germany; klaus.hackl@rub.de, dennis.kochmann@rub.de

Abstract The plastic deformation of crystalline solids often gives rise to complex microstructural patterns, which can be observed experimentally and which have been reasoned to form as minimizers of non-quasiconvex energy potentials. We model the time-continuous evolution of laminate microstructures in finite-strain elastoplasticity via incrementally solving the stationarity conditions of the underlying minimum principles, replacing the nonconvex potentials by relaxed forms that account for laminate microstructures of first order. Here, we apply a time-incremental approach to investigate the orientation of the forming laminates dependent on the active slip system orientation. We present results for monotonic as well as cyclic stress-strain behavior. Moreover, we compare results from the present approach to those obtained by the well-established method using the so-called condensed energy for monotonic loading.

Key words: Microstructure, plasticity, finite strains, relaxation

1 Introduction

The macroscopic response of a material body deforming plastically under the action of external forces is a direct result of physical mechanisms occurring on the body's microscale. Hence, microstructural mechanisms are crucial in order to understand the macroscopic material behavior. The accommodation of plastic deformation by huge numbers of such defects is dictated by the concepts of free energy and dissipation. In elasto-plasticity the free energy is commonly assumed to stem from the uniform deformation of the periodic crystal lattice and aims at restoring the material body in a zero-stress state. It therefore comprises only energy due to elastic deformation. Dissipation arises as a consumption of energy due to irreversible changes of the microstructure. The internal state of the material's microstructure is commonly described in terms of so-called internal or history variables, a change of which causes dissipation. In crystalline solids dissipation may occur as a consequence of dislocation motion through gliding or climbing, or as a result of grain boundary motion, phase transformations or sources of internal friction.

R. de Borst and E. Ramm (eds.), *Multiscale Methods in Computational Mechanics*, Lecture Notes in Applied and Computational Mechanics 55, DOI 10.1007/978-90-481-9809-2_7,

Given the free energy and the dissipation of a deformed material body as functions of the displacement field and of the set of all internal variables, the principles of minimum potential energy and the principle of maximum dissipation or of minimum dissipation potential [1] determine the actual configuration of the body. In rate-independent plasticity models the incremental evolution of the internal variables is often described by a Lagrange functional consisting of the sum of elastic power and dissipation due to changes of the internal state of the material [2, 3].

Experimental evidence indicate that very often it is favorable for a material to accommodate an imposed deformation gradient not by a homogeneous deformation field but rather by forming microstructural patterns which accommodate the overall deformation by mixing different homogeneous states of minimum energy. Microstructure formation has been observed e.g. in the context of phase transformations and for deformation twin structures, or for dislocation walls in single-crystals. An interesting feature of all these microstructures is that they tend to form similar spatial patterns, hinting at a universal underlying mechanism. Ericksen was among the first to conclude microstructural patterns as a direct result of energy minimum principles [4], whose idea was transformed later into a more complex theory predominantly for the treatment of phase transformations [5, 6] and problems in elasto-plasticity [2,3]. In all of these examples the free energy is non-quasiconvex and hence multi-welled in nature. As a consequence the material body, aiming to reduce its energy, does not respond by means of a homogeneous deformation state but breaks up into multiple phases at local energy minima in such way that it is compatible with the overall imposed deformation field or any given boundary conditions. Solutions to describe these phase mixtures were developed by employing the theory of relaxation to find the quasiconvex hull of the free energy. This approach is based on small-scale fluctuations related to probability distributions of deformation gradients, so-called Young measures. These small-scale fluctuations correspond to the observed material microstructures. The particular features of those, like orientation or volume fractions, can be calculated via relaxed potentials.

Approaches to model microstructural patterns must account for the two crucial problems of the initiation of a forming microstructure and the subsequent evolution of the newly-formed structure. The initiation of a forming microstructure usually follows from a loss of stability. At a material point, local instability corresponds to a state where the crystal can reduce its energy by breaking up the homogeneous deformation state into fine structures. By considering associated potentials in a time-incremental setting, several authors have investigated the initiation of microstructures [3, 7–10] using a so-called condensed energy functional. This method was successfully applied to the evolution of inelastic materials, see [11, 12]. However, application of the condensed energy is based on the assumption that no microstructure is present at the beginning of the time step to be modeled. Therefore, it is suitable to model the onset of microstructure formation. Yet in order to model the time-continuous evolution of microstructures we need to account for microstructural reorientation required to change from one state in time to the next. Hence, the incremental update of the internal variables will depend on the set of internal variables at the end of the preceding time step so that the use of the condensed energy functional may be of high mathematical interest but also at risk to loose physical meaning. To overcome this problem, we developed an incremental formulation based on laminated Young measures [13,14]. Here we review this incremental strategy first and then discuss novel results from its application to monotonic and cyclic loading.

2 Non-Convex Potentials and Relaxation

In an isothermal setting, the state of a general inelastic material is defined by its deformation gradient $\mathbf{F} = \nabla \boldsymbol{\phi}$, where $\boldsymbol{\phi}(\mathbf{X})$ represents the displacement field, and a collection of internal variables $\mathbf{K}(\mathbf{X})$. Denoting the specific Helmholtz free energy by $\Psi(\mathbf{F}, \mathbf{K})$, we introduce thermodynamically conjugate stresses by defining

$$\mathbf{P} = \frac{\partial \Psi}{\partial \mathbf{F}}, \qquad \mathbf{Q} = -\frac{\partial \Psi}{\partial \mathbf{K}}. \tag{1}$$

Here, $\mathbf{P}$ denotes the first Piola–Kirchhoff stress tensor and $\mathbf{Q}$ captures the driving forces of the internal variables $\mathbf{K}$. The evolution of these internal variables is governed either by a so-called inelastic potential $J(\mathbf{K}, \mathbf{Q})$ or its Legendre-transform [2], the dissipation functional

$$\Delta(\mathbf{K}, \dot{\mathbf{K}}) = \sup \left\{ \dot{\mathbf{K}}{:}\mathbf{Q} - J(\mathbf{K}, \mathbf{Q}) \,\middle|\, \mathbf{Q} \right\}. \tag{2}$$

Evolution equations for the internal variables are then given in the two equivalent forms

$$\dot{\mathbf{K}} \in \frac{\partial J}{\partial \mathbf{Q}}, \qquad \mathbf{Q} \in \frac{\partial \Delta}{\partial \dot{\mathbf{K}}}. \tag{3}$$

Note that, along with $(1)_2$, the latter form simply rephrases the well-known Biot equation of standard generalized materials [15–18].

The complete evolution problem can now be described in terms of two minimum principles where we follow ideas presented in [2, 3, 19]. We denote the total free energy of the body by

$$\mathcal{I}(t, \boldsymbol{\phi}, \mathbf{K}) = \int_\Omega \Psi(\nabla\boldsymbol{\phi}, \mathbf{K})\, \mathrm{d}v - \ell(t, \boldsymbol{\phi}), \tag{4}$$

where $\ell(t, \boldsymbol{\phi})$ represents the potential of external forces, Ω is the body's volume and $\partial\Omega$ denotes a subset of its boundary. The actual displacement field then follows from the principle of minimum potential energy

$$\boldsymbol{\phi} = \operatorname{argmin}\left\{ \mathcal{I}(t, \boldsymbol{\phi}, \mathbf{K}) \,\middle|\, \boldsymbol{\phi} = \boldsymbol{\phi}_0 \text{ on } \Gamma_\mathbf{u} \right\}, \tag{5}$$

where given boundary conditions $\boldsymbol{\phi}_0$ are prescribed on the subset $\Gamma_\mathbf{u}$ of the body's boundary.

Introducing the Lagrange functional [3]

$$\mathcal{L}(\boldsymbol{\phi}, \mathbf{K}, \dot{\mathbf{K}}) = \frac{\mathrm{d}}{\mathrm{d}t}\, \Psi(\nabla\boldsymbol{\phi}, \mathbf{K}) + \Delta(\mathbf{K}, \dot{\mathbf{K}}), \tag{6}$$

i.e., the sum of the energy rate and the dissipation potential, we can write the evolution equation (3) in the form

$$\dot{\mathbf{K}} = \operatorname{argmin}\{\, \mathcal{L}(\boldsymbol{\phi}, \mathbf{K}, \dot{\mathbf{K}}) \mid \dot{\mathbf{K}}\,\}, \tag{7}$$

where the dot denotes differentiation with respect to time. Noting that

$$\mathcal{L} = \frac{\mathrm{d}}{\mathrm{d}t}\Psi(\mathbf{K}, \mathbf{K}) + \Delta(\mathbf{K}, \dot{\mathbf{K}}) = \frac{\partial \Psi}{\partial \mathbf{F}} : \dot{\mathbf{F}} + \frac{\partial \Psi}{\partial \mathbf{K}} \cdot \dot{\mathbf{K}} + \Delta(\mathbf{K}, \dot{\mathbf{K}}), \tag{8}$$

it becomes obvious that minimization with respect to $\dot{\mathbf{K}}$ ensures that solutions automatically satisfy the Biot equation [15–18]. Also, it has been shown [1] that for dissipation potentials which are homogeneous of degree n the above principle of minimum dissipation potential is equivalent to the principle of maximum dissipation (i.e., of maximum entropy production). For a thorough investigation of this principle and its relation to the principle of maximum dissipation, see [1].

The dissipation distance

$$D(\mathbf{K}_0, \mathbf{K}_1) = \inf\left\{ \int_0^1 \Delta(\mathbf{K}(s), \dot{\mathbf{K}}(s))\, \mathrm{d}s \mid \mathbf{K}(0) = \mathbf{K}_0, \mathbf{K}(1) = \mathbf{K}_1 \right\} \tag{9}$$

has been introduced [19] as a measure of the total dissipated energy during a finite time increment $[t_n, t_{n+1}]$, during which the internal variables change from $\mathbf{K}_0$ to $\mathbf{K}_1$. In rate-independent plasticity an approximate formulation for the above minimization problem is then given in terms of the so-called condensed energy

$$\Psi^{\text{cond}}_{\mathbf{K}_n}(\mathbf{F}) = \inf\{\, \Psi(\mathbf{F}, \mathbf{K}) + D(\mathbf{K}_n, \mathbf{K}) \mid \mathbf{K}\,\}. \tag{10}$$

The condensed energy has been used in the literature to calculate the onset of microstructures [3, 7–10]. This approach, however, is based on the assumption that the material is homogeneous and does not exhibit a microstructure at the beginning of the time increment. Hence, it is not suitable to describe the evolution of already existing microstructures since, for each time step, the internal variables already exhibit a microstructure at the beginning of the time increment as a result of a relaxation process in the preceding time-increment. Here, we will make use of laminated Young measures to describe the microstructural characteristics and model their evolution in an incremental setting. Then, the energy and dissipation potentials can be replaced by the corresponding relaxed potentials to render the minimization problem well-posed. In the sequel we follow the basic strategy outlined in [13, 14].

3 First-Order Laminate Microstructures

Relaxation of the aforementioned potentials involves computation of the quasiconvex envelope. Although it was derived successfully for some model cases, the quasiconvex envelope is hard to determine in general due to its non-local nature. One possible approximation is via lamination [2,3,7,9,13,14,20]. We will adopt this approach here in principle but for conciseness restrict our considerations to first-order laminates in the following.

We make use of the crystal plasticity model introduced originally in [2]. We study the evolution of microstructures by employing a Neo-Hookean material with free energy density

$$\Psi(\mathbf{F}_e) = \frac{1}{2}\mu[\operatorname{tr}\mathbf{F}_e^T\mathbf{F}_e - 3] + \frac{\lambda}{4}\left[j^2 - 2\left(1 + 2\frac{\mu}{\lambda}\right)\ln j - 1\right], \tag{11}$$

where $\mathbf{F}_e$ denotes the elastic deformation gradient tensor, $j = \det\mathbf{F}_e$, and $\mu > 0$ and $\lambda > 0$ are the elastic moduli. The split of the deformation gradient into an elastic part $\mathbf{F}_e$ and an irreversible plastic part $\mathbf{F}_p$ is based on the standard multiplicative decomposition

$$\mathbf{F} = \mathbf{F}_e\mathbf{F}_p. \tag{12}$$

To account for material hardening, we make use of an internal hardening history variable p. In most of the subsequent analysis we will limit our considerations to incompressible material behavior so that $j = \det\mathbf{F}_e = 1$. Note that incompressibility requires that not only $\det\mathbf{F}_p = 1$ but also $\det\mathbf{F}_e = \det\mathbf{F} = 1$. The assumption of incompressibility is, of course, rather strong albeit necessary to allow for the closed-form analytical solution for the semi-relaxed energy as derived in the sequel. This model is primarily intended to model the behavior of metallic materials, where it is presumed that the elastic strains are small compared to the plastic strains, so that the error made by the assumption that $\det\mathbf{F}_e = 1$ is assumed rather small.

Then, the energy (11) reduces to

$$\Psi(\mathbf{F}_e, p) = \frac{1}{2}\mu(\operatorname{tr}\mathbf{F}_e^T\mathbf{F}_e - 3) + \kappa\, p^\alpha, \qquad \det\mathbf{F} = 1, \tag{13}$$

where $\kappa > 0$ is a hardening modulus of the material and α is commonly 2 (linear hardening) or 4 (higher-order hardening law).

Plastic deformation is accommodated by dislocation slip on specific active slip systems. A slip system is characterized by vectors $\mathbf{s}$ and $\mathbf{m}$ ($|\mathbf{s}| = |\mathbf{m}| = 1, \mathbf{s}\cdot\mathbf{m} = 0$), where $\mathbf{s}$ characterizes the slip direction, and $\mathbf{m}$ denotes the unit vector normal to the corresponding slip plane. If we have n active slip systems, the flow rule can be given in the form

$$\dot{\mathbf{F}}_p\mathbf{F}_p^{-1} = \sum_i^n \dot{\gamma}_i\, \mathbf{s}_i \otimes \mathbf{m}_i \tag{14}$$

with plastic slip rates $\dot{\gamma}_i$ and the initial conditions $\gamma_i(0) = 0$. Time-integration of (14) reduces to a neat analytical form, if we assume that all slip directions lie in the same slip plane ($\mathbf{m}_i = \mathbf{m}$). Then, by time-integration of (14) we infer [2]

$$\mathbf{F}_p^{-1} = \mathbf{I} - \sum_i^n \gamma_i\, \mathbf{s}_i \otimes \mathbf{m}. \tag{15}$$

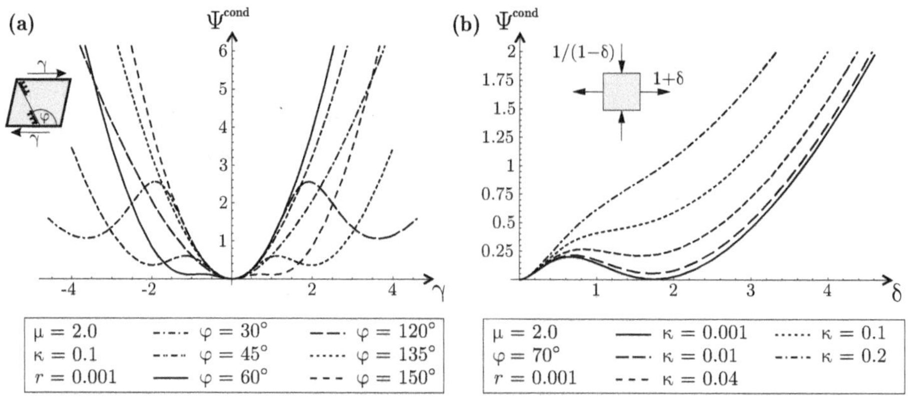

Fig. 1. Unrelaxed condensed energy of a homogeneous single crystal with one active slip system. The energy is clearly not convex. (a) Condensed energy during a simple shear test for fixed hardening and varying slip system orientation, (b) condensed energy during a tension-compression test for fixed slip system orientation and varying hardening parameter values.

In general, depending on the crystallographic orientations, there is a variety of possible choices for slip systems in metals, which do not have to lie within a common plane. Also, our numerical examples in the following will be reduced to single-slip only, a rather strong physical assumption. Note that, however, for simplicity we here assume infinite latent hardening [8, 11], i.e., as soon as one slip system becomes active, it is unconvenient to activate any of the remaining slip systems, since, even at moderate latent hardening, the well structure of the energy density suggests to seek a minimum energy state by developing regions of single slip only [3].

The flow rule for the hardening variables depends on the number of active slip systems. In the simplest case of infinite latent hardening and only one active slip system in the laminate we consider [2]

$$\dot{p} = |\dot{\gamma}| \tag{16}$$

with the initial condition $p(0) = 0$. If several active slip systems are present (16) must be modified appropriately [3, 14]. A simplest choice would be to introduce p_i in each system and require $\dot{p}_i = |\dot{\gamma}_i|$. In general, however, cross hardening will be desirable so that a more general formulation becomes necessary [3]. For the dissipation functional due to plastic slip we assume

$$\Delta(\dot{\gamma}) = r \sum_{i}^{n} |\dot{\gamma}_i|, \tag{17}$$

with a positive constant r (the critical resolved shear stress), see [2, 13, 14] for details.

The condensed energy functional (10) modified for the present approach with a single active slip system ($n = 1$) for a homogeneously deformed body (i.e. without microstructure) now takes the form

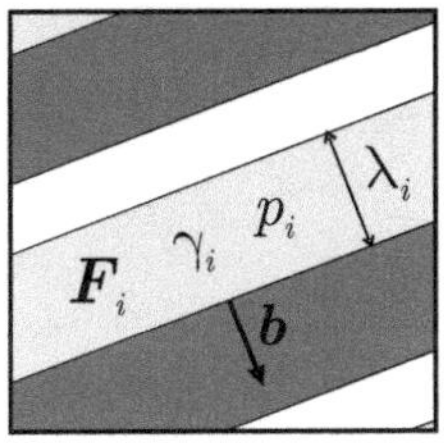

Fig. 2. Sketch of a first-order laminate for $N = 3$ with normal vector $\mathbf{b}$.

$$\psi^{\text{cond}}(\mathbf{F}) = \inf\left\{\Psi(\mathbf{F}(\mathbf{I} - \gamma\mathbf{s}\otimes\mathbf{m}), p) + r|\gamma| \,\middle|\, \gamma \,:\, p = |\gamma|\right\}. \tag{18}$$

The unrelaxed energy path is illustrated in Figure 1 for a plane-strain simple shear test and a tension-compression test where the in-plane orientation of the active slip system is defined by $\mathbf{s} = (\cos\varphi, \sin\varphi, 0)^T$. Note that the appearing non-convexity depends on the choice of the active slip system. In Figure 1a we illustrate the non-convex energy for various slip system orientations for a simple shear test ($\mathbf{F} = \mathbf{I} + \gamma\,\mathbf{e}_1 \otimes \mathbf{e}_2$). It becomes apparent that the non-convexity appears during positive loading ($\gamma > 0$) if $\varphi > 90°$, and during negative loading if $\varphi < 90°$. The influence of the hardening parameter κ is demonstrated in Figure 1b for tension-compression ($\mathbf{F} = (1+\delta)\,\mathbf{e}_2 \otimes \mathbf{e}_2 + \frac{1}{1+\delta}\mathbf{e}_1 \otimes \mathbf{e}_1$) with $\alpha = 4$ (in both graphics). With increasing hardening constant κ the second energy well is lifted considerably. In all of these cases the energy is clearly non-convex giving rise to instability and hence to the formation of microstructure as has already been discussed e.g. in [2, 9, 13, 14].

Due to the non-convex condensed energy, microstructures arise as energy minimizers. Let us assume a specific microstructure, viz. a laminate of first order which is characterized by N volume fractions λ_i separated by parallel planes with normal vector $\mathbf{b}$, as sketched in Figure 2. To every volume fraction i there correspond values of the internal variables γ_{ij} and p_{ij} for each active slip system j. Moreover, in every volume fraction we have a deformation gradient which we write in the form

$$\mathbf{F}_i = \mathbf{F}(\mathbf{I} + \mathbf{a}_i \otimes \mathbf{b}) \tag{19}$$

so that deformation gradients differ only by tensors of rank one, enforcing compatibility at laminate interfaces. We impose the correct volume average of the deformation gradient by

$$\sum_{i=1}^{N} \lambda_i \mathbf{F}_i = \mathbf{F} \quad \Leftrightarrow \quad \sum_{i=1}^{N} \lambda_i \mathbf{a}_i = \mathbf{0}. \tag{20}$$

To ensure incompressibility of each laminate phase, we must enforce that for every phase i we have

$$\det \mathbf{F}_i = 1 \quad \Leftrightarrow \quad \mathbf{a}_i \cdot \mathbf{b} = 0. \tag{21}$$

We consider changes of the normal vector $\mathbf{b}$ to be of dissipative type since any change of $\mathbf{b}$ would require a reorientation of the internal variables and thus lead to dissipation. The amplitudes $\mathbf{a}_i$ on the other hand can be changed purely elastically. Taking the two constraints (20) and (21) into account and using (13), we write the relaxed energy as

$$\Psi^{\mathrm{rel}}(\mathbf{F}, \lambda_i, \gamma_{ij}, p_{ij}, \mathbf{b}) = \tag{22}$$
$$\frac{\mu}{2}\sum_i^N \lambda_i \left[\operatorname{tr}\mathbf{C}_{\mathrm{e},i} - 2\mathbf{\Lambda}\cdot\mathbf{a}_i - 2\rho_i\mathbf{a}_i\cdot\mathbf{b} + \kappa\sum_j^n p_{ij}^{\alpha} \right] - \frac{3}{2}\mu$$

where $\mathbf{C}_{\mathrm{e},i} = \mathbf{F}_{\mathrm{e},i}^{\mathsf{T}}\mathbf{F}_{\mathrm{e},i}$ denotes the elastic right Cauchy–Green tensor in phase i with, following (15) and (19),

$$\mathbf{F}_{\mathrm{e},i} = \mathbf{F}(\mathbf{I} + \mathbf{a}_i \otimes \mathbf{b})\left(\mathbf{I} - \sum_j^n \gamma_{ij}\mathbf{s}_j \otimes \mathbf{m}\right), \tag{23}$$

and $\mathbf{\Lambda}$ and ρ_i are Lagrange multipliers to enforce constraints (20) and (21). Minimization with respect to the elastic amplitude vectors $\mathbf{a}_i$ yields the relaxed energy [13]:

$$\begin{aligned}\Psi^{\mathrm{rel}}(\mathbf{F}, \lambda_i, \gamma_{ij}, p_{ij}, \mathbf{b}) &= \kappa\sum_i^N \lambda_i \sum_j^n p_{ij}^{\alpha} + \frac{\mu}{2}(\operatorname{tr}\mathbf{C} - 3)\\ &+ \frac{\mu}{2}\left[\frac{1}{\sum_i^N \frac{\lambda_i}{\mathbf{b}_i\cdot\mathbf{b}}}\left(\sum_j^N\sum_k^N \frac{\lambda_j\lambda_k\mathbf{b}_j\cdot\mathbf{C}\mathbf{b}_k}{\mathbf{b}_j\cdot\mathbf{b}\,\mathbf{b}_k\cdot\mathbf{b}} - \frac{1}{\mathbf{b}\cdot\mathbf{C}^{-1}\mathbf{b}}\right)\right.\\ &\left.+\sum_i^N \lambda_i\left(\frac{\mathbf{b}_i\cdot\mathbf{b}}{\mathbf{b}\cdot\mathbf{C}^{-1}\mathbf{b}} - \frac{\mathbf{b}_i\cdot\mathbf{C}\mathbf{b}_i}{\mathbf{b}_i\cdot\mathbf{b}}\right) + \sum_i^N \lambda_i\sum_j^n\left(\gamma_{ij}^2\mathbf{s}_j\cdot\mathbf{C}\mathbf{s}_j - 2\gamma_{ij}\mathbf{s}_j\cdot\mathbf{C}\mathbf{m}\right)\right].\end{aligned} \tag{24}$$

Depending on the geometric problem formulation the analytical solution (24) is exact under some circumstances:

The most general case will be a laminate microstructure with several active slip systems which may differ between phases. In this case the solution (24) holds if all active slip systems within one phase lie within the same glide plane (i.e. $\mathbf{m}_{ij} = \mathbf{m}_i$ in each phase i) so that

$$\mathbf{b}_i = \mathbf{b} - \sum_j^n \gamma_{ij}(\mathbf{b}\cdot\mathbf{m}_i\,\mathbf{s}_{ij} + \mathbf{b}\cdot\mathbf{s}_{ij}\,\mathbf{m}_i) + \sum_j^n \gamma_{ij}^2\mathbf{b}\cdot\mathbf{s}_{ij}\,\mathbf{s}_{ij}. \tag{25}$$

From this general case several simple examples can be deduced. On the one hand, if only one slip system $(\mathbf{s}_i, \mathbf{m}_i)$ is active within each phase i of the laminate but the active slip system may differ between phases, then the above solution holds with

$$\mathbf{b}_i = \mathbf{b} - \gamma_i(\mathbf{b}\cdot\mathbf{m}_i\,\mathbf{s}_i + \mathbf{b}\cdot\mathbf{s}_i\,\mathbf{m}_i) + \gamma_i^2\mathbf{b}\cdot\mathbf{s}_i\,\mathbf{s}_i. \tag{26}$$

This is the typical case of infinite latent hardening in each phase but differing orientations between phases. The interfaces are then formed as so-called dipolar dislocation

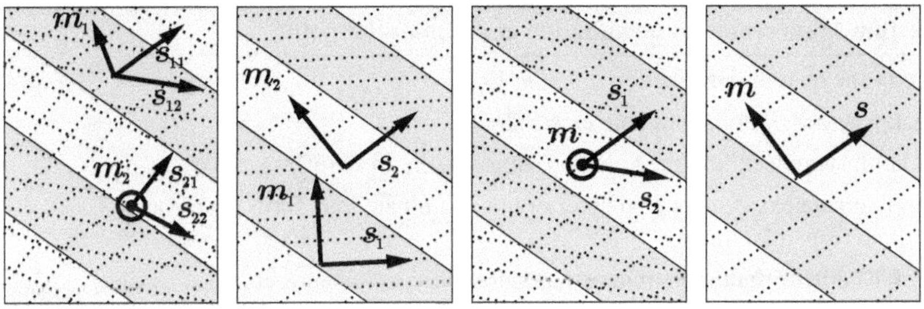

Fig. 3. Orientation of (i) two active slip systems which differ between phases but lie within the same plane within each phase, (ii) one active slip system within each phase, (iii) two active slip systems with a common plane in the entire crystal, (iv) only one active slip system in the entire crystal.

walls. Finally, the case of only one single active slip system $(\mathbf{s}, \mathbf{m})$ throughout the entire crystal (i.e. equal active slip systems in all laminate phases) has been studied [13]. This assumption reduces the above form to

$$\mathbf{b}_i = \mathbf{b} - \gamma_i(\mathbf{b} \cdot \mathbf{m}\,\mathbf{s} + \mathbf{b} \cdot \mathbf{s}\,\mathbf{m}) + \gamma_i^2 \mathbf{b} \cdot \mathbf{s}\,\mathbf{s}. \tag{27}$$

Examples of these cases are illustrated in Figure 3. If several slip systems which do not lie in the same glide plane are active within one phase, (15) is no longer valid so that the above solution will be rendered more complex [21].

For simplicity of the following examples we reduce the present model to a two-phase laminate ($N = 2$) and define the volume fraction of phase 2 as λ such that by taking into account (17) the dissipation potential for a single active slip system may be written in the form [13]

$$\Delta^*(\lambda, \gamma_i, \dot{\lambda}, \dot{\gamma}_i) = r\left(\left|\dot{\lambda}(\gamma_1 - \gamma_2)\right| + (1-\lambda)\,|\dot{\gamma}_1| + \lambda\,|\dot{\gamma}_2|\right), \tag{28}$$

and the Lagrange functional now becomes

$$\mathcal{L}(\mathbf{F}, \lambda, \gamma_i, p_i, \dot{\lambda}, \dot{\gamma}_i, \mathbf{b}) = \frac{\mathrm{d}}{\mathrm{d}t}\,\Psi^{\mathrm{rel}}(\mathbf{F}, \lambda, \gamma_i, p_i, \mathbf{b}) + \Delta^*(\lambda, \gamma_i, \dot{\lambda}, \dot{\gamma}_i). \tag{29}$$

Here, one of the major differences of the present model to previous approaches becomes apparent from the first term of (28). A change of the volume fractions causes dissipation, and we correctly account for the transformation of some part of one phase into a part of the other phase. Therefore, the dissipation (and thus the evolution of the internal variables) depends on the microstructure at the beginning of each time step.

4 Incremental Numerical Scheme

Via the principle given in (7) we now arrive at evolution equations for the internal variables from the above Lagrange functional, which for single-slip plasticity reads

Algorithm 1: Incremental formulation at the material point

- specify the overall deformation gradient $\mathbf{F}_{n+1}$
- check whether or not a laminate exists already
- if a laminate exists: find the volume fraction λ_{n+1} from (30)
- if no laminate exists: check whether or not a laminate may form via (34), and if so form a laminate.
- check whether or not the laminate has vanished
- find the hardening parameters $p_{i,n+1}$ via energetic averaging
- find the plastic slips $\gamma_{i,n+1}$ from (31) and (32)
- update the hardening parameters $p_{i,n+1}$ due to changes of the plastic slips
- store the new internal variables and compute the stresses

$$-r\,|\gamma_1-\gamma_2|\,\text{sign}\,\dot{\lambda} \in \frac{\partial \Psi^{\text{rel}}}{\partial \lambda}, \tag{30}$$

$$-r(1-\lambda)\text{sign}\,\dot{\gamma}_1 \in \frac{\partial \Psi^{\text{rel}}}{\partial \gamma_1} + \frac{\partial \Psi^{\text{rel}}}{\partial p_1}\text{sign}\,\dot{\gamma}_1, \tag{31}$$

$$-r\,\lambda\text{sign}\,\dot{\gamma}_2 \in \frac{\partial \Psi^{\text{rel}}}{\partial \gamma_2} + \frac{\partial \Psi^{\text{rel}}}{\partial p_2}\text{sign}\,\dot{\gamma}_2. \tag{32}$$

Instead of solving the stationarity conditions one can also minimize the Lagrange functional

$$\begin{aligned}\mathcal{L}(\mathbf{F},\lambda,\gamma_i,p_i,\dot{\lambda},\dot{\gamma}_i,\dot{p}_i,\mathbf{b}) &= \frac{\partial \Psi^{\text{rel}}}{\partial \lambda}\dot{\lambda} + \sum_{i=1}^{2}\left(\frac{\partial \Psi^{\text{rel}}}{\partial \gamma_i}\dot{\gamma}_i + \alpha\,\kappa\,p_i^{\alpha-1}|\dot{\gamma}_i|\right)\\ &+ r\,\left[\dot{\lambda}(\gamma_1-\gamma_2) + (1-\lambda)\,|\dot{\gamma}_1| + r\,\lambda\,|\dot{\gamma}_2|\right],\end{aligned} \tag{33}$$

numerically with respect to the rates of all internal variables.

With our goal of computing the evolution of plastic microstructures in mind, we found an incremental formulation to be solved numerically, using finite deformation increments $\left[\mathbf{F}_n,\mathbf{F}_{n+1}\right]$ with known initial conditions $\mathbf{F}_n$, $\gamma_{i,n}$, λ_n, $p_{i,n}$ at the beginning of each time step $[t_n,t_{n+1}]$ and the known deformation gradient $\mathbf{F}_{n+1}$ at the end of the time step. Equations (30)–(32) can then be used to compute the updates $\Delta\lambda = \lambda_{n+1}-\lambda_n$, $\Delta\gamma_i = \gamma_{i,n+1}-\gamma_{i,n}$, $\Delta p_i = p_{i,n+1}-p_{i,n}$.

Our numerical procedure solves the evolution equations (30)–(32) for each load increment in a staggered manner [13, 14]. As long as no microstructure exists, we numerically determine the condensed relaxed energy

$$\begin{aligned}\Psi_{n+1}^{\text{rel,cond}}(\mathbf{F}_{n+1}) = \min\Big\{\Psi^{\text{rel}}(\mathbf{F}_{n+1},\gamma_{1,n},\gamma_2,\lambda,\mathbf{b}) + r\,|\lambda(\gamma_2-\gamma_{1,n})|\,\Big|\,\gamma_2,\lambda,\mathbf{b}\;:\\ p_2 = |\gamma_2|,\ 0\le\lambda\le 1,\ |\mathbf{b}|=1\Big\}\end{aligned} \tag{34}$$

and check whether or not $\Psi^{\text{rel,cond}}_{n+1}(\mathbf{F}_{n+1}) \leq \Psi^{\text{cond}}_{n+1}(\mathbf{F}_{n+1})$, i.e. if the formation of a laminate can reduce the energy below that of the homogeneous crystal. If $\Psi^{\text{rel,cond}}_{n+1}(\mathbf{F}_{n+1}) \leq \Psi^{\text{cond}}_{n+1}(\mathbf{F}_{n+1})$, then a laminate forms with $(\gamma_2, \lambda, \mathbf{b}) = \operatorname{argmin} \Psi^{\text{rel,cond}}_{n+1}(\mathbf{F}_{n+1})$ from (34). A rotation of the existing laminate microstructure is understood to be admissible only, if the amount of released energy is bigger than the dissipation required for the rotation [13], i.e. if

$$\Psi^{\text{rel}}(\mathbf{F}_n, \lambda_n, \gamma_{i,n}, p_{i,n}, \mathbf{b}_{n+1}) - \Psi^{\text{rel}}(\mathbf{F}_n, \lambda_n, \gamma_{i,n}, p_{i,n}, \mathbf{b}_n) + 2r\lambda_n(1-\lambda_n)\left|\gamma_{1,n} - \gamma_{2,n}\right| \leq 0.$$

Of crucial importance is the updating procedure for the hardening variables upon changes of the volume fractions. We follow the approach outlined in [13, 14] and update the hardening variables by means of an energetic averaging of the corresponding values in the mixed phases. The entire algorithm is outlined in a concise form in Algorithm 1.

5 Results

5.1 Evolution of the Internal Variables and Laminate Orientation

The presented incremental strategy to compute the characteristics of the developing laminate will be applied to a variety of examples. In the sequel we restrict ourselves for conciseness to only one active slip system. Let us first demontrate the results obtained from Algorithm 1 for a simple shear deformation with the macroscopic deformation gradient $\mathbf{F} = \gamma \mathbf{e}_1 \otimes \mathbf{e}_2$. Results for the evolution of all internal variables, the energy and the Cauchy shear stress are summarized in Figure 4 both without ($\kappa = 0$) and with hardening ($\kappa = 0.1$). At sufficiently high shear strain the second phase arises with already finite plastic slip. During the course of straining the volume fraction of the second phase rises to about 35% and then decreases to finally vanish, so that the crystal becomes homogeneous upon the recovery of the convexity of the condensed energy. The effect of relaxation becomes apparent in particular from the energy path and the shear stress in comparison with the unrelaxed solution. The orientation of the forming laminate is also sketched for both cases.

The orientation of the forming laminate microstructure depends considerably on the orientation of the active slip system, as has already been pointed out in [3]. The influence of the slip system on the laminate normal orientation $\mathbf{b}$ is illustrated in Figure 5. It becomes obvious that laminate interfaces prefer orientations close to perpendicular to the slip direction so that dislocations can pile-up to constitute the interface boundaries. Interestingly, the illustrations indicate another consistent relation between the orientation of the active slip system and the orientation of the laminate: for $\varphi < 135°$ the deviation of the interface normal with respect to the negative slip direction is clockwise, for $\varphi > 135°$ it appears counter-clockwise.

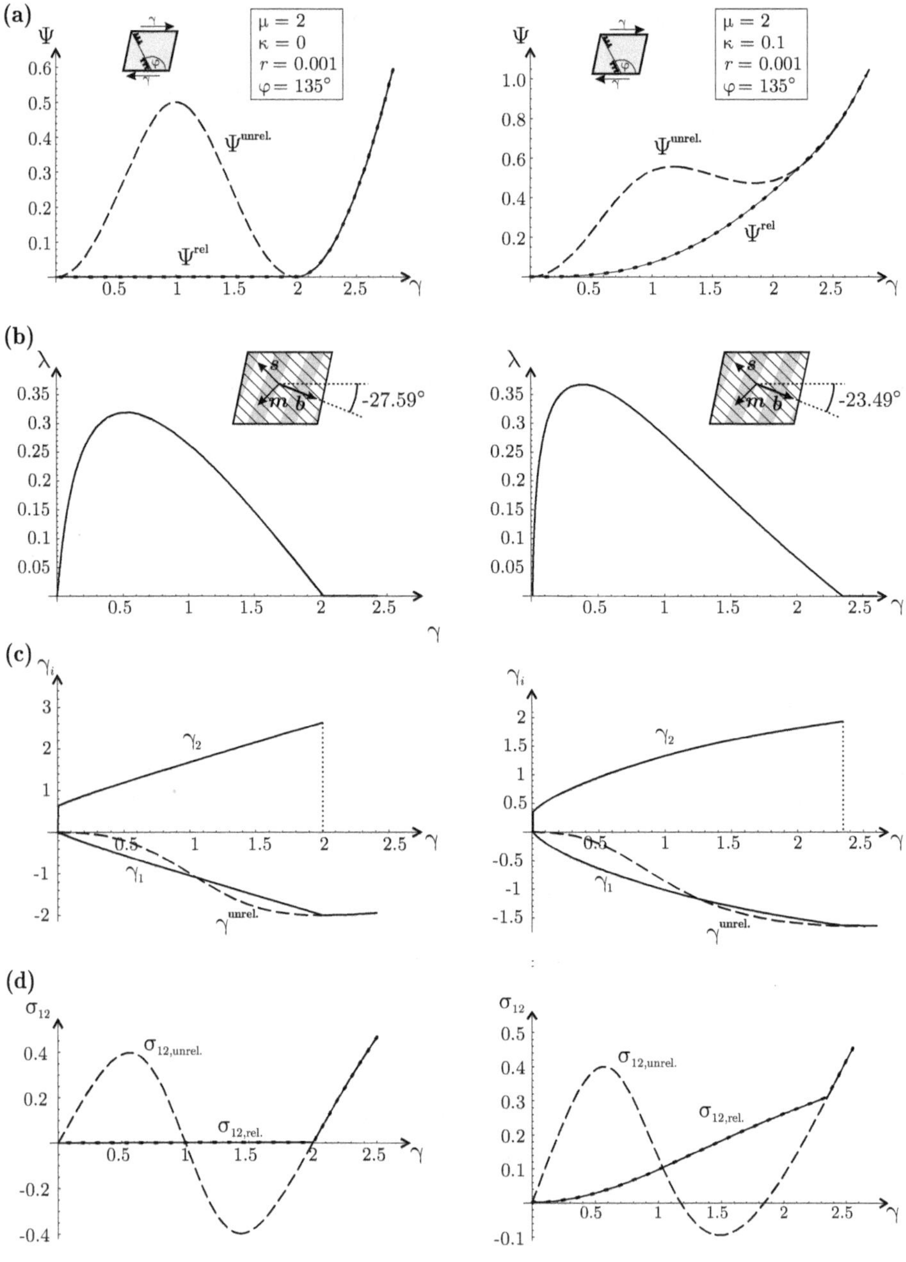

Fig. 4. Evolution of (a) the energy, (b) the volume fraction of phase 2, (c) the plastic slips, (d) the Cauchy shear stress for a plane simple shear test ($\mathbf{F} = \gamma \mathbf{e}_1 \otimes \mathbf{e}_2$).

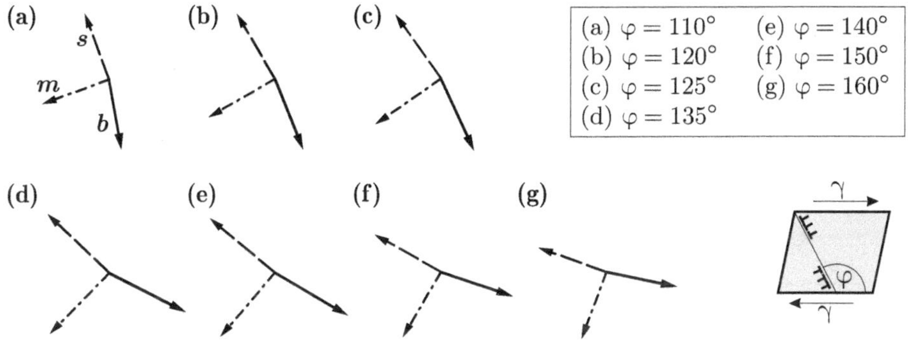

Fig. 5. Orientation of the forming laminate (interface normal **b**) for various orientations of the slip system (**s**, **m**) for $\mu = 2$, $\kappa = 0.01$ and $r = 0.001$.

5.2 Energy Reduction

The incremental method outlined in this contribution accounts for the existing microstructure at the beginning of each time step, in contrast to many earlier approaches which made use of the condensed energy functional [3, 7–10, 22]. Therefore, it is worth comparing results from the present method with those obtained from a condensed energy functional formulation. In Figure 6 we summarize the energy path, the Cauchy tensile stress, the developing plastic slips and hardening variables as well as the evolving volume fraction of the second phase for a tension-compression-test with the macroscopic deformation gradient

$$\mathbf{F} = \begin{pmatrix} 1+\delta & 0 & 0 \\ 0 & 1/(1+\delta) & 0 \\ 0 & 0 & 1 \end{pmatrix}.$$

The unrelaxed solution for the energy, $\Psi^{\text{cond},1}$, and the corresponding relaxed solution $\Psi^{\text{cond,rel},1}$, obtained from the condensed energy for a first-order laminate, i.e.

$$\Psi^{\text{cond,rel}}(\mathbf{F}) = \inf\Big\{(1-\lambda)\Psi^{\text{cond}}(\mathbf{F}_1) + \lambda\Psi^{\text{cond}}(\mathbf{F}_2)\,\Big|\,\lambda, \mathbf{F}_1, \mathbf{F}_2 : \\ 0 \le \lambda \le 1,\ (1-\lambda)\mathbf{F}_1 + \lambda\mathbf{F}_2 = \mathbf{F},\ \text{rank}(\mathbf{F}_1 - \mathbf{F}_2) \le 1\Big\}$$

are compared to the analogous path Ψ^{rel} obtained from the present incremental approach. The energy from the present approach indicates a considerable reduction in comparison with the condensed solution; a similar deviation becomes obvious from the stress-strain curve. This difference can be explained by considering the evolution of the plastic slips γ_i and the hardening variables p_i (see Figure 6c). The condensed energy functional is based on the assumption that the hardening variables evolve

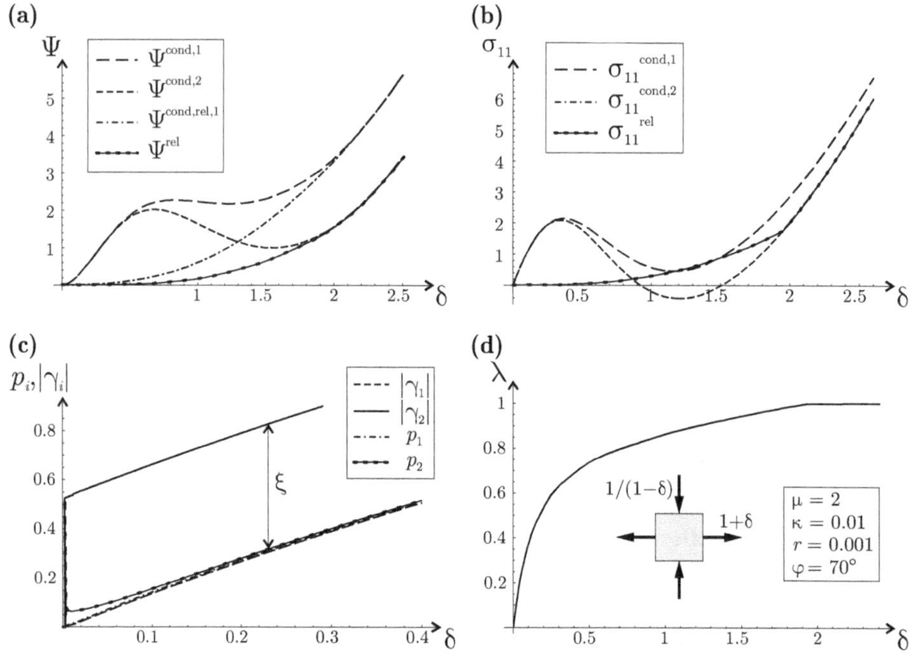

Fig. 6. Comparison of results from energy condensation with those from the present incremental technique: (a) path of energy, (b) Cauchy tensile stress, (c) plastic slips and hardening variables, and (d) volume fraction of phase 2 from the present approach.

with the plastic slips such that for each time step $p_i = |\gamma_i|$. The updating procedure employed here for the hardening variables (energetic averaging upon changes of the volume fractions) yields a spontaneous decrease of p_2 after the formation of the second laminate phase so that $p_2 = |\gamma_2| - \xi$ ($\xi > 0$). In the example in Figure 6 we see that after some straining p_2 approaches $p_1 = |\gamma_1|$. At the end of the non-convex regime only the second laminate phase remains with $p_2 = |\gamma_2| - \xi$. Accounting for this difference in the condensed energy functional yields the unrelaxed solution $\Psi^{\text{cond},2}$ which perfectly matches the incremental solution.

5.3 Cyclic Behavior

The present method yields interesting results when applied to deformation cycles [13]. In particular, the stress-strain behavior changes with an increasing number of load cycles due to specific microstructural mechanisms as discussed in the sequel. In Figure 7 we illustrate the evolving internal variables as well as the stress-strain curve and the energy path for a cyclic simple shear test ($\mathbf{F} = \mathbf{I} + \gamma\, \mathbf{e}_1 \otimes \mathbf{e}_2$) with given material parameters. The first cycle begins with elastic and soon plastic deformation of the homogeneous single crystal. At about 1.2% strain a laminate forms with high

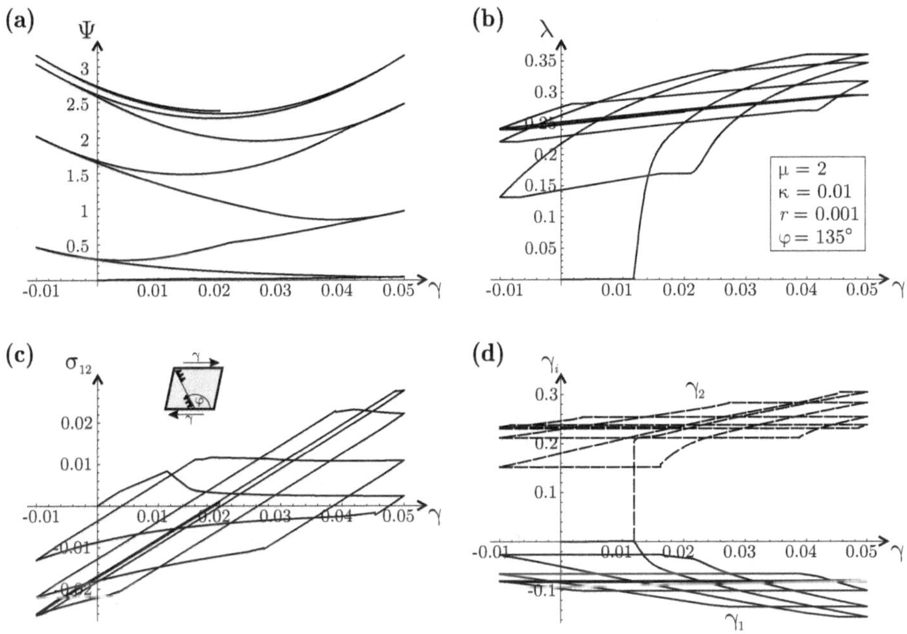

Fig. 7. Cyclic simple shear test: (a) energy path, (b) volume fraction of phase 2, (c) stress-strain hysteresis, (d) plastic slips.

plastic strain γ_2 already. The volume fraction of phase 2, λ, gradually rises to about 35%. Unloading in every cycle is first purely elastic again followed by plastic flow and an inverse lamination effect: The volume fraction of the second phase and the plastic slips in both phases decrease. With an increasing number of cycles the behavior changes: The broad hysteresis of the first cycle gradually decays so that after less than four cycles the stress-strain behavior reaches an almost steady cycle that hardly changes during subsequent load cycles and ressembles almost pure elastic loading and unloading. The cyclic paths of the internal variables for each cycle also become more and more shallow. The big changes of the volume fraction and the plastic slips during the first cycle are not repeated during subsequent cycles but the changes of the internal variables with increasing number of cycles gradually decay. The final behavior results from an almost permanent laminate microstructure with hardly changing volume fractions (λ resides between 25 and 29%) and plastic slips (almost constant values $\gamma_1 \approx -0.08$ and $\gamma_2 \approx 0.232$). The energy curve in Figure 7a illustrates the increasing amount of stored energy due to the rise of hardening.

Such a cyclic stress-strain behavior has been observed, at least in principle, in experiments on copper single-crystals. Of course, the number of loading cycles in experiments before the elastic shakedown is considerably higher than in this simple example: On the one hand, numerical simulations are not performed with the exact material data, and numerical experiments indicate that the choice of the material properties highly affects the observed stress-strain response. On the other hand, the simulation is based on simplifying assumptions such as single-slip plasticity, so that it can be presumed that the presence and interaction of multiple active slip systems will essentially affect the microstructure transformation into the observed steady state.

6 Discussion and Conclusions

In this contribution we reviewed an incremental strategy to model not only the origin but in particular the subsequent time-continuous evolution of laminate microstructures in finite-strain plasticity. The non-convexity of the free energy gives rise to microstructure formation. A relaxation of the free energy density by means of an approximation as a first-order laminate provides the analytical framework to simulate the mechanical behavior of such laminates. The analytical relaxation is applicable to a single active slip system in all phases, different single active slip systems in all phases, as well as multiple active slip systems as long as they lie within the same glide plane.

The incremental method is applied to several examples where we have laid emphasis on a comparison with approaches from the literature that make use of a condensed energy functional, and we have demonstrated the specific laminate behavior during cyclic loading. In comparison with those solutions obtained from using a condensed energy functional the present method gives rise to a considerable energy reduction mainly due to the updating procedure of the internal hardening variables (in contrast to the common approach of energy condensation we do not assume that $p_i = |\gamma_i|$ in every phase). Several examples illustrated the evolving laminate characteristics. The cyclic behavior obtained from the present approach is of particular interest: without hardening, the stress-strain curve follows the same hysteresis loop during subsequent load cycles [13]. When plastic hardening is accounted for, the hysteresis loop is clearly cycle-dependent. The broad hysteresis loop during the first load cycles gradually decreases to eventually yield an almost steady microstructure – the internal variables reach a steady state and hardly change during subsequent cycles, and the stress-strain curve exhibits an almost constant slope with a vanishing hysteresis. The microstructural essentials that come along with this behavior have been outlined.

In conclusion we have demonstrated the applicability of a novel incremental method to model the time-continuous evolution of laminate microstructures in finite plasticity to several interesting problems.

References

1. Hackl, K. and Fischer, F.D., On the relation between the principle of maximum dissipation and inelastic evolution given by dissipation. *Proc. Roy. Soc. London A*, 464:117–132, 2008.

2. Carstensen, C., Hackl, K., and Mielke, A., Non-convex potentials and microstructures in finite-strain plasticity. *Proc. R. Soc. London A*, 458:299–317, 2002.
3. Ortiz, M. and Repetto, E.A.. Nonconvex energy minimization and dislocation structures in ductile single crystals. *J. Mech. Phys. Solids*, 47:397–462, 1999.
4. Ericksen, J.L., Equilibrium of bars. *J. Elasticity*, 5:191–201, 1975.
5. Ball, J.M. and James, R.D., Fine phase mixtures as minimizers of energy. *Arch. Rat. Mech. Anal.*, 100:13–52, 1987.
6. Govindjee, S., Mielke, A., and Hall, G.J., The free energy of mixing for n-variant martensitic phase transformations using quasi-convex analysis. *J. Mech. Phys. Solids*, 51:1–26, 2003.
7. Bartels, S., Carstensen, C., Hackl, K., and Hoppe, U., Effective relaxation for microstructure simulation: algorithms and applications. *Comp. Meth. Appl. Meth. Eng.*, 193:5143–5175, 2004.
8. Conti, S. and Theil, F., Single-slip elastoplastic microstructures. *Arch. Rat. Mech. Anal.*, 178:125–148, 2005.
9. Lambrecht, M., Miehe, C., and Dettmar, J., Energy relaxation of non-convex incremental stress potentials in a strain-softening elastic-plastic bar. *Int. J. Solids Struct.*, 40:1369–1391, 2003.
10. Mielke, A., Deriving new evolution equations for microstructures via relaxation of variational incremental problems. *Comp. Meth. Appl. Meth. Eng.*, 193:5095–5127, 2004.
11. Conti, S. and Ortiz, M., Minimum principles for the trajectories of systems governed by rate problems. *J. Mech. Phys. Solids*, 56:1885–1904, 2008.
12. Mielke, A. and Ortiz, M., A class of minimum principles for characterizing the trajectories and the relaxation of dissipative systems. *ESAIM Control Optim. Calc. Var.*, 14:494–516, 2007.
13. Hackl, K. and Kochmann, D.M., Relaxed potentials and evolution equations for inelastic microstructures. In: B. Daya Reddy (Ed.), *IUTAM Symposium on Theoretical, Computational and Modelling Aspects of Inelastic Media*. Springer, Dordrecht, pp. 27–39, 2008.
14. Kochmann, D.M. and Hackl, K., Time-continuous evolution of microstructures in finite plasticity. In: K. Hackl (Ed.), *IUTAM Symposium on Variational Concepts with Applications to the Mechanics of Materials*. Springer, Dordrecht, 2010.
15. Biot, M.A., *Mechanics of Incremental Deformations*. John Wiley & Sons, New York, 1965.
16. Germain, P., *Cours de Mécanique des Milieux Continus*. Masson et Cie, Paris, 1973.
17. Nguyen, Q.S., *Stability and Nonlinear Solid Mechanics*. John Wiley & Sons, New York, 2000.
18. Ziegler, H. and Wehrli, C., The derivation of constitutive relations from the free energy and the dissipation function. In: T.Y. Wu and J.W. Hutchinson (Eds.), *Advances in Applied Mechanics IV*. Academic Press, 1987.
19. Mielke, A., Finite elastoplasticity, Lie groups and geodesics on SL(d). In: P. Newton, A. Weinstein, and P. Holmes (Eds.), *Geometry, Dynamics, and Mechanics*. Springer, Berlin, 2002.
20. Miehe, C., Schotte, J., and Lambrecht, M., Homogenization of inelastic solid materials at finite strains based on incremental minimization principles. Application to the texture analysis of polycrystals. *J. Mech. Phys. Solids*, 50:2123–2167, 2002.
21. Hackl, K., Mielke, A., and Mittenhuber, D., Dissipation distances in multiplicative elastoplasticity. In: W.L. Wendlang and M. Efendiev (Eds.), *Analysis and Simulation of Multifield Problems*. Springer, New York, 2003.

22. Carstensen, C., Conti, S., and Orlando, A., Mixed analytical-numerical relaxation in finite single-slip crystal plasticity. *Contin. Mech. Thermodyn.*, 20:275–301, 2008.
23. Hackl, K., Schmidt-Baldassari, M., and Zhang, W., A micromechanical model for polycrystalline shape-memory alloys. *Mat. Sci. Eng. A*, 378:503–506, 2003.

The Micromorphic versus Phase Field Approach to Gradient Plasticity and Damage with Application to Cracking in Metal Single Crystals

Ozgur Aslan and Samuel Forest

MINES ParisTech, Centre des Matériaux, CNRS UMR 7633, BP 87,
91003 Evry Cedex, France; samuel.forest@ensmp.fr

Abstract The micromorphic approach consists in introducing additional degrees of freedom in the continuum mechanical framework, that are constitutively related to mechanical variables like strain or internal variables for hardening and damage. We show the strong links between the micromorphic theory originated from Erigen and Mindlin's work, and phase field models for which the Landau–Ginzburg equation can be interpreted as a balance of generalized forces. The application deals with the formulation of a micromorphic crystal model involving plasticity and damage according to crystallographic planes. Finite element simulations of crack propagation under mode I conditions are presented.

Key words: Micromorphic theory, phase field, crystal plasticity, damage mechanics, crack

1 Generalized Continua and Material Microstructure

The mechanics of generalized continua represents a way of introducing, in the continuum description of materials, some characteristic length scales associated with their microstructure [24]. Such intrinsic lengths and generalized constitutive equations can be identified in two ways. Direct identification is possible from experimental curves exhibiting clear size effects in plasticity or fracture or from full-field strain measurements of strongly heterogeneous fields [13]. The effective properties of such generalized continua can also be derived from scale transition and homogenization techniques by prescribing appropriate boundary conditions on a representative volume of material with microstructure [5].

The multiplication of generalized continuum model formulations from Cosserat to strain gradient plasticity in literature may leave an impression of disorder and inconsistency. Recent accounts have shown, on the contrary, that unifying presentations of several classes of generalized continuum theories are possible [11, 18]. One of them, called the micromorphic approach, encompasses most theories incorporating additional degrees of freedom from the well-established Cosserat, microstretch and micromorphic continua [9] up to Aifantis and Gurtin strain gradient plasticity theories.

R. de Borst and E. Ramm (eds.), *Multiscale Methods in Computational Mechanics*, Lecture Notes in Applied and Computational Mechanics 55, DOI 10.1007/978-90-481-9809-2_8,

The objective of this chapter is to present this systematic approach for incorporating intrinsic lengths in non-linear continuum mechanical models and to illustrate the efficiency of the method in the case of an anisotropic plasticity and damage model. The so-called *microdamage* model takes the crystallography of plasticity and fracture in metal single crystals.

The micromorphic approach is exposed in Section 2 together with the closely related phase field approach. Differences and similarities between the micromorphic framework and the phase field approach are pointed out following the general framework provided in [17]. A single crystal plasticity and damage model is explored in Section 3. The finite element implementation and its validation for monotonic crack growth are shown in the two last sections.

The notations used are the same as the ones given in [11]. For the sake of conciseness, the theory and applications are presented within the framework of small deformation.

2 Micromorphic Approach

2.1 Thermomechanics with Additional Degrees of Freedom

One starts from an elastoviscoplasticity model formulation within the framework of the classical Cauchy continuum and classical continuum thermodynamics according to Germain et al. [16] and Maugin [22]. The material behaviour is characterized by the reference sets of degrees of freedom and state variables

$$DOF\,0 = \{\underline{\boldsymbol{u}}\}, \quad STATE\,0 = \{\underset{\sim}{\boldsymbol{\varepsilon}}, \quad T, \quad \alpha\} \tag{1}$$

on which the free energy density function ψ may depend. The displacement vector is $\underline{\boldsymbol{u}}$. The strain tensor is denoted by $\underset{\sim}{\boldsymbol{\varepsilon}}$ whereas α represents the whole set of internal variables of arbitrary tensorial order accounting for nonlinear processes at work inside the material volume element, like isotropic and kinematic hardening variables. The absolute temperature is T.

Additional degrees of freedom $^{\chi}\phi$ are then introduced in the previous original model. They may be of any tensorial order and of different physical nature (deformation, plasticity or damage variable). The notation $^{\chi}$ indicates that these variables eventually represent some microstructural features of the material so that we will call them micromorphic variables or microvariables (microdeformation, microdamage, etc.). The *DOF* and *STATE* spaces are extended as follows:

$$DOF = \{\underline{\boldsymbol{u}}, \quad {}^{\chi}\phi\}, \quad STATE = \{\underset{\sim}{\boldsymbol{\varepsilon}}, \quad T, \quad \alpha, \quad {}^{\chi}\phi, \quad \nabla^{\chi}\phi\} \tag{2}$$

Depending on the physical nature of $^{\chi}\phi$, it may or may not be a state variable. For instance, if the microvariable is a microrotation as in the Cosserat model, it is not a state variable for objectivity reasons and will appear in *STATE* only in combination with the macrorotation. In contrast, if the microvariable is a microplastic equivalent strain, as in Aifantis model, it can then explicitly appears in the state space.

The virtual power of internal forces is then extended to the power done by the micromorphic variable and its first gradient:

$$\mathcal{P}^{(i)}(\underline{v}^{\star}, {}^{\chi}\dot{\phi}^{\star}) = -\int_{\mathcal{D}} p^{(i)}(\underline{v}^{\star}, {}^{\chi}\dot{\phi}^{\star})\, dV$$

$$p^{(i)}(\underline{v}^{\star}, {}^{\chi}\dot{\phi}^{\star}) = \underset{\sim}{\sigma} : \nabla \underline{v}^{\star} + a^{\chi}\dot{\phi}^{\star} + \underline{b} \cdot \nabla^{\chi}\dot{\phi}^{\star} \tag{3}$$

where $\mathcal{D}$ is a subdomain of the current configuration Ω of the body. The Cauchy stress is $\underset{\sim}{\sigma}$ and a and $\underline{b}$ are generalized stresses associated with the micromorphic variable and its first gradient. Similarly, the power of contact forces must be extended as follows:

$$\mathcal{P}^{(c)}(\underline{v}^{\star}, {}^{\chi}\dot{\phi}^{\star}) = \int_{\mathcal{D}} p^{(c)}(\underline{v}^{\star}, {}^{\chi}\dot{\phi}^{\star})\, dV, \quad p^{(c)}(\underline{v}^{\star}, {}^{\chi}\dot{\phi}^{\star}) = \underline{t} \cdot \underline{v}^{\star} + a^{c\,\chi}\dot{\phi}^{\star} \tag{4}$$

where $\underline{t}$ is the traction vector and a^c a generalized traction. For conciseness, we do not extend the power of forces acting at a distance and keep the classical form:

$$\mathcal{P}^{(e)}(\underline{v}^{\star}, {}^{\chi}\dot{\phi}^{\star}) = \int_{\mathcal{D}} p^{(e)}(v^{\star}, {}^{\chi}\dot{\phi}^{\star})\, dV, \quad p^{(e)}(\underline{v}^{\star}, {}^{\chi}\dot{\phi}^{\star}) = \rho \underline{f} \cdot \underline{v}^{\star} \tag{5}$$

where $\rho \underline{f}$ accounts for given simple body forces. Following Germain [15], given body couples and double forces working with the gradient of the velocity field, could also be introduced in the theory. The generalized principle of virtual power with respect to the velocity and micromorphic variable fields, is presented here in the static case only:

$$\mathcal{P}^{(i)}(\underline{v}^{\star}, {}^{\chi}\dot{\phi}) + \mathcal{P}^{(e)}(\underline{v}^{\star}, {}^{\chi}\dot{\phi}^{\star}) + \mathcal{P}^{(c)}(\underline{v}^{\star}, {}^{\chi}\dot{\phi}) = 0, \quad \forall \mathcal{D} \subset \Omega, \quad \forall \underline{v}^{\star}, {}^{\chi}\dot{\phi} \tag{6}$$

The method of virtual power according to Maugin [21] is used then to derive the standard local balance of momentum equation:

$$\operatorname{div} \underset{\sim}{\sigma} + \rho \underline{f} = 0, \quad \forall \underline{x} \in \Omega \tag{7}$$

and the generalized balance of micromorphic momentum equation:

$$\operatorname{div} \underline{b} - a = 0, \quad \forall \underline{x} \in \Omega \tag{8}$$

The method also delivers the associated boundary conditions for the simple and generalized tractions:

$$\underline{t} = \underset{\sim}{\sigma} . \underline{n}, \quad a^c = \underline{b} . \underline{n}, \quad \forall \underline{x} \in \partial \mathcal{D} \tag{9}$$

The local balance of energy is also enhanced by the generalized micromorphic power already included in the power of internal forces (3):

$$\rho \dot{\epsilon} = p^{(i)} - \operatorname{div} \underline{q} + \rho r \tag{10}$$

where ϵ is the specific internal energy, $\underline{q}$ the heat flux vector and r denotes external heat sources. The entropy principle takes the usual local form:

$$-\rho(\dot{\psi} + \eta\dot{T}) + p^{(i)} - \frac{\underline{q}}{T}.\nabla T \geq 0 \tag{11}$$

where it is assumed that the entropy production vector is still equal to the heat vector divided by temperature, as in classical thermomechanics according to Coleman and Noll [6]. Again, the enhancement of the theory goes through the enriched power density of internal forces (3). The entropy principle is exploited according to classical continuum thermodynamics to derive the state laws. At this stage it is necessary to be more specific on the dependence of the state functions $\psi, \eta, \underset{\sim}{\sigma}, a, \underline{b}$ on state variables and to distinguish between dissipative and non-dissipative mechanisms. The introduction of dissipative mechanisms may require an increase in the number of state variables. These different situations are considered in the following subsections.

2.2 Non-Dissipative Contribution of Generalized Stresses and Micromorphic Model

Dissipative events are assumed here to enter the model only via the classical mechanical part. Total strain is split into elastic and plastic parts:

$$\underset{\sim}{\varepsilon} = \underset{\sim}{\varepsilon}^e + \underset{\sim}{\varepsilon}^p \tag{12}$$

The following constitutive functional dependencies are then introduced:

$$\psi = \hat{\psi}(\underset{\sim}{\varepsilon}^e, T, \alpha, {}^\chi\phi, \nabla^\chi\phi), \quad \underset{\sim}{\sigma} = \hat{\underset{\sim}{\sigma}}(\underset{\sim}{\varepsilon}^e, T, \alpha, {}^\chi\phi, \nabla^\chi\phi), \quad \eta = \hat{\eta}(\underset{\sim}{\varepsilon}^e, T, \alpha, {}^\chi\phi, \nabla^\chi\phi)$$

$$a = \hat{a}(\underset{\sim}{\varepsilon}^e, T, \alpha, {}^\chi\phi, \nabla^\chi\phi), \quad \underline{b} = \hat{\underline{b}}(\underset{\sim}{\varepsilon}^e, T, \alpha, {}^\chi\phi, \nabla^\chi\phi) \tag{13}$$

The entropy inequality (11) can be expanded as

$$\left(\underset{\sim}{\sigma} - \rho\frac{\partial\hat{\psi}}{\partial\underset{\sim}{\varepsilon}^e}\right) : \dot{\underset{\sim}{\varepsilon}}^e + \rho\left(\eta + \frac{\partial\hat{\psi}}{\partial T}\right)\dot{T} + \left(a - \rho\frac{\partial\hat{\psi}}{\partial{}^\chi\phi}\right){}^\chi\dot{\phi} + \left(\underline{b} - \rho\frac{\partial\hat{\psi}}{\partial\nabla{}^\chi\phi}\right)\cdot\nabla{}^\chi\dot{\phi}$$

$$+\underset{\sim}{\sigma} : \dot{\underset{\sim}{\varepsilon}}^p - \rho\frac{\partial\hat{\psi}}{\partial\alpha}\dot{\alpha} - \frac{\underline{q}}{T}\cdot\nabla T \geq 0 \tag{14}$$

Assuming that no dissipation is associated with the four first terms of the previous inequality, the following state laws are found:

$$\underset{\sim}{\sigma} = \rho\frac{\partial\hat{\psi}}{\partial\underset{\sim}{\varepsilon}^e}, \quad \eta = -\frac{\partial\hat{\psi}}{\partial T}, \quad X = \rho\frac{\partial\hat{\psi}}{\partial\alpha} \tag{15}$$

$$a = \rho\frac{\partial\hat{\psi}}{\partial{}^\chi\phi}, \quad \underline{b} = \rho\frac{\partial\hat{\psi}}{\partial\nabla{}^\chi\phi} \tag{16}$$

and the residual dissipation is

$$D^{res} = W^p - X\dot{\alpha} - \frac{\underline{q}}{T}.\nabla T \geq 0 \tag{17}$$

where W^p represents the (visco-)plastic power and X the thermodynamic force associated with the internal variable α. The existence of a convex dissipation potential, $\Omega(\underset{\sim}{\sigma}, X)$ depending on the thermodynamic forces can then be assumed from which the evolution rules for internal variables are derived, that identically fulfill the entropy inequality, as usually done in classical continuum thermomechanics [16]:

$$\dot{\underset{\sim}{\varepsilon}}^p = \frac{\partial \Omega}{\partial \underset{\sim}{\sigma}}, \quad \dot{\alpha} = \frac{\partial \Omega}{\partial X} \tag{18}$$

Micromorphic Model

After presenting the general approach, we readily give the most simple example which provides a direct connection to several existing generalized continuum models. We consider first cases where ϕ and ${}^{\chi}\phi$ are observer invariant quantities. The free energy density function ψ is chosen as a function of the generalized relative strain variable e defined as:

$$e = \phi - {}^{\chi}\phi \tag{19}$$

thus introducing a coupling between macro and micromorphic variables. Assuming isotropic material behavior for brevity, the additional contributions to the free energy can be taken as quadratic functions of e and $\nabla^{\chi}\phi$:

$$\psi(\underset{\sim}{\varepsilon}, T, \alpha, {}^{\chi}\phi, \nabla^{\chi}\phi) = \psi^{(1)}(\underset{\sim}{\varepsilon}, T, \alpha) + \psi^{(2)}(e = \phi - {}^{\chi}\phi, \nabla^{\chi}\phi, T), \quad \text{with} \tag{20}$$

$$\rho\psi^{(2)}(e, \nabla^{\chi}\phi, T) = \frac{1}{2}H_{\chi}(\phi - {}^{\chi}\phi)^2 + \frac{1}{2}A\nabla^{\chi}\phi \cdot \nabla^{\chi}\phi \tag{21}$$

After inserting the state laws (16)

$$a = \rho\frac{\partial \psi}{\partial {}^{\chi}\phi} = -H_{\chi}(\phi - {}^{\chi}\phi), \quad \underline{b} = \rho\frac{\partial \psi}{\partial \nabla^{\chi}\phi} = A\nabla^{\chi}\phi \tag{22}$$

into the additional balance equation (8), the following partial differential equation is obtained, at least for a homogeneous material under isothermal conditions:

$$\phi = {}^{\chi}\phi - \frac{A}{H_{\chi}}\Delta^{\chi}\phi \tag{23}$$

where Δ is the Laplace operator. This type of equation is encountered at several places in the mechanics of generalized continua especially in the linear micromorphic theory [7, 9, 23] and in the so-called implicit gradient theory of plasticity and damage [8, 25, 26]. Note however that this equation corresponds to a special quadratic potential and represents the simplest micromorphic extension of the classical theory. It involves a characteristic length scale defined by

$$l_c^2 = \frac{A}{H_{\chi}} \tag{24}$$

This length is real for positive values of the ratio A/H_{χ}. The additional material parameters H_{χ} and A are assumed to be positive in this work. This does not exclude a softening material behaviour that can be induced by the proper evolution of the internal variables (including $\phi = \alpha$ itself).

2.3 Viscous Generalized Stress and Phase Field Model

Generalized stresses can also be associated with dissipation by introducing the viscous part a^v of a:

$$\underset{\sim}{\boldsymbol{\varepsilon}} = \underset{\sim}{\boldsymbol{\varepsilon}}^e + \underset{\sim}{\boldsymbol{\varepsilon}}^p, \quad a = a^e + a^v \tag{25}$$

The entropy inequality (11) now becomes

$$\left(\underset{\sim}{\boldsymbol{\sigma}} - \rho\frac{\partial\hat{\psi}}{\partial\underset{\sim}{\boldsymbol{\varepsilon}}^e}\right) : \dot{\underset{\sim}{\boldsymbol{\varepsilon}}}^e + \rho\left(\eta + \frac{\partial\hat{\psi}}{\partial T}\right)\dot{T} + \left(a^e - \rho\frac{\partial\hat{\psi}}{\partial\,{}^\chi\phi}\right){}^\chi\dot{\phi} + \left(\underline{\boldsymbol{b}} - \rho\frac{\partial\hat{\psi}}{\partial\boldsymbol{\nabla}\,{}^\chi\phi}\right)\cdot\boldsymbol{\nabla}\,{}^\chi\dot{\phi}$$

$$+\underset{\sim}{\boldsymbol{\sigma}} : \dot{\underset{\sim}{\boldsymbol{\varepsilon}}}^p - \rho\frac{\partial\hat{\psi}}{\partial\alpha}\dot{\alpha} + a^v\,{}^\chi\dot{\phi} - \frac{\underline{\boldsymbol{q}}}{T}\cdot\boldsymbol{\nabla}T \geq 0 \tag{26}$$

Assuming that no dissipation is associated with the four first terms of the previous inequality, the following state laws are found:

$$\underset{\sim}{\boldsymbol{\sigma}} = \rho\frac{\partial\hat{\psi}}{\partial\underset{\sim}{\boldsymbol{\varepsilon}}^e}, \quad \eta = -\frac{\partial\hat{\psi}}{\partial T}, \quad X = \rho\frac{\partial\hat{\psi}}{\partial\alpha} \tag{27}$$

$$a^e = \rho\frac{\partial\hat{\psi}}{\partial\,{}^\chi\phi}, \quad \underline{\boldsymbol{b}} = \rho\frac{\partial\hat{\psi}}{\partial\boldsymbol{\nabla}\,{}^\chi\phi} \tag{28}$$

and the residual dissipation is

$$D^{res} = \underset{\sim}{\boldsymbol{\sigma}} : \dot{\underset{\sim}{\boldsymbol{\varepsilon}}}^p - X\dot{\alpha} + a^v\,{}^\chi\dot{\phi} - \frac{\underline{\boldsymbol{q}}}{T}\cdot\boldsymbol{\nabla}T \geq 0 \tag{29}$$

Evolution rules for viscoplastic strain, internal variables, and the additional degrees of freedom can be derived from a dissipation potential $\Omega(\underset{\sim}{\boldsymbol{\sigma}}, X, a^v)$:

$$\dot{\underset{\sim}{\boldsymbol{\varepsilon}}}^p = \frac{\partial\Omega}{\partial\underset{\sim}{\boldsymbol{\sigma}}}, \quad \dot{\alpha} = \frac{\partial\Omega}{\partial X}, \quad {}^\chi\dot{\phi} = \frac{\partial\Omega}{\partial a^v} \tag{30}$$

Convexity of the dissipation potential then ensures positivity of dissipation rate for any process.

Note that no dissipative part has been assigned to the generalized stress $\underline{\boldsymbol{b}}$ since then exploitation of second principle does not seem to be straightforward. Instead, the total gradient $\boldsymbol{\nabla}\,{}^\chi\phi$ can be split into elastic and plastic parts, as it will be done in Section 2.4.

Phase Field Model

The dissipation potential can be decomposed into the various contributions due to all thermodynamic forces. Let us assume for instance that the contribution of the viscous generalized stress a^v is quadratic:

$$\Omega(\underset{\sim}{\boldsymbol{\sigma}}, X, a^v) = \Omega_1(\underset{\sim}{\boldsymbol{\sigma}}, X) + \Omega_2(a^v), \quad \Omega_2(a^v) = \frac{1}{2\beta}a^{v2} \tag{31}$$

The use of the flow rule (30) and of the additional balance equation (8) then leads to

$$\beta^\chi \dot{\phi} = a^v = a - a^e = a - \rho \frac{\partial \hat{\psi}}{\partial^\chi \phi} = \operatorname{div} \rho \frac{\partial \hat{\psi}}{\partial \nabla^\chi \phi} - \rho \frac{\partial \hat{\psi}}{\partial^\chi \phi} \tag{32}$$

One recognizes the Landau–Ginzburg equation that arises in phase field theories. The previous derivation of Landau–Ginzburg equation is due to Gurtin [1, 17]. The coupling with mechanics is straightforward according to this procedure and more general dissipative mechanics can be put forward.

2.4 Elasto-Plastic Decomposition of Generalized Strains

Instead of the previous decomposition of generalized stresses, the introduction of additional dissipative mechanisms can rely on the split of all strain measures into elastic and plastic parts:

$$\underset{\sim}{\varepsilon} = \underset{\sim}{\varepsilon}^e + \underset{\sim}{\varepsilon}^p, \quad {}^\chi\phi = {}^\chi\phi^e + {}^\chi\phi^p, \quad \underline{\kappa} = \nabla^\chi\phi = \underline{\kappa}^e + \underline{\kappa}^p \tag{33}$$

The objectivity of ${}^\chi\phi$ is required for such a unique decomposition to be defined. We do not require here that

$$\underline{\kappa}^e = \nabla^\chi\phi^e, \quad \kappa^p = \nabla^\chi\phi^p \tag{34}$$

although such a model also is possible, as illustrated by the *gradient of strain theory* put forward in [12]. The Clausius–Duhem inequality then writes

$$\left(\underset{\sim}{\sigma} - \rho \frac{\partial \hat{\psi}}{\partial \underset{\sim}{\varepsilon}^e}\right) : \dot{\underset{\sim}{\varepsilon}}^e + \rho \left(\eta + \frac{\partial \hat{\psi}}{\partial T}\right) \dot{T} + \left(a - \rho \frac{\partial \hat{\psi}}{\partial^\chi \phi}\right) {}^\chi\dot{\phi}^e + \left(\underline{b} - \rho \frac{\partial \hat{\psi}}{\partial \nabla^\chi \phi}\right) \cdot \dot{\underline{\kappa}}^e$$

$$+ \underset{\sim}{\sigma} : \dot{\underset{\sim}{\varepsilon}}^p - \rho \frac{\partial \hat{\psi}}{\partial \alpha} \dot{\alpha} + a^\chi \dot{\phi}^p + \underline{b} \cdot \dot{\underline{\kappa}}^p - \frac{\boldsymbol{q}}{T} \cdot \nabla T \geq 0 \tag{35}$$

Assuming that no dissipation is associated with the four first terms of the previous inequality, the following state laws are found:

$$\underset{\sim}{\sigma} = \rho \frac{\partial \hat{\psi}}{\partial \underset{\sim}{\varepsilon}^e}, \quad \eta = -\frac{\partial \hat{\psi}}{\partial T}, \quad X = \rho \frac{\partial \hat{\psi}}{\partial \alpha} \tag{36}$$

$$a = \rho \frac{\partial \hat{\psi}}{\partial^\chi \phi^e}, \quad \underline{b} = \rho \frac{\partial \hat{\psi}}{\partial \underline{\kappa}^e} \tag{37}$$

and the residual dissipation is

$$D^{res} = \underset{\sim}{\sigma} : \dot{\underset{\sim}{\varepsilon}}^p - X\dot{\alpha} + a^\chi \dot{\phi}^p + \underline{b} \cdot \dot{\underline{\kappa}}^p - \frac{\boldsymbol{q}}{T} \cdot \nabla T \geq 0 \tag{38}$$

Evolution rules for viscoplastic strain, internal variables, and the additional degrees of freedom can be derived from a dissipation potential $\Omega(\underset{\sim}{\sigma}, X, a, \underline{b})$:

$$\dot{\underset{\sim}{\varepsilon}}^p = \frac{\partial \Omega}{\partial \underset{\sim}{\sigma}}, \quad \dot{\alpha} = \frac{\partial \Omega}{\partial X}, \quad {}^{\chi}\dot{\phi}^p = \frac{\partial \Omega}{\partial a}, \quad \dot{\underline{\kappa}}^{\,p} = \frac{\partial \Omega}{\partial \underline{b}} \tag{39}$$

As a result of the additional balance equation (8) combined with the previous state laws, the type of derived partial differential equation can be made more specific by adopting a quadratic free energy potential for $\underline{b}$ (modulus A) and a quadratic dissipation potential with respect to a (parameter β). We obtain:

$$\beta\, {}^{\chi}\dot{\phi} = a + \beta\, {}^{\chi}\dot{\phi}^e = \operatorname{div} A\underline{\kappa} - \operatorname{div} A\underline{\kappa}^{\,p} + \beta\, {}^{\chi}\dot{\phi}^e \tag{40}$$

Decompositions of stresses and strains can also be mixed, for instance in the following way:

$$\underset{\sim}{\varepsilon} = \underset{\sim}{\varepsilon}^e + \underset{\sim}{\varepsilon}^p, \quad a = a^e + a^v, \quad \underline{\kappa} = \nabla^{\chi}\phi = \underline{\kappa}^{\,e} + \underline{\kappa}^{\,p} \tag{41}$$

based on which a constitutive theory can be built.

3 Continuum Damage Model for Single Crystals and Its Regularization

We present here a constitutive model for damaging viscoplastic single crystal behaviour aiming at simulating crack initiation and propagation. The micromorphic approach is then applied to this model in order to obtain a regularized continuum damage formulation with a view to simulating mesh-independent crack propagation in single crystals.

3.1 Constitutive Equations

In the proposed crystal plasticity model taken from [20], viscoplasticity and damage are coupled by introducing an additional damage strain variable $\underset{\sim}{\varepsilon}^d$, into the strain rate partition equation:

$$\dot{\underset{\sim}{\varepsilon}} = \dot{\underset{\sim}{\varepsilon}}^e + \dot{\underset{\sim}{\varepsilon}}^p + \dot{\underset{\sim}{\varepsilon}}^d \tag{42}$$

where $\dot{\underset{\sim}{\varepsilon}}^e$ and $\dot{\underset{\sim}{\varepsilon}}^p$ are the elastic and the plastic strain rates, respectively. The flow rule for plastic part is written at the slip system level by means of the orientation tensor $\underset{\sim}{m}^s$:

$$\underset{\sim}{m}^s = \frac{1}{2}(\underline{n}^{\,s} \otimes \underline{l}^{\,s} + \underline{l}^{\,s} \otimes \underline{n}^{\,s}) \tag{43}$$

where $\underline{n}^{\,s}$ is the normal to the plane of slip system s and $\underline{l}^{\,s}$ stands for the corresponding slip direction. Then, plastic strain rate reads:

$$\dot{\underset{\sim}{\varepsilon}}^p = \sum_{s=1}^{N_{slip}} \dot{\gamma}^s \underset{\sim}{m}^s \tag{44}$$

The flow rule on slip system s is a classical Norton rule with threshold.

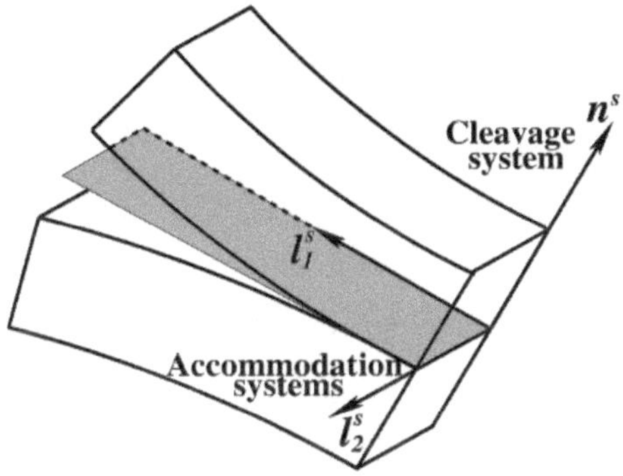

Fig. 1. Illustration of the cleavage and two accommodation systems to be associated to the crystallographic planes.

$$\dot{\gamma}^s = \left\langle \frac{|\tau^s - x^s| - r^s}{K} \right\rangle^n sign(\tau^s - x^s) \tag{45}$$

where r^s and x^s are the variables for isotropic and kinematic hardening respectively and K and n are material parameters to be identified from experimental curves.

Material separation is assumed to take place w.r.t. specific crystallographic planes, like cleavage planes in single crystals. The word *cleavage* is written in a more general sense that its original meaning in physical metallurgy associated with brittle fracture of non-f.c.c. crystals. In the continuum mechanical model, *cleavage* means cracking along a specific crystallographic plane as it is often observed in low cycle fatigue of f.c.c. crystals like single crystal nickel-base superalloys. The damage strain $\dot{\underset{\sim}{\boldsymbol{\varepsilon}}}^d$ is decomposed in the following crystallographic contributions:

$$\dot{\underset{\sim}{\boldsymbol{\varepsilon}}}^d = \sum_{s=1}^{N_{damage}} \dot{\delta}_c^s \underline{\boldsymbol{n}}_d^s \otimes \underline{\boldsymbol{n}}_d^s + \dot{\delta}_1^s \underline{\boldsymbol{n}}_d^s \overset{sym}{\otimes} \underline{\boldsymbol{l}}_{d_1}^s + \dot{\delta}_2^s \underline{\boldsymbol{n}}_d^s \overset{sym}{\otimes} \underline{\boldsymbol{l}}_{d_2}^s \tag{46}$$

where $\dot{\delta}^s$, $\dot{\delta}_1^s$ and $\dot{\delta}_2^s$ are the strain rates for mode I, mode II and mode III crack growth, respectively and N_{damage}^d stands for the number of damage planes which are fixed crystallographic planes depending on the crystal structure. Cleavage damage is represented by the opening δ^s of crystallographic cleavage planes with the normal vector $\underline{\boldsymbol{n}}^s$. Additional damage systems must be introduced for the in-plane accommodation along orthogonal directions $\underline{\boldsymbol{l}}_1^s$ and $\underline{\boldsymbol{l}}_2^s$, once cleavage has started (Figure 1). Three damage criteria are associated to one cleavage and two accommodation systems:

$$f_c^s = \left|\underline{\boldsymbol{n}}_d^s \cdot \underset{\sim}{\boldsymbol{\sigma}} \cdot \underline{\boldsymbol{n}}_d^s\right| - Y_c^s \tag{47}$$

$$f_i^s = \left|\underline{\boldsymbol{n}}_d^s \cdot \underset{\sim}{\boldsymbol{\sigma}} \cdot \underline{\boldsymbol{l}}_{d_i}^s\right| - Y_i^s \quad (i = 1, 2) \tag{48}$$

The critical normal stress Y^s for damage decreases as δ increases:

$$Y_c^s = Y_0^s + H\delta_c^s\,, \quad Y_i^s = Y_0^s + H\delta_i^s \tag{49}$$

where Y_0^s is the initial damage stress (usually coupled to plasticity) and H is a negative modulus which controls material softening due to damage. Finally, evolution of damage is given by the following equations:

$$\dot{\delta}_c^s = \left\langle \frac{f_c^s}{K_d} \right\rangle^{n_d} \text{sign}\left(\underline{\boldsymbol{n}}_d^s \cdot \underset{\sim}{\boldsymbol{\sigma}} \cdot \underline{\boldsymbol{n}}_d^s\right) \tag{50}$$

$$\dot{\delta}_i^s = \left\langle \frac{f_i^s}{K_d} \right\rangle^{n_d} \text{sign}\left(\underline{\boldsymbol{n}}_d^s \cdot \underset{\sim}{\boldsymbol{\sigma}} \cdot \underline{\boldsymbol{l}}_{d_i}^s\right) \tag{51}$$

where K_d and n_d are material parameters.

These equations hold for all conditions except when the crack is closed ($\delta_c^s < 0$) and compressive forces are applied ($\underline{\boldsymbol{n}}_d^s \cdot \underset{\sim}{\boldsymbol{\sigma}} \cdot \underline{\boldsymbol{n}}_d^s < 0$). In this case, damage evolution stops ($\dot{\delta}_c^s = \dot{\delta}_i^s = 0$), corresponding to the unilateral damage conditions.

Note that the damage variables δ introduced in the model differ from the usual corresponding variables of standard continuum damage mechanics that vary from 0 to 1. In contrast, the δs are strain-like quantities that can ever increase.

Coupling between plasticity and damage is generated through initial damage stress Y_0 in (49) which is controlled by cumulative slip variable γ_{cum}:

$$\dot{\gamma}_{cum} = \sum_{s=1}^{N_{slips}} |\dot{\gamma}^s| \tag{52}$$

Then, Y_0 takes the form:

$$Y_0^s = \sigma_n^c \, e^{-d\gamma_{cum}} + \sigma_{ult} \tag{53}$$

This formulation suggests an exponential decaying regime from a preferably high initial cleavage stress value σ_n^c, to an ultimate stress, σ_{ult} which is close to but not equal to zero for numerical reasons and d is a material constant.

This model, complemented by the suitable constitutive equations for viscoplastic strain, has been used for the simulation of crack growth under complex cyclic loading at high temperature [19]. Significant mesh dependency of results was found [20].

In the present work, the model is further developed by switching from classical to microdamage continuum in order to assess the regularization capabilities of a higher order theory. The coupling of the model with microdamage theory is achieved by introducing a cumulative damage variable calculated from the damage systems and a new threshold function $Y_0(\delta, \gamma_{cum})$:

$$\dot{\delta}_{cum} = \sum_{s=1}^{N_{planes}} \dot{\delta}^s \ , \ \text{where} \ \dot{\delta}^s = |\dot{\delta}_c^s| + |\dot{\delta}_1^s| + |\dot{\delta}_2^s| \tag{54}$$

$$Y_0 = \sigma_n^c \, e^{-d\gamma_{cum} - H\delta_{cum}} + \sigma_{ult} \tag{55}$$

3.2 Microdamage Continuum

The micromorphic medium introduced by Eringen and Suhubi [10] possesses a full microdeformation field $\underset{\sim}{\chi}$, in addition to the classical displacement field $\underline{u}$. Containing additional degrees of freedom and balance equations, the micromorphic continuum approach can be considered as the main framework for most generalized continuum models [11]. Alternative micromorphic variables other than the full strain tensor can be chosen [2, 11]. The strain gradient effect can for instance be limited to the damage strain $\underset{\sim}{\varepsilon}^d$ gradient and more specifically to the damage variable δ_{cum} introduced in the previous model and noted here δ for conciseness.

In microdamage theory, the introduced microvariable is a scalar microdamage parameter ${}^{\chi}\delta$:

$$DOF = \{\underline{u}\,,\ {}^{\chi}\delta\} \qquad STRAIN = \{\underset{\sim}{\varepsilon},\ {}^{\chi}\delta, \nabla\,{}^{\chi}\delta\} \tag{56}$$

The power of internal forces is extended as

$$p^{(i)} = \underset{\sim}{\sigma} : \dot{\underset{\sim}{\varepsilon}} + a\,{}^{\chi}\dot{\delta} + \underline{b}\,.\nabla\,{}^{\chi}\dot{\delta} \tag{57}$$

where generalized stresses a, $\underline{b}$ have been introduced. The generalized balance equations are:

$$\operatorname{div}\underset{\sim}{\sigma} = 0, \qquad a = \operatorname{div}\underline{b} \tag{58}$$

The free energy density is taken as a quadratic potential in the elastic strain, damage δ, relative damage $\delta - {}^{\chi}\delta$ and microdamage gradient $\nabla{}^{\chi}\delta$:

$$\rho\psi = \frac{1}{2}\underset{\sim}{\varepsilon}^e : \underset{\approx}{c} : \underset{\sim}{\varepsilon}^e + \frac{1}{2}\sum_{s=1}^{N_{damage}} H\delta_s^2 + \frac{1}{2}\,{}^{\chi}H(\delta - {}^{\chi}\delta)^2 + \frac{1}{2}A\nabla\,{}^{\chi}\delta\cdot\nabla\,{}^{\chi}\delta \tag{59}$$

where H, ${}^{\chi}H$ and A are scalar material constants. Then, the elastic response of the material becomes

$$\underset{\sim}{\sigma} = \rho\frac{\partial\psi}{\partial\underset{\sim}{\varepsilon}^e} = \underset{\approx}{c} : \underset{\sim}{\varepsilon}^e \tag{60}$$

The generalized stresses read

$$a = \rho\frac{\partial\psi}{\partial\,{}^{\chi}\delta} = -\,{}^{\chi}H(\delta - {}^{\chi}\delta)\,, \qquad \underline{b} = A\nabla\,{}^{\chi}\delta \tag{61}$$

and the driving force for damage can be derived as

$$Y^s = \rho\frac{\partial\psi}{\partial\delta^s} = H\delta^s + {}^{\chi}H(\delta^s - {}^{\chi}\delta) \tag{62}$$

The damage criterion is now

$$f^s = \left|\underline{n}^s\cdot\underset{\sim}{\sigma}\cdot\underline{n}^s\right| - Y_0 - Y^s = 0 \tag{63}$$

Substituting the linear constitutive equations for generalized stresses into the additional balance equation (58), assuming homogeneous material properties, leads to the following partial differential equation for the microdamage:

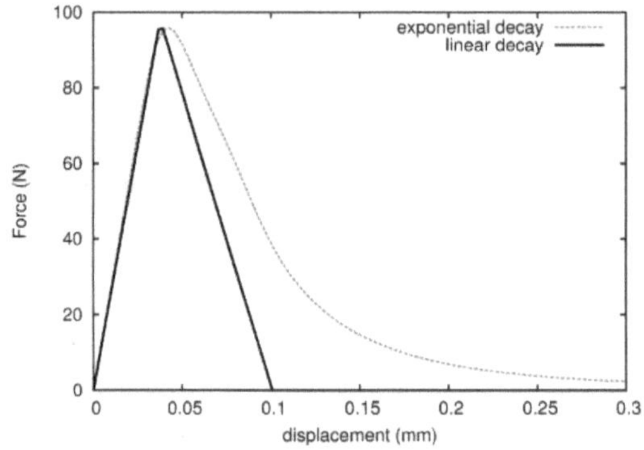

Fig. 2. Comparison between force vs. displacement diagram of a 1D softening rod for linear and exponential decay.

$$
{}^{\chi}\delta - \frac{A}{{}^{\chi}H}\Delta {}^{\chi}\delta = \delta \tag{64}
$$

where the macrodamage δ acts as a source term. Exactly this type of Helmholtz equation has been postulated in the so-called implicit gradient theory of plasticity and damage [8, 14, 25, 26], where the microvariables are called non local variables and where the generalized stresses a and $\underline{b}$ are not explicitly introduced (see [7, 11] for the analogy between this latter approach and the micromorphic theory).

The solution of the problem of failure of a 1D bar in tension/compression was treated in [2]. The characteristic size of the damage zone was shown to be controlled by the characteristic length corresponding to the inverse of

$$
\omega = \sqrt{|H\, {}^{\chi}H / A(H + {}^{\chi}H)|} \tag{65}
$$

In comparison with the standard strain gradient approaches [14, 25], microdamage theory eliminates the final fracture problem without any modification to the damage function, since there exists no direct coupling between the force stress $\underset{\sim}{\sigma}$ and the generalized stresses, a and $\underline{b}$. For a better representation of a cracked element, we suggest an exponential drop both for damage threshold Y_0 and the modulus A, since the element should become unable to store energy due to the generalized stresses when broken (see Figure 2):

$$
Y_0 = \sigma_n^c\, e^{-H\delta} + \sigma_{ult}\ , \quad \underline{b} = A e^{-H\delta} \nabla\, {}^{\chi}\delta \tag{66}
$$

4 Finite Element Implementation

4.1 Variational Formulation and Discretization

The variational formulation of the microdamage approach can be derived directly from the principle of virtual power (57):

$$\int_{\Omega} p^{(i)} dV + \int_{\partial\Omega} p^{(c)} dS = 0 \quad (67)$$

$$\int_{\Omega} (\underset{\sim}{\sigma} : \dot{\underset{\sim}{\varepsilon}} + a\, {}^{\chi}\dot{\delta} + \underline{b} . \nabla\, {}^{\chi}\dot{\delta}) dV + \int_{\partial\Omega} (\underline{t} . \dot{\underline{u}} + a\, {}^{\chi}\dot{\delta}) dS = 0 \quad (68)$$

Finite element discretization of the displacement field $\underline{u}$ and the microdamage field ${}^{\chi}\delta$ takes the following form:

$$u = N_u d_u, \quad \nabla u = B_u d_u, \quad {}^{\chi}\delta = N_\delta d_\delta, \quad \nabla^{\chi}\delta = B_\delta d_\delta \quad (69)$$

where d_u and d_δ are the nodal degrees of freedom. N_u and N_δ represent the shape functions and B_u and B_δ stand for their partial derivatives with respect to the coordinates. In this work we use isoparametric quadratic elements for both types of degrees of freedom.

Finally, the discretized equilibrium equations read:

$$\int_{\Omega} B_u^T \underset{\sim}{\sigma} dV = \int_{\Omega} {N_u}^T \underline{f}\, dV + \int_{\Gamma} {N_u}^T \underline{t}\, dS \quad (70)$$

$$\int_{\Omega} ({N_\delta}^T a + {B_\delta}^T \underline{b}) dV = \int_{\Gamma} {N_\delta}^T a_c dS \quad (71)$$

4.2 Implicit Incremental Formulation

A fully implicit Newton–Raphson incremental formulation is developed for solving (70, 71). The corresponding time discretization is now introduced. Using the known values of the state variables $\underset{\sim}{\varepsilon}^e(t)$, $\upsilon^s(t)$ (integrated from $\dot{\upsilon}^s = |\dot{\gamma}^s|$),$\delta^s_{c,i}(t)$, $\delta^s_{cum}(t)$ for the current time step, the values at $t + \Delta t$ are estimated by a straight forward linearization procedure.

$$\underset{\sim}{\varepsilon}^e(t + \Delta t) = \underbrace{\Delta t\, \dot{\underset{\sim}{\varepsilon}}^e(t + \Delta t)}_{\Delta \underset{\sim}{\varepsilon}^e} + \underset{\sim}{\varepsilon}^e(t) \quad (72)$$

$$\upsilon^s(t + \Delta t) = \Delta t\, \dot{\upsilon}^s(t + \Delta t) + \upsilon^s(t) \quad (73)$$

$$\delta^s_{c,i}(t + \Delta t) = \Delta t\, \dot{\delta}^s_{c,i}(t + \Delta t) + \delta^s_{c,i}(t) \quad (74)$$

$$\delta^s_{cum}(t + \Delta t) = \Delta t\, \dot{\delta}^s_{cum}(t + \Delta t) + \delta^s_{cum}(t) \quad (75)$$

The model is implemented into the FE code ZeBuLoN [4], using a θ-method for the local integration. In order to calculate the state variable increments, the residuals and their Jacobian are written as follows:

$$R_{\underset{\sim}{\varepsilon}^e} = \Delta \underset{\sim}{\varepsilon}^e + \Delta \underset{\sim}{\varepsilon}^p + \Delta \underset{\sim}{\varepsilon}^d - \Delta \underset{\sim}{\varepsilon} \quad (76)$$

$$= \Delta\varepsilon^e + \sum_{s=1}^{N_{slip}} \underset{\sim}{m}^s \Delta\upsilon^s sign(\tau^s - x^s) + \sum_{s=1}^{N_{planes}} \Delta\delta^s_c \underline{n}^s_d \otimes \underline{n}^s_d + \Delta\delta^s_i \underline{n}^s_d \otimes \underline{l}^s_{d_i} \quad (77)$$

$$R_{\upsilon^s} = \Delta\upsilon^s - \Delta t \left\langle \frac{\Phi^s}{K} \right\rangle^n \tag{78}$$

$$R_{\delta_c^s} = \Delta\delta_c^s - \Delta t \left\langle \frac{f_c^s}{K_d} \right\rangle^{n_d} sign(\underline{\boldsymbol{n}}_d^s \cdot \underset{\sim}{\boldsymbol{\sigma}} \cdot \underline{\boldsymbol{n}}_d^s) \tag{79}$$

$$R_{\delta_i^s} = \Delta\delta_c^i - \Delta t \left\langle \frac{f_i^s}{K_d} \right\rangle^{n_d} sign(\underline{\boldsymbol{n}}_d^s \cdot \underset{\sim}{\boldsymbol{\sigma}} \cdot \underline{\boldsymbol{l}}_{d_i}^s) \tag{80}$$

$$R_{\delta_{cum}^s} = \Delta\delta_{cum} - \Delta \left(\sum_{s=1}^{N_{planes}} |\delta_c^s| + |\delta_1^s| + |\delta_2^s| \right) \tag{81}$$

$$[J] = \frac{\partial\{R\}}{\partial\{\Delta\nu\}} = 1 - \Delta t \left. \frac{\partial\{\dot{\nu}\}}{\partial\{\Delta\nu\}} \right|_{t+\Delta t} \tag{82}$$

where $\{R\}^T = \left[R_{\varepsilon^e},\ R_{\upsilon^s},\ R_{\delta_c^s},\ R_{\delta_i^s},\ R_{\delta_{cum}} \right]$ and ν stands for the internal state variables to be integrated locally. Then, the Jacobian matrix becomes

$$[J] = \begin{pmatrix} \frac{\partial R_{\varepsilon^e}}{\partial\Delta\varepsilon^e} & \frac{\partial R_{\varepsilon^e}}{\partial\Delta\upsilon^s} & \frac{\partial R_{\varepsilon^e}}{\partial\Delta\delta_c^s} & \frac{\partial R_{\varepsilon^e}}{\partial\Delta\delta_i^s} & \frac{\partial R_{\varepsilon^e}}{\partial\Delta\delta_{cum}} \\ \frac{\partial R_{\upsilon^s}}{\partial\Delta\varepsilon^e} & \frac{\partial R_{\upsilon^s}}{\partial\Delta\upsilon^e} & \frac{\partial R_{\upsilon^s}}{\partial\Delta\delta_c^s} & \frac{\partial R_{\upsilon^s}}{\partial\Delta\delta_i^s} & \frac{\partial R_{\upsilon^s}}{\partial\Delta\delta_{cum}} \\ \frac{\partial\delta_c^s}{\partial\Delta\varepsilon^e} & \frac{\partial\delta_c^s}{\partial\Delta\upsilon^e} & \frac{\partial\delta_c^s}{\partial\delta_c^s} & \frac{\partial\delta_c^s}{\partial\delta_{cum}} & \frac{\partial\delta_c^s}{\partial\delta_{cum}} \\ \frac{\partial\delta_i^s}{\partial\Delta\varepsilon^e} & \frac{\partial\delta_i^s}{\partial\Delta\upsilon^e} & \frac{\partial\delta_i^s}{\partial\delta_c^s} & \frac{\partial\delta_i^s}{\partial\delta_i^s} & \frac{\partial\delta_i^s}{\partial\delta_{cum}} \\ \frac{\partial\delta_{cum}}{\partial\Delta\varepsilon^e} & \frac{\partial\delta_{cum}}{\partial\Delta\upsilon^e} & \frac{\partial\delta_{cum}}{\partial\Delta\delta_i^s} & \frac{\partial\delta_{cum}}{\partial\Delta\delta_c^s} & \frac{\partial\delta_{cum}}{\partial\Delta\delta_{cum}} \end{pmatrix} \tag{83}$$

After convergence, the θ-method allows the calculation of the tangent matrix of the behaviour. R can be decomposed into two parts as

$$\{R\} = \{R_i\} - \{R_e\} \tag{84}$$

where R_e corresponds to the applied load. After the convergence (i.e. $\{R\} \approx \{0\}$) an infinitesimal variation can be applied to the residual equation such as

$$\delta\{R\} = \{0\} = \delta\{R_i\} - \delta\{R_e\} \tag{85}$$

which can be rewritten in the form:

$$\delta\Delta\nu_{int} = [J]^{-1}\,\delta\{R_e\} \tag{86}$$

For the calculation of elastic strain increment, the above relation reads

$$\delta\Delta\underset{\sim}{\boldsymbol{\varepsilon}}^e = \underset{\approx}{\boldsymbol{J}}_e\delta\Delta\underset{\sim}{\boldsymbol{\varepsilon}}, \quad \delta\Delta\underset{\sim}{\boldsymbol{\sigma}} = \underset{\approx}{\boldsymbol{C}} : \underset{\approx}{\boldsymbol{J}}_e\delta\Delta\underset{\sim}{\boldsymbol{\varepsilon}} \tag{87}$$

Note that a consistent tangent matrix can directly be obtained from $[\underset{\approx}{\boldsymbol{C}} : \underset{\approx}{\boldsymbol{J}}_e]$. A more comprehensive presentation of this systematic numerical integration method can be found in [3].

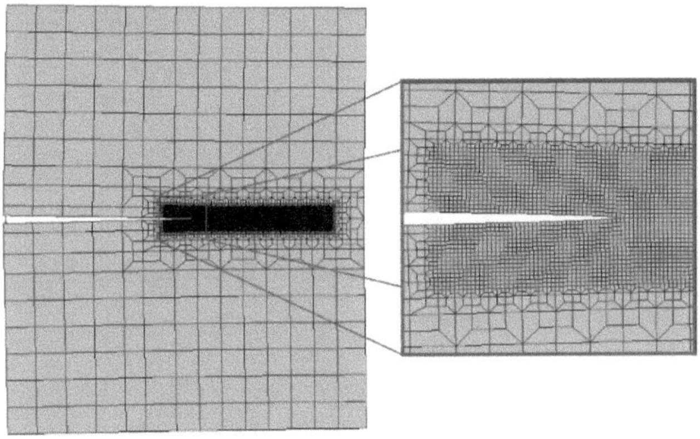

Fig. 3. FE mesh of a CT-like specimen created by ZeBuLoN GUI.

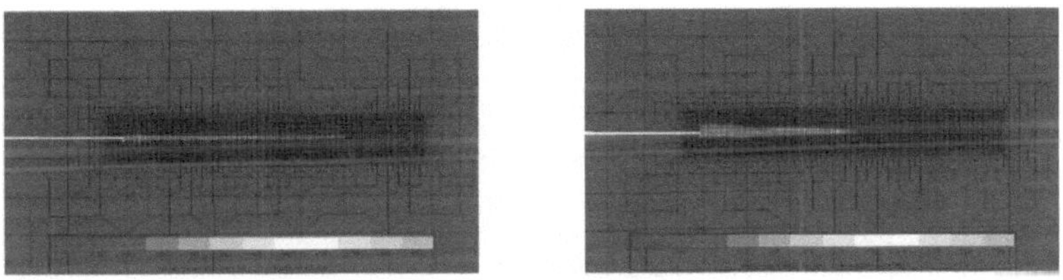

Fig. 4. Crack growth in a 2D single crystal CT-like specimen with a single cleavage plane aligned through the horizontal axis under vertical tension. Field variable δ. (Left) A = 100 MPa·mm^2, $H = -20000$ MPa, $^\chi H = 30000$ MPa; (Right) A = 150 MPa·mm^2, $H = -10000$ MPa, $^\chi H = 30000$ MPa.

5 Numerical Examples

As a 2D example, a single crystal CT-like specimen under tension is analysed. The corresponding FE mesh is given in Figure 3. Analyses are performed for two different crack widths, obtained by furnishing different material parameters which control the characteristic length. The propagation of a crack, corresponding stress fields and the comparison with classical elastic solutions are given in Figures 4 and 5. This comparison shows that the microdamage model is able to reproduce the stress concentration at the crack tip except very close to the crack tip where finite stress values are predicted.

Another 2D example, namely a plate under uniaxial tension with several cleavage planes, is investigated (see Figure 6). In order to trigger localization, an initial geometric defect is created on the left edge. First, a cleavage plane is oriented at 30° from the horizontal axis. FEA results show that localization path is perfectly matching with the cleavage plane and the size of the localization band is controlled by ω in (65) (Figure 6, left). Second, two orthogonal cleavage planes are placed with an orientation of 45° from the horizontal axis representing $\{111\}$ planes. For the former case, damage is coupled with plasticity and two localization bands merge together

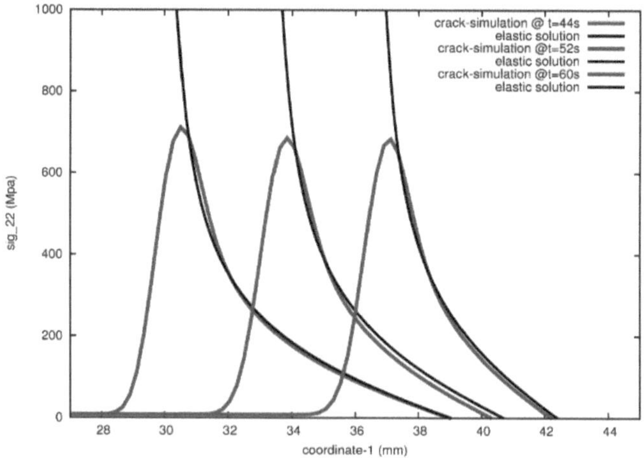

Fig. 5. Evolution of the crack and the stress fields in a CT-like specimen compared with corresponding elastic solutions.

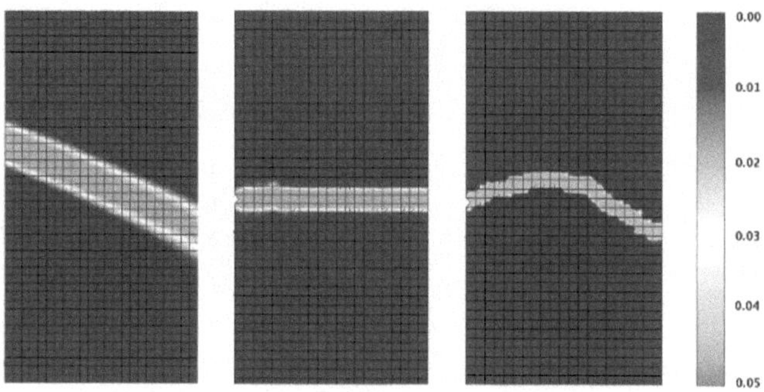

Fig. 6. Crack growth in a 2D single crystal block with a single inclined cleavage plane (left) and two orthogonal planes oriented at 45 degrees (middle and right) under vertical tension with 10% strain. Field variable δ.

and form a straight crack path which can be considered as a type of ductile crack (Figure 6, middle). For the latter case, plasticity is excluded from the calculation and crack path is allowed to choose its path between the orthogonal planes which results in a brittle type of crack propagation (Figure 6, right).

In [2], FEA of a CT-like fracture mechanics specimen under tension was considered. The analysis was done by creating a cleavage plane parallel to the horizontal axis and the loading was performed from the center of the pin. For a given characteristic length (associated with parameters $A = 200$ MPa·mm^2, $H = -16000$ MPa, ${}^{\chi}H = 50000$ MPa), mesh refinement of the specimen lead to a unique fracture curve and a finite size crack width, as shown in Figure 7 and in the loading curves in [2].

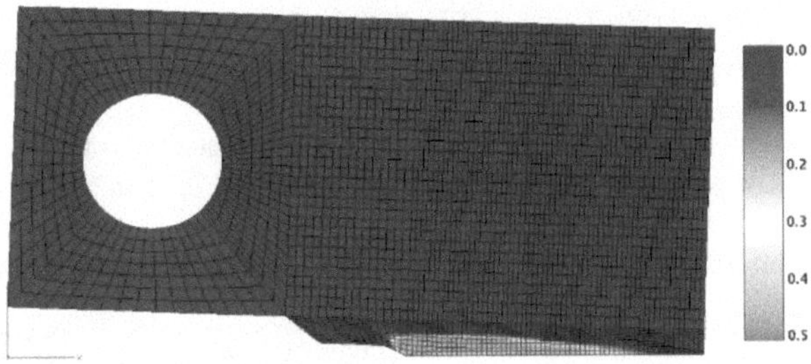

Fig. 7. Crack growth in a CT-like specimen under tension under vertical tension with 8% strain. Field variable δ.

6 Conclusion

The proposed systematic treatment of the thermomechanics of continua with additional degrees of freedom leads to model formulations ranging from micromorphic to phase field models. In particular, a general framework for the introduction of dissipative processes associated with the additional degrees of freedom has been proposed. If internal constraints are enforced on the relation between macro and microvariables in the micromorphic approach, standard second gradient and strain gradient plasticity models can be retrieved as shown in [11].

As a variant of micromorphic continuum, microdamage continuum and its regularization capabilities for the modelling of crack propagation in single crystals have been studied. First, a crystallographic constitutive model which accounts for continuum damage with respect to fracture planes has been presented. Then, the theory has been extended from classical continuum to microdamage continuum, respectively. It has been shown that the approach can be a good candidate for solving mesh dependency and the prediction of final fracture. Analytical fits and numerical results showed that the theory is well suited for FEA and possesses a great potential for the future modelling aspects. Comparison with available data on crack growth especially under cyclic loading in nickel-based superalloys, will be decisive to conclude on the ability of the approach to reach realistic prediction of component failure.

References

1. K. Ammar, B. Appolaire, G. Cailletaud, F. Feyel, and F. Forest. Finite element formulation of a phase field model based on the concept of generalized stresses. *Computational Materials Science*, 45:800–805, 2009.
2. O. Aslan and S. Forest. Crack growth modelling in single crystals based on higher order continua. *Computational Materials Science*, 45:756–761, 2009.
3. J. Besson, G. Cailletaud, J.-L. Chaboche, S. Forest, and M. Blétry. *Non-Linear Mechanics of Materials*. Springer, 2009.
4. J. Besson, R. Leriche, R. Foerch, and G. Cailletaud. Object-oriented programming applied to the finite element method. Part II: Application to material behaviours. *Rev. Eur. Elem. Finis*, 7:567–588, 1998.

5. G. Cailletaud, S. Forest, D. Jeulin, F. Feyel, I. Galliet, V. Mounoury, and S. Quilici. Some elements of microstructural mechanics. *Computational Materials Science*, 27:351–374, 2003.
6. B.D. Coleman and W. Noll. The thermodynamics of elastic materials with heat conduction and viscosity. *Arch. Rational Mech. and Anal.*, 13:167–178, 1963.
7. T. Dillard, S. Forest, and P. Ienny. Micromorphic continuum modelling of the deformation and fracture behaviour of nickel foams. *European Journal of Mechanics A/Solids*, 25:526–549, 2006.
8. R.A.B. Engelen, M.G.D. Geers, and F.P.T. Baaijens. Nonlocal implicit gradient-enhanced elasto-plasticity for the modelling of softening behaviour. *International Journal of Plasticity*, 19:403–433, 2003.
9. A.C. Eringen. *Microcontinuum Field Theories*. Springer, New York, 1999.
10. A.C. Eringen and E.S. Suhubi. Nonlinear theory of simple microelastic solids. *Int. J. Engng Sci.*, 2:189–203, 389–404, 1964.
11. S. Forest. The micromorphic approach for gradient elasticity, viscoplasticity and damage. *ASCE Journal of Engineering Mechanics*, 135:117–131, 2009.
12. S. Forest and R. Sievert. Elastoviscoplastic constitutive frameworks for generalized continua. *Acta Mechanica*, 160:71–111, 2003.
13. M.G.D. Geers, R. de Borst, W.A.M. Brekelmans, and R.H.J. Peerlings. On the use of local strain fields for the determination of the intrinsic length scale. *Journal de Physique IV*, 8:Pr8–167–174, 1998.
14. N. Germain, J. Besson, and F. Feyel. Simulation of laminate composites degradation using mesoscopic non-local damage model and non-local layered shell element. *Modelling and Simulation in Materials Science and Engineering*, 15:S425–S434, 2007.
15. P. Germain. La méthode des puissances virtuelles en mécanique des milieux continus, première partie: Théorie du second gradient. *J. de Mécanique*, 12:235–274, 1973.
16. P. Germain, Q.S. Nguyen, and P. Suquet. Continuum thermodynamics. *Journal of Applied Mechanics*, 50:1010–1020, 1983.
17. M.E. Gurtin. Generalized Ginzburg–Landau and Cahn–Hilliard equations based on a microforce balance. *Physica D*, 92:178–192, 1996.
18. C.B. Hirschberger and P. Steinmann. Classification of concepts in thermodynamically consistent generalized plasticity. *ASCE Journal of Engineering Mechanics*, 135:156–170, 2009.
19. N. Marchal, S. Flouriot, S. Forest, and L. Remy. Crack-tip stress-strain fields in single crystal nickel-base superalloys at high temperature under cyclic loading. *Computational Materials Science*, 37:42–50, 2006.
20. N. Marchal, S. Forest, L. Rémy, and S. Duvinage. Simulation of fatigue crack growth in single crystal superalloys using local approach to fracture. In D. Steglich, J. Besson, and D. Moinereau (Eds.), *Local Approach to Fracture*, 9th European Mechanics of Materials Conference, Euromech-Mecamat, Moret-sur-Loing, France. Presses de l'Ecole des Mines de Paris, pp. 353–358, 2006.
21. G.A. Maugin. The method of virtual power in continuum mechanics: Application to coupled fields. *Acta Mechanica*, 35:1–70, 1980.
22. G.A. Maugin. *Thermomechanics of Nonlinear Irreversible Behaviors*. World Scientific, 1999.
23. R.D. Mindlin. Micro-structure in linear elasticity. *Arch. Rat. Mech. Anal.*, 16:51–78, 1964.
24. H.B. Mühlhaus. *Continuum Models for Materials with Microstructure*. Wiley, 1995.

25. R.H.J. Peerlings, M.G.D. Geers, R. de Borst, and W.A.M. Brekelmans. A critical comparison of nonlocal and gradient-enhanced softening continua. *Int. J. Solids Structures*, 38:7723–7746, 2001.
26. R.H.J. Peerlings, T.J. Massart, and M.G.D. Geers. A thermodynamically motivated implicit gradient damage framework and its application to brick masonry cracking. *Comput. Methods Appl. Mech. Engrg.*, 193:3403–3417, 2004.

Homogenization and Multiscaling of Granular Media for Different Microscopic Constraints

C. Miehe, J. Dettmar and D. Zäh

Institute of Applied Mechanics, Chair I, University of Stuttgart, Pfaffenwaldring 7, 70550 Stuttgart, Germany; cm@mechbau.uni-stuttgart.de

Abstract We outline a method for quasi-static homogenization of granular microstructures and its embedding into two-scale simulations. A consistent extension of classical stiff, soft and periodic boundary conditions from continuous to granular microstructures induces new classes of micro-to-macro transitions for granular aggregates. These boundary conditions include constraints not only for particle center displacements but also for particle rotations at a driving boundary frame. The stiff and soft constraints at the driving frame of the particle aggregate induce upper and lower bounds of the particle aggregate stiffness. We outline a unified implementation of the displacement- and rotational constraints by penalty methods that proves to be convenient for straightforward integration into discrete element codes. Finally, we embed granular microstructures into a coarse graining discrete-to-finite element model, where they govern the micromechanical behavior of a two-scale simulation.

Key words: Multiscale method, homogenization, granular materials, particles

1 Introduction

As shown in Figure 1, we consider a homogenized macro-continuum with locally attached microstructure, representing an aggregate of discrete solid granules which possibly come into contact. Conceptual approaches to the modeling of such two-scale scenarios are outlined in [1, 3, 6, 8, 9, 11–13, 16, 19, 23], see also the references cited therein. In a recent work by Miehe and Dettmar [16], the authors outlined a framework of homogenization of granular materials for *periodic* (P) microstructures. The formulation was based on the introduction of a characteristic driving frame, consisting of particles with non-trivial periodic center translations and rotations. It was shown that such a framework yields for quasi-static homogenization of particle aggregates *symmetric* homogenized Cauchy stresses suitable for the application of standard macro-continuum formulations. We now extend this formulation to other classes of microscopic boundary constraints. These constraints transform classical continuum-based definitions of linear displacements (D) and uniform stress (S) to particle aggregates. This includes the definition of a characteristic Hill–Mandel-type *microheterogeneity condition for particle aggregates*. It is shown that the proposed

R. de Borst and E. Ramm (eds.), *Multiscale Methods in Computational Mechanics*, Lecture Notes in Applied and Computational Mechanics 55, DOI 10.1007/978-90-481-9809-2_9,

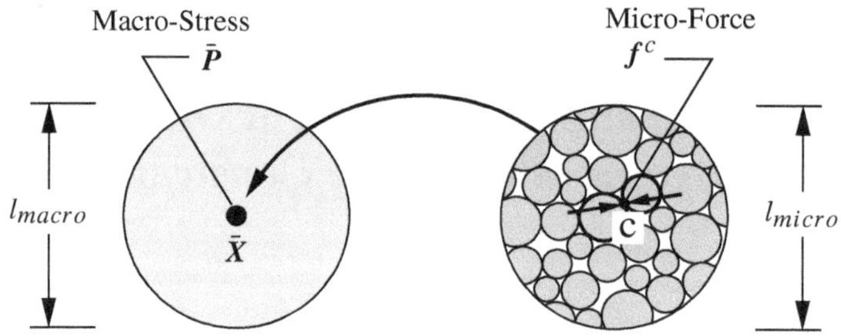

Fig. 1. The notion of a continuum with granular micro-structure. Associated with a typical point $\bar{X} \in \bar{\mathcal{V}}$ of a homogenized macro-continuum $\bar{\mathcal{V}} \subset \mathcal{R}^3$ is a micro-structure $\mathcal{V} \subset \mathcal{R}^3$ that characterizes a cell of an aggregate of solid particles which possibly come into contact. A distinct averaging of the contact micro-forces f^c in $\mathcal{V}$ determines the overall macro-stress $\bar{P}$ at $\bar{X} \in \bar{\mathcal{V}}$.

three types $I = D, P, S$ of boundary conditions satisfy a priori this condition. Condition (D) defines linear center translations and zero rotations of particles at the boundary frame, providing a *stiff* microstructural response and an *upper bound* of the aggregate's stiffness. Condition (S) defines uniform center forces and free rotations of particles at the boundary frame, yielding a *soft* microstructural response and a *lower bound* of the stiffness. The stiffness obtained for periodic boundary conditions (P) lies between these bounds. The theoretical investigations and numerical experiments underline the result of Miehe and Dettmar [16], who considered the periodicity conditions (P) to be an optimal choice. They are expected to yield the most realistic microstructural stiffness for coarse microstructures.

Following conceptually the recent work by Miehe and Bayreuther [15], we outline a unified implementation of the boundary constraints by penalty methods. This defines for the three classes $I = D, P, S$ of *stiff, periodic* and *soft* constraints microstructural support forces and couples. It is shown that the support forces define the overall nominal stress of the aggregate. The representations (28), (29) and (30) allow a unified and simple implementation of deformation-driven homogenization in particle codes based on the discrete element technique (DE). Of particular importance is the non-standard implementation of the soft condition (S) in the deformation-driven scenario. It allows a comparative study of the overall responses for $I = D, P, S$ in deformation-driven processes. In the numerical studies, we consider micromechanical models of particle interaction for two- and three-dimensional spherical granules. The quasi-static response of the granular microstructures is realized by applying a dynamic relaxation technique in typical load increments. We demonstrate the performance of the proposed method by a full two-scale analysis of a granular medium with discrete-to-finite element (DE-FE) coupling, where the proposed homogenization analysis is applied at typical integration points of a discretized macro-structure.

2 Quasi-Static Homogenization of Granular Aggregates

2.1 Deformation-Driven Homogenization of Microstructures

Definition of Particle Microstructures

Figure 2 shows deformations of a typical granular microstructure, where *driving frames* are highlighted. The microstructural domain $\mathcal{V} \subset \mathcal{R}^3$ is filled by a representative aggregate of particles. Its definition is not restricted to rectangular shapes or those with straight edges. The total number of particles of the representative volume $|\mathcal{V}|$ is decomposed into M particles that represent the *boundary frame* of the assembly and the remaining N particles in the *interior domain*. The centroids of the particles $\mathcal{P}_p$, $p = 1 \dots N$ in the interior domain have the positions

$$\boldsymbol{X}_p \in \mathcal{V} \quad \text{for} \quad p = 1 \dots N \tag{1}$$

in the reference configuration. The centroids of the particles $\mathcal{P}_q$, $q = 1 \dots M$ of the boundary frame, those that are dark shaded in Figures 2 and 3, have the positions

$$\boldsymbol{X}_q \in \partial\mathcal{V} \quad \text{for} \quad q = 1 \dots M \, . \tag{2}$$

They define a quasi-surface of the granular microstructure and provide a frame for the macroscopic driving of the granular aggregate in the interior domain of the representative elementary volume. For a precise definition of the *special case of periodic microstructures*, we refer to the work by Miehe and Dettmar [16]. The kinematics of rigid particles is governed by the particle deformation maps

$$\boldsymbol{\varphi}_i(\boldsymbol{X}, t) = \boldsymbol{x}_i(t) + \boldsymbol{q}_i(t)[\boldsymbol{X} - \boldsymbol{X}_i] \quad \text{for} \quad i = 1 \dots N + M \, , \tag{3}$$

in terms of the *particle center translation* $\boldsymbol{x}_i(t)$ and the *particle rotation* $\boldsymbol{q}_i(t) \in SO(3)$, e.g. parametrized by Euler angles. The motion of the particles is driven by contact forces as commented on in Section 3 below. In what follows, we derive a framework for homogenization of particle aggregates that provides a straightforward modification of the formulations of computational homogenization for continuous microstructures, such as outlined in [14]. In particular, it may be viewed as a *discrete counterpart* of the continuous setting. This discretization is conceptually performed by the transition

$$\int_{\partial\mathcal{V}_q} \boldsymbol{N} \, dA \to \boldsymbol{A}_q \quad \text{and} \quad \int_{\partial\mathcal{V}_q} \boldsymbol{t} \, dA \to \boldsymbol{a}_q \quad \text{for} \quad q = 1 \dots M \tag{4}$$

of infinitesimal area vectors to *finite area vectors* $\boldsymbol{A}_q$ and infinitesimal forces to *finite forces* $\boldsymbol{a}_q$ at the center points $q = 1 \dots M$ of the boundary frame. Here, $\partial\mathcal{V}_q \subset \partial\mathcal{V}$ is the part of the surface associated with the discrete center point q, see Figure 2.

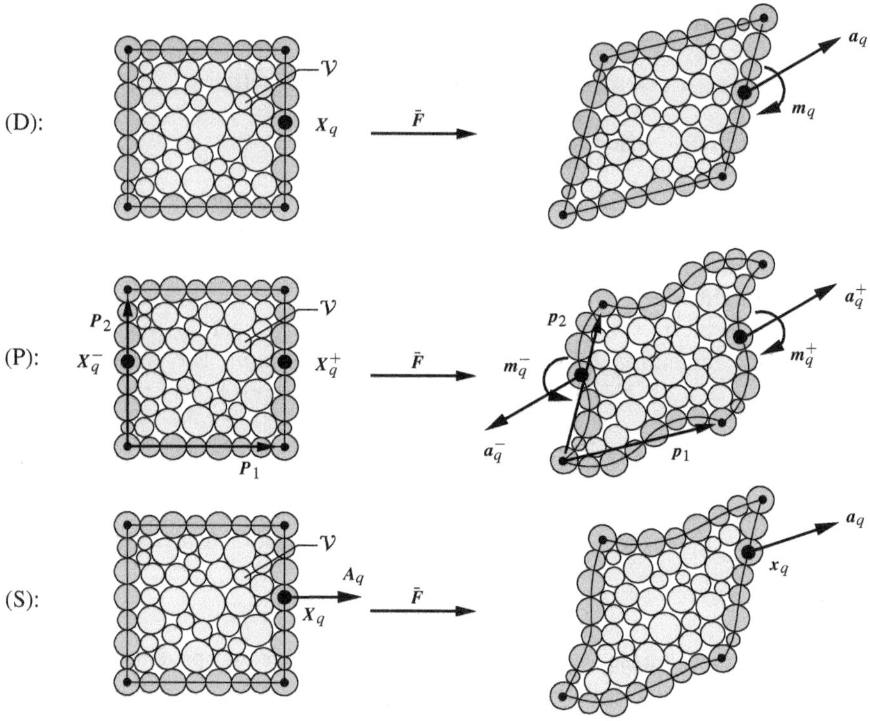

Fig. 2. Three types $I = D, P, S$ of boundary constraints acting on a granular microstructure, which are consistent with a Hill-type microheterogeneity condition for particle aggregates. Fictitious support forces $\boldsymbol{a}_q$ and couples $\boldsymbol{m}_q$ are defined by the penalty terms (28) and (29) for linear displacements/zero rotations (D), periodic displacements/rotations and uniform tractions/free rotations (S) at the boundary frame, yielding the overall stress $\bar{\boldsymbol{P}}$ of the aggregate in (30).

Microscopic Boundary Conditions

The basic difficulty of the transition from continuous to particle microstructures is the non-consistency of the microscopic particle rotations with a homogenized standard macro-continuum under consideration. This non-consistency can be resolved by making particular assumptions with respect to the rotations of the particles at the boundary frame. In what follows, we propose *stiff* and *soft* microscopic boundary constraints (D) and (S) as well as *periodic* conditions (P). These assumptions affect the stiffness of the particle aggregate and are considered to define reasonable bounds for its overall stiffness. A first definition in terms of a prescribed macroscopic deformation gradient $\bar{\boldsymbol{F}}(t)$ assumes zero fluctuations of particle center translations and particle rotations on the boundary, i.e.

$$(D): \quad \boldsymbol{x}_q = \bar{\boldsymbol{F}}\boldsymbol{X}_q \quad \text{and} \quad \boldsymbol{q}_q = \boldsymbol{1} \quad \text{for} \quad q = 1 \ldots M\,. \tag{5}$$

These *stiff boundary conditions* require that the particles at the boundary frame do not rotate. Clearly, this constraint yields an unequal distribution of support couples

$$\boldsymbol{m}_q \neq \boldsymbol{0} \quad \text{for} \quad q = 1 \dots M \tag{6}$$

on the boundary frame as indicated in Figure 2a. A non-trivial periodicity of the deformation of the granular aggregate may be described by periodic fluctuations of the particle center translations accompanied by periodic particle rotations on the boundary, i.e.

$$(P): \quad [\![\boldsymbol{x}_q]\!] = \bar{\boldsymbol{F}}\,[\![\boldsymbol{X}_q]\!] \quad \text{and} \quad [\![\boldsymbol{q}_q]\!] = \boldsymbol{0} \quad \text{for} \quad q = 1 \dots M^+ , \tag{7}$$

where $M^+ = M/2$. These *periodicity constraints* for granular materials have been investigated in [16]. Consistent with (7) are for periodic aggregates the anti-periodicity of the particle support forces and couples

$$\langle \boldsymbol{a}_q \rangle = \boldsymbol{0} \quad \text{and} \quad \langle \boldsymbol{m}_q \rangle = \boldsymbol{0} \quad \text{for} \quad q = 1 \dots M^+ , \tag{8}$$

on the boundary frame as indicated in Figure 2b. Finally, a traction-type *soft boundary constraint* for particle aggregates is proposed to be of the form

$$\boldsymbol{a}_q = \bar{\boldsymbol{P}} \boldsymbol{A}_q \quad \text{for} \quad q = 1 \dots M \tag{9}$$

as the dual counterpart to $(5)_1$. It determines for given macroscopic stress $\bar{\boldsymbol{P}}$ the discrete support force $\boldsymbol{a}_q$ in terms of the discrete area vector $\boldsymbol{A}_q$ introduced in (4). In order to obtain a formulation equivalent to (9) in terms of a prescribed macroscopic deformation gradient $\bar{\boldsymbol{F}}$, we define the constraint

$$(S): \quad \bar{\boldsymbol{F}} = \frac{1}{|\mathcal{V}|} \sum_{q=1}^{M} \boldsymbol{x}_q \otimes \boldsymbol{A}_q , \tag{10}$$

on the particle center translations $\boldsymbol{x}_q$ at the boundary frame, which is considered as a weak counterpart to (9). Scalar contraction of this equation with a Lagrange parameter $\boldsymbol{\lambda}(t)$ induces the global constraint $c(\{\boldsymbol{x}_q\}_{q=1\dots M}, \boldsymbol{\lambda}; \bar{\boldsymbol{F}}) = -\frac{1}{|\mathcal{V}|} \sum_{q=1}^{M} (\boldsymbol{\lambda} \cdot \boldsymbol{A}_q) \cdot \boldsymbol{x}_q + \boldsymbol{\lambda} : \bar{\boldsymbol{F}} = 0$. Taking the derivative with respect to the macro-deformation $\bar{\boldsymbol{F}}$ identifies the Lagrange parameter by the macro-stress $\bar{\boldsymbol{P}} := \partial_{\bar{\boldsymbol{F}}} c = \boldsymbol{\lambda}$. The variation of the constraint gives the surface term of prescribed forces $\delta c = -\frac{1}{|\mathcal{V}|} \sum_{q=1}^{M} (\boldsymbol{\lambda} \cdot \boldsymbol{A}_q) \cdot \delta \boldsymbol{x}_q$. Thus the weak constraint in equation (10) enforces the force-type boundary condition (9) for prescribed macro-deformations $\bar{\boldsymbol{F}}(t)$. Observe that no constraints on the particle rotations were introduced. Thus the particles are free to rotate according to

$$\boldsymbol{m}_q = \boldsymbol{0} \quad \text{for} \quad q = 1 \dots M \tag{11}$$

as indicated in Figure 2c, providing the softest response of the driving frame.

Microscopic Equilibrium State

For the above defined discrete granular microstructures, where the particles split into those on the boundary frame and those in the bulk, three types of equilibrium conditions may be defined: (i) for particles in the interior domain, (ii) for particles which build the boundary frame and finally (iii) for the total aggregate. This is indicated in Figure 3. The *particles $\mathcal{P}_p$ in the interior domain* of the microstructure interact among each other through discrete contact forces $\boldsymbol{f}_p^c$ at discrete contact points $\boldsymbol{x}_p^c$ on the surfaces of the rigid particles, as shown in Figure 3b. The equilibrium conditions read

$$\sum_{c\in\mathcal{A}_p^c} \boldsymbol{f}_p^c = \boldsymbol{0} \quad \text{and} \quad \sum_{c\in\mathcal{A}_p^c} (\boldsymbol{x}_p^c - \boldsymbol{x}_p) \times \boldsymbol{f}_p^c = \boldsymbol{0} \quad \text{for} \quad p = 1 \ldots N\,. \qquad (12)$$

Note that the symbol $\mathcal{A}_p^c$ denotes the currently active set of contacts for the particular particle $\mathcal{P}_p$. Clearly, the above equations represent the local equilibrium conditions for both forces and couples on each particle $\mathcal{P}_p$, $p = 1, \ldots, N$ with regard to their displaced centroids $\boldsymbol{x}_p$. The *particles $\mathcal{P}_q$ on the defined boundary frame* of the granular assembly interact with those in the interior domain through discrete contact forces $\boldsymbol{f}_q^c$ at contact points $\boldsymbol{x}_q^c$ on the deformed surfaces of the rigid particles. The driving of these particles is performed by *fictitious support forces $\boldsymbol{a}_q$* and *fictitious support couples $\boldsymbol{m}_q$* that act on the displaced and rotated centroids of the boundary frame particles $\mathcal{P}_q$, see Figure 3a. In analogy to equation (12), the discrete equilibrium conditions for the particles on the boundary read

$$\sum_{c\in\mathcal{A}_q^c} \boldsymbol{f}_q^c = -\boldsymbol{a}_q \quad \text{and} \quad \sum_{c\in\mathcal{A}_q^c} (\boldsymbol{x}_q^c - \boldsymbol{x}_q) \times \boldsymbol{f}_q^c = -\boldsymbol{m}_q \quad \text{for} \quad q = 1 \ldots M\,. \qquad (13)$$

These equilibrium conditions are defined with regard to the displaced and rotated centroids $\boldsymbol{x}_q$ of the particles $\mathcal{P}_q, q = 1 \ldots M$ on the boundary frame. In the above equation (13) the contact forces $\boldsymbol{f}_q^c$ that act between the boundary- and the bulk particles are considered to be *real forces*, while the externally applied support forces and couples $\boldsymbol{a}_q$ and $\boldsymbol{m}_q$ were introduced as artificial quantities. These forces are considered to be *fictitious forces*, which replace the mechanical action of the adjacent particle microstructures. Clearly, the equilibrium conditions for the whole particle aggregate can be expressed solely in terms of the support forces $\boldsymbol{a}_q$ and couples $\boldsymbol{m}_q$ as

$$\sum_{q=1}^{M} \boldsymbol{a}_q = \boldsymbol{0} \quad \text{and} \quad \sum_{q=1}^{M} (\boldsymbol{x}_q \times \boldsymbol{a}_q + \boldsymbol{m}_q) = \boldsymbol{0}\,. \qquad (14)$$

Obviously, all interior contact forces drop out and the equilibrium conditions appear solely in terms of microscopic surface data.

Macroscopic Equilibrium State

Let $\bar{\boldsymbol{\varphi}}_t : \bar{\mathcal{V}} \to \mathcal{R}^3$ denote the deformation map of the homogenized macrostructure $\bar{\mathcal{V}} \subset \mathcal{R}^3$ at time $t \in \mathcal{R}$, which is considered as a standard continuum. A static

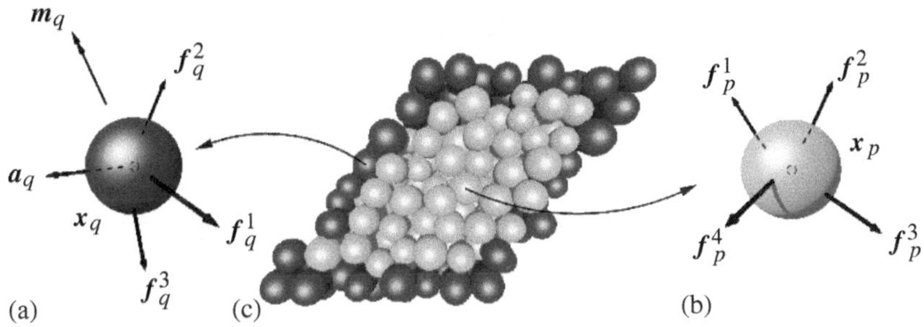

Fig. 3. Quasi-static equilibrium of particle aggregates: (a) for particles at boundary frame with support forces $\boldsymbol{a}_q$ and couples $\boldsymbol{m}_q$, (b) for granules in the interior domain, and (c) for total granular microstructure.

equilibrium state of this macro-continuum is governed by the field equations

$$\operatorname{tr}[\nabla_{\bar{X}} \bar{\boldsymbol{P}}] = \boldsymbol{0} \quad \text{and} \quad \bar{\boldsymbol{F}} \cdot \bar{\boldsymbol{P}}^T - \bar{\boldsymbol{P}} \cdot \bar{\boldsymbol{F}}^T = \boldsymbol{0} \tag{15}$$

in $\bar{\mathcal{V}}$. Here, the macroscopic stress tensor $\bar{\boldsymbol{P}}$ is related to the macroscopic deformation gradient $\bar{\boldsymbol{F}} := \nabla_{\bar{X}} \bar{\boldsymbol{\varphi}}_t(\bar{\boldsymbol{X}})$ and determined by a homogenization procedure of contact forces in the granular microstructure. We define the nominal macro-stress of the particle aggregate by

$$\bar{\boldsymbol{P}} = \frac{1}{|\mathcal{V}|} \sum_{q=1}^{M} \boldsymbol{a}_q \otimes \boldsymbol{X}_q \tag{16}$$

in terms of the support forces $\boldsymbol{a}_q$ at the boundary frame. Recall that these support forces are determined by the microscopic equilibrium conditions (12), (13) and (14) of the granular aggregate including the microscopic contact interactions of the particles. Clearly, the overall response of the microstructure is also influenced by the microscopic boundary conditions (5), (7) and (10). In line with the definition of (16), we postulate a Hill–Mandel-type microheterogeneity condition for granular particle assemblies by

$$\bar{\boldsymbol{P}} : \dot{\bar{\boldsymbol{F}}} = \frac{1}{|\mathcal{V}|} \sum_{q=1}^{M} \boldsymbol{a}_q \cdot \dot{\boldsymbol{x}}_q \tag{17}$$

in terms of the fictitious support forces $\boldsymbol{a}_q$ on the boundary frame. Rewriting this identity in the format

$$\frac{1}{|\mathcal{V}|} \sum_{q=1}^{M} \boldsymbol{a}_q \cdot \dot{\boldsymbol{x}}_q - \bar{\boldsymbol{P}} : \dot{\bar{\boldsymbol{F}}} - \bar{\boldsymbol{P}} : \dot{\bar{\boldsymbol{F}}} + \bar{\boldsymbol{P}} : \dot{\bar{\boldsymbol{F}}} = 0$$

and inserting the alternative representations

$$\bar{\boldsymbol{P}} : \dot{\bar{\boldsymbol{F}}} = \frac{1}{|\mathcal{V}|} \sum_{q=1}^{M} (\bar{\boldsymbol{P}} \cdot \boldsymbol{A}_q) \cdot \dot{\boldsymbol{x}}_q = \frac{1}{|\mathcal{V}|} \sum_{q=1}^{M} \boldsymbol{a}_q \cdot (\dot{\bar{\boldsymbol{F}}} \cdot \boldsymbol{X}_q) = \frac{1}{|\mathcal{V}|} \sum_{q=1}^{M} (\bar{\boldsymbol{P}} \cdot \boldsymbol{A}_q) \cdot (\dot{\bar{\boldsymbol{F}}} \cdot \boldsymbol{X}_q)$$

based on (10) and (16), the above Hill-type condition can be recast into the statement

$$\frac{1}{|\mathcal{V}|}\sum_{q=1}^{M}(\boldsymbol{a}_q - \bar{\boldsymbol{P}}\cdot\boldsymbol{A}_q)\cdot(\dot{\boldsymbol{x}}_q - \dot{\bar{\boldsymbol{F}}}\cdot\boldsymbol{X}_q) = \boldsymbol{0}\,. \tag{18}$$

Obviously, this constraint is satisfied by the above outlined three classes of microstructural boundary conditions (5), (7) and (10) of the particle aggregate, which characterize the above outlined homogenization methods to be *consistent* with the Hill–Mandel-type condition (17).

With known macroscopic nominal stress tensor $\bar{\boldsymbol{P}}$ we may obtain other macroscopic stress measures based on standard geometric relationships, such as the Kirchhoff and Cauchy stresses

$$\bar{\boldsymbol{\tau}} = \bar{\boldsymbol{P}}\bar{\boldsymbol{F}}^T \quad \text{and} \quad \bar{\boldsymbol{\sigma}} = \bar{\boldsymbol{\tau}}/\det[\bar{\boldsymbol{F}}]\,. \tag{19}$$

Here, the determination of $\bar{\boldsymbol{F}}$ governs the volume change $|v| = \det[\bar{\boldsymbol{F}}]\ |\mathcal{V}|$ of the granular microstructure. Insertion of (16) into (19) identifies the (true) Cauchy stress of the aggregate

$$\bar{\boldsymbol{\sigma}} = \frac{1}{|v|}\sum_{q=1}^{M}\boldsymbol{a}_q \otimes (\bar{\boldsymbol{F}}\boldsymbol{X}_q). \tag{20}$$

It has already been shown in [16] that this expression is *symmetric for periodic boundary conditions* (P). Setting $\bar{\boldsymbol{F}}\boldsymbol{X}_q = \boldsymbol{x}_q - \boldsymbol{w}_q$ with periodic fluctuations $[\![\boldsymbol{w}_q]\!] = \boldsymbol{0}$ and taking into account anti-periodic support forces $\langle\boldsymbol{a}_q\rangle = \boldsymbol{0}$, we may recast (20) for the periodicity condition into the form

$$\bar{\boldsymbol{\sigma}} = \frac{1}{|v|}\sum_{q=1}^{M}\boldsymbol{a}_q \otimes \boldsymbol{x}_q \tag{21}$$

in terms of the current position $\boldsymbol{x}_q$ at the boundary. For the equilibrium condition (14) it is clear that $\sum_{q=1}^{M}\boldsymbol{a}_q \otimes \boldsymbol{x}_q$ is *symmetric* due to the anti-periodic support couples $\langle\boldsymbol{m}_q\rangle = \boldsymbol{0}$ as stated in $(8)_2$. Hence, for the modeling of true periodicity, the support couples play a crucial role. The symmetry cannot be proved for the stiff and soft boundary conditions (D) and (S), respectively. Here, the proposed homogenization technique does not necessarily give a symmetric macro-stress. However, the skew part of the stress is considered to be a *boundary-condition caused effect*. This is consistent with observations made by Bardet and Vardoulakis [2], who for certain non-periodic boundary conditions characterized the stress asymmetry in granular materials as a boundary effect. Clearly, symmetry may a posteriori be enforced in these situations by a *symmetrization* of the Kirchhoff stress, inducing the modification

$$\bar{\boldsymbol{P}} \Leftarrow \text{sym}[\bar{\boldsymbol{P}}\bar{\boldsymbol{F}}^T]\bar{\boldsymbol{F}}^{-T} \tag{22}$$

of the first Piola (nominal) stress. Then, $\bar{\boldsymbol{\tau}}$ and $\bar{\boldsymbol{\sigma}}$ computed by (19) are symmetric.

Note carefully that the particle rotations do not directly enter the above definition of overall stresses and microheterogeneity. This is due to the fact that the macrostructure is a priori defined to be a standard continuum without rotational degrees. The presence of a possible asymmetry of the Cauchy stresses (20) for the boundary constraints (D) and (S) is a *boundary effect* whose influence decreases for increasingly large numbers of particles in the interior domain, i.e. $M/N \to 0$. It does not motivate higher order macro-theories, where couple stresses appear as independent fields for which additional constitutive assumptions are needed, such as considered for example in [4,5,10,20] or [8]. This is underlined by the fact that support couples do not violate the symmetry of the macroscopic Cauchy stresses (21) in the case of periodic boundary constraints.

Macroscopic Boundary Conditions

Boundary conditions of the homogenized macro-continuum are defined in a standard manner, consisting of Dirichlet-type and Neumann-type conditions of the form

$$\bar{\varphi} = \bar{\varphi}_D \;\text{ on } \partial\bar{\mathcal{V}}_\varphi \qquad \text{and} \qquad \bar{\boldsymbol{P}} \cdot \bar{\boldsymbol{N}} = \bar{\boldsymbol{t}}_N \;\text{ on } \partial\bar{\mathcal{V}}_t \,, \tag{23}$$

where the surface $\partial\bar{\mathcal{V}}$ of the macro-continuum with outward normal $\bar{\boldsymbol{N}}$ is decomposed according to $\partial\bar{\mathcal{V}} = \partial\bar{\mathcal{V}}_\varphi \cup \partial\bar{\mathcal{V}}_t$.

2.2 Penalty-Type Implementation of Boundary Constraints

We now outline details of the implementation of the different boundary conditions for the particle microstructure. The method is motivated by a fictitious elastic response of the particle interaction governed by a variational principle with an overall potential. Details of the variational based implementation of continuous microstructures were recently outlined in [15] based on Lagrange and penalty methods. Here, we conceptually apply these formulations to particle aggregates. Assuming an fictitious elastic response of the microstructure governed by an overall potential Π, we start from a variational problem

$$\Pi(\{\boldsymbol{x}_i\}_{i=1}^{N+M}, \{\boldsymbol{q}_i\}_{i=1}^{N+M}) \to \text{Min!}\,, \tag{24}$$

subject to the microstructural boundary conditions (D), (P) or (S) defined in (5), (7) and (10). This variational principle is assumed to determine the center translations $\{\boldsymbol{x}_i\}_{i=1}^{N+M}$ and the rotations $\{\boldsymbol{q}_i\}_{i=1}^{N+M}$ of the granules, for a prescribed macroscopic deformation $\bar{\boldsymbol{F}}$ as schematically visualized in Figure 2. An implementation of the boundary constraints by a penalty method is determined by the variational principle

$$\bar{W}_I^\epsilon(\bar{\boldsymbol{F}}) = \inf_{\boldsymbol{x}_i, \boldsymbol{q}_i} \Big[\, \Pi(\{\boldsymbol{x}_i\}_{i=1}^{N+M}, \{\boldsymbol{q}_i\}_{i=1}^{N+M}) - p_I(\{\boldsymbol{x}_q\}_{q=1}^{M}, \{\boldsymbol{q}_q\}_{q=1}^{M}); \bar{\boldsymbol{F}}) \,\Big] \tag{25}$$

in terms of the penalty terms p_I for $I = D, P, S$. For the assumed elastic response, $\bar{W}_I^\epsilon(\bar{\boldsymbol{F}})$ is considered as a potential for the homogenized nominal stresses $\bar{\boldsymbol{P}}$. This characteristic feature of constraint variational frameworks of homogenization was outlined in [15]. Hence, for the equilibrium state consistent with the above minimization, we have

$$\bar{\boldsymbol{P}}_I^{\epsilon} = \partial_{\bar{\boldsymbol{F}}} \bar{W}_I^{\epsilon}(\bar{\boldsymbol{F}}) = -\partial_{\bar{\boldsymbol{F}}} p_I(\{\boldsymbol{x}_q\}_{q=1}^M, \{\boldsymbol{q}_q\}_{q=1}^M); \bar{\boldsymbol{F}}) . \tag{26}$$

Thus, the overall macro-stress is obtained by taking the derivative of the penalty terms with regard to the driving macroscopic deformation gradient $\bar{\boldsymbol{F}}$. For the above introduced boundary constraints, we introduce the quadratic penalty functions

$$p_I = \begin{cases} \dfrac{1}{2|\mathcal{V}|} \displaystyle\sum_{q=1}^{M} \{ \epsilon_f | \boldsymbol{x}_q - \bar{\boldsymbol{F}} \boldsymbol{X}_q |^2 + \epsilon_c | \boldsymbol{q}_q - \boldsymbol{1} |^2 \} , \\ \dfrac{1}{2|\mathcal{V}|} \displaystyle\sum_{q=1}^{M^+} \{ \epsilon_f | [\![\boldsymbol{x}_q]\!] - \bar{\boldsymbol{F}} [\![\boldsymbol{X}_q]\!] |^2 + \epsilon_c | [\![\boldsymbol{q}_q]\!] |^2 \} , \\ \dfrac{\epsilon_f}{2} \left| \dfrac{1}{|\mathcal{V}|} \displaystyle\sum_{q=1}^{M} \boldsymbol{x}_q \otimes \boldsymbol{A}_q - \bar{\boldsymbol{F}} \right|^2 . \end{cases} \tag{27}$$

The penalty parameters ϵ_f and ϵ_c affect the translational and rotational parts of the constraints, respectively. Note that the penalty method induces *support forces and couples* at the boundary frame of the microstructure. The *support forces* are defined by taking the derivative of the penalty term with regard to the center translations $\boldsymbol{a}_q^I := -\partial_{\boldsymbol{x}_q} p_I(\{\boldsymbol{x}_q\}_{q=1}^M, \{\boldsymbol{q}_q\}_{q=1}^M); \bar{\boldsymbol{F}})$, yielding

$$\boldsymbol{a}_q^I := \begin{cases} -\dfrac{\epsilon_f}{|\mathcal{V}|} (\boldsymbol{x}_q - \bar{\boldsymbol{F}} \boldsymbol{X}_q) , \\ \mp \dfrac{\epsilon_f}{|\mathcal{V}|} ([\![\boldsymbol{x}_q]\!] - \bar{\boldsymbol{F}} [\![\boldsymbol{X}_q]\!]) , \\ -\epsilon_f \left(\dfrac{1}{|\mathcal{V}|} \displaystyle\sum_{q=1}^{M} (\boldsymbol{A}_q \cdot \boldsymbol{A}_q) \boldsymbol{x}_q - \bar{\boldsymbol{F}} \boldsymbol{A}_q \right) . \end{cases} \tag{28}$$

The *support couples* defined follow by taking the derivative of the penalty terms with regard to the particle rotations $\boldsymbol{m}_q^I := -\partial_{\boldsymbol{q}_q} p_I(\{\boldsymbol{x}_q\}_{q=1}^M, \{\boldsymbol{q}_q\}_{q=1}^M); \bar{\boldsymbol{F}})$, yielding

$$\boldsymbol{m}_q^I := \begin{cases} -\dfrac{\epsilon_c}{|\mathcal{V}|} (\boldsymbol{q}_q - \boldsymbol{1}) , \\ \mp \dfrac{\epsilon_c}{|\mathcal{V}|} [\![\boldsymbol{q}_q]\!] , \\ \boldsymbol{0} \end{cases} \tag{29}$$

for the different boundary conditions $I = D, P, S$. These support forces and couples enforce for large but finite values of the penalty parameters ϵ_f and ϵ_c the different boundary conditions. Finally, the exploitation of the potential relationship (26) gives representations of the macroscopic overall stresses of the particle aggregate for the penalty formulation under consideration

$$\bar{\boldsymbol{P}}_I^\epsilon = \begin{cases} \dfrac{1}{|\mathcal{V}|} \displaystyle\sum_{q=1}^{M} \epsilon_f \left(\boldsymbol{x}_q - \bar{\boldsymbol{F}} \boldsymbol{X}_q \right) \otimes \boldsymbol{X}_q \,, \\ \dfrac{1}{|\mathcal{V}|} \displaystyle\sum_{q=1}^{M^+} \epsilon_f \left([\![\boldsymbol{x}_q]\!] - \bar{\boldsymbol{F}} [\![\boldsymbol{X}_q]\!] \right) \otimes [\![\boldsymbol{X}_q]\!] \,, \\ \epsilon_f \left(\dfrac{1}{|\mathcal{V}|} \displaystyle\sum_{q=1}^{M} \boldsymbol{x}_q \otimes \boldsymbol{A}_q - \bar{\boldsymbol{F}} \right) . \end{cases} \tag{30}$$

This summarizes the proposed unified implementation of the boundary constraints of the microstructure and the evaluation of the overall stresses by a penalty method. The formulations for the support forces and couples (28) and (29) as well as the definition of overall stresses (30) are in what follows also applied to *dissipative microstructures without an underlying variational principle.* This provides a simple implementation of all three constraints into discrete particle codes. The implementation of the Dirichlet-type constraints $I = D, P$ can alternatively be performed in a straightforward manner in displacement-type codes as conceptually outlined in [18]. However, the implementation of the soft constraint $I = S$ needs the above formulation.

3 Microstructural Modeling of Granular Materials

3.1 Micromechanical Model for Interparticle Contact

Typical micromechanical models consider two contacting particles in the current configuration as visualized in Figure 4, referred to as *master* $\mathcal{P}_M$ and *slave* $\mathcal{P}_S$ *particles*. They are governed by contact-*normal-force* and frictional-*tangential-force* mechanisms. Here, we do not point out details of those models. In the subsequent numerical simulations, we use the quasi-static frictional contact model outlined in [16] for spherical particles and in [17] for elliptical particles. The micromechanical model of two contacting master and slave particles depicted in Figure 4 is governed by a *generalized displacement vector*, containing the particle center translations and rotations, and a *generalized force vector*, containing the particle center forces and couples

$$\boldsymbol{d}^c := [\, \boldsymbol{x}_M \; \boldsymbol{\vartheta}_M \; \boldsymbol{x}_S \; \boldsymbol{\vartheta}_S \,]^T \quad \text{and} \quad \boldsymbol{p}^c := [\, \boldsymbol{f}_M \; \boldsymbol{m}_M \; \boldsymbol{f}_S \; \boldsymbol{m}_S \,]^T \,, \tag{31}$$

where $c \in \mathcal{A}(t) := \{c := (M, S) \,|\, \mathcal{P}_M \cap \mathcal{P}_S$ denotes an element of the currently active set of contact points in the particle aggregate. Here, we have parametrized the particle rotations $\boldsymbol{q}_i(t) \in SO(3)$ by Euler angles $\boldsymbol{\vartheta}_i(t) \in \mathcal{R}^3$. The micromechanical constitutive model

$$\boldsymbol{p}^c = \hat{\boldsymbol{p}}^c(\boldsymbol{d}^c, \dot{\boldsymbol{d}}^c; \mathcal{H}^c) \tag{33}$$

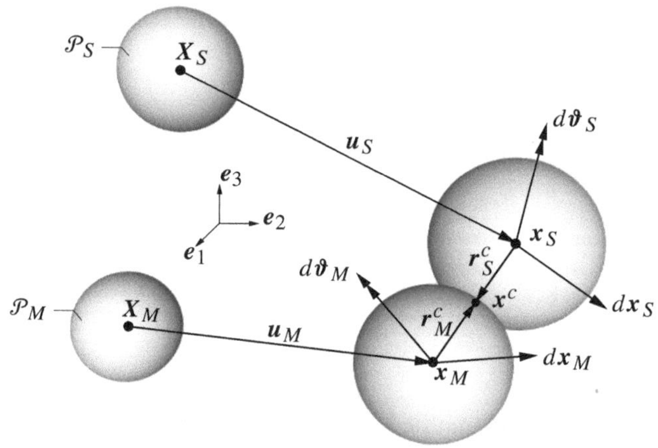

Fig. 4. Micromechanical particle contact: Master and slave particles $\mathcal{P}_M$ and $\mathcal{P}_S$ at current positions $(\boldsymbol{x}_M, \boldsymbol{\vartheta}_M)$ and $(\boldsymbol{x}_S, \boldsymbol{\vartheta}_S)$, respectively. The incremental slip governing the frictional contact force at $\boldsymbol{x}^c$ is determined by the increments $(d\boldsymbol{x}_M, d\boldsymbol{x}_S)$ and $(d\boldsymbol{\vartheta}_M, d\boldsymbol{\vartheta}_S)$.

defines the generalized force vector $\boldsymbol{p}^c$ in terms of the current generalized displacement vector $\boldsymbol{d}^c$, its rate $\dot{\boldsymbol{d}}^c$ and a set of history variables $\mathcal{H}^c$ at the contact point. It is defined in terms of the contact normal and tangential forces P^c and T^c, respectively, i.e. $\boldsymbol{f}_M = P^c\boldsymbol{n}+T^c\boldsymbol{t}$ and $\boldsymbol{m}_M = \boldsymbol{r}^c \times (P^c\boldsymbol{n}+T^c\boldsymbol{t})$. Details of the algorithmic formulation of the model for finite time increments are outlined in [16]. This micromechanical model is described by three fundamental micro-force-potential functions $\psi_p(\varepsilon_e^c)$, $\psi_t(\gamma_e^c)$, $\psi_h(\alpha)$ and a Coulomb-type slip criterion function $\phi(T^c, P^c, \beta)$.

3.2 Dynamic Relaxation of the Microstructural Response

Problems inherent in the above outlined quasi-static approach to granular microstructures are the multiple *stability problems* in particle clusters and the *loss of contacts* of individual particles or clusters of particles. An artificial dynamic relaxation analysis circumvents these problems, see for example [1,21]. In this section we consider an explicit numerical model for the dynamic regularization of the quasi-static analysis of granular assemblies, where both global and local artificial damping mechanisms are introduced.

A dynamic generalization of the quasi-static equilibrium equations in a typical load increment $[t_n, t_{n+1}]$ of the microstructure assumes the form

$$\mathfrak{m}\ddot{\mathfrak{d}}(T) + \mathfrak{c}\dot{\mathfrak{d}}(T) + \mathfrak{p}(\mathfrak{d}(T), state(t_n)) = \mathbf{0}\,. \tag{34}$$

$\mathfrak{d}(T)$ and $\mathfrak{p}(T)$ converge within the relaxation analysis to the quasi-static values $\boldsymbol{d}(t_{n+1})$ and $\boldsymbol{p}(t_{n+1})$. Here, $\boldsymbol{d} := \mathsf{A}_{c\in\mathcal{A}}\boldsymbol{d}^c$ is the global generalized displacement array and $\boldsymbol{p} := \mathsf{A}_{c\in\mathcal{A}}\boldsymbol{p}^c$ the global generalized force array, obtained form their local counterparts (31) by a standard assembly procedure. $state(t_n)$ denotes the history variables at t_n. $\mathfrak{m}$ and $\mathfrak{c}$ are artificial global mass and damping matrices, respectively. They are assumed to be diagonal with contributions

$$\mathfrak{m}^p := \text{diag}[\ m_p, m_p, \theta_p\ ] \quad \text{and} \qquad \mathfrak{c}^p := \alpha_g \mathfrak{m}^p\ , \tag{35}$$

where m_p represents a fictitious mass and θ_p a fictitious mass moment of inertia of the particle p. They are computed with a fictitious particle density ρ. The damping matrix is assumed to be related to the mass matrix by the parameter α_g. The artificial *initial-boundary value problem of particle motion* within the relaxation analysis of a typical load driving step $[t_n, t_{n+1}]$ is parametrized by the fictitious time T. It consists of the dynamic equations (34), the initial conditions

$$\mathfrak{d}(T=0) = \boldsymbol{d}(t_n) \quad \text{and} \qquad \dot{\mathfrak{d}}(T=0) = \dot{\boldsymbol{d}}(t_n) \tag{36}$$

for the generalized particle displacement and the alternative penalty-type boundary constraints (28) and (29) for support forces and couples at the boundary frame. The discretization of the dynamic system (34) for constant relaxation time steps $\Delta T := T_{k+1} - T_k$ is performed by a standard central difference scheme. Here, the acceleration and velocity at time T_k are approximated by

$$\dot{\mathfrak{d}}_k = \frac{1}{2}[\dot{\mathfrak{d}}_{k+\frac{1}{2}} + \dot{\mathfrak{d}}_{k-\frac{1}{2}}] \quad \text{and} \quad \ddot{\mathfrak{d}}_k = \frac{1}{\Delta T}[\dot{\mathfrak{d}}_{k+\frac{1}{2}} - \dot{\mathfrak{d}}_{k-\frac{1}{2}}]\ , \tag{37}$$

where $T_{k-\frac{1}{2}}$ and $T_{k+\frac{1}{2}}$ refer to time instances in the middle of the previous and actual relaxation steps $[T_{k-1},\ T_k]$ and $[T_k,\ T_{k+1}]$, respectively. Insertion into (34) gives the simple update equation of the velocity

$$\dot{\mathfrak{d}}_{k+\frac{1}{2}} = \left[(1 - \frac{1}{2}\alpha_g \Delta T)\dot{\mathfrak{d}}_{k-\frac{1}{2}} - \Delta T \mathfrak{m}^{-1} \mathfrak{p}(\mathfrak{d}_k, state(t_n))\right] / (1 + \frac{1}{2}\alpha_g \Delta T)\ . \tag{38}$$

Standard integration then yields the current generalized displacement $\mathfrak{d}_{k+1} = \mathfrak{d}_k + \Delta T \dot{\mathfrak{d}}_{k+\frac{1}{2}}$ at T_{k+1}. The relaxation analysis is terminated if the fictitious mean kinetic energy

$$\bar{E}_{k+\frac{1}{2}} := \frac{1}{2}\dot{\mathfrak{d}}^T_{k+\frac{1}{2}} \mathfrak{m} \dot{\mathfrak{d}}_{k+\frac{1}{2}} / (N+M) \leq tol_E \tag{39}$$

goes below a certain limit. Global damping is introduced in the update equation (38) by the global damping parameter α_g. A second kind of damping, referred to as the contact damping, defines additional viscous forces P_v^c and T_v^c at the points where two particles are in contact, see [16] for details.

4 Numerical Examples and Comparative Study

In what follows, the performance of the proposed homogenization techniques is demonstrated. To this end, we investigate a representative three-dimensional test example of spherical particles in deformation-driven scenarios, which underline the bound character of the three constraints $I = D, P, S$. We then perform a full two-scale analysis of a biaxial laboratory test of a granular medium. This couples a macroscopic continuum problem with locally attached discrete microstructures.

4.1 Specification of Basic Micromechanical Functions

For a comprehensive presentation of the numerical results, the micromechanical force-potential functions need to be specified. We assume simple quadratic functions

$$\psi_p = \frac{1}{2} c_p (\varepsilon^c)^2 \,, \quad \psi_t = \frac{1}{2} c_t (\gamma_e^c)^2 \,, \quad \psi_h = \frac{1}{2} h\alpha^2 \,, \tag{40}$$

yielding linear force-displacement relations. The potentials contain the three physical parameters c_p, c_t and h. They govern in connection with the interparticle friction angle φ in the slip criterion function

$$\phi(T^c, P^c, \beta) = |T^c| - P^c \tan[\varphi] - \beta \le 0 \quad \text{with} \quad \beta = \psi_h'(\alpha) \,. \tag{41}$$

the quasi-static micromechanical response outlined in Section 3.1. The potential functions for the artificial local damping mechanism are also assumed to be quadratic, i.e.

$$\psi_{pv} = \frac{1}{2} \alpha_c c_p (\dot{\varepsilon}^c)^2 \,, \quad \psi_{tv} = \frac{1}{2} \alpha_c c_t (\dot{\gamma}_e^c)^2 \,. \tag{42}$$

The local damping parameter α_c, a global damping function $\hat{\alpha}_g(n_r)$ depending on the number n_r of relaxation steps and the fictitious density ρ govern the dynamic relaxation processes in typical load increments of the microstructures. For more details of this model, we refer to [17].

4.2 Compression-Shear Mode for Cubic Microstructures

We considers an aggregate of 4076 spherical particles as shown in Figure 6. The initial volume of the microstructure is $|\mathcal{V}| = 2531.5 \cdot 10^6$ square units and the volume density of the assembly is characterized by $\tilde{\rho} = 0.52$, adding up to a porosity of $\tilde{p} = 0.48$. The void ratio follows as $\tilde{e} = 0.93$. The physical and technical parameters used in this simulation are specified in Table 1. The force penalty parameter is set at $\epsilon_f = 10^8$ for the simulations (D), (P) with the Dirichlet-type boundary conditions and at $\epsilon_f = 10^1$ for the Neumann-type boundary condition (S). The global damping function is specified by the parameters $n_{r0} = 50$, $\alpha_{g0} = 20.0$ and $\bar{\alpha} = 1.0$. They ensure that global damping $\alpha_g = (n_r - n_{r0})^2/\bar{\alpha}$ is applied once 50 relaxation steps with $\Delta T = 2.0 \cdot 10^{-3}$ are performed. The deformation is applied in 1100 constant load steps of $\Delta t = 0.1$. At first, a triaxial compression mode $\bar{\boldsymbol{F}}_c$ is performed in 100 load steps up to $t_0 = 10.0$. Then the compression state is held constant and a simple shear mode $\bar{\boldsymbol{F}}_s$ is applied in the remaining 1000 load steps. These two deformation modes read

$$\bar{\boldsymbol{F}}_c = \boldsymbol{1} - t \begin{bmatrix} 0.002 & 0 & 0 \\ 0 & 0.002 & 0 \\ 0 & 0 & 0.002 \end{bmatrix} , \quad \bar{\boldsymbol{F}}_s = \boldsymbol{1} - \begin{bmatrix} 0.2 & 0 & (t_0 - t)0.01 \\ 0 & 0.2 & 0 \\ 0 & 0 & 0.2 \end{bmatrix} . \tag{43}$$

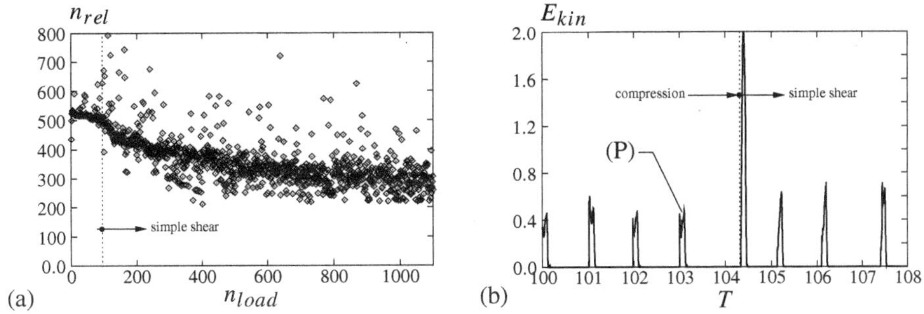

Fig. 5. Cubic microstructure with 4076 particles. (a) Number of relaxation steps n_{rel} versus load steps n_{load} for the case of (P) periodic boundary conditions, (b) mean kinetic energy E_{kin} versus simulation time, showing the dynamic relaxations within incremental load steps.

Table 1. Physical and technical parameters

	Physical Parameters		
normal stiffness	c_p	=	$1.4 \cdot 10^7$
tangential stiffness	c_t	=	$1.4 \cdot 10^7$
linear hardening	h	=	0
interparticle friction angle	φ	=	25°
	Technical Parameters		
fictitious density	ρ	=	1.0
contact damping	α_c	=	0.01
initial global damping	α_{g0}	=	20.0
low–damping relaxation steps	n_{r0}	=	50
intensity of relaxation function	$\bar{\alpha}$	=	1.0
penalty force factor	ϵ_f	=	10^8 / 10^1
penalty couple factor	ϵ_c	=	10^8
load increment	Δt	=	0.1
relaxation step length	ΔT	=	$2.0 \cdot 10^{-3}$

Each load step is followed by n_{rel} relaxation steps which relax the fictitious incremental dynamic response to a quasi-static state.

Figure 5a depicts a varying number n_{rel} of relaxation steps executed within some load steps, which are required to reach this static state. Here, the global relaxation procedure outlined in Section 3.2 is applied in combination with the energetic convergence criterion (39). Figure 5b depicts the mean kinetic energy $\bar{E}_{kin}$ of the assembly for some representative load steps in the time interval $T \in [100,\ 108]$. This curve is obtained from the simulation (P) with periodic boundary constraints. It is evident that at the end of each load step the mean kinetic energy reaches almost zero and hence quasi-static states are gained.

In Figure 6 three microstructural deformation states are displayed: firstly at the end of the compression mode or the beginning of shear deformation at $\bar{F}_{13} = 0.0$ and $\bar{F}_{11} = -0.2$, secondly at the intermediate deformation at $\bar{F}_{13} = 0.5$ and finally

(D):

(P):

(S):

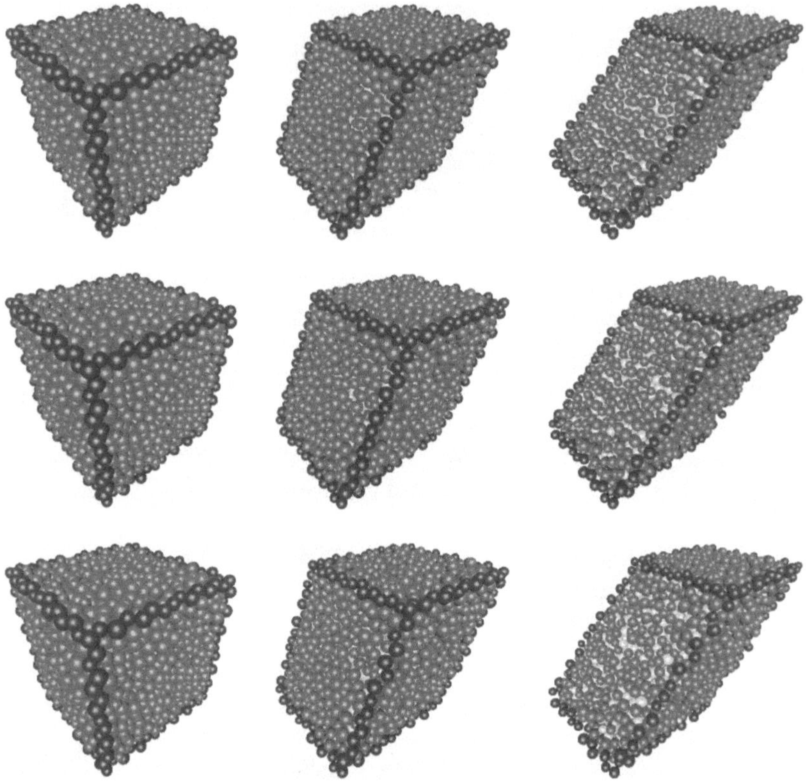

Fig. 6. Deformation of a cubic microstructure with 4076 spherical particles under a biaxial compression mode followed by a simple shearing. Displayed are the deformation states for the three boundary constraints $I = D, P, S$, where (D) and (S) provide stiff and soft bounds of the periodic conditions (P).

at the end of the driving process $\bar{F}_{13} = 1.0$. The results are displayed for the three alternative homogenization approaches based on linear center translations/zero rotations (D), periodic center translations/rotations (P) and uniform center forces/free rotations (S) of particles at the boundary frame. The plotted deformations of the microstructure underline the *stiff* character of (D) and the *soft* character of (S).

Figure 7 depicts the homogenized macroscopic normal Cauchy stresses $\bar{\sigma}_{11}$ versus the initial macroscopic compression $\Delta \bar{F}_{11}$ and the prescribed macroscopic shear deformation $\bar{F}_{13}$. These curves underline the bound character of the *stiff* boundary conditions of linear center translations/zero rotations (D), and the *soft* boundary conditions of uniform center forces/free rotations (S) of particles at the boundary frame. These Dirichlet-type and Neumann-type conditions act as a upper and lower bounds for the macroscopic stress-deformation response. The result for periodic translations/rotations (P) lies between these bounds and is considered to be the most appropriate choice for computational homogenization of granular aggregates.

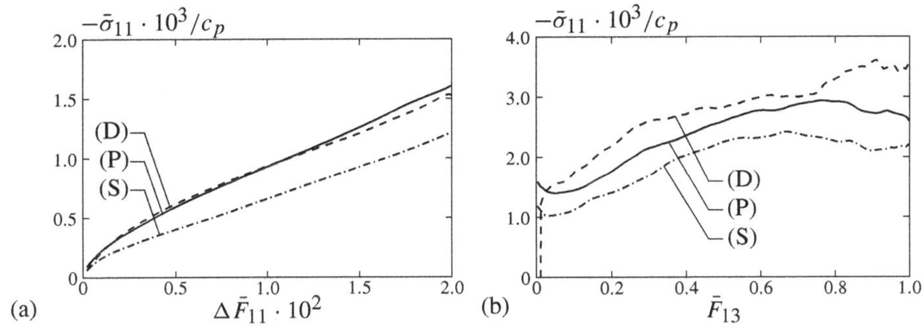

Fig. 7. Cubic microstructure with 4076 spherical particles. Homogenized Cauchy stresses $-\bar{\sigma}_{11}$ versus (a) initial macroscopic compression $\Delta\bar{F}_{11}$ and (b) simple shearing $\bar{F}_{13}$ for the three boundary constraints $I = D, P, S$, where (D) and (S) provide stiff and soft bounds of the periodic condition (P).

5 Multiple Scale Simulation of a Granular Medium

5.1 Two-Scale Simulations Based on DE-FE Coupling

In the previous section, we demonstrated the driving of *single* microstructures for prescribed macroscopic deformation modes. These computations serve as *virtual experiments* for the investigation of local stress-strain curves. We now outline the basic setup of *true two-scale analyses* of a granular medium. This couples full macroscopic continuum problems with locally attached discrete microstructures as depicted schematically in Figure 1. Here, the stress response at local points of the macro-continuum is determined by the computational homogenization methods outlined in the previous sections.

Figure 8 illustrates our concept of a deformation-driven two-scale analyses of a granular medium. The macrostructural boundary-value problem outlined in Section 2.1 is discretized in a standard manner by a continuum *finite element method* (FE). The local constitutive response at a typical integration point of the macro-elements is determined by the microstructural boundary problem of a particle aggregate, governed by the *discrete element technique* (DE) outlined in Section 3. The discrete-to-finite element coupling DE-FE of the two-scale simulations is performed in a deformation-driven process by the following *algorithmic cycle* within a typical global solution step. First, we compute the kinematic deformation variables $\{\bar{\boldsymbol{F}}_g\}_{g=1,n_g}$ at all n_g integration points of the macroscopic FE mesh. For these macro-deformations, the associated macro-stresses $\{\bar{\boldsymbol{P}}_g\}_{g=1,n_g}$ are obtained from DE computations accompanied by the proposed alternative homogenization procedures (D), (P) or (S). With those macro-stresses at hand, discrete equilibrium of the macroscopic FE mesh determines an updated deformation of the granular medium. As a representative two-scale simulation, we consider a classical biaxial test depicted in Figure 9.

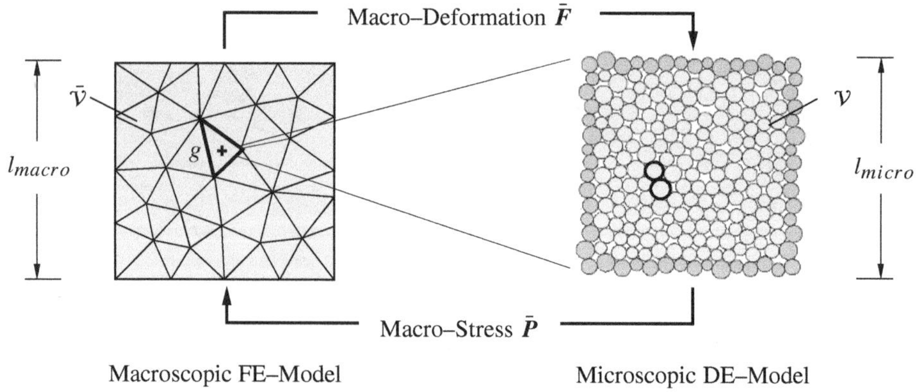

Fig. 8. Deformation-driven two-scale simulations of granular media. The coupling of the macrostructural FE problem with the microstructural DE models is realized by the sequence $\bar{F}_g \to \bar{P}_g$, i.e. by computing homogenized stresses for prescribed deformations at typical intergation points $g = 1 \dots n_g$ of the macrostructural mesh.

5.2 Simulation of a Biaxial Compression Test of a Soil

Experimental Setup

The biaxial apparatus selected for the computational simulation is the device that was built and extensively applied in a Grenoble laboratory. The apparatus is described in [22], a comprehensive review of many test results is given in [7]. A prismatic test specimen of granular material is placed into a pressure cell, see Figure 9a. First, the specimen is subjected to a uniform hydrostatic pressure $\bar{\sigma}_3$. The soil specimen is encased by a latex membrane, the hydrostatic pressure imposed through a fluid, e.g. silicone oil, into which the specimen is placed. Next, a strain-controlled axial compressive load is applied by pressing down a rigid die as shown in Figure 9a. The contact surface of the die is lubricated with silicone grease in order to minimize friction. The hydrostatic lateral pressure $\bar{\sigma}_3$ is kept constant during axial loading. The aim here is to simulate the test named `shf03` of [7]. This test concerns a specimen of *dry Hostun sand* with dimensions 112 × 251 mm, that yields doubly *symmetric* localized zones, i.e. shear bands which develop cross-diagonally through the test specimen.

Coupled FE-DE Two-Scale Model

The FE mesh of the macrostructure is depicted in Figure 9b. Bearing the experimental observation in mind, we exploit the symmetry and model only one fourth of the test specimen. The upper left quarter is modeled by a FE mesh of 401 triangular elements. Vertical downward displacements at the top nodes are prescribed. On the microscopic side, a granular DE microstructure with 219 circular plane particles is employed, yielding a total of about one million microscopic degrees of freedom of

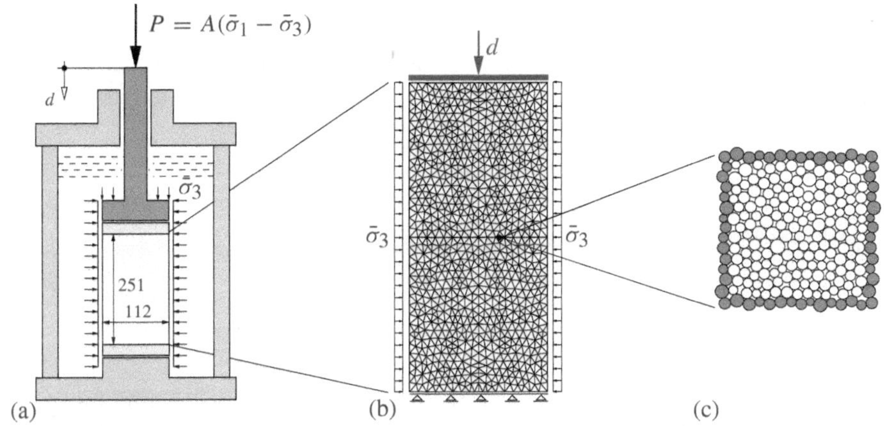

Fig. 9. Biaxial compression test. (a) Cross section through test apparatus, (b) macroscopic FE model with 1604 triangular finite elements, and (c) representative microscopic DE structure with 219 circular plane particles. A denotes the cross-sectional horizontal area of the specimen and $\bar{\sigma}_3$ the hydrostatic pressure in which the specimen is placed.

the coupled DE-FE problem. Periodic boundary constraints (P) are enforced on the microstructures. The applied macroscopic loading step is $\Delta t = 1 \cdot 10^{-2}$, the prescribed forces (lateral stresses) and subsequent vertical displacements are imposed in 8 and 40 load steps, respectively. On the microscopic side, the penalty parameters are taken to be $c_p = c_t = 1 \cdot 10^8$. Hardening is not considered. The interparticle friction angle is $\varphi = 30°$. The fictitious density of the particles is taken to be $\rho = 100$ and the damping parameters are $\alpha_c = \alpha_{g0} = 0.01$. In order to match the localized pattern of test shf03, an imperfection needs to be incorporated. To this end, the microscopic interparticle friction angle is slightly reduced in the finite elements where the shear band is expected to develop. Through this reduction, the onset of the localization is triggered. The lateral stress is increased during the first 8 load steps up to its final value $\bar{\sigma}_3 = 8/3 \cdot 10^4$.

Results and Discussion

In Figure 10 the deformed macroscopic structure at a top displacement of $d = 4.0$ is plotted with six representative granular microstructures at selected macroscopic sampling points. The localized zone has successfully been triggered and the shear bands have already started to develop inwardly. The contour plot shows the norm $|\bar{\boldsymbol{R}}|$ of the rotational part of the local macro-deformation $\bar{\boldsymbol{F}}$.

In Figure 11a the relation of the Roscoe planar stress measure $\bar{t}/\bar{s}$ against the logarithmic axial strain $\bar{\varepsilon}$ is reported and compared with the experimental results of Desrues and Viggiani [7]. These overall stress and strain measures of the specimen are defined as

$$\bar{s} := \frac{1}{2}(\bar{\sigma}_1 + \bar{\sigma}_3) > 0 \,, \quad \bar{t} := \frac{1}{2}(\bar{\sigma}_1 - \bar{\sigma}_3) \geq 0 \quad \text{and} \quad \bar{\varepsilon} := -\ln\frac{H}{H_0} \,, \tag{44}$$

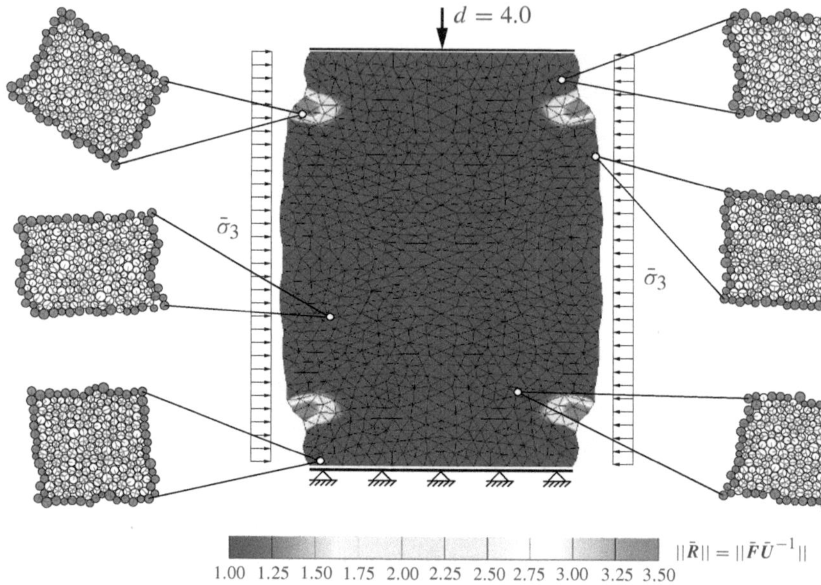

Fig. 10. Coupled two-scale simulation of a biaxial compression test. Deformed configuration at $d = 4.0$ of the macroscopic FE mesh with six selected DE microstructures of 256 circular particles. The constant hydrostatic pressure in which the test specimen was placed is simulated by a constant lateral stress.

where H and H_0 are the actual and initial heights of the macroscopic specimen and $\bar{\sigma}_1$ and $\bar{\sigma}_3$ denote the major principal stress and hydrostatic overall stress in the specimen, respectively. $\bar{t}$ denotes the deviatoric and $\bar{s}$ the mean planar stresses. The effective mobilized friction angle φ_m follows from $\sin(\varphi_m) = \bar{t}/\bar{s}$. The first part of the load-displacement curve in Figure 11a shows a stiffer behavior than the experimentally observed one. The softer behavior of the experiment may be attributed to the lubrication of the top die that minimizes friction between the die and the specimen. That lubrication lets the top particles move not only vertically but also horizontally. In the simulation, the top nodes were fixed in both directions which results in the initially stiffer behavior. However, the influence of this effect vanishes with increasing load. Figure 11b plots the macroscopic mobilized friction angles φ_m of the simulation and the experiment versus the axial strain $\bar{\varepsilon}$, where the lateral stress $\bar{\sigma}_3$ has reached its maximum. Initially, the simulated φ_m is higher, probably due to the lateral flexibility in the experiment mentioned above. As the vertical strain advances, the simulation and experiment reach almost similar values of about $\varphi_m \approx 33°$. Clearly, the two-scale model considered here still represents a strong idealization of the reality and gives due to several simplifications essentially qualitative results. However, it outlines a multiscale simulation of granular media on a precise conceptual level.

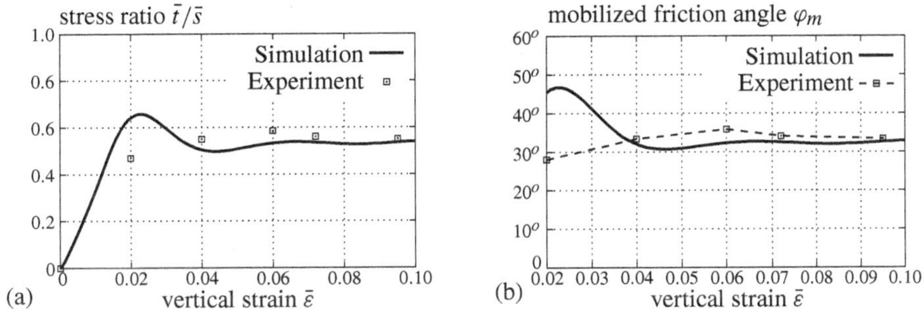

Fig. 11. Biaxial compression test. (a) Roscoe stress measure $\bar{t}/\bar{s}$ and (b) effective mobilized friction angle φ_m versus axial strain $\bar{\varepsilon}$ obtained from the two-scale simulation and comparison with experimental results of Desrues and Viggiani [7].

6 Conclusion

We outlined methods for quasi-static homogenization of granular microstructures and their embedding into full two-scale simulations. A consistent transition of classical stiff, soft and periodic boundary conditions from continuous to granular microstructures induced new classes of micro-to-macro transitions for granular aggregates. These include constraints not only for particle center displacements but also for particle rotations at a boundary frame that drives the aggregate. The proposed stiff and soft constraints yield upper and lower bounds of the aggregate's stiffness. On the computational side, we outlined a unified implementation of the displacement and rotational constraints by penalty methods that prove to be convenient for straightforward implementation into discrete element codes. The periodic boundary conditions provide the most suitable setting for two-scale modeling with coarse microstructures. The best choice for coarse non-periodic microstructures are the soft stress-driven conditions, which avoid over-constrained solutions inherent in the stiff displacement-driven constraints. Finally, we embedded granular microstructures into a coarse graining discrete-to-finite element coupling and performed a true two-scale simulation of a granular medium.

Acknowledgement

Partial support for this research was provided by the German Research Foundation DFG under grant Mi 295/10-2.

References

1. J.P. Bardet and J. Proubet. Adaptative dynamic relaxation for statics of granular materials. *Computers and Structures*, 39:221–229, 1991.

2. J.P. Bardet and I. Vardoulakis. The asymmetry of stress in granular media. *International Journal of Solids and Structures*, 38:353–367, 2001.
3. R.I. Borja and J.R. Wren. Micromechanics of granular media Part I: Generation of overall constitutive equation for assemblies of circular disks. *Computer Methods in Applied Mechanics and Engineering*, 127:13–36, 1995.
4. C.S. Chang and C.L. Liao. Constitutive relation for a particulate medium with the effect of particle rotation. *International Journal of Solids and Structures*, 26(4):437–453, 1990.
5. C.S. Chang and L. Ma. A micromechanical-based micropolar theory for deformation of granular solids. *International Journal of Solids and Structures*, 28(1):67–86, 1991.
6. P.A. Cundall and O.D.L. Strack. A discrete numerical model for granular assemblies. *Géotechnique*, 29(1):47–65, 1979.
7. J. Desrues and G. Viggiani. Strain localization in sand: an overview of the experimental results obtained in grenoble using stereophotogrammetry. *International Journal for Numerical and Analytical Methods in Geomechanics*, 28:279–321, 2004.
8. W. Ehlers, E. Ramm, S. Diebels, and G.A. D'Addetta. From particle ensembles to cosserat continua: homogenization of contact forces towards stresses and couple stresses. *International Journal of Solids and Structures*, 40:6681–6702, 2003.
9. H.J. Herrmann and S. Luding. Modeling granular media on the computer. *Continuum Mechanics and Thermodynamics*, 10:189–231, 1998.
10. K. Kanatani. A micropolar continuum theory for the flow of granular materials. *International Journal of Engineering Science*, 17:419–432, 1979.
11. K. Kaneko, K. Terada, T. Kyoya, and Y. Kishino. Global-local analysis of granular media in quasi-static equilibrium. *International Journal of Solids and Structures*, 40:4043–4069, 2003.
12. S. Luding. Micro-macro transition for anisotropic, frictional granular packings. *International Journal of Solids and Structures*, 41:5821–5836, 2004.
13. H.A. Meier, P. Steinmann, and E. Kuhl. Towards multiscale computation of confined granular media – Contact forces, stresses and tangent operators. *Technische Mechanik*, 28(1):32–42, 2008.
14. C. Miehe. Computational micro-to-macro transitions for discretized micro-structures of heterogeneous materials at large strains based on the minimization of averaged incremental energy. *Computer Methods in Applied Mechanics and Engineering*, 192:559–591, 2002.
15. C. Miehe and C. Bayreuther. On multiscale fe analyses of heterogeneous structures: From homogenization to multigrid solvers. *International Journal for Numerical Methods in Engineering*, 71:1135–1180, 2007.
16. C. Miehe and J. Dettmar. A framework for micro-macro transitions in periodic particle aggregates of granular materials. *Computer Methods in Applied Mechanics and Engineering*, 193:225–256, 2004.
17. C. Miehe, J. Dettmar, and D. Zäh. Homogenization and two-scale simulations of granular materials for different microstructural constraints. *International Journal of Numerical Methods in Engineering*, submitted, 2009.
18. C. Miehe, J. Schröder, and J. Schotte. Computational homogenization analysis in finite plasticity. simulation of texture development in polycrystalline materials. *Computer Methods in Applied Mechanics and Engineering*, 171:387–418, 1999.
19. J.J. Moreau. Some numerical methods in multibody dynamics: Application to granular materials. *European Journal of Mechanics A/Solids*, 13(4):93–114, 1994.
20. H.-B. Mühlhaus and I. Vardoulakis. The thickness of shear bands in granular materials. *Géotechnique*, 37(3):271–283, 1987.

21. P. Underwood. Dynamic relaxation. In T. Belytschko and T.J.R. Hughes (Eds.), *Computational Methods for Transient Analysis*. Elsevier Science Publisher, pp. 245–265, 1983.
22. I. Vardoulakis and M. Goldscheider. Biaxialgerät zur Untersuchung der Festigkeit und Dilatanz von Scherfugen in Böden (A biaxial apparatus for testing of strength and dilatancy of shear bands in solids). *Geotechnik*, 3:19–31, 1980.
23. J.R. Wren and R.I. Borja. Micromechanics of granular media Part II: Overall tangential moduli and localization model for periodic assemblies of circular disks. *Computer Methods in Applied Mechanics and Engineering*, 141:221–246, 1997.

Effective Hydraulic and Mechanical Properties of Heterogeneous Media with Interfaces

L. Dormieux[1], L. Jeannin[2] and J. Sanahuja[3]

[1]ENPC, Laboratoire Navier, Champs/Marne, France; dormieux@lmsgc.enpc.fr
[2]GDFSuez, Paris, France
[3]EDF R&D, Moret-sur-Loing, France

Abstract The role of interfaces on the effective behavior of composites is supported by strong experimental evidences. The purpose of the paper is to adapt the standard closed-form homogenization schemes for heterogeneous media with random microstructure in order to incorporate the effect of interfaces. The Mori–Tanaka and self-consistent schemes which are respectively used for matrix-inclusion and polycrystal-like morphologies are revisited. A polarization-based variational approach is also considered.

Key words: Interfaces, polarization method, self-consistent scheme, Mori–Tanaka scheme, permeability, elasticity

1 Introduction

An interface in a composite can be regarded as a two dimensional geometrical model of a thin layer which is expected to affect significantly the effective properties. The purpose of the present paper is to illustrate this on hydraulic properties (in the sense of Darcy's law) and mechanical properties (in the linear elastic regime).

Depending on the morphology of the microstructure, interfaces can separate inclusions from the matrix in the case of an inclusionary phase, or they can separate the grains to one another in the case of a granular medium.

Cement-based materials like mortar or concrete provide examples of the matrix-inclusion microstructure, sand grains or aggregates being embedded in a cement matrix. But in fact, the properties of the latter, for instance its porosity, are modified in the vicinity of the inclusions (see e.g. [7]) which is generally referred to as interfacial transition zone. The latter is often described as a thin layer with high permeability and low stiffness as compared to the matrix.

The polycrystal-like morphology characterizes the second important family of microstructure in which the role of interfaces can be decisive. Referring again to geomaterials, sandstones constitute an important example of cemented granular rock in which both hydraulic and mechanical properties are of great practical interest since sandstones are the host rock of many oil and gas reservoirs. Modelling the cement or

R. de Borst and E. Ramm (eds.), *Multiscale Methods in Computational Mechanics*, Lecture Notes in Applied and Computational Mechanics 55, DOI 10.1007/978-90-481-9809-2_10,

the joint between two grains as an interface is an appealing alternative as compared to the three dimensional thin layer model.

The purpose of the paper is to investigate how to adapt the standard schemes of homogenization of heterogeneous media with random microstructure in order to incorporate the presence of interfaces. Section 2 is devoted to the determination of an effective permeability coefficient. Both the matrix-inclusion and the polycrystal morphology will be considered. A variational approach providing an upper-bound of the effective permeability coefficient is also presented. Section 3 derives self-consistent estimates of the effective elastic properties of a porous granular rock.

2 Hydraulic Model for a Porous Matrix with Impermeable Inclusionary Phase

We consider a representative elementary volume (*rev*) Ω made up of a porous matrix Ω_m with permeability k_m and non-porous impermeable grains G_i which play the role of an inclusionary phase: $\Omega_g = \cup_i G_i$. Each grain is surrounded by an interface with permeability η. The purpose of this section is to quantify how the interfaces S_i between the grains and the matrix contribute to the homogenized permeability k^{hom}.

We define the microscopic scale as the one which reveals the heterogeneity of the microstructure, namely the grains, the matrix and the interfaces in between. In contrast, at the macroscopic scale, the *rev* is regarded as a homogenized body. If $a(z)$ is a field defined on Ω at the microscopic scale, its average is denoted by $\overline{a}$:

$$\overline{a} = \frac{1}{|\Omega|} \int_{\Omega} a(z)\, dV \tag{1}$$

Similarly, we also introduce the intrinsic averages $\overline{a}^{\alpha}$ (with $\alpha = m$ or I) on Ω_α:

$$\overline{a}^{\alpha} = \frac{1}{|\Omega_\alpha|} \int_{\Omega_\alpha} a(z)\, dV \tag{2}$$

Following their definitions, $\overline{a}$ as well $\overline{a}^{\alpha}$ are macroscopic quantities. Darcy's law is assumed to be relevant for describing the fluid flow in the matrix at the microscopic scale. It relates the filtration vector $\boldsymbol{v}$ to the fluid pressure gradient:

$$z \in \Omega_m : \quad \boldsymbol{v} = -k_m\, \mathbf{grad}\, p \tag{3}$$

Note that $p(z)$ as well as $\boldsymbol{v}(z)$ are defined at the microscopic scale. Let $\Sigma \subset \Omega_m$ denote a surface with normal $\boldsymbol{n}$. The fluid flow across Σ reads:

$$D = \int_{\Sigma} \boldsymbol{v} \cdot \boldsymbol{n}\, dS \tag{4}$$

In the interface S_i between grain G_i and the matrix, the fluid flow is again defined by a filtration vector $\boldsymbol{q}$ which is proportional to the fluid pressure gradient in the interface:

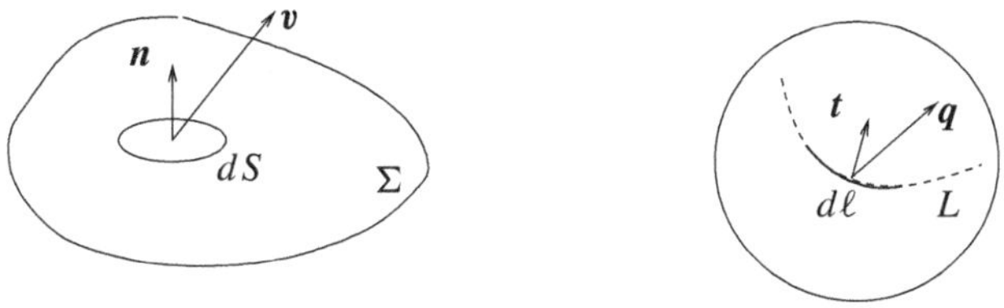

Fig. 1. Filtration vector in the matrix ($\boldsymbol{v}$) and in the interface ($\boldsymbol{q}$)

$$z \in S_i : \quad \boldsymbol{q} = -\eta\, \mathbf{grad}_S\, p \tag{5}$$

Following its definition, $\boldsymbol{q}$ lies within the tangent plane of S_i. Let $L \subset S_i$ denote a line which normal in the tangent plane is $\boldsymbol{t}$. The fluid flow across L reads:

$$D = \int_L \boldsymbol{q} \cdot \boldsymbol{t}\, d\ell \tag{6}$$

The physical definition of $\boldsymbol{v}$ and $\boldsymbol{q}$ (resp. k_m and η) implies that both the ratios $\boldsymbol{q}/\boldsymbol{v}$ and η/k_m have the dimension of a length. A macroscopic fluid pressure gradient $\nabla \boldsymbol{P}$ is applied to the *rev* in the sense of Hashin:

$$z \in \partial\Omega : \quad p(z) = \nabla \boldsymbol{P} \cdot z \tag{7}$$

Any pressure field which meets the above boundary condition is said to be *admissible* with $\nabla \boldsymbol{P}$. In turn, the average flow in the *rev* represents the macroscopic filtration vector. It involves both the contribution of the matrix and that of the interfaces:

$$\overline{\boldsymbol{v}} = (1-c)\overline{\boldsymbol{v}}^m + \frac{1}{|\Omega|} \sum_i \int_{S_i} \boldsymbol{q}\, dS \tag{8}$$

where c is the grain volume fraction. We note from (3) that $\overline{\boldsymbol{v}}^m = -k_m \overline{\mathbf{grad}\, p}^m$:

$$\overline{\boldsymbol{v}} = -(1-c)k_m \overline{\mathbf{grad}\, p}^m + \frac{1}{|\Omega|} \sum_i \int_{S_i} \boldsymbol{q}\, dS \tag{9}$$

The intrinsic definition of k^{hom} yields:

$$\overline{\boldsymbol{v}} = -k^{hom} \nabla \boldsymbol{P} \tag{10}$$

Eventually, estimating the homogenized permeability amounts to estimating the average pressure gradient in the matrix and the fluid flow in the interfaces.

2.1 Mori–Tanaka Estimate

The Mori–Tanaka estimate of the homogenized permeability is derived from an auxiliary problem in which a single grain G surrounded by its interface with the matrix

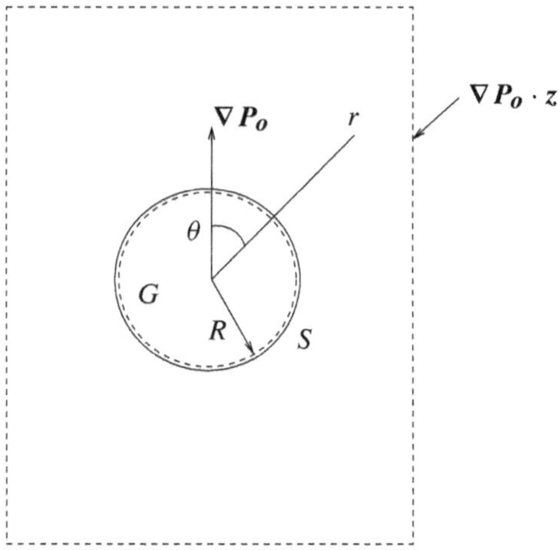

Fig. 2. Generalized Eshelby problem (Mori–Tanaka approach).

is embedded in an infinite medium with permeability k_m (Figure 2). At infinity, the boundary conditions on the fluid pressure read $p \to \nabla \boldsymbol{P_o} \cdot z$, where $\nabla \boldsymbol{P_o}$ is an auxiliary macroscopic pressure gradient, which is going to be related later to the true macroscopic pressure gradient $\nabla \boldsymbol{P}$. In the Mori–Tanaka approach, the average pressure gradient $\overline{\mathbf{grad}\,\mathbf{p}}^m$ over the matrix Ω_m is estimated from the average pressure gradient in the medium surrounding the grain G in the auxiliary problem (Figure 2), which is clearly equal to $\nabla \boldsymbol{P_o}$:

$$\overline{\mathbf{grad}\,p}^m = \nabla \boldsymbol{P_o} \tag{11}$$

Similarly, the average pressure gradient in G is taken as an estimate for the average pressure gradient in Ω_g. The latter can be derived analytically in the case of a spherical grain. Let $\nabla \boldsymbol{P_o} = \nabla P_o \boldsymbol{e}_3$. Using spherical coordinates defined w.r.t. the center of the grain with $\theta = 0$ corresponding to the axis colinear to $\boldsymbol{e}_3$, the pressure field is sought in the form

$$\begin{aligned} r < R &: p = Ar\cos\theta \\ r > R &: p = \left(\alpha r + \frac{\beta}{r^2}\right)\cos\theta \end{aligned} \tag{12}$$

in which R denotes the grain radius. Of course, no fluid flow occurs in the grain. Accordingly, the fluid pressure for $r < R$ must be interpreted as a limit when the grain permeability tends towards 0. The continuity of the fluid pressure at $r = R$ reads

$$AR = \alpha R + \frac{\beta}{R^2} \tag{13}$$

As regards the flow in the interface, the fluid mass balance requires that

$$\mathrm{div}_S \boldsymbol{q} = -v_{r\,|r=R^+} \tag{14}$$

Owing to (5) and (12), the fluid flow in the interface is found to be characterized by

$$\boldsymbol{q} = \eta A \sin\theta \, \boldsymbol{e}_\theta \tag{15}$$

Similarly, Darcy's law in the domain $r > R$ together with (12) yields

$$r > R: \quad \boldsymbol{v} = -k_m \left(\left(\alpha - \frac{2\beta}{r^3} \right) \cos\theta \, \boldsymbol{e}_r - \left(\alpha + \frac{\beta}{r^3} \right) \sin\theta \, \boldsymbol{e}_\theta \right) \tag{16}$$

Accordingly, (14) reads

$$k_m \left(\alpha - \frac{2\beta}{R^3} \right) = 2A \frac{\eta}{R} \tag{17}$$

Finally, the boundary condition at infinity prescribes

$$\alpha = \nabla P_o \tag{18}$$

The unknown A and β are then derived from (13) and (17). Introducing $\delta = \eta/(k_m R)$:

$$A = \frac{3}{2(1+\delta)} \nabla P_o \quad ; \quad \beta = \frac{R^3}{2} \frac{1-2\delta}{1+\delta} \nabla P_o \tag{19}$$

Note that $A\boldsymbol{e}_3$ is equal to the (uniform) pressure gradient in G. As stated before, it is used as an estimate for $\overline{\textbf{grad}\, p}^g$:

$$\overline{\textbf{grad}\, p}^g = \frac{3}{2(1+\delta)} \nabla \boldsymbol{P_o} \tag{20}$$

We still have to relate $\nabla \boldsymbol{P_o}$, which has been interpreted as the average pressure in the matrix to the macroscopic pressure gradient $\nabla \boldsymbol{P}$. Recalling that $\overline{\textbf{grad}\, p} = \nabla \boldsymbol{P}$, it is readily seen from (11) and (20) that

$$\nabla \boldsymbol{P} = \left(1 - c + \frac{3c}{2(1+\delta)} \right) \nabla \boldsymbol{P_o} \tag{21}$$

We are left with the determination of the homogenized permeability. Going back to (9), the contribution of the interfaces to $\overline{\boldsymbol{v}}$ is estimated from (15):

$$\frac{1}{|\Omega|} \sum_i \int_{S_i} \boldsymbol{q} \, dS \approx N \int_S \eta A \sin\theta \, \boldsymbol{e}_\theta \, dS = -\frac{8}{3} \pi R^2 N \eta A \, \boldsymbol{e}_3 \tag{22}$$

where N is the grain density (number of grains per unit volume), which is related to the grain volume fraction by $c = 4\pi R^3 N/3$:

$$\frac{1}{|\Omega|} \sum_i \int_{S_i} \boldsymbol{q} \, dS \approx -\frac{\eta}{R} \frac{3c}{1+\delta} \nabla \boldsymbol{P_o} \tag{23}$$

Introducing (11) and (23) together with (21) into (9) and comparing with (10) eventually yields the Mori–Tanaka (MT) estimate of the homogenized permeability:

$$k_{MT}^{hom} = 2k_m \frac{1 - c + \delta(1+2c)}{2 + c + 2\delta(1-c)} \tag{24}$$

2.2 The Variational Approach

The purpose of the present section is to establish a rigorous upper bound of the homogenized permeability under the assumption of an isotropic distribution of the phases (grains and matrix). The method is inspired of the polarization technique as presented in [6]. In the context of Darcy's law, the starting point is the following inequality:

$$\frac{k}{2}(\mathbf{grad}\, p)^2 \leq \frac{k_o}{2}(\mathbf{grad}\, p)^2 - \boldsymbol{w} \cdot \mathbf{grad}\, p - \frac{1}{2(k-k_o)}\boldsymbol{w}^2 \tag{25}$$

The latter holds irrespective of the values of $\boldsymbol{w}$ and $\mathbf{grad}\, p$ provided that $k \leq k_o$. Let us now recall the variational definition of the homogenized permeability k^{hom}. For a given macroscopic gradient $\nabla \boldsymbol{P}$:

$$\frac{k^{hom}}{2}\nabla \boldsymbol{P}^2 = \frac{1}{|\Omega|} \inf_{p\,\text{adm.}\,\nabla \boldsymbol{P}} \left(\int_\Omega \frac{k(z)}{2}(\mathbf{grad}\, p)^2\, dV + \sum_i \int_{S_i} \frac{\eta}{2}(\mathbf{grad}_S\, p)^2\, dS \right) \tag{26}$$

where $k(z) = k_m$ in Ω_m and $k(z) = 0$ in Ω_g. For the forthcoming application, it is convenient to model the interface as a thin (spherical) layer I_i with infinitesimal width e having the permeability η/e. Denoting as before the grain volume fraction by c, we note for later use that the volume fraction of the interface layers is $3ce/R$. The matrix volume fraction is $1 - c(1 + 3e/R)$. (26) then reads:

$$\frac{k^{hom}}{2}\nabla \boldsymbol{P}^2 = \frac{1}{|\Omega|} \inf_{p\,\text{adm.}\,\nabla \boldsymbol{P}} \left(\int_{\Omega_m \cup \Omega_g} \frac{k(z)}{2}(\mathbf{grad}\, p)^2\, dV + \sum_i \int_{I_i} \frac{\eta}{2e}(\mathbf{grad}\, p)^2\, dV \right) \tag{27}$$

Let us now choose $k_o = \eta/e$ in (25). Applying this remark for $k = k_m$ and for $k = 0$, it is found that

$$\begin{aligned} \frac{k^{hom}}{2}\nabla \boldsymbol{P}^2 \leq \frac{1}{|\Omega|} \Bigg(& \inf_{p\,\text{adm.}\,\nabla \boldsymbol{P}} \left(\int_\Omega \frac{k_o}{2}(\mathbf{grad}\, p)^2 - \boldsymbol{w} \cdot \mathbf{grad}\, p\, dV \right) \\ & - \int_\Omega \frac{1}{2(k(z)-k_o)}\boldsymbol{w}^2\, dV \Bigg) \end{aligned} \tag{28}$$

The validity of (28) is subjected to the condition $k_m \leq k_o = \eta/e$, which holds since $e \to 0$. Note that the choice of the so-called polarization field $\boldsymbol{w}(z)$ in (28) is absolutely arbitrary.

It is readily seen that the infimum in (28) is attained by the solution of the problem defined by

$$\begin{aligned} &\text{div}\, \boldsymbol{v} = 0 \\ &\boldsymbol{v} = -k_o\, \mathbf{grad}\, p + \boldsymbol{w}(z) \\ &p = \nabla \boldsymbol{P} \cdot z \quad \text{on } \partial\Omega \end{aligned} \tag{29}$$

The latter satisfies

$$\mathbf{grad}\, p = \nabla \boldsymbol{P} + \Gamma_o \boldsymbol{w} \tag{30}$$

where Γ_o is the (linear) so-called Green operator of the domain Ω with uniform permeability k_o. Since $\Gamma_o \boldsymbol{w}$ is the pressure gradient associated with the solution to (29) with $\nabla \boldsymbol{P} = 0$, we note that

$$\int_{\Omega} (-k_o \Gamma_o \boldsymbol{w} + \boldsymbol{w}(z)) \cdot \Gamma_o \boldsymbol{w} \, dV = 0 \tag{31}$$

Besides, from the average rule $\nabla \boldsymbol{P} = \overline{\mathbf{grad}\, p}$, it follows that $\overline{\Gamma_o \boldsymbol{w}} = 0$. Making use of these remarks, the use of (30) in (28) yields

$$\frac{k^{hom}}{2} \nabla \boldsymbol{P}^2 \leq \frac{k_o}{2} \nabla \boldsymbol{P}^2 - \overline{\boldsymbol{w}} \cdot \nabla \boldsymbol{P} - \frac{1}{2} \overline{\boldsymbol{w} \cdot \Gamma_o \boldsymbol{w}} - \frac{1}{|\Omega|} \int_{\Omega} \frac{1}{2(k(z) - k_o)} \boldsymbol{w}^2 \, dV \tag{32}$$

which again holds for any polarization field $\boldsymbol{w}(z)$. We now select a piecewise constant field of the form

$$\boldsymbol{w}(z) = \boldsymbol{w}_m \chi_m(z) + \sum_i \boldsymbol{w}_g \chi_{G_i}(z) \tag{33}$$

where $\boldsymbol{w}_m$ and $\boldsymbol{w}_g$ are two arbitrary vectors, χ_m and χ_{G_i} are the characteristic functions of Ω_m and G_i respectively. The polarization field of (33) is therefore equal to 0 in the interface layer I_i separating the matrix from the grain G_i. It follows that

$$\overline{\boldsymbol{w}} = (1 - c(1 + 3\frac{e}{R}))\boldsymbol{w}_m + c\boldsymbol{w}_g \tag{34}$$

Since $\Gamma_o \boldsymbol{W} = 0$ for a uniform polarization field ($\boldsymbol{w}(z) = \boldsymbol{W}$) and recalling $\overline{\Gamma_o \boldsymbol{w}} = 0$, it is convenient to observe that

$$\overline{\boldsymbol{w} \cdot \Gamma_o \boldsymbol{w}} = \overline{(\boldsymbol{w} - \boldsymbol{W}) \cdot \Gamma_o (\boldsymbol{w} - \boldsymbol{W})} \tag{35}$$

Selecting $\boldsymbol{W} = \boldsymbol{w}_m$, it appears that

$$\overline{\boldsymbol{w} \cdot \Gamma_o \boldsymbol{w}} = \overline{\boldsymbol{w}' \cdot \Gamma_o \boldsymbol{w}'} \tag{36}$$

with $\boldsymbol{w}' = -\boldsymbol{w}_m \sum_i \chi_{G_i \cup I_i}(z) + \boldsymbol{w}_g \sum_i \chi_{G_i}(z)$, that is,

$$\begin{aligned} \overline{\boldsymbol{w} \cdot \Gamma_o \boldsymbol{w}} = & \sum_{i,j} \overline{\chi_{G_i \cup I_i} \boldsymbol{w}_m \cdot \Gamma_o(\chi_{G_j \cup I_j} \boldsymbol{w}_m)} \\ & -2 \sum_{i,j} \overline{\chi_{G_i} \boldsymbol{w}_g \cdot \Gamma_o(\chi_{G_j \cup I_j} \boldsymbol{w}_m)} \\ & + \sum_{i,j} \overline{\chi_{G_i} \boldsymbol{w}_g \cdot \Gamma_o(\chi_{G_j} \boldsymbol{w}_g)} \end{aligned} \tag{37}$$

where the symmetry $\overline{\chi_{G_i} \boldsymbol{w}_g \cdot \Gamma_o(\chi_{G_j \cup I_j} \boldsymbol{w}_m)} = \overline{\chi_{G_j \cup I_j} \boldsymbol{w}_m \cdot \Gamma_o(\chi_{G_i} \boldsymbol{w}_g)}$ has been used. We now introduce the assumption of isotropic spatial distribution. Reasoning as in the derivation of the Hashin-Shtrikman bounds [6], we only keep the "squared" terms ($i = j$) in the righthand side of (37) which reduces to

$$
\begin{aligned}
\overline{\boldsymbol{w} \cdot \Gamma_o \boldsymbol{w}} = & \sum_i \overline{\chi_{G_i \cup I_i} \boldsymbol{w}_m \cdot \Gamma_o(\chi_{G_i \cup I_i} \boldsymbol{w}_m)} \\
& -2 \sum_i \overline{\chi_{G_i} \boldsymbol{w}_g \cdot \Gamma_o(\chi_{G_i \cup I_i} \boldsymbol{w}_m)} \\
& + \sum_i \overline{\chi_{G_i} \boldsymbol{w}_g \cdot \Gamma_o(\chi_{G_i} \boldsymbol{w}_g)}
\end{aligned}
\tag{38}
$$

This amounts to saying that it is sufficient to determine the restriction of $\Gamma_o(\chi_{G_i} \boldsymbol{w}_g)$ (resp. $\Gamma_o(\chi_{G_i \cup I_i} \boldsymbol{w}_m)$) to G_i (resp. $G_i \cup I_i$). Since the Green operator Γ_o is not known analytically, it is useful to take advantage of an approximation involving the Green operator Γ_o^∞ of the infinite medium which can be justified under the no long-range order hypothesis [9]

$$
\Gamma_o \boldsymbol{w} = \Gamma_o^\infty(\boldsymbol{w}) - \boldsymbol{P}_o^\Omega \cdot \overline{\boldsymbol{w}} \tag{39}
$$

$\boldsymbol{P}_o^\Omega$ is the (second order) Hill tensor of the *rev*, which reduces to $1/(3k_o)\mathbf{1}$ in the case of a spherical *rev*. The interest of (39) lies in the fact that, for any spherical domain G with uniform polarization vector $\boldsymbol{W}$ embedded in an infinite medium, the pressure gradient $\Gamma_o^\infty(\chi_G \boldsymbol{W})$ is the well known solution of the so-called inclusion Eshelby problem: its restriction to G is uniform and equal to $\boldsymbol{W}/(3k_o)$ (see e.g. [3]). Hence, introducing (39) into (38) yields

$$
\overline{\boldsymbol{w} \cdot \Gamma_o \boldsymbol{w}} = \frac{ec}{3\eta}\left(-2\boldsymbol{w}_m \cdot \boldsymbol{w}_g + \boldsymbol{w}_g^2 + \boldsymbol{w}_m^2\left(1 + \frac{3e}{R}\right)\right) - \frac{ec^2}{3\eta}\left(\boldsymbol{w}_g - \left(1 + \frac{3e}{R}\right)\boldsymbol{w}_m\right)^2 \tag{40}
$$

In addition, it is readily seen that

$$
\frac{1}{|\Omega|}\int_\Omega \frac{1}{2(k(\underline{z}) - k_o)} \boldsymbol{w}^2 \, dV = \frac{1}{2(k_m - \eta/e)} \boldsymbol{w}_m^2 \left(1 - c\left(1 + \frac{3e}{R}\right)\right) - \frac{c}{2\eta/e} \boldsymbol{w}_g^2 \tag{41}
$$

We are left with introducing (34),(40) and (41) into (32). This reveals that the right hand side in (32) is a quadratic form of the vectors $\boldsymbol{w}_g$ and $\boldsymbol{w}_m$. The optimal values of the latter are obtained by differentiation and take the form $\mathcal{F}(k_m, \eta/R, e/R, c)\nabla \boldsymbol{P}$ with a leading term of the order of $(e/R)^{-1}$. Eventually, the limit $e/R \to 0$ in the optimal upper bound yields

$$
k^{hom} \le k_m\,(1 - c + 2c\delta) \tag{42}
$$

It appears as the classical Voigt bound of a porous medium comprising an impermeable phase of volume fraction c, corrected by a term taking the influence of the interfaces into account.

2.3 The Self-Consistent Approach

As opposed to the morphology described in Section 2, we now consider a polycrystal-like porous medium, in which the *rev* is made up of impermeable grains separated by thin interfaces and of a pore space Ω_p (volume fraction ϕ). The macroscopic filtration vector is therefore the sum of these two contributions:

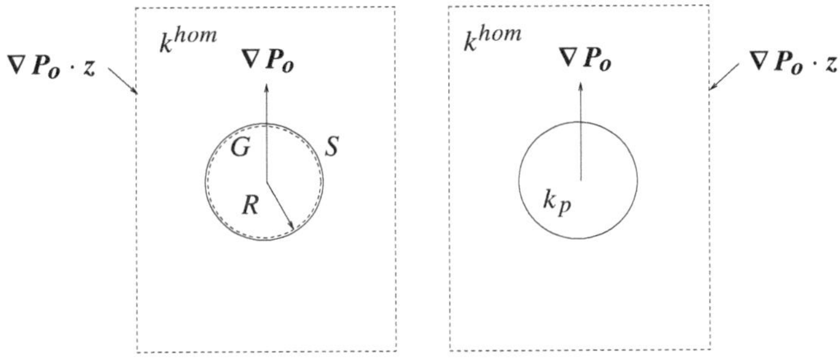

Fig. 3. Two Eshelby problems (self-consistent scheme).

$$\overline{\boldsymbol{v}} = \phi \overline{\boldsymbol{v}}^p + \frac{1}{|\Omega|} \sum_i \int_{S_i} \boldsymbol{q} \, dS \tag{43}$$

where $\overline{\boldsymbol{v}}^p$ denotes the average velocity over the pore space. The characteristic size of the pores is assumed to be very large as compared to the width of the interfaces. Accordingly, the pore space is regarded as a phase with infinite permeability. The homogenized permeability is therefore expected to be controlled by the porosity ϕ and the interface permeability η. In practice, a Darcy's law of the form $\boldsymbol{v} = k_p \, \mathbf{grad} \, p$ is adopted for the flow in the pores. The limit $k_p \to \infty$ is taken in the result of the homogenization.

In order to deal with the polycrystal morphology, a self-consistent scheme is proposed. The idea is to resort to two auxiliary Eshelby-type problems respectively related to the flow in the pores and in the interfaces (Figure 3). In view of estimating the flow in the interfaces, we first consider a spherical grain G (radius R) surrounded by its own interface, embedded in an infinite homogeneous medium with the same boundary conditions as in Section 2.1. However, as opposed to Section 2.1, the permeability of this infinite medium is now equal to the homogenized permeability k^{hom} which is the unknown of the problem. Equations (19) and (22) are still valid up to the fact that δ is now equal to $\eta/(k^{hom} R)$:

$$\frac{1}{|\Omega|} \sum_i \int_{S_i} \boldsymbol{q} \, dS \approx -\frac{3(1-\phi)}{1+\delta} \frac{\eta}{R} \nabla \boldsymbol{P}_o \tag{44}$$

where it has been used that $8\pi R^2 N/3 = 2(1-\phi)/R$. In order to estimate the average flow $\overline{\boldsymbol{v}}^p$ in the pore space, a second Eshelby problem is considered. A spherical pore with permeability k_p is embedded in the infinite medium with permeability k^{hom}. The same boundary conditions at infinity are used. This case exactly corresponds to the classical Eshelby problem (see e.g. [3]). The uniform pressure gradient in the pore is taken as an estimate for the fluid pressure gradient in the pore space:

$$\overline{\mathbf{grad} \, p}^p = \left(1 + \frac{k_p - k^{hom}}{3k^{hom}}\right)^{-1} \nabla \boldsymbol{P}_o \tag{45}$$

Accordingly, the average $\overline{\boldsymbol{v}}^p$ of the filtration vector in Ω_p is estimated as $-k_p\overline{\mathbf{grad}\,p}^p$. The limit when $k_p \to \infty$ reads

$$\overline{\boldsymbol{v}}^p = -3k^{hom}\nabla \boldsymbol{P}_o \quad ; \quad \overline{\mathbf{grad}\,p}^p \to 0 \tag{46}$$

The average rule $\nabla \boldsymbol{P} = (1-\phi)\overline{\mathbf{grad}\,p}^g$ is used in order to relate $\nabla \boldsymbol{P}_o$ to the true macroscopic pressure gradient. Equation (20) yields

$$\nabla \boldsymbol{P} = \frac{3(1-\phi)}{2(1+\delta)}\nabla \boldsymbol{P}_o \tag{47}$$

Eventually, (43) and (46) together with (47) are introduced into (43). Comparing the result to the homogenized Darcy's law (10), it is found that

$$k^{hom} = \frac{2}{1-3\phi}\frac{\eta}{R} \tag{48}$$

Note that the self-consistent model predicts that $k^{hom} \to \infty$ when $\phi \to 1/3^-$.

3 Mechanical Model for a Granular Cemented Rock

3.1 General Framework

Cemented granular rocks like sandstones can be considered as an assemblage of solid grains held up by a cementing material, while voids between grains constitute the porous phase. The idea of the proposed mechanical modelling is to introduce a composite elementary brick made up of a solid grain surrounded by an interface representing the mechanical role of the cement. The sandstone itself is now described as a disordered composite material made up of such bricks and of the complementary pore space. The pore volume fraction classically referred to as porosity is as before denoted by ϕ.

In view of performing an homogenization of this heterogeneous material, we again consider a representative elementary volume (*rev*), denoted by Ω. As before (Section 2), we define the microscopic scale as the one which reveals the heterogeneity of the microstructure, and the macroscopic scale at which the *rev* is regarded as a homogenized body. At the microscopic scale, the stresses and strains are characterized by tensorial fields denoted respectively by $\boldsymbol{\sigma}(z)$ and $\boldsymbol{\varepsilon}(z)$, where z is the position vector defining the location in the *rev*. In contrast, at the macroscopic scale stresses and strains are defined by the tensors $\boldsymbol{\Sigma}$ and $\boldsymbol{E}$ related to the fields $\boldsymbol{\sigma}(z)$ and $\boldsymbol{\varepsilon}(z)$ by the average rules $\boldsymbol{\Sigma} = \overline{\boldsymbol{\sigma}}$ and $\boldsymbol{E} = \overline{\boldsymbol{\varepsilon}}$ with the notation introduced in (1). Furthermore, the loading to which the *rev* is subjected is defined according to the so-called Hashin uniform strain boundary conditions:

$$z \in \partial\Omega : \quad \boldsymbol{\xi}(z) = \boldsymbol{E}\cdot z \tag{49}$$

where $\boldsymbol{\xi}$ denotes the microscopic displacement field. The first step of homogenization consists in solving the boundary value problem defined on Ω by (49), according to the relevant constitutive behaviors of the solid grains and the interfaces. In particular, the stress field $\boldsymbol{\sigma}(z)$ is sought as a function of $\boldsymbol{E}$. The macroscopic stress associated to $\boldsymbol{E}$ is then derived from the stress average rule:

$$\boldsymbol{\Sigma} = \overline{\boldsymbol{\sigma}} = \sum_i f_i \overline{\boldsymbol{\sigma}}^{G_i} \tag{50}$$

where f_i is the volume fraction of the solid grain G_i ($\sum_i f_i = 1 - \phi$) and $\overline{a}^{G_i}$ denotes the average of $a(z)$ over G_i:

$$\overline{a}^{G_i} = \frac{1}{|G_i|} \int_{G_i} a(z) \mathrm{d}V \tag{51}$$

A linear elastic behavior of both the solid and the cement is assumed. Let $\boldsymbol{C}^s$ denote the stiffness tensor of the solid. In the solid grain G_i, the linear state equation reads

$$z \in G_i : \quad \boldsymbol{\sigma}(z) = \boldsymbol{C}^s : \boldsymbol{\varepsilon}(z) \tag{52}$$

The idea of the interface as a geometrical and mechanical model of grain-to-grain interaction in the presence of cement consists in introducing a displacement jump $[\boldsymbol{\xi}]$ taking place at the boundary ∂G_i of the grain G_i. It can be splitted into a normal component $\xi_n = [\boldsymbol{\xi}] \cdot \boldsymbol{n}$ and a tangential one $[\boldsymbol{\xi}]_t = [\boldsymbol{\xi}] - \xi_n \boldsymbol{n}$, where $\boldsymbol{n}$ denotes the outward unit normal to ∂G_i. According to the assumption of linear elastic behavior of the cement, the stress vector acting on the interface $\boldsymbol{T} = \boldsymbol{\sigma} \cdot \boldsymbol{n}$ is linearly related to $[\boldsymbol{\xi}]$:

$$\begin{aligned} T_n &= \boldsymbol{T} \cdot \boldsymbol{n} = k_n \xi_n \\ \boldsymbol{T}_t &= \boldsymbol{T} - T_n \boldsymbol{n} = k_t [\boldsymbol{\xi}]_t \end{aligned} \tag{53}$$

which also reads

$$\boldsymbol{T} = \boldsymbol{k} \cdot [\boldsymbol{\xi}] \quad \text{with} \quad \boldsymbol{k} = k_n(\boldsymbol{n} \otimes \boldsymbol{n}) + k_t(\boldsymbol{1} - \boldsymbol{n} \otimes \boldsymbol{n}) \tag{54}$$

It is convenient to introduce a generalized strain field $\boldsymbol{\varepsilon}^{gen}$ defined as the sum of the smooth strain field $\boldsymbol{\varepsilon}$ in the solid grains and of the contribution of the displacement jumps.

$$\boldsymbol{\varepsilon}^{gen}(z) = \boldsymbol{\varepsilon}(z) + \sum_i [\boldsymbol{\xi}]_i(z) \overset{s}{\otimes} \boldsymbol{n}_i(z) \, \delta_{\partial G_i} \tag{55}$$

in which $\boldsymbol{n}_i$ is the unit normal to the grain boundary ∂G_i and $\delta_{\partial G_i}$ is the Dirac distribution the support of which is this grain boundary:

$$\int \delta_{\partial G_i}(z) f(z) \, dV = \int_{\partial G_i} f(z) \, dS \tag{56}$$

In particular, the strain average rule in fact involves the generalized strain field $\boldsymbol{\varepsilon}^{gen}$ and reads

$$\boldsymbol{E} = \overline{\boldsymbol{\varepsilon}^{gen}} = \frac{1}{|\Omega|} \left(\sum_i \int_{G_i} \boldsymbol{\varepsilon} \, dV + \sum_i \int_{\partial G_i} [\boldsymbol{\xi}]_i \overset{s}{\otimes} \boldsymbol{n}_i \, dS + \int_{\Omega_p} \boldsymbol{\varepsilon} \, dV \right) \tag{57}$$

where Ω_p denotes the pore space in Ω. In the linear case, the local response at the microscopic scale and in particular the local stress field $\boldsymbol{\sigma}(\underline{z})$ are expected to depend linearly on the loading parameter $\boldsymbol{E}$. According to the stress average rule (50), the macroscopic stress is itself linearly related to $\boldsymbol{E}$:

$$\boldsymbol{\Sigma} = \boldsymbol{C}_{hom} : \boldsymbol{E} \tag{58}$$

where $\boldsymbol{C}_{hom}$ is the homogenized stiffness tensor. Section 3.2 will provide a self-consistent estimate of the latter.

3.2 The Self-Consistent Homogenization Scheme

Following [2], a self-consistent approach seems to be the appropriate homogenization scheme for dealing with a polycrystal-like granular morphology. In short, the principle of the self-consistent reasoning in the mechanical framework consists in assuming that each phase (solid composite brick or pore) reacts to the mechanical loading applied to the *rev* as if it were an inclusion embedded in a homogeneous material which is the sought homogenized medium. It thus appears as the mechanical counterpart of Section 2.3. For simplicity, a spherical shape is adopted for the "inclusion" representing both a solid composite brick and a pore. Accordingly, an isotropic self-consistent estimate $\boldsymbol{C}_{ac}$ of $\boldsymbol{C}_{hom}$, characterized by the bulk and shear moduli k_{ac} and μ_{ac}, is expected:

$$\boldsymbol{C}_{ac} = 3k_{ac}\boldsymbol{J} + 2\mu_{ac}\boldsymbol{K} \tag{59}$$

The classical Eshelby-based self-consistent scheme and its application to porous media has been extensively described by many authors (see e.g. [3, 5, 8, 13]). In the present case, the new feature lies in the fact that the solid inclusion is not homogeneous but is made up of an elastic sphere *surrounded by an elastic interface*. The elastic sphere together with its interface will be denoted by $\mathcal{S}$ in the following. In the line of [4], the heterogeneous nature of this inclusion requires to introduce a generalized Eshelby problem which can be viewed as a particular example of the morphologically representative pattern technique [14]. On this basis, a generalized self-consistent approach can be derived [2]. As usual, an auxiliary macroscopic strain $\boldsymbol{E}_0$ is introduced. The strain average rule will be used later in order to relate $\boldsymbol{E}_0$ to the macroscopic strain $\boldsymbol{E}$. Two elementary problems are considered. In each case, an inclusion is embedded in an infinite medium with stiffness $\boldsymbol{C}_{ac}$, with the boundary conditions $\underline{\xi} \rightarrow \boldsymbol{E}_0 \cdot \underline{z}$ at infinity. Problem I is the classical Eshelby problem in which the inclusion is a spherical pore $\mathcal{P}$. $\boldsymbol{S}^{ac}_{sph}$ denoting the Eshelby tensor of the sphere, the average strain in the pore $\mathcal{P}$ reads

$$\overline{\boldsymbol{\varepsilon}}^{\mathcal{P}} = \left(\boldsymbol{I} - \boldsymbol{S}^{ac}_{sph}\right)^{-1} : \boldsymbol{E}_0 \tag{60}$$

In problem II, the inclusion is a solid sphere with radius R and stiffness tensor $\boldsymbol{C}^s$, surrounded by an interface (normal and tangential stiffness coefficients k_n and k_t). This problem takes the following form:

$$\begin{cases} r < R & \boldsymbol{\sigma}_{II} = \boldsymbol{C}_s : \boldsymbol{\varepsilon}_{II} \\ r = R & \boldsymbol{\sigma}_{II} = \mathbf{k}[\boldsymbol{\xi}_{II}] \\ r > R & \boldsymbol{\sigma}_{II} = \boldsymbol{C}_{hom} : \boldsymbol{\varepsilon}_{II} \\ |z| \to \infty & \boldsymbol{\xi}_{II} = \boldsymbol{E}_0 \cdot \boldsymbol{z} \end{cases} \tag{61}$$

with $[\boldsymbol{\xi}] = \boldsymbol{\xi}(R^+) - \boldsymbol{\xi}(R^-)$. The average stress $\overline{\boldsymbol{\sigma}}^{\mathcal{S}}$ in the composite solid inclusion reads

$$\overline{\boldsymbol{\sigma}}^{\mathcal{S}} = \frac{3}{4\pi R^3} \int_{r<R} \boldsymbol{\sigma}_{II} \, dV = \frac{3}{4\pi R^2} \int_{r=R} \boldsymbol{e}_r \otimes \boldsymbol{\sigma}_{II} \cdot \boldsymbol{e}_r \, dS \tag{62}$$

while the average strain $\overline{\boldsymbol{\varepsilon}}^{\mathcal{S}}$ involves the displacement $\boldsymbol{\xi}_{II}(R^+)$ on the external lip of the interface ($r = R^+$):

$$\overline{\boldsymbol{\varepsilon}}^{\mathcal{S}} = \frac{3}{4\pi R^3} \int_{r=R} \boldsymbol{\xi}_{II}(R^+) \overset{s}{\otimes} \boldsymbol{e}_r \, dS \tag{63}$$

Since the displacement and stress fields $\boldsymbol{\xi}_{II}$ and $\boldsymbol{\sigma}_{II}$ clearly linearly depend on $\boldsymbol{E}_0$, this also holds for the averages $\overline{\boldsymbol{\varepsilon}}^{\mathcal{S}}$ and $\overline{\boldsymbol{\sigma}}^{\mathcal{S}}$:

$$\overline{\boldsymbol{\varepsilon}}^{\mathcal{S}} = \boldsymbol{A}_0^{\mathcal{S}} : \boldsymbol{E}_0 \quad ; \quad \overline{\boldsymbol{\sigma}}^{\mathcal{S}} = \boldsymbol{B}_0^{\mathcal{S}} : \boldsymbol{E}_0 \tag{64}$$

The idea of the homogenization scheme is to adopt $\overline{\boldsymbol{\varepsilon}}^{\mathcal{S}}$ (resp. $\overline{\boldsymbol{\sigma}}^{\mathcal{S}}$) as an estimate for the average strain (resp. stress) in the whole solid phase of the *rev*. In particular, this amounts to assuming that the average $\overline{\boldsymbol{\sigma}}^{G_i}$ is independent of the considered grain: $\forall i,\ \overline{\boldsymbol{\sigma}}^{G_i} \approx \overline{\boldsymbol{\sigma}}^{\mathcal{S}}$. (50) then further implies that

$$\boldsymbol{\Sigma} = (1-\phi)\overline{\boldsymbol{\sigma}}^{\mathcal{S}} = (1-\phi)\boldsymbol{B}_0^{\mathcal{S}} : \boldsymbol{E}_0 \tag{65}$$

We still have to relate the auxiliary strain $\boldsymbol{E}_0$ to the macroscopic strain $\boldsymbol{E}$. Adopting $\overline{\boldsymbol{\varepsilon}}^{\mathcal{P}}$ (see (60)) as an estimate of the average strain in the pore space, the strain average rule reads

$$\boldsymbol{E} = \phi\overline{\boldsymbol{\varepsilon}}^{\mathcal{P}} + (1-\phi)\overline{\boldsymbol{\varepsilon}}^{\mathcal{S}} = \left(\left(\boldsymbol{I} - \boldsymbol{S}_{sph}^{ac}\right)^{-1} + \boldsymbol{A}_0^{\mathcal{S}}\right) : \boldsymbol{E}_0 \tag{66}$$

Eventually, a combination of (65) and (66) yields the macroscopic state equation in the form

$$\boldsymbol{\Sigma} = (1-\phi)\boldsymbol{B}_0^{\mathcal{S}} : \left(\left(\boldsymbol{I} - \boldsymbol{S}_{sph}^{ac}\right)^{-1} + \boldsymbol{A}_0^{\mathcal{S}}\right)^{-1} : \boldsymbol{E} \tag{67}$$

which amounts to the derivation of the following self-consistent estimate of the effective stiffness:

$$\boldsymbol{C}_{ac} = (1-\phi)\boldsymbol{B}_0^{\mathcal{S}} : \left(\left(\boldsymbol{I} - \boldsymbol{S}_{sph}^{ac}\right)^{-1} + \boldsymbol{A}_0^{\mathcal{S}}\right)^{-1} \tag{68}$$

Comparison of (68) with (59) yields two coupled equations to be solved by the unknowns k_{ac} and μ_{ac}. Note that, in practice, the latter can be obtained in considering successively a purely spherical auxiliary strain tensor $\boldsymbol{E}_0 = E_0\mathbf{1}$ and a purely deviatoric one [2]. More precisely, the first equation can be solved for k_{ac} in the form

$$k_{ac} = \frac{4(1-\phi)\mu_{ac} R k_n k_s}{12\mu_{ac} k_s + 4\mu_{ac} k_n R + 3\phi R k_n k_s} \tag{69}$$

The shear modulus μ_{ac} is the solution of a third order equation:

$$A\mu_{ac}^3 + B\mu_{ac}^2 + C\mu_{ac} + D = 0 \tag{70}$$

where A, B, C, D are polynomial functions of $k_s, \mu_s, Rk_n, Rk_t, \phi$.

In the sequel, we shall restrict to the case of incompressible grains ($k_s \sim \infty$) and incompressible interfaces ($k_n \sim \infty$). The link (69) between the homogenized bulk and shear moduli then reduces to

$$k_{ac} = \frac{4(1-\phi)}{3\phi}\mu_{ac} \tag{71}$$

The homogenized behavior now depends on μ_s, on the non-dimensional parameter $\kappa = k_t R/\mu_s$ and on the porosity ϕ. Furthermore, the homogenized estimate of μ_{ac} proves to be solution to the following polynomial equation [2]:

$$16(5+\kappa)(3-\phi)\left(\frac{\mu_{ac}}{\mu_s}\right)^2 + [(9+77\phi)\kappa + 114(3\phi-1)]\frac{\mu_{ac}}{\mu_s} + 57\kappa(2\phi-1) = 0 \tag{72}$$

We denote by $M(\kappa, \phi)$ the positive solution of (72):

$$\mu_{ac} = M(\kappa,\phi)\mu_s \quad ; \quad k_{ac} = \frac{4(1-\phi)}{3\phi} M(\kappa,\phi)\mu_s \tag{73}$$

At least theoretically, (73) suggests that the value of κ could be derived from the comparison between the effective linear elastic properties of the sandstone and that of the solid making up the grains. In practice, the determination of κ from the elastic properties is more complicated since the latter are significantly affected by the confining pressure.

The limit case $\kappa \to \infty$ corresponds to the usual situation of a polycrystal made up of linear elastic grains with perfect interfaces (no displacement discontinuity at the grain-to-grain boundary). The usual self-consistent solution is retrieved:

$$\lim_{\kappa \to \infty} M = 3\frac{1-2\phi}{3-\phi} \tag{74}$$

The limit case $\kappa \to 0$ corresponds to the situation of smooth (frictionless) interfaces. The conclusion then depends on the porosity. If $\phi < 1/3$, the effective stiffness is still of the order of the grain stiffness. Indeed, it is found that

$$\phi < 1/3 : \quad \lim_{\kappa \to 0} M = \frac{57}{40}\frac{1-3\phi}{3-\phi} \tag{75}$$

However, for higher values of the porosity, the effective stiffness vanishes as compared to the grain stiffness . In fact, for $\phi \in]1/3, 1/2[$, the effective stiffness is of the order of $k_t R \ll \mu_s$, which amounts to saying that M is of the order of κ:

$$\phi > 1/3 : \quad M = \frac{\kappa}{2}\frac{1-2\phi}{3\phi-1} + O(\kappa) \tag{76}$$

4 Concluding Remarks

In both the hydraulic and the mechanical frameworks, the interface contribution appears through non-dimensional parameters ($\delta = \eta / k_m R$ and $\kappa = k_t R / \mu_s$). This emphasizes the fact that the effect of interfaces introduces an internal length, namely the radius of the interfaces (in the spherical case). As far as the mechanical properties are concerned, this suggests an explanation for experimental evidence showing size effect. A typical example is given by the brittle failure of granular rocks like sandstones [10, 11].

For the sake of brevity, the matrix-inclusion situation is not included in the mechanical part of this paper (Section 3). Still, it can be easily addressed from the solution to the generalized Eshelby problem considered in Section 3.2. From a practical point of view, mortar is an interesting illustration of the influence of interfaces on elastic properties. If perfect cohesion at the contact between sand grains and cement paste is assumed, then the standard Hashin–Shtrikman bound (using the cement paste as reference medium) should provide a lower bound of the effective properties which is invalidated by the experimental data [12]. This clearly indicates that the perfect cohesion assumption is not realistic and emphasizes the interest of an interface-based model of the cement-paste-to-grain contact.

Another important example is provided by the nanocomposites (see e.g. [1]). In this kind of materials, the existence of an interface around an inclusion embedded in a matrix implies a discontinuity of the stress vector. The latter is the mechanical counterpart of the discontinuity of the filtration velocity in the hydraulic problem. The variational approach developed in Section 2.2 can be readily extended to the derivation of bounds for the elastic properties of nanocomposites.

References

1. Duan, H.L., Wang, J., Huang, Z.P., and Karihaloo, B.L., Size-dependent effective elastic constants of solids containing nano-inhomogeneities with interface stress. *J. Mech. Phys. Solids*, 53:1574–1596, 2005.
2. Dormieux, L., Sanahuja, J., and Maalej, Y., Résistance d'un polycristal avec interfaces imparfaites. *C. R. Mécanique*, 335:25–31, 2007.
3. Dormieux, L., Kondo, D., and Ulm, F.-J., *Microporomechanics*. Wiley, 2006.
4. Hashin, Z., The spherical inclusion with imperfect interface. *J. Applied Mech.*, 58:444–449, 1991.
5. Kröner, E., Self-consistent scheme and graded disorder in polycrystal elasticity. *J. Phys. F: Met. Phys.*, 8:2261–2267, 1978.
6. Ponte-Castanada, P. and Willis, J.R., The effect of spatial distribution on the effective behavior of composite materials and cracked media. *J. Mech. Phys. Solids*, 43:1919–1951, 1995.
7. Scrivener, K.L., Crumbie, A.K., and Laugesen, P., The interfacial transition zone (ITZ) between cement paste and aggregate in concrete. *Interface Science*, 12(4):411–421, 2004.
8. Suquet, P., Effective behavior of non linear composites. In *Continuum Micromechanics*, P. Suquet (Ed.). Springer, 1997.

9. Willis, J.R., Variational and related methods for the overall properties of composites. *Advances of Applied Mechanics*, 21:1–78, 1981.
10. Wong, T.F., David, C., and Zhu, W., The transition form brittle to cataclastic flow: Mechanical deformation. *Journal of Geophysical Research*, 102(B2-10):3009–3025.
11. Wong, T.F., David, C., and Menendez, B., Mechanical compaction. In *Mechanics of Fluid-Saturated Rocks*, Y. Gueguen and M. Bouteca (Eds.). Elsevier, pp. 55–114, 2004.
12. Yang, C.C., Effect of the transition zone on the elastic moduli of mortar. *Cement and Concrete Research*, 28(5):727–736, 1998.
13. Zaoui, A., Continuum micromechanics: Survey. *Journal of Engineering Mechanics*, 128(8):808–816, 2002.
14. Zaoui, A., Structural morphology and constitutive behavior of microheterogeneous materials. In *Continuum micromechanics*, P. Suquet (Ed.). Springer, 1997.

An Extended Finite Element Method for the Analysis of Submicron Heat Transfer Phenomena

Pilhwa Lee[1], Ronggui Yang[1] and Kurt Maute[1,2]

[1]Department of Mechanical Engineering, UCB 427, University of Colorado, Boulder, CO 80309-0427, USA; pilhwa.lee@colorado.edu, ronggui.yang@colorado.edu
[2]Center for Aerospace Structures, Department of Aerospace Engineering Sciences, UCB 429, University of Colorado, Boulder, CO 80309-0429, USA; kurt.maute@colorado.edu

Abstract Manipulating the spatial layout of heterogeneous materials at submicron scales allows for the design of novel nano-engineered material with unique thermal properties. To analyze heat conduction at submicron scales of geometrically complex nano-structured materials, an extended finite element method (XFEM) is presented. Appropriate for both diffusive and ballistic domains, heat conduction is described by the phonon Boltzmann transport equations. Specifically, the gray phonon model is used along with a diffusive scattering model describing the transmission and reflection of phonons at material interfaces. The geometry of the material interfaces is described by a level-set approach. The phonon distribution is discretized in the velocity space by a discrete ordinate approach and in the spatial domain by a stabilized Galerkin finite element method. Discontinuities of the phonon distribution across material interfaces are captured via enriched shape functions. To enforce interface scattering conditions and boundary conditions, a stabilized Lagrange multiplier method is presented. The proposed method is verified through comparison with benchmark results. The utility of the XFEM approach is demonstrated through the thermal analysis of experimentally characterized material samples and computer-designed nano-composites.

Key words: Extended finite element method, gray phonon model, diffusive mismatch model, phonon Boltzmann transport equation, stabilized Lagrange multiplier formulation

1 Introduction

The ability to design, manipulate, and realize materials and devices at submicron and nanometer scales is likely to lead to revolutionary advances across a wide spectrum of technologies. Early successes have been achieved in nano-electronics, nano-materials, and nano-medicine. Despite these advances, the design at submicron scales is typically done in an ad-hoc manner. While scientific understanding and fabrication techniques for nano-scale materials and devices have dramatically advanced and have become increasingly accessible in recent years, appropriate formal design

R. de Borst and E. Ramm (eds.), *Multiscale Methods in Computational Mechanics*, Lecture Notes in Applied and Computational Mechanics 55, DOI 10.1007/978-90-481-9809-2_11,

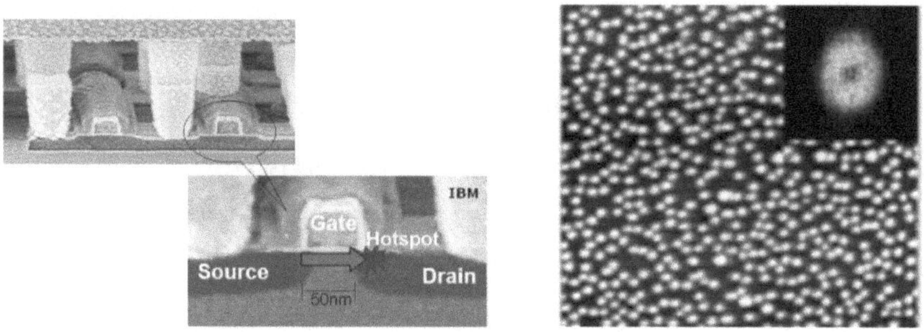

(a) (b)

Fig. 1. Heat transfer problems at submicron scale: (a) a nano-scale transistor that generates heat in a $\approx$ 10 nm region, (b) thermoelectric PbSeTe/PbTe quantum-dot super-lattice [18].

approaches that take advantage of these advances are lacking. In particular, analysis methods are needed that allow the efficient analysis and development of engineered designs. Traditional atomistic modeling methods predominately developed for advancing the scientific understanding of nano-structures are often computationally too costly to be employed in an iterative design process.

This paper is concerned with the development of an analysis tool that allows efficiently and accurately analyzing heat transfer processes at submicron scales. Managing the heat transfer at such small scales plays an increasingly important role in a broad range of applications. For example, in recent years the size of transistors has dramatically decreased and the number of transistors on a chip increased, requiring novel thermal management strategies [14, 28]. In particular the emergence of nano-scale hot-spots (see Figure 1a) requires design concepts that mitigate the thermal impact on the electronic performance. Another example is the development of nano-composites for thermoelectric energy conversion [9]. The thermoelectric efficiency can be increased by simultaneously increasing the electric conductivity and decreasing the thermal conductivity. The latter effect has been utilized by quantum-dot composites and super-lattice structures (see Figure 1b).

The analysis of heat transfer phenomena at submicron scales requires new engineering analysis tools as the commonly used Fourier model of heat conduction fails at such small length scales [17, 22, 36]. Instead, the thermal properties of a material need to be derived from quantum mechanics and statistical mechanics. In this study we assume that heat is solely propagated via the vibration of atoms, i.e. via phonons. The assumption holds for dielectric materials and semiconductors. However, in electric conductors electrons are the main heat carriers. Therefore this class of materials is not considered in the sequel but the proposed numerical framework can be easily extended to include electrons as carriers for heat conduction.

The vibration of a periodic lattice of atoms is typically described by the dispersion characteristics which defines the relation between the lattice vibration frequency $\omega_{q,p}$ and the wave vector $\boldsymbol{k}$. The group velocity, with which the energy is propagated, is given by the slope of dispersion curve: $\boldsymbol{v}_{q,p} = d\omega_{q,p}/d\boldsymbol{k}$. Dispersion diagrams for

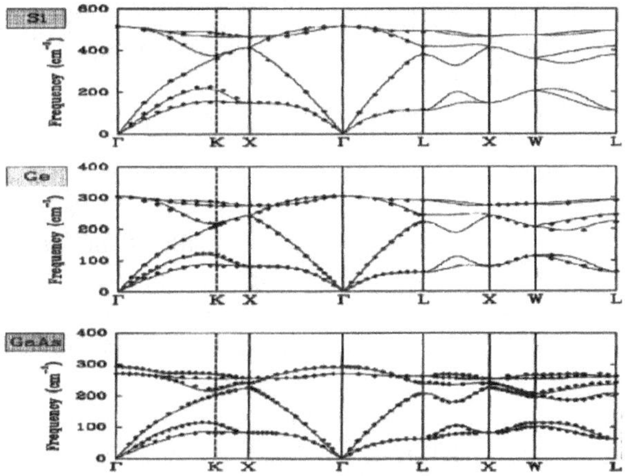

Fig. 2. Dispersion diagrams for Si, Ge, and GeAs.

typical semiconductor materials are shown in Figure 2. We can distinguish acoustic and optical branches, denoted by the subscript q, as well as different polarization p depending on the type and orientation of the vibration modes. A phonon is a quantized mode of vibration associated with the lattice vibration frequency $\omega_{q,p}$.

Recalling the wave-particle duality in quantum mechanics, we can consider the energy transport both as a wave phenomenon and a particle transport process, depending on length scale of interest. As heat conduction in solids is dominated by modes with a wavelength of $\approx 1\ nm$ and the feature sizes in nano-structures is typically larger than $1\ nm$, we conveniently describe the heat transport by a quasi-particle model. The associated energy density $e(\omega_{q,p})$ of a phonon is

$$e(\omega_{q,p}) = n_{q,p}(\omega_{q,p})\ \hbar\ \omega_{q,p}\ D_{q,p}(\omega_{q,p}) \tag{1}$$

where $n_{q,p}$ is the particle density, $\hbar$ the Planck constant, and $D_{q,p}$ the number of modes per unit volume.

The energy transport via phonons is described by kinetic theory, in particular the Boltzmann transport equations:

$$\frac{\partial n_{q,p}}{\partial t} + \boldsymbol{v}_{q,p} \cdot \frac{\partial n_{q,p}}{\partial \boldsymbol{x}} = C(n_{q,p}) \tag{2}$$

For designing nano-composites it is most often sufficient to consider only steady-state conditions. In this case the time derivative term vanishes. The collision term $C(n_{q,p})$ is typically formulated via a relaxation time approximation which is similar to the Bathnagar–Gross–Krook (BGK) collision operator [4] frequently used in rarefied gas dynamics:

$$C(n_{q,p}) = \frac{n^{\text{eq}}(n_{q,p}, \boldsymbol{v_{q,p}}) - n_{q,p}}{\tau_{q,p}} \tag{3}$$

where n^{eq} is the equilibrium distribution and $\tau_{q,p}$ the relaxation time. The latter depends on the mean free path length $\Lambda_{q,p}$ and the group velocity of the phonons:

$$\Lambda_{q,p} = \|\boldsymbol{v}_{q,p}\|\tau_{q,p} \tag{4}$$

Depending on the ratio of the mean free path length over the characteristic length of the heat transport problem, that is the Knudsen number, we distinguish two regimes of heat transport. For Knudsen numbers smaller than 0.1, diffusive effects dominate and Fourier's law is valid. When Knudsen numbers are larger than one, i.e. the feature size is smaller or equal the mean free path length of the phonons, a phonon model is needed to capture ballistic phenomena, which is crucial to predict correctly heat conduction.

A fundamental difference between the diffusive and ballistic regime is the transport of heat across material interfaces. Unlike in the diffusive regime, the scattering of phonons at material interfaces leads to an interfacial thermal resistance [36] and a discontinuity in the phonon distribution. Accurately predicting the scattering of phonons requires modeling nonlocal and non-equilibrium phenomena by resolving the lattice vibrations [8]. In this study, we approximate the phonon scattering effects at material interfaces via a diffusive mismatch model [36].

The Boltzmann transport equations provide a convenient framework for analyzing the heat transport by phonons as standard discretization schemes for partial differential equations can be applied. The phonon Boltzmann transport equations have been solved by Direct Monte Carlo Simulation [19, 37], finite volume schemes in combination with discrete ordinate methods [23, 27], and finite element methods [13]. Finite element formulations have also been applied to solving the neutron Boltzmann transport equations [11, 32]. However, these numerical schemes have been developed primarily to study rather simple material layouts, such as spherical or square inclusions in nano-composites and super-lattices.

In this study we propose a novel computational framework based on a finite element approach. To conveniently analyze complex geometries and vary the geometries in the design process, we describe the material layout by a level-set method and enforce the boundary and interface conditions via an immersed boundary method [16, 33]. This approach avoids the often cumbersome and lengthy generation of unstructured meshes aligned with the material interfaces. In particular, we adopt the extended finite element method (XFEM) that was originally proposed for modeling the propagation of cracks [3, 12, 26]. Recently, XFEM has been applied, for example, to phase change problems [10] and fluid-structure interaction problems [15, 38]. For solving the Boltzmann transport equations we locally enrich the spatial discretization of the phonon distribution to capture discontinuities at material interfaces and enforce interface scattering conditions and boundary conditions with a stabilized Lagrange multiplier method. We apply this computational approach to a simplified version of the phonon Boltzmann transport equations based on the gray-phonon model, verify its accuracy through comparison with results in the literature, and demonstrate its versatility with the analysis of a typical nano-composite and a conceptual design study.

The remainder of this paper is structured as follows: In Section 2 we briefly outline the level-set approach for describing the geometry of the material interfaces. The governing equations of the gray phonon model are summarized in Section 3. The methods for discretizing the gray phonon distribution in the velocity space and the spatial domain as well as the Lagrange multiplier method for enforcing the boundary and interface conditions are described in Section 4. We report on verification, analysis, and design studies in Section 5.

2 Level-Set Description of Material Layout

The geometry of the material layout is described implicitly by the level set method. This approach was first proposed by Osher and Sethian [31] and has been successfully applied in image processing and to describe the evolution of interfaces in systems governed by partial differential equations, such as multi-phase flows [35]. Recently level-set method have become popular for shape and topology optimization purposes [1]. For general overviews of level-set methods and their applications the reader is referred to [29, 30, 34].

The geometry of a two-phase composite can be described via the level set function $\phi(\boldsymbol{x})$ as follows:

$$\Omega_1 = \{x \,|\, \phi(\boldsymbol{x}) > 0\} \tag{5}$$

$$\Omega_2 = \{x \,|\, \phi(\boldsymbol{x}) < 0\} \tag{6}$$

$$\Gamma = \{x \,|\, \phi(\boldsymbol{x}) = 0\} \tag{7}$$

where phase 1 occupies subdomain Ω_1 and phase 2 subdomain Ω_2.

A level set representation of a Si-Ge nano-composite material is shown in Figure 3. Based on the gray intensity of an image of the composite a level-set function is initialized. To reduce the noise in the image and to simplify the geometry, the level set function is smoothed. The contour line of the zero level set is plotted representing the interface of the material phases. This level-set description of the material layout can be used by numerical analysis methods in combination with immersed boundary techniques, such as XFEM, to enforce interface and boundary conditions (see Section 4).

3 Gray Phonon Model

The Boltzmann transport equations (2) describe the evolution of the phonon distributions $n_{q,p}$ associated with the lattice frequencies $\omega_{q,p}$ of all branches q and polarization p. Depending on the dispersion characteristics, a large number of phonon distributions need to be modeled to accurately predict the total energy transport. Instead, one can often simplify the phonon model by considering only one phonon distribution carrying the total energy density $\bar{e}$ and traveling with an average group velocity $\bar{v}$:

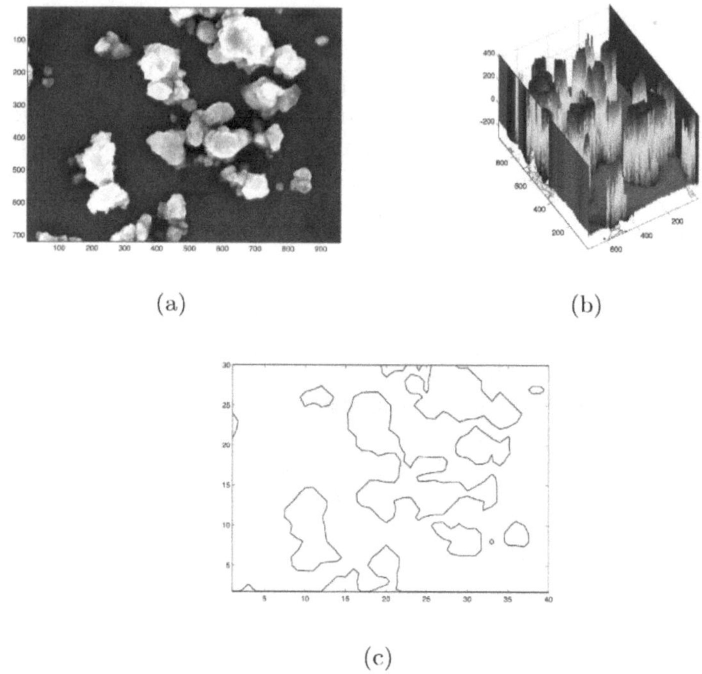

Fig. 3. Level set description: (a) SEM image of Si-Ge nano-composite – brighter color marks Si nano-particles [21], (b) smoothed level set function, and (c) contour plot of zero level set.

$$\bar{e} = \sum_{q,p} \int_0^{\omega_{max}} n_{q,p}(\omega_{q,p})\, \hbar\, \omega_{q,p}\, D_{q,p}(\omega_{q,p})\, d\omega_{q,p} \tag{8}$$

The mode density $D_{q,p}$ and the cut-off frequency ω_{max} can be approximated by the Debye model [42]. The simplified model is referred to gray phonon model.

Introducing the gray phonon model into the Boltzmann transport equations (2) it is convenient to rewrite the equations with $\bar{e}$ being the total energy density distribution function. Restricting this study to steady-state problems, the governing equations are

$$\boldsymbol{s} \cdot \frac{\partial \bar{e}}{\partial \boldsymbol{x}} = \frac{\bar{e}^{\mathrm{eq}} - \bar{e}}{\Lambda} \tag{9}$$

where $\boldsymbol{s}$ is a unit vector describing the direction of the phonon motion:

$$\boldsymbol{s} = (\cos(\pi\mu), \sin(\pi\mu)), \quad \mu \in [-1, 1) \tag{10}$$

We assume that the phonons at equilibrium follow the Bose–Einstein statistics:

$$n_{q,p}^{eq} = \frac{1}{e^{\frac{\hbar\omega_{q,p}}{k_{\mathrm{B}}T}} - 1} \tag{11}$$

where k_B is the Boltzmann constant and T is the macroscopic temperature. The equilibrium distribution $\bar{e}^{eq}$ can be determined analogously to (8) with $n_{q,p} = n^{eq}_{q,p}$. Summing over all branches and polarizations and integrating over all lattice frequencies $\omega_{q,p}$, the equilibrium distribution $\bar{e}^{eq}$ is only a function of the macroscopic temperature T which is defined via the total energy carried by all phonons $\bar{e}$:

$$\frac{1}{2}\int_{-1}^{1} \bar{e}\, d\mu = C\left(T - T_{\text{ref}}\right) \tag{12}$$

where C is the heat capacity and T_{ref} is a reference temperature. The equilibrium distribution can be expressed as an integral of the directional distributions function $\bar{e}$ as follows:

$$\bar{e}^{eq} = \frac{1}{2}\int_{-1}^{1} \bar{e}\, d\mu \tag{13}$$

To analyze the thermal properties of materials, one of the following boundary conditions is frequently imposed:

$$\bar{e}(\boldsymbol{x},\mu)|_L - \bar{e}(\boldsymbol{x},\mu)|_R = \text{constant} \tag{14}$$

$$\text{For } \boldsymbol{s}\cdot\boldsymbol{n} < 0,\ x \in \partial D,\ \ \bar{e}(\boldsymbol{x},\mu) = \bar{e}(\boldsymbol{x},-\mu) \tag{15}$$

$$\text{For } \boldsymbol{s}\cdot\boldsymbol{n} > 0,\ x \in \partial D,\ \ \bar{e}(\boldsymbol{x},\mu) = \bar{e}^{*}(\boldsymbol{x},\mu) \tag{16}$$

$$\text{For } x \in \partial D,\ \ T(\bar{e}) = T^{*} \tag{17}$$

where $\boldsymbol{n}$ is outwards pointing normal, and ∂D is the domain boundary. Equations (14) and (15) represent periodic and reflective boundary conditions, respectively. Equation (16) enforces an emitting heat flux and Eq. (17) prescribes the temperature.

While more accurate and complex models for predicting phonon scattering at material interfaces are available, for the sake of simplicity, we assume in this study that the phonon scattering at material interfaces is diffusive. The interface scattering is described by the diffusive mismatch model (DMM) which assumes that phonons emerging from an interface are independent of their origin. The transmission and reflection of phonons at the material interfaces are governed by energy balance conditions. In the case of a two-phase composite, these conditions can be written as follows:

For $\boldsymbol{s}\cdot\boldsymbol{n} > 0$

$$\bar{e}^{-}(\boldsymbol{x},\mu) = -\epsilon_{21}\int_{\boldsymbol{s}\cdot\boldsymbol{n}<0} \boldsymbol{s}\cdot\boldsymbol{n}\,\bar{e}^{-}\, d\mu + \alpha_{12}\int_{\boldsymbol{s}\cdot\boldsymbol{n}>0} \boldsymbol{s}\cdot\boldsymbol{n}\,\bar{e}^{+}\, d\mu \tag{18}$$

$$\bar{e}^{+}(\boldsymbol{x},\mu) = -\alpha_{21}\int_{\boldsymbol{s}\cdot\boldsymbol{n}<0} \boldsymbol{s}\cdot\boldsymbol{n}\,\bar{e}^{-}\, d\mu + \epsilon_{12}\int_{\boldsymbol{s}\cdot\boldsymbol{n}>0} \boldsymbol{s}\cdot\boldsymbol{n}\,\bar{e}^{+}\, d\mu \tag{19}$$

where

$$\bar{e}^{-}(\boldsymbol{x},\mu) = \lim_{\epsilon\to 0,\epsilon>0} \bar{e}(\boldsymbol{x}+\epsilon\boldsymbol{n}) \tag{20}$$

$$\bar{e}^{+}(\boldsymbol{x},\mu) = \lim_{\epsilon\to 0,\epsilon>0} \bar{e}(\boldsymbol{x}-\epsilon\boldsymbol{n}) \tag{21}$$

and α_{ij} and ϵ_{ij} denote the transmission coefficient and reflection coefficient from phase i to phase j. The unit vector $\boldsymbol{n}$ is normal to the interface and points towards phase 2. The transmission coefficient is determined by

$$\alpha_{ij} = \frac{C_j \bar{v}_j}{\sum_k C_k \bar{v}_k} \tag{22}$$

where $\bar{v}_k$ is the average phonon group velocity for phase k. The specific heat capacity of phase k is denoted to C_k. Energy conservation at the interface implies that $\alpha_{ij} + \epsilon_{ij} = 1$. The assumption that scattered phonons bear no relationship to their origin, i.e. whether they are transmitted or reflected, requires that α_{ij} is equal to ϵ_{ji}.

4 Discretization Methods

In the gray phonon model, a single state variable $\bar{e}$ defines the phonon distribution at point $\boldsymbol{x}$ traveling with the velocity $\bar{v}\boldsymbol{s}$. To develop a numerical analysis method, the phonon distribution $\bar{e}$ is discretized both in velocity and spatial domains. For the discretization in the velocity domain, spherical harmonics and discrete ordinate methods are typically used [24, 40]. Spherical harmonics approximations are based on global, periodic functions and are efficient for smooth phonon distributions, typically for diffusive energy transport. In the ballistic regime, the phonon distribution in the velocity space often lacks smoothness. Therefore we apply a discrete ordinate method which is based on a local, discontinuous approximation but requires a fine resolution of the velocity space. Note only the angular space needs to be discretized as the gray phonon model uses a single, average group velocity.

The spatial discretization can be performed by finite difference methods, finite volume methods, and Galerkin finite element methods. In this study we focus on finite element approaches. Due to the convective operator in the Boltzmann transport equations (2) and (9), the numerical solution needs to be stabilized either by a discontinuous Galerkin formulation [13, 41], a Galerkin least-square approach [6], or a streamline upwind Petrov–Galerkin (SUPG) stabilization [7]. The latter approach is adopted in this study due to its computational efficiency. To allow for a convenient analysis of complex geometries, we embed the SUPG approach into an XFEM formulation and enforce boundary and interface conditions via a Lagrange multiplier method.

4.1 Discrete Ordinate Method

The velocity space is discretized uniformly with equal weights for all discrete directions. We resolve the velocity domain by N directions:

$$s_i \cdot \nabla e_i = \frac{1}{\Lambda}\left(\sum_{j=1}^{N} w_j e_j - e_i\right) \quad \text{for } i = 1, ..., N \tag{23}$$

$$s_i = \left(\cos\left(2\pi\left(\frac{i}{N} - \frac{1}{2N}\right)\right), \sin\left(2\pi\left(\frac{i}{N} - \frac{1}{2N}\right)\right)\right) \tag{24}$$

$$w_i = \frac{1}{N} \tag{25}$$

where s_i is the discrete directional unit vector and the weighting factors w_i are chosen to be uniform. This discretization leads to the semi-discrete form of the gray phonon Boltzmann equations which can be written in a weak formulation as follows:

$$R_i = \int_{\Omega} \delta e_i \left\{ s_i \cdot \nabla e_i - \frac{1}{\Lambda}\left(\sum_{j=1}^{N} w_j e_j - e_i\right)\right\} d\Omega + R_{i,\text{stab}} + R_{i,\lambda} = 0 \tag{26}$$

where $R_{i,\text{stab}}$ and $R_{i,\lambda}$ denote contributions from the SUPG stabilization and the interface constraints, respectively. Both terms will be discussed in detail below.

4.2 Extended Finite Element Method

The main idea of XFEM is to enrich shape functions in elements where a boundary/interface resides in order to capture discontinuities in the solution. The enrichment function depends on the order of the discontinuity. In the phonon scattering problem, there is a jump in the phonon distribution at the material interface. Assuming again a two-phase composite, we capture this jump by enriching the shape functions with a Heaviside function as follows:

$$\bar{e}(\boldsymbol{x}) = \sum_{I} N_I(\boldsymbol{x})(\tilde{e}_I + \psi(\boldsymbol{x})\hat{e}_I) \tag{27}$$

$$\psi(\boldsymbol{x}) = H(\boldsymbol{x}) = \begin{cases} +1 & \forall \boldsymbol{x} \in \Omega_1 \\ 0 & \forall \boldsymbol{x} \in \Omega_2 \end{cases} \tag{28}$$

where $N_I(\boldsymbol{x})$ are standard shape functions, and $N_I(\boldsymbol{x})\psi(\boldsymbol{x})$ are the enriched shape functions. The standard degrees of freedom and the enriched degrees of freedom are denoted by $\tilde{e}_I$ and $\hat{e}_I$, and the domains of phase 1 and 2 are denoted by Ω_1 and Ω_2, respectively. In this study, $N_I(\boldsymbol{x})$ are bilinear shape functions. Following an isoparametric approach, the element geometry is also discretized by the shape functions $N_I(\boldsymbol{x})$.

To increase the numerical accuracy, the intersected elements are integrated piecewise based on a Delaunay triangulization which is aligned with the intersection:

$$\int_{\Omega_{\text{ELE}}} R(\boldsymbol{x}) d\Omega = \sum_{i} \int_{\Omega_i} R(\boldsymbol{x}) d\Omega \tag{29}$$

where Ω_{ELE} is the domain of an extended element, and Ω_i is the ith triangular subdomain of Ω_{ELE}. The triangular subdomains are shown in Figure 4b.

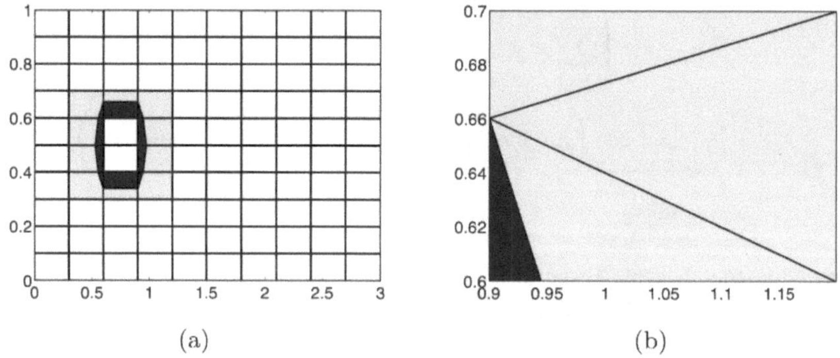

Fig. 4. XFEM mesh and Delaunay triangulization; (a) the elements intersected by the interface are enriched with Heaviside function. (b) the extended element is partitioned for the triangular integration.

An SUPG approach is used to stabilize the convective term leading to the additional residual term $R_{i,\text{stab}}$:

$$R_{i,\text{stab}} = \int_{\Omega} \tau_{\text{stab}}\, \boldsymbol{s}_i \cdot \nabla \delta e \left\{ \boldsymbol{s}_i \cdot \nabla e_i - \frac{1}{\Lambda} \left(\sum_{j=1}^{N} w_j e_j - e_i \right) \right\} d\Omega \tag{30}$$

where τ_{stab} is the stabilization factor. Owing to the simplicity of the convective term, the stabilization factor is equal the element size, i.e. $\tau_{\text{stab}} = h$.

4.3 Lagrange Multiplier Method

The boundary conditions and interface conditions are enforced by a Lagrange multiplier method. With $c(\bar{e}_i) = 0$ representing a generic boundary and interface condition, the contributions of the Lagrange multiplier formulation to the weak form of the governing equations (26) are

$$R_{i,\lambda} = \int_{\Gamma} \delta\lambda \; c \; d\Gamma + \int_{\Gamma} \delta e_i \; \frac{\partial c}{\partial e_i} \lambda \; d\Gamma \tag{31}$$

where the Lagrange multiplier $\lambda(\boldsymbol{x})$ is discretized by standard shape functions.

The Lagrange multiplier method leads to mixed finite element formulation. The orders of the approximation of λ and $\bar{e}$ need to satisfy the Ladysenskaja–Babuska–Brezzi inf-sup condition [5]. This condition can be easily satisfied when imposing constraints along element edges as the relation between the edge length and the size of the element is known a-priori. However, satisfying the inf-sup condition along an interface in an XFEM formulation is more challenging as the length of the interface within an element varies in the mesh (see Figure 5a). This issue was first observed by Dolbow et al. who noticed non-physical oscillations in the solution along the interface [20]. The instability problem can be resolved by projecting the approximation

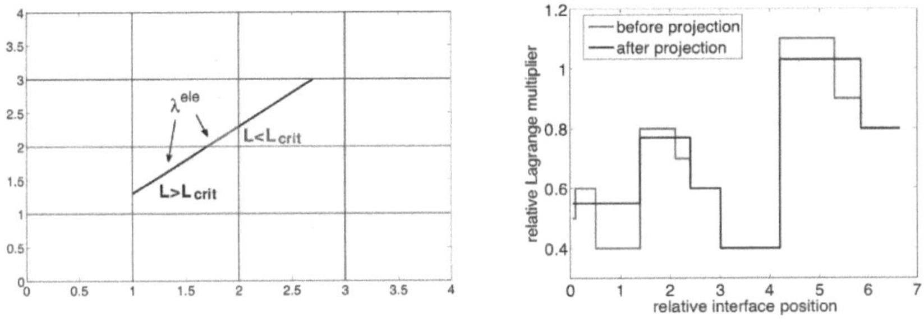

(a) (b)

Fig. 5. (a) Discretization of Lagrange multiplier along material interface, (b) projection of Lagrange multiplier onto the lower dimensional space.

of the Lagrange multiplier onto an appropriate lower-order subspace that satisfies the inf-sup condition [2, 25].

In this study, we approximate the Lagrange multiplier in each element by piecewise constant shape functions. As the interface length becomes less than a critical value, L_{crit}, this approximation violates the inf-sup condition. We therefore project the Lagrange multiplier approximation onto a lower-order subspace by imposing additional constraints on the Lagrange multiplier degrees of freedom of adjacent interface segments:

$$\boldsymbol{P}\,\hat{\boldsymbol{\lambda}} = 0 \tag{32}$$

The construction of the projection $\boldsymbol{P}$ is illustrated in Figure 5a. To satisfy the inf-sup condition we collect n_λ adjacent interfaces such that

$$\sum_{i}^{n_\lambda} L_i \geq L_{\text{crit}} \tag{33}$$

and require $\hat{\lambda}_i = \hat{\lambda}_j,\ i, j = 1 \ldots n_\lambda$. In other words, we assign one Lagrange multiplier degree of freedom to all segments i. The effect of this projection on the discretization of the Lagrange multiplier along a material interface is illustrated in Figure 5b showing the original piecewise constant Lagrange multiplier discretization before projection and the lower dimensional approximation after projection.

5 Numerical Examples

We study the accuracy and versatility of the proposed computational framework by a verification example, the analysis of a real nano-composite specimen and a conceptual design study.

5.1 Verification Example

The accuracy of the proposed XFEM approach is verified through comparison with the results of Yang and Chen [39], which were obtained by a combination of finite volume and discrete ordinate methods for solving the gray phonon Boltzmann transport equations. We consider a periodic Ge-Si nano-composite made of a square Si inclusion embedded into a Ge host. We study the thermal conductivity for two inclusion sizes. For the small inclusion ballistic effects dominate while for the large inclusion heat conduction is dominated by diffusive effects.

The mean free path of Ge and Si are 198.6 and 268.2 nm, respectively. The average group velocities are 1042 m/s for Ge and 1804 m/s for Si at room temperature. The transmission coefficient from Ge to Si is 0.65 following the diffusive mismatch model. For simulating heat conduction in the ballistic regime, a square Si inclusion of size 10 nm is bedded into a Ge host. The length of the unit cell is 23.5 nm. To simulate predominantly diffusive heat conduction, a square Si inclusion of size 268 nm is considered. The length of the unit cell is 629.8 nm.

We predict the phonon distribution and the macroscopic temperature fields applying periodic boundary conditions between the left and right edge of the computational domain by Eq. (14) and imposing a temperature difference of 1 K. Reflective boundary conditions are applied at the upper and lower sides. Transmission and reflection conditions defined by Eq. (21) are enforced at the material interface.

The simulation results for a 64×64 mesh are shown in Figure 6. Note the material interface is not aligned with the mesh. The temperature distribution in the computational domain and along $y^* = 0.5, 0.7, 0.85$ are shown, where y^* denotes the vertical position of the horizontal cuts relative to the length of the computational domain. The XFEM results are in good agreement with those of Yang and Chen [39]. Minor differences along the interfaces are likely to stem from differences in the spatial and angular resolutions.

The convergence of the proposed XFEM approach is studied by refining discretization of the velocity space, increasing the mesh size, and varying critical interface length L_{crit} in ballistic regime. In Figure 7 we show the dependency of the effective thermal conductivity computed from the phonon distribution on the discretization parameters. The numerical results suggest for the present example that the velocity space needs to be resolved by about 10 directions. The thermal conductivity converges monotonically as the mesh is refined. A critical interface length of $L_{\text{crit}} = 0.6$–0.7 relative to the edge length of the element leads to a stable and accurate enforcement of the interface conditions.

5.2 Analysis of Nano-Composites

The utility of the proposed XFEM approach is illustrated with the analysis of a Ge-Si nano-composite sample shown in Figure 8, which is taken from the study of Joshi et al. [21]. The size of the sample is 1000 nm × 750 nm. Converting the image of the sample into a level-set representation, as described in Section 2, the proposed XFEM

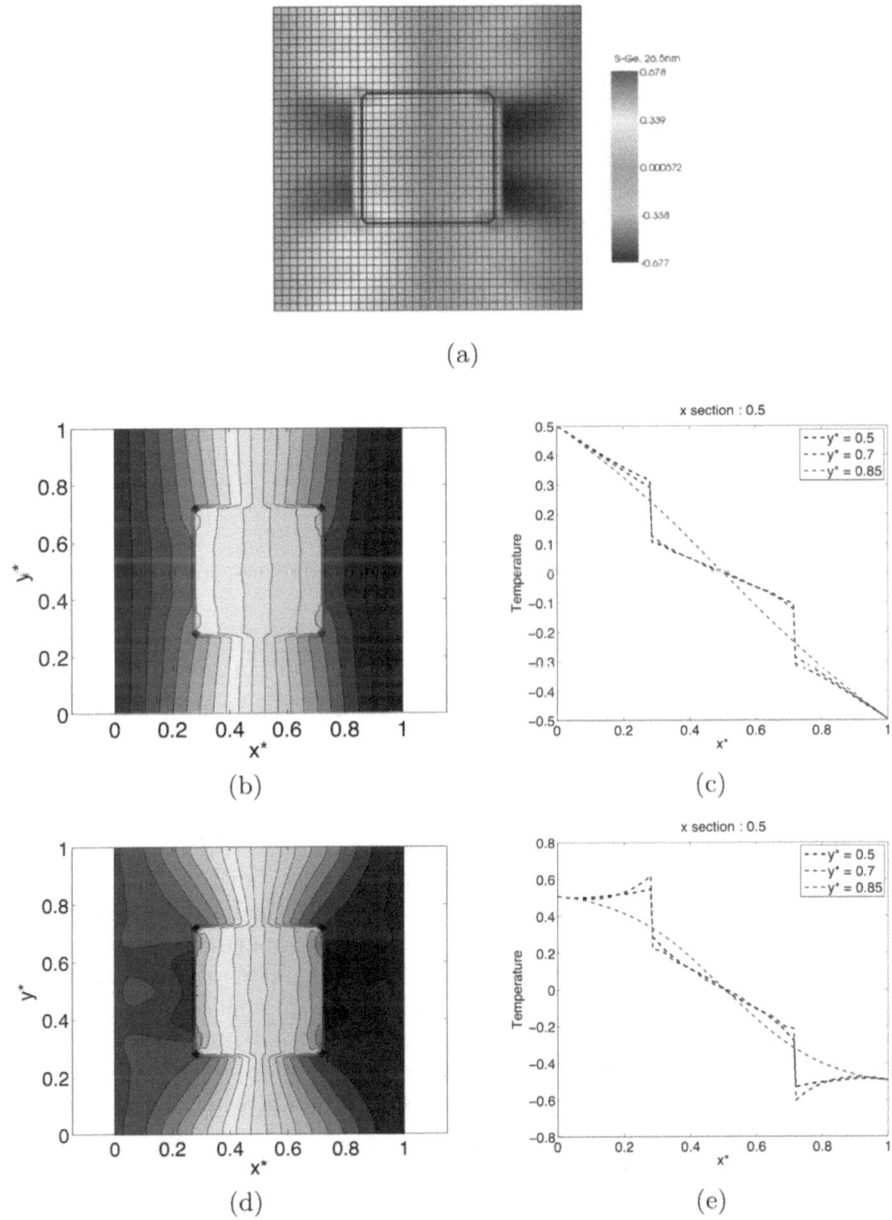

Fig. 6. XFEM verification example: (a) geometry and mesh; (b) diffusive regime: temperature contour; (c) diffusive regime: temperature distribution at $y^* = 0.5$ (black), $y^* = 0.7$ (red), and $y^* = 0.85$ (blue); (d) ballistic regime: temperature contour; (e) ballistic regime: temperature distribution at $y^* = 0.5$ (black), $y^* = 0.7$ (red), and $y^* = 0.85$ (blue).

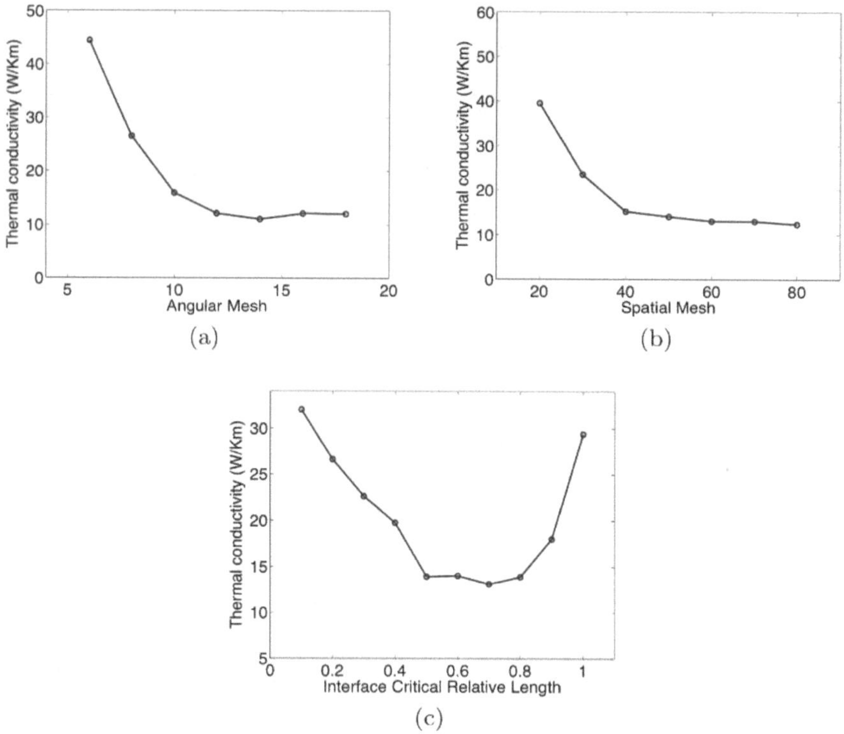

Fig. 7. Convergence of thermal conductivity.

approach analyzes the problem without any additional steps, such as constructing a spline representation of the interfaces and generating body-fitted meshes.

We analyze the thermal properties of the sample by applying periodic boundary conditions at the left and right edges and reflective boundary condition at the upper and lower edges of the computational domain. Again, transmission and reflection conditions based on a diffusive mismatch model are applied along the Ge-Si interfaces. In Figure 8 the temperature contours for a 64×48 mesh are shown. The resolution of the velocity space is 16, and the threshold for projection L_{crit} is 0.7. The computed effective thermal conductivity is 3.176 W/mK which agrees well with the experimental results of Joshi et al. [21].

5.3 Design Study

The proposed XFEM approach also facilitates conceptual design studies and can be conveniently integrated into formal optimization procedures. This feature is illustrated with a simple example. We consider again a Ge host material and study the influence of the shape and layout of Si inclusions on the effective thermal conductivity. To illustrate the importance of accurate modeling ballistic effects at submicron scales, we compare temperature predictions and thermal conductivity calculations

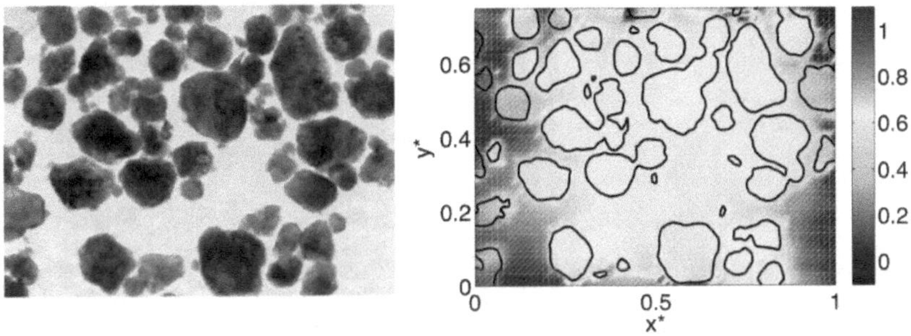

(a) (b)

Fig. 8. (a) Real nano-composite material image, (b) the computed temperature contour and interface applying the level set of (a).

Table 1. Effective thermal conductivity for different material layouts, W/mK.

Domain	2 circles	2 circles & 1 clover
Diffusive	80.45	85.21
Ballistic	38.23	15.66

based on Fourier's law (diffusive regime) and the gray phonon model. In the latter case, we consider a computational domain of 23.5×23.5 nm and apply a diffusive mismatch model to predict phonon scattering at the material interfaces.

First we consider two circular inclusions and then add a clover-shaped inclusion. The temperature contours and Si-Ge interfaces are shown in Figure 9 and the predictions of the effective thermal conductivity are summarized in Table 1. Note the thermal conductivities predicted by the diffusive and ballistic models differ considerably. Furthermore, the diffusive model predicts that the thermal conductivity increases when inserting the clover-shaped inclusion. This seems to be intuitive as Si has a higher thermal conductivity than Ge. However, the phonon model predicts a drop in thermal conductivity as the addition of the clover increases the scattering of phonons at material interfaces.

6 Conclusions

We have presented an extended finite element method for the analysis of heat conduction phenomena at submicron scales. Our approach combines a level-set representation of the geometry, a SUPG stabilized finite element formulation of the gray phonon Boltzmann transport equations, a local enrichment strategy to represent discontinuous phonon distributions at material interfaces, and a stabilized Lagrange multiplier method to enforce boundary conditions and interface conditions.

As illustrated by numerical examples, this computational framework allows for the efficient, accurate and convenient analysis of complex layouts of composite ma-

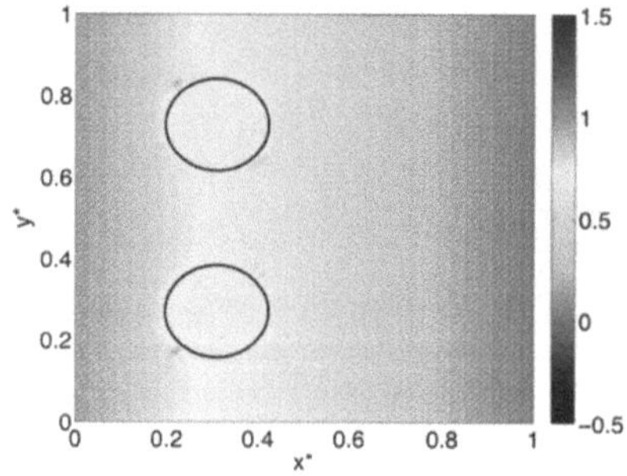

(a) two circles, diffusive, k=80.45W/mK

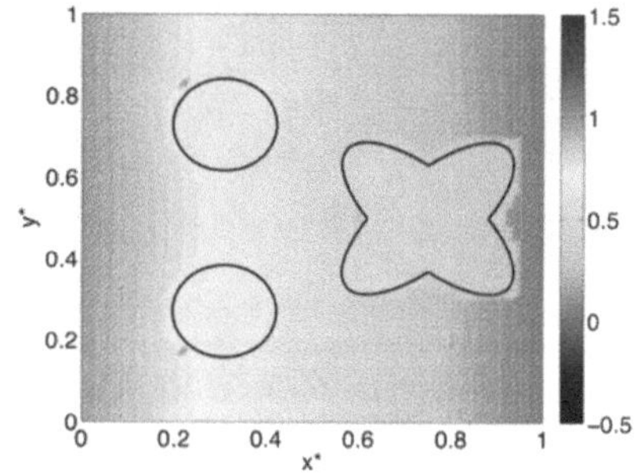

(b) two circles and one clover, diffusive, k=85.21W/mK

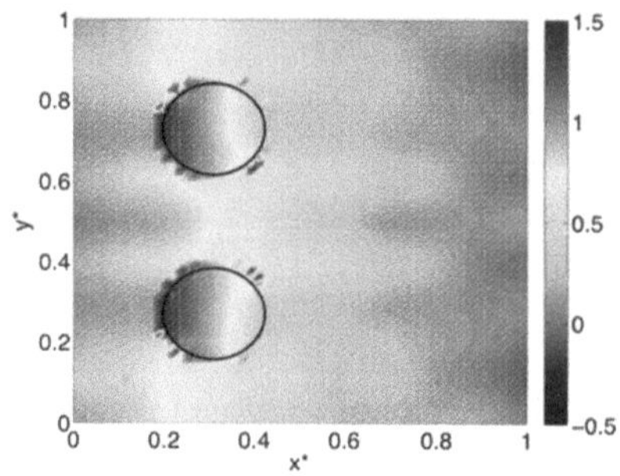

(c) two circles, ballistic, k=38.23W/mK

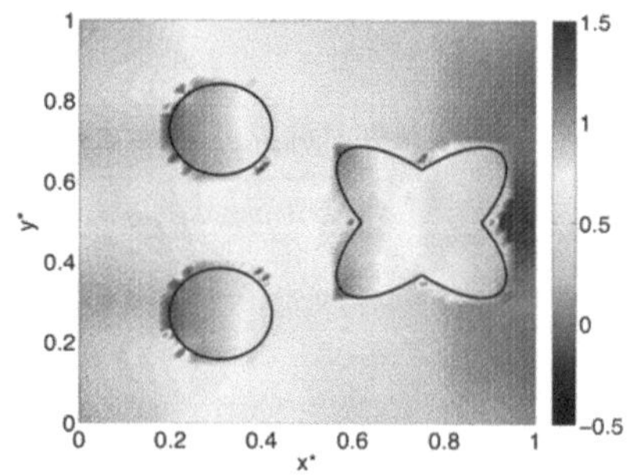

(d) two circles and one clover, ballistic, k=15.66W/mK

Fig. 9. Design study of Si inclusions and Ge host in ballistic and diffusive regimes.

terials. These features render the proposed method an attractive computational tool for scientist to quickly analyze new materials and for design engineers to develop novel materials. Furthermore, the proposed computational framework is applicable to other problems that can be modeled by the Boltzmann transport equations, such as radiation problems and hydrodynamic flows.

Acknowledgments

The authors acknowledge the support of the National Science Foundation under grant CMMI-0729520 and the Air Force Office of Scientific Research under grant FA9550-08-1-0078. The opinions and conclusions presented in this chapter are those of the authors and do not necessarily reflect the views of the sponsoring organization.

References

1. G. Allaire, F. Jouve, and A.M. Toader. A level-set method for shape optimization. *C.R. Math.*, 334:1125–1130, 2002.
2. E. Bechet, N. Moes, and B. Wohlmuth. A stable Lagrange multiplier space for stiff interface conditions within the extended finite element method. *Int. J. Numer. Meth. Engng.*, 78:931–954, 2009.
3. T. Belytschko and T. Black. Elastic crack growth in finite elements with minimal remeshing. *Int. J. Numer. Methods Eng.*, 45:601–620, 1999.
4. P.L. Bhatnagar, E.P. Gross, and M. Krook. A model for collision processes in gases. I. Small amplitude processes in charged and neutral one-component systems. *Phys. Rev.*, 94(3):511–525, 1954.
5. F. Brezzi and M. Fortin. *Mixed and Hybrid Finite Element Methods*. Springer-Verlag, 1991.
6. M.O. Bristeau, O. Pironneau, and R. Glowinski. On the numerical solution of nonlinear problems in fluid dynamics by least squares and finite element methods. (I) Least square formulations and conjugate gradient solution of the continuous problems. *Comput. Methods Appl. Mech. Engrg.*, 17:619–657, 1979.
7. A.N. Brooks and T.J.R. Hughes. Streamline upwind/Petrov–Galerkin formulations for convection dominated flows with particular emphasis on the incompressible Navier-Stokes equations. *Comput. Methods Appl. Mech. Engrg.*, 32:199–259, 1982.
8. G. Chen. Nonlocal and nonequilibrium heat conduction in the vicinity of nanoparticles. *J. Heat Transfer*, 118:539–545, 1996.
9. G. Chen, M.S. Dresselhaus, G. Dresselhaus, J.-P. Fleurial, and T. Caillat. Recent developments in thermoelectric materials. *Int. Mater. Rev.*, 48(1):45–66, 2003.
10. J. Chessa, P. Smolinski, and T. Belytschko. The extended finite element method (XFEM) for solidification problems. *Int. J. Numer. Methods Eng.*, 53:1959–1977, 2002.
11. S.V. Criekingen. A 2-D/3-D Cartesian geometry non-comforming spherical harmonic neutron transport solver. *Ann. Nucl. Energy*, 34:177–187, 2007.
12. J.E. Dolbow. An extended finite element method with discontinuous enrichment for applied mechanics. PhD Thesis, Northwestern University, 1999.
13. A. Evgrafov, K. Maute, R.G. Yang, and M.L. Dunn. Topology optimization for nano-scale heat transfer. *Int. J. Numer. Meth. Engng*, 77(285-300), 2009.
14. K. Fushinobu, A. Majumdar, and K. Hijikata. Heat generation and transport in submicron semiconductor devices. *J. Heat Transfer*, 117:25–31, February 1995.
15. A. Gerstenberger and W.A. Wall. An extended finite element method/Lagrange multiplier based approach for fluid-structure interaction. *Comput. Methods Appl. Mech. Engrg.*, 197:1699–1714, 2008.
16. R. Glowinski, T.-W. Pan, and T.I. Hesla. A distributed Lagrange multiplier/fictitious domain method for partiulate flows. *Int. J. Multiphase Flow*, 25:755–794, 1999.
17. K.E. Goodson. Thermal conduction in nonhomogeneous CVD diamond layers in electronic microstructures. *J. Heat Transfer*, 118:279–286, 1996.
18. T.C. Harman, P.J. Taylor, M.P. Walsh, and B.E. LaForge. Quantum dot superlattice thermoelectric materials and devices. *Science*, 297:2229–2232, 2002.
19. M.-S. Jeng, R. Yang, D.W. Song, and G. Chen. Modeling the thermal conductivity and phonon transport in nanoparticle composites using Monte Carlo simulation. *ASME J. Heat Transfer*, 130(042410):1–11, 2008.
20. H. Ji and J.E. Dolbow. On strategies for enforcing interfacial constraints and evaluating jump conditions with the extended finite element method. *Int. J. Numer. Meth. Engng.*, 61:2508–2535, 2004.

21. G. Joshi, H. Lee, Y. Lan, X. Wang, G. Zhu, D. Wang, R. W. Gould, D. C. Cuff, M. Y. Tang, M. S. Dresselhaus, G. Chen, and Z. Ren. Enhanced thermoelectric figure-of-merit in nanostructured p-type silicon germanium bulk alloys. *Nano Lett.*, 8(12):4670–4674, 2008.
22. Y.S. Ju and K.E. Goodson. Phonon scattering in silicon films with thickness of order 100 nm. *Appl. Phys. Lett.*, 74(20):3005–3007, 1999.
23. S.R. Mathur and J.Y. Murthy. Radiative heat transfer in periodic geometries using a finite volume scheme. *J. Heat Transfer*, 121:357–364, May 1999.
24. M.F. Modest. *Radiative Heat Transfer*. McGraw-Hill, 1993.
25. N. Moes, E. Bechet, and M. Tourbier. Imposing Dirichlet boundary conditions in the extended finite element method. *Int. J. Numer. Meth. Engng.*, 67:1641–1669, 2006.
26. N. Moes, J. Dolbow, and T. Belytschko. A finite element method for crack growth without remeshing. *Int. J. Numer. Methods Eng.*, 46:131–150, 1999.
27. J.Y. Murthy and S.R. Mathur. Computation of sub-micron thermal transport using an unstructured finite volume method. *Trans. ASME*, 124:1176–1181, 2002.
28. S.V.J. Narumanchi, J.Y. Murthy, and C.H. Amon. Boltzmann transport equation-based thermal modeling approaches for hotspots in microelectronics. *Heat Mass Transfer*, 42:478–491, 2006.
29. S. Osher and N. Paragios (Eds.), *Geometric Level Set Methods in Imaging, Vision, and Graphics*. Springer, 2003.
30. S.J. Osher and R. P. Fedkiw. *Level Set Methods and Dynamic Implicit Surfaces*. Springer, 2002.
31. S.J. Osther and J.A. Sethian. Fronts propagating with curvature-dependent speed: Algorithms based on Hamilton–Jacobi formulations. *J. Comp. Phys.*, 79:12–49, 1988.
32. C.C. Pain, M.D. Eaton, R.P. Smedley-Stevenson, A.J.H. Goddard, M.D. Piggott, and C.R.E. de Oliveira. Streamline upwind Petrov–Galerkin methods for the steady-state Boltzmann transport equation. *Comput. Methods Appl. Mech. Engrg.*, 195:4448–4472, 2006.
33. C.S. Peskin. The immersed boundary method. *Acta Numer.*, 2:479–517, 2002.
34. J.A. Sethian. *Level Set Methods and Fast Marching Methods: Evolving Interfaces in Computational Geometry, Fluid Mechanics, Computer Vision, and Materials Science*. Cambridge University Press, 1999.
35. M. Sussman, P. Smereka, and S. Osher. A level set approach for computing solutions to incompressible two-phase flow. *J. Comp. Phys.*, 114:146–159, 1994.
36. E.T. Swartz and R.O. Pohl. Thermal boundary resistance. *Rev. Mod. Phys.*, 61(3):606–668, 1989.
37. W. Tian and R. Yang. Phonon transport and thermal conductivity percolation in random nanoparticle composites. *Comput. Model Eng. Sci.*, 24:123–141, 2008.
38. G.J. Wagner, N. Moes, W.K. Liu, and T. Belytschko. The extended finite element method for rigid particles in Stokes flow. *Int. J. Numer. Methods Eng.*, 51:293–313, 2001.
39. R. Yang and G. Chen. Thermal conductivity modeling of periodic two-dimensional nanocmposites. *Phys. Rev. B*, 69(195316):1–10, 2004.
40. R. Yang, G. Chen, M. Laroche, and Y. Taur. Multidimensional transient heat conduction at nanoscale using the ballistic-diffusive equations and the Boltzmann equation. *ASME J. Heat Transfer*, 127:298–306, 2005.
41. O.C. Zienkiewicz, R.L. Taylor, S.J. Sherwin, and J. Peiro. On discontinuous Galerkin methods. *Int. J. Numer. Meth. Engng.*, 58:1119–1148, 2003.
42. J.M. Ziman. *Electrons and Phonons*. Oxford University Press, 1960.

PART 3

Composites, Laminates, and Structures: Optimization

Multiscale Modeling and Simulation of Composite Materials and Structures

Jacob Fish

Multiscale Science and Engineering Center, Rensselaer Polytechnic Institute, Troy, NY 12211, USA; fishj@rpi.edu

Abstract We describe various spatial and temporal multiscale approaches for composite materials and structures. Spatial multiscale approaches are grouped into two categories: information-passing and concurrent. In the concurrent multiscale methods in space multiple scales are simultaneously resolved, whereas in the information-passing schemes, the fine scale is modeled and its gross response is infused into the continuum scale. The issue of appropriate scale selection is discussed. Among the temporal multiscale application we describe block cycle and temporal homogenization approaches with application to fatigue life prediction of composites.

Key words: Multiscale, homogenization, composites, concurrent, information-passing

1 Introduction

With recent developments in computational science and engineering and continued increase in processing power on one hand and rapidly increasing use of composite materials on the other hand, the time is ripe for the emergence of *truly predictive* computational tools that would improve our fundamental understanding of composites behavior and hence ensure safety without heavy reliance on experiments. Further reduction of testing and thus time and cost of development as well as insertion of composites into primary structures is hindered by their *lack of predictability*. This lack of predictability of existing analysis and design tools in composites has often resulted in overdesign and thus limited their use since the overdesigned component may not yield any design advantages. There are several reasons to this state of affairs:

(a) *Existence of multiple spatial and temporal scales and multiple physical processes*: The behavior of the composites is determined by phenomena operating on a range of time and spatial scales. To enable design of composites to their potential, simplistic empirical approaches (such as knockdown factors) must be replaced by multiscale-multiphysics approaches, which can reliably predict behavior of these architectures. Commonly used phenomenological model cannot extend beyond the database.

R. de Borst and E. Ramm (eds.), *Multiscale Methods in Computational Mechanics*, Lecture Notes in Applied and Computational Mechanics 55, DOI 10.1007/978-90-481-9809-2_12,

(b) *Lack of integrated design tools*: An experimentally validated multiscale design system (MDS) that would account for phenomena at multiple spatial and temporal scales to predict the behavior of composite platforms has long been the dream of engineers. Such a design system would be indispensable in systematic exploration of alternative designs at the material and structural scales. Development of such a design system would advance the state-of-the-art in the field far beyond what an equivalent investment in its comprising components (materials, mechanics, testing and computations).

A typical composite structure, such as Airbus A380, is meters long, while the diameter of the fastener hole is of few millimeters. In order to determine the force redistribution between various structural components, joints must be accurately modeled at a scale which is several orders of magnitude smaller than the structure. The useful life and cost of maintenance of composite structures depends on the quality of modeling at each level and the ability of reliable transfer of the appropriate information between various modeling levels.

In analyzing a typical composite structure the current practice today is to carry out a sequence of uncoupled analyses corresponding to different length scales including:

(i) the level of a structural component;
(ii) the level of a joint; and
(iii) the level of a composite material microstructure.

On the structural level, structural components are explicitly represented, while details of interconnects are not resolved except perhaps for their overall contribution. At the level of the interconnect the details of subcomponents are explicitly accounted for, while microstructural details are not accounted for. Note that the three analysis steps are currently performed independently in the sense that output from one is used as an input to the second. The obvious question arises as to the validity of such uncoupled or information-passing approach [1, 2]. Is there a need for concurrent multiscale (coupled) approach that will consider phenomena simultaneously at several different levels? If the answer is positive, is it feasible to devise a concurrent multiscale approach that could be successfully utilized by naval designers in day-to-day operations?

In order to answer the first question it is important to recall that composite structures are often designed for temporary overloads in which case inelastic deformation is permitted. The key question then becomes: can the appropriate information extracted from the nonlinear analysis of the joint be incorporated into structural analysis where the joint details are excluded in order to reduce the size of the structural problem? While for linear problems this seems to be a viable approach, for nonlinear history dependent systems such an idealization is highly questionable since instantaneous material properties in the vicinity of the interconnect undergoing irreversible processes may vary in space and time due to their history dependence. Thus what is ultimately needed is explicit representation of material microstructure in the critical regions of the joint and smeared (or homogenized) material properties away from the

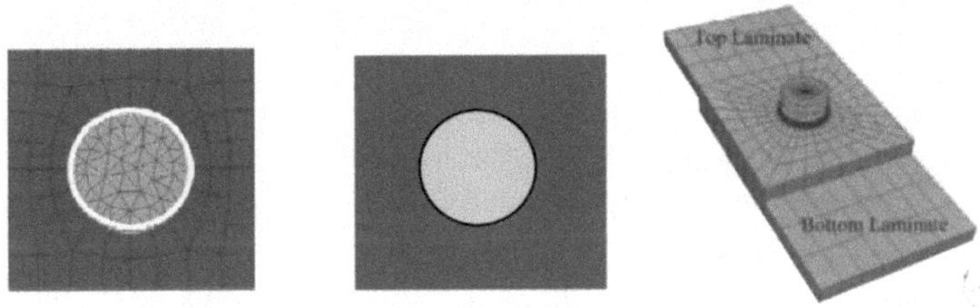

Micro model Meso model Macro model

Fig. 1. Linking micromechanical and macromechanical problems through a mesomechanical model.

hot spots. In the following we describe such state-of-the-art multiscale technologies and discuss their cons and pros.

The use of computational homogenization methods that can predict the overall behavior of the composite material from a detailed finite element analysis of a unit cell [3–6, 32] have so far seen limited use in practice. There are several reasons that can be attributed to this:

1. Material characterization and calibration to experimental data are often available on the macroscale only;
2. Significant computational cost;
3. Lack of accuracy in the vicinity of high gradient regions; and
4. Lack of mechanistic approach to fatigue of composites.

The *first barrier* is concerned with scale-specific measurements of in-situ properties, uncertainty quantification and indirect calibration by inverse methods [7]. Multiscale model calibration (or parameter identification) in composites is a significantly more challenging task than in single scale material models, since material parameters are defined at a finer scale, often orders of magnitude smaller than the scale at which experiments are conducted. For details we refer to [8] for calibration of the reduced order model and to [9] for multiscale sensitivity analyses needed in calibration at multiple scales.

The *second barrier* is purely due to computational complexity. The cost of solving just a two-scale nonlinear problem is tremendous. To illustrate the computational complexity involved, consider a macro-problem with N_{cells} Gauss points, n load increments in the macroscale, and I_{coarse} and I_{fine} average iterations in the macro- and micro-scales, respectively. The total number of linear solves of a micro-problem is $N_{\text{cells}} \cdot n \cdot I_{\text{coarse}} \cdot I_{\text{fine}}$ – a formidable computational cost if the number of unit cells and degrees-of-freedom in a unit cell is substantial. This tyranny of spatial scales can be effectively addressed by a combination of parallel methods and by introducing an intermediate mesomechanical model as shown in Figure 1.

While parallelization in space is natural since unit cell problems are fully parallelizable in space (see for instance [10]), parallelization in time remains an outstanding issue. Some promising results on parallelization in time have been obtained by utilizing waveform relaxation scheme [11] and space-time variational multigrid method [12].

Development of mesomechanical (or reduced order) models for periodic heterogeneous continua has been an active research area in the past decade. Perhaps, one of the oldest mesomechanical approaches is based on purely kinematical Taylor's hypothesis (closely related to Cauchy–Born rule) which assumes a uniform deformation in the fine scale; it satisfies compatibility but fails to account for equilibrium across microconstituents boundaries. A major progress in mesomechanical modeling (at the expense of computational cost) has been made by utilizing the Voronoi cell method [13], the spectral method [14], the network approximation method [15], the Fast Fourier Transforms [16, 17], and the Transformation Field Analysis (TFA) [18, 19]. Despite significant progress, the need for flexible low-cost mesomechanical approach, which can be easily adapted to meet accuracy needs, still remains.

The TFA [20], which has its roots in early works of Laws [21], Willis [22] and Dvorak [23], is based on a brilliant idea that allows precomputing certain information (localization operators, concentration tensors, transformation influence functions) in the preprocessing phase prior to nonlinear analysis, which consequently can be carried out with a small subset of unknowns. By this approach the effect of eigenstrains, representing inelastic strains, thermal strains or phase transformation strains, is accounted for by solving a sequence of linear elasticity problems. The salient feature of TFA based approaches is that unit cell equilibrium equations, which have to be solved for $N_{\text{cells}} \cdot n \cdot I_{\text{coarse}} \cdot I_{\text{fine}}$ times in the direct homogenization approach (see earlier discussion) are satisfied a priori, in the preprocessing stage.

Yet, despite the promise, TFA has found limited success in practice. There are several reasons to this:

1. it does not account for interface failure,
2. it is not hierarchical in the sense that model improvement is not possible when higher accuracy is needed such in the regions of boundary layers,
3. it assumes periodicity (or classical uniform displacement or traction boundary conditions),
4. it is limited to two scales, and
5. it lacks a rigorous mathematical framework

The above deficiencies have been recently circumvented by the eigendeformation based homogenization [24–27]. The eigendeformation based homogenization has the following characteristics:

1. it accounts for interface failure in addition to failure of microconstituents; interface failure is modeled using so-called eigenseparations - a concept similar to eigenstrains used for modeling inelastic deformation of phases;
2. it is equipped with a hierarchical model improvement capability, i.e., it incorporates a hierarchical sequence of computational homogenization models where the most inexpensive member of the sequence represents uniform deformation within each microphase (inclusion, matrix and interface), whereas the most com-

prehensive model of the hierarchical sequence coincides with a direct homogenization approach; and

3. like TFA it constructs residual-free stress fields and thus de facto eliminating the need for solving discrete equilibrium equations in the unit cell. This removes a major computational bottleneck and thus permits accounting for arbitrary micromechanical details at a computational cost comparable to phemenological modeling of nonlinear heterogeneous medium.

The *third barrier*, the lack of accuracy in the vicinity of high gradient regions, is concerned with the principal limitations of the homogenization approach: periodicity and uniformity of macroscale fields. Various hierarchical techniques [28, 29], higher order homogenization [30], and nonperiodic homogenization methods [31, 32] have been proposed to extend the range of validity of the computational homogenization approaches. While higher order theories [33–35] are equipped with a capability of subjecting the unit cell to "true" macroscopic deformation, they have not found their way into commercial arena because they require consideration of coupled tractions and C^1 continuous formulation. To extend the range of applicability of the mathematical homogenization theory to problems where scale separation may not exist, such as in the case of nonperiodic solutions or problems where the coarse solution may rapidly vary over the domain of a unit cell, Multiscale Enrichment based on Partition of Unity (MEPU) [36, 37] has been developed. MEPU is a synthesis of the mathematical homogenization and Partition of Unity [38–41] methods.

The *fourth barrier*, the lack of mechanistic approach to fatigue of composites, is by at large responsible for the fact that to-date life prediction of critical components is solely based on experiments. While in academia Paris-like laws of fatigue life prediction in composites (see for instance [42]) are continued to be pursued very few if any complex composite components have been designed that way. On other hand, simulation based approach for fatigue life prediction of composites [43–46], is being increasingly used in industry. By this approach fatigue of composites is viewed as a multiscale phenomenon in space and time (see Sections 2 and 3 for details).

The chapter is organized as follows. Section 2 outlines information-passing multiscale approaches for composites in space including direct and reduced order homogenization. Concurrent multiscale strategy in space is presented in Section 3. Section 4 is devoted to temporal scale linking with application to fatigue.

2 Information-Passing Multiscale Approaches in Space

2.1 Direct Mathematical Homogenization for Nonlinear Problems

We consider a strong form of the initial-boundary value problem on a domain Ω_X with boundaryΓ_X given as

$$\frac{\partial P^{\zeta}_{ij}(\mathbf{F}^{\zeta})}{\partial X_j} + b^{\zeta}_{Xi} - \rho^{\zeta}_{X}\frac{\partial^2 u^{\zeta}_i}{\partial t^2} = 0 \quad \text{on} \quad \Omega_X$$

$$F_{ik}^{\zeta} = \delta_{ik} + \frac{\partial u_i^{\zeta}}{\partial X_k} \tag{1}$$

with boundary conditions:

$$\begin{aligned} &P_{ij}^{\zeta} N_{Xj} = \bar{t}_{Xi} \quad \text{on} \quad \Gamma_X^t \\ &u_i^{\zeta} = \bar{u}_i \quad \text{on} \quad \Gamma_X^u \\ &\Gamma_X^t \cup \Gamma_X^u = \Gamma_X \quad \text{and} \quad \Gamma_X^t \cap \Gamma_X^u = 0 \end{aligned} \tag{2}$$

and initial conditions:

$$\begin{aligned} &u_i^{\zeta} = \bar{u}_i^0 \quad \text{at} \quad t = 0 \\ &\frac{\partial u_i^{\zeta}}{\partial t} = \bar{v}_i^0 \quad \text{at} \quad t = 0 \end{aligned} \tag{3}$$

where $\mathbf{P}^{\zeta}$ denotes the first Piola–Kirchhoff stress tensor; $\mathbf{F}^{\zeta}$ the deformation gradient; $\mathbf{b}_X$ the body force vector, $\mathbf{N}_X$ the unit normal to the boundary Γ_X and ρ_X the initial mass density. Lower case subscripts i and j denote spatial dimensions. Subscripts X and x refer to the initial and deformed configurations, respectively. The superscript ζ denotes dependence of a variable on the fine (micro) scale features. $\mathbf{X}$ is a material coordinate in the initial macroscopic domain Ω_X and $\mathbf{x}$ is the corresponding deformed (spatial) coordinate. Summation convention over repeated subscripts is employed except for the subscripts X and x.

In the classical two-scale mathematical homogenization, displacements are expanded as

$$u_i^{\zeta}(X,t) = u_i^0(X,t) + \zeta u_i^1(X,Y,t) + \zeta^2 u_i^2(X,Y,t) + O(\zeta^3) \tag{4}$$

where Y represents material coordinates in the initial microscopic (unit cell) domain Θ_Y and y is the deformed (spatial) coordinate in the unit cell domain Θ_y. The two scale coordinates are related by $Y \equiv X/\zeta$ with $0 < \zeta \ll 1$.

For large unit cell distortions, the macro displacement $u_i^0(X,t)$ is no longer constant over the unit cell domain and therefore equation has to be modified using Taylor's series expansion around the centroid of the unit cell (for more details see [47]). The modified asymptotic expansion of the displacement field may be written as

$$u_i^{\zeta}(X,t) = u_i^0(\hat{X},t) + \zeta \underbrace{[(\bar{F}_{ij}(\hat{X},t) - \delta_{ij}) + u_i^1(\hat{X},Y,t)]}_{\hat{u}_i^1(\hat{X},Y,t)} + O(\zeta^2) \tag{5}$$

where $\bar{F}_{ij}(\hat{X},t)$ is the overall (macro) deformation gradient, and $\hat{u}_i^1(\hat{X},Y,t)$ is the micro displacement. Using the modified asymptotic expansion, the original problem may be decomposed into the coarse- and fine- scale problems, which are two-way coupled. The two-scale problem is summarized below:

(a) *Fine-scale problem*

$$\text{Given } \bar{F}_{ik}(\hat{X},t)\text{, find } \hat{u}_i^1(\hat{X},Y,t) \text{ on } \Theta_Y \text{ such that:}$$

$$\frac{\partial P_{ij}\left(F\left(\hat{X},Y,t\right)\right)}{\partial Y_j} = 0 \quad \text{on} \quad \Theta_Y$$

$$g(\hat{u}^1) = 0 \quad \text{on} \quad \Gamma_Y \tag{6}$$

(b) *Coarse-scale problem*

$$\text{Given } \bar{P}_{ij}\text{, find } \hat{u}_i^0(\hat{X},t) \text{ on } \Omega_X \text{ such that}$$

$$\frac{\partial \bar{P}_{ij}}{\partial X_j} + \bar{b}_{Xi} = \bar{\rho}_X \frac{\partial^2 \hat{u}_i^0}{\partial t^2} \quad \text{on } \Omega_X$$

$$\bar{P}_{ij} N_{Xj} = \bar{t}_{Xi} \quad \text{on } \Gamma_X^t; \quad \hat{u}_i^0 = \bar{u}_i \quad \text{on } \Gamma_X^u$$

$$\hat{u}_i^0 = \bar{u}_i^0 \quad \text{at } t = 0; \quad \frac{\partial \hat{u}_i^0}{\partial t} = \bar{v}_i^0 \quad \text{at } t = 0 \tag{7}$$

where $\mathbf{g}(\hat{\mathbf{u}}^1) = 0$ denotes the boundary conditions (periodic or nonperiodic [47]) for the unit cell problem. $\bar{P}_{ij}$, $\bar{b}_{Xi}$ and $\bar{\rho}_X$ are the overall stress, body force and mass density, respectively.

Both the coarse- and fine-scale problems are solved using finite element method. In the discretization phase, the finite element equations can be reformulated in terms of Cauchy stress for both problems.

2.2 Eigendeformation-Based Reduced Order Homogenization

Here we consider small deformations in the material coordinate system (i.e. small strains, but large rotations). $\mathbf{x}, \mathbf{y}$ will denote the material coordinate in macroscopic domain Ω and unit cell domain Θ, respectively. We specifically consider constitutive equations and kinematic equations in the form

$$\sigma_{ij}^{\zeta}(\mathbf{x}) = L_{ijkl}^{\zeta}(\mathbf{x}) \left[\varepsilon_{kl}^{\zeta}(\mathbf{x}) - \sum_I {}^I\mu_{kl}^{\zeta}(\mathbf{x}) \right], \quad \mathbf{x} \in \Omega \tag{8}$$

$$\varepsilon_{ij}^{\zeta}(\mathbf{x}) = u_{(i,j)}^{\zeta}(\mathbf{x}) \equiv \frac{1}{2}(u_{i,j}^{\zeta} + u_{j,i}^{\zeta}), \quad \mathbf{x} \in \Omega \tag{9}$$

The constitutive relation described in Eq. (8) assumes an additive decomposition of total strains ε_{ij}^{ζ} into elastic and inelastic components (or more generally stated as eigenstrains ${}^I\mu_{kl}^{\zeta}$); the left superscript I indicates different types of eigenstrains, such as inelastic deformation (damage, plasticity, etc.), thermal changes, moisture effects and phase transformation.

Using the constitutive (8) and kinematic (9) equations, the equilibrium equation at the fine scale can be expressed as

$$\left\{L_{ijkl}(\mathbf{y})\left[\bar{\varepsilon}_{\mathbf{kl}}(\mathbf{x})+\mathbf{u}^{\mathbf{1}}_{(\mathbf{k},\mathbf{y_l})}(\mathbf{x},\mathbf{y})-\sum_{\mathbf{I}}{}^{\mathbf{I}}\mu_{\mathbf{kl}}(\mathbf{x},\mathbf{y})\right]\right\}_{,y_j}=0 \tag{10}$$

where $\bar{\varepsilon}_{kl}(\mathbf{x})$ is an overall strain field. Following [48] the displacement field is decomposed using the elastic influence function H_{ikl}, the eigenstrain influence function h^{μ}_{ikl} and the eigenseparation influence function $h^{\delta}_{i\hat{n}}$ as

$$\begin{aligned} u^1_i(\mathbf{x},\mathbf{y}) &= u^{1e}_i(\mathbf{x},\mathbf{y})+\sum_I {}^I u^{1\mu}_i(\mathbf{x},\mathbf{y})+u^{1\delta}_i(\mathbf{x},\mathbf{y}) \\ &= H_{ikl}(\mathbf{y})\bar{\varepsilon}_{kl}(\mathbf{x}) \\ &\quad+\int_{\Theta} h^{\mu}_{ikl}(\mathbf{y},\hat{\mathbf{y}})\sum_I {}^I\mu_{kl}(\mathbf{x},\hat{\mathbf{y}})d\Theta \\ &\quad+\int_S h^{\delta}_{i\hat{n}}(\mathbf{y},\hat{\mathbf{y}})\delta_{\hat{n}}(\mathbf{x},\hat{\mathbf{y}})dS \end{aligned} \tag{11}$$

where the influence (or Green's) functions H_{ikl}, h^{μ}_{ikl} and $h^{\delta}_{i\hat{n}}$ are computed by solving a sequence of elastic boundary value problems independent of and prior to nonlinear macro analysis. These functions are chosen to satisfy the equilibrium equation (1) for arbitrary $\bar{\varepsilon}_{kl}(\mathbf{x})$, ${}^I\mu_{kl}(\mathbf{x},\hat{\mathbf{y}})$ and $\delta_{\hat{n}}(\mathbf{x},\hat{\mathbf{y}})$. The subscript $\hat{n}$ in the eigenseparation influence function denotes the component in the local Cartesian coordinate system of the interface.

To reduce the computational complexity of solving the unit cell problem the eigenstrains are discretized in terms of piecewise constant shape functions $N^{(\alpha)}(\mathbf{y})$ as

$${}^I\mu_{ij}(\mathbf{x},\mathbf{y})=\sum_{\alpha=1}^{n} N^{(\alpha)}(\mathbf{y})\,{}^I\mu^{(\alpha)}_{ij}(\mathbf{x}) \tag{12}$$

where

$$N^{(\alpha)}(\mathbf{y})=\begin{cases}1 & \mathbf{y}\in\Theta^{(\alpha)}\\ 0 & \mathbf{y}\notin\Theta^{(\alpha)}\end{cases} \tag{13}$$

$${}^I\mu^{(\alpha)}_{ij}(\mathbf{x})=\frac{1}{|\Theta^{(\alpha)}|}\int_{\Theta^{(\alpha)}} {}^I\mu_{ij}(\mathbf{x},\mathbf{y})d\Theta \tag{14}$$

in which the total volume of the unit cell is partitioned into n nonoverlapping subdomains denoted by $\Theta^{(\alpha)}$. Partitions at various scales are denoted by a superscript enclosed in parenthesis: (α), (β) denote the phase (volume) partitions; (ξ), (η) denote the interface partitions.

The eigenseparation $\delta_{\hat{n}}(\mathbf{x},\tilde{\mathbf{y}})$ is discretized in terms of a C^0 continuous interface partition shape function $N^{(\xi)}(\tilde{\mathbf{y}})$ as

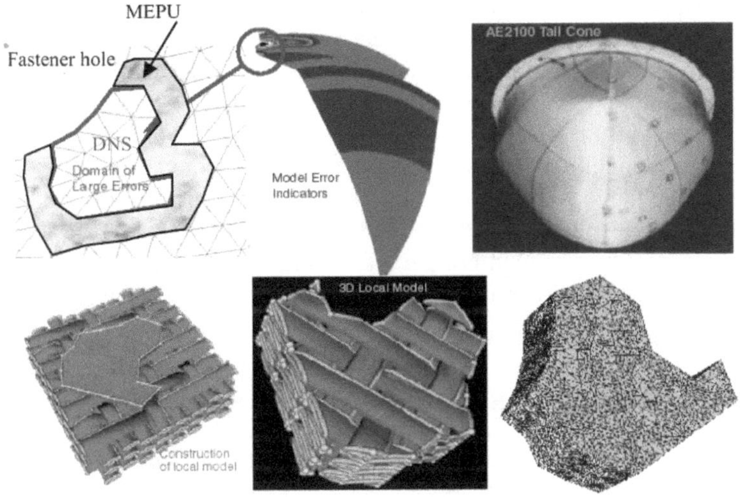

Fig. 2. Concurrent multiscale modeling of the engine tailcone.

$$\delta_{\hat{n}}(\mathbf{x}, \tilde{\mathbf{y}}) = \sum_{\xi=1}^{m} N^{(\xi)}(\tilde{\mathbf{y}})\delta_{\hat{n}}^{(\xi)}(\mathbf{x}) \tag{15}$$

where $N^{(\xi)}(\tilde{\mathbf{y}})$ is a linear combination of piecewise linear finite element shape functions defined over partition ξ

$$N^{(\xi)}(\tilde{\mathbf{y}}) = \begin{cases} \sum\limits_{a \in S^{(\xi)}} N_a(\tilde{\mathbf{y}}) & \tilde{\mathbf{y}} \in S^{(\xi)} \\ 0 & \tilde{\mathbf{y}} \notin S^{(\xi)} \end{cases} \tag{16}$$

$$\delta_{\hat{n}}^{(\xi)}(\mathbf{x}) = \frac{1}{|S^{(\xi)}|} \int_{S^{(\xi)}} \delta_{\hat{n}}(\mathbf{x}, \tilde{\mathbf{y}}) dS \tag{17}$$

in which the total interface in the unit cell is divided into m partitions denoted by $S^{(\xi)}$, and $N_a(\tilde{\mathbf{y}})$ is a linear shape function associated with a finite element mesh node a along the interface.

Equations (12)–(17) together with the constitutive equations for the eigenstrains and eigenseparations form a reduced system of equations from which ${}^{I}\mu_{ij}^{(\alpha)}(\mathbf{x})$ and $\delta_{\hat{n}}^{(\xi)}(\mathbf{x})$ are computed at each Gauss point. For more details we refer to [48]. In the present book chapter perfect interfaces are considered in the reduced order model.

3 Concurrent Multiscale Methods in Space

Figure 2 depicts concurrent domain decomposition based multiscale strategy employed in [49] for modeling of Rolls Royce engine tail cone. By this approach model error estimators (see Section 3.2) guide the construction of the nearly optimal concurrent multiscale models where the direct numerical simulation (DNS) is used in the critical regions of the fastener holes, the reduced order model (ROM) is used away from the hot spots and the multiscale enrichment based on partition of unity (MEPU) is used at the interface between the two models. Formulation of the consistent interface conditions between the DNS and ROM models is necessary to accurately resolve the local fields at the interface. Prior experience [49] indicates that simply enforcing displacement compatibility condition between DNS and ROH induces spurious stresses, which pollute the solution in the vicinity of the interface. An alternative, multigrid based concurrent multiscale strategy has been developed in [50, 51].

3.1 Multiscale Enrichment Based on Partition of Unity (MEPU)

To convey the basic idea of MEPU, we start by stating the key result from the mathematical homogenization theory for periodic elastic heterogeneous media. Consider a three-term double-scale asymptotic expansion of the solution, $u_i = u_i^0(\mathbf{x}) + \zeta \mathbf{u}_i^1(\mathbf{x}, \mathbf{y}) + \zeta^2 \mathbf{u}_i^2(\mathbf{x}, \mathbf{y})$, where $\mathbf{x}$ and $\mathbf{y} = \mathbf{x}/\zeta$ are the coarse and fine scale position vectors, respectively; $0 < \zeta = l/L \ll 1$, and l, L denote the characteristic size of the fine and coarse scale, respectively. The first and the second order terms are decomposed as

$$\begin{aligned} \mathbf{u}_i^1(\mathbf{x}, \mathbf{y}) &= \chi_{ikl}(\mathbf{y})\bar{\varepsilon}_{kl}(\mathbf{x}) \\ \mathbf{u}_i^2(\mathbf{x}, \mathbf{y}) &= \eta_{ijmn}(\mathbf{y})\bar{\varepsilon}_{mn,x_j}(\mathbf{x}) \end{aligned} \tag{18}$$

where $\chi_{ikl}(\mathbf{y})$ is the first order influence function (symmetric with respect to kl indices); $\eta_{ijmn}(\mathbf{y})$ the second order influence function used for the second order enrichment. It is convenient to replace the pair of subscripts kl in χ_{ikl} denoting the first order enrichment modes by a single uppercase subscript, A. Likewise, the subscripts jmn in η_{ijmn} denoting the second order enrichment modes by another single uppercase subscript, B. In MEPU, the approximation of solution field is constructed by replacing $\bar{\varepsilon}_A$ and $\bar{\varepsilon}_{B,\alpha}$ by a set of independent degrees-of-freedom $a_{A\alpha}$ and $b_{B\alpha}$. The resulting enriched solution approximation states

$$u_i = \bar{N}_\beta(\mathbf{x})d_{i\beta} + N_\alpha(\mathbf{x})[\chi_{iA}a_{A\alpha} + \eta_{iB}(\mathbf{x})b_{B\alpha}] \tag{19}$$

where summation convention is employed for the repeated indices; Greek subscripts denote finite element nodes. Note that the shape functions used for the coarse scale discretization, $\bar{N}_\beta(\mathbf{x})$, which might be different from those used in the partition of unity decomposition $N_\alpha(\mathbf{x})$.

With the second order enrichment in place, MEPU is equipped with the deformation modes capable of capturing linear variation of coarse scale strain gradients over the unit cell domain. Such an enrichment is necessary in the high gradients region, such as in the vicinity of cracks or cutouts where the characteristic size of the unit cell is comparable to the coarse scale features. Nevertheless, the second order enrichment involves additional degrees-of-freedom, and therefore should be used when the first order enrichment is inaccurate. For linear problems, the influence functions $\chi_{ikl}(\mathbf{y})$ and $\eta_{ijmn}(\mathbf{y})$ in Eq. (19) can be precomputed by solving a first and second order unit cell problem(s) (see [37] for more details).

3.2 Adaptive Model Selection

The question that needs to be answered is how accurate are the reduced order and MEPU models and where they should be placed? For example, the use of the reduced order model, which is based on the first homogenization, may lead to inadequate resolution of the overall behavior due to three major sources of errors: (i) truncation error, (ii) model reduction error, and (iii) the approximation of the boundary conditions in the unit cell problem.

To estimate the three sources of error a hierarchical modeling strategy [52] is often employed. Hierarchical models are a sequence of mathematical models, which include increasingly more sophisticated effects. The modeling error associated with any member of the sequence is assessed by comparing it to the most-sophisticated member of the hierarchical sequence. For example, to estimate the truncation error, higher order multiple scale expansions can be considered. Truncation error in both the local and nonlocal reduced order model can be obtained by estimating the ratio between the first term neglected in the asymptotic expansion to those taken into account. The model reduction error can be estimated by progressively increasing the number of partitions in the unit cell and comparing the results in the quantities of interest to the model with fewer partitions. The boundary condition error can be assessed by increasing the unit cell size. Similar strategies can be employed to estimate the modeling error in the two MEPU models.

3.3 Numerical Example

Figure 3 shows the circular tube model, the mesh of a unit cell and the microstructure of the braided composites without matrix. The inner diameter of the tube is 5.96 cm, and the thickness is 0.19 cm. The length of the tube is 18.0 cm. The tube bottom is constrained in the vertical direction. The top of the tube is subjected to compression by a rigid body with a constant velocity 25 cm/s. Only one quarter of the tube is modeled in the simulation with the consideration of symmetry. Adaptive strategy is employed where direct homogenization is used in the critical regions identified by the magnitude of truncation error whereas reduced order model is used elsewhere. The unit cell is partitioned into four phases: one for matrix, one for axial tow and two for bias tows. The tow is modeled as elastic material with isotropic damage. Figure 4 compares the multiscale simulation and experiment results for load-displacement history. Good agreement with the experiment result can be observed.

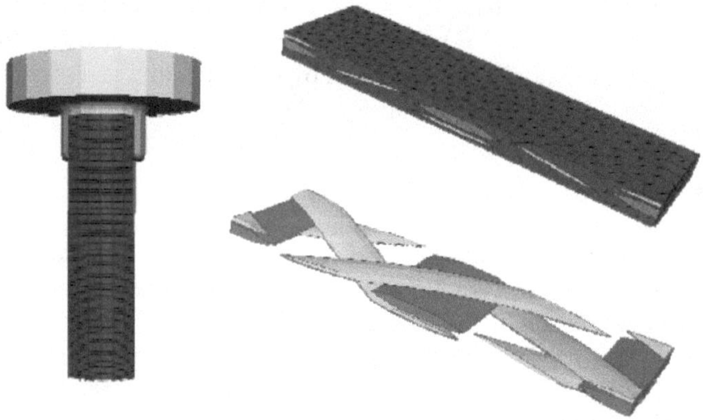

Fig. 3. Finite element model of a circular tube (left), unit cell mesh (top right), weave architecture (bottom right).

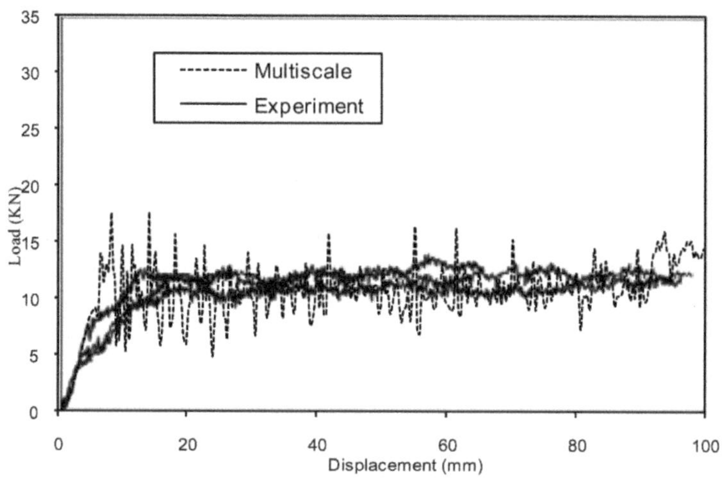

Fig. 4. Comparison of load-displacement history.

4 Temporal Multiscale Model for Fatigue Life Prediction

One of the most effective simulation based approaches for fatigue life prediction of composite materials and structures is based on so-called adaptive cycle block technique for temporal upscaling [43]. The temporal scales exist due to slow degradation of material properties on the one hand and rapidly oscillatory loading on the other. For temporal upscaling, an efficient time integrator has been developed by discretizing the cyclic loading history into a series of load cycle blocks where each block consists of several load cycles. Since the increment of damage accumulation in a single load cycle is typically very small, the derivative of the damage parameter with

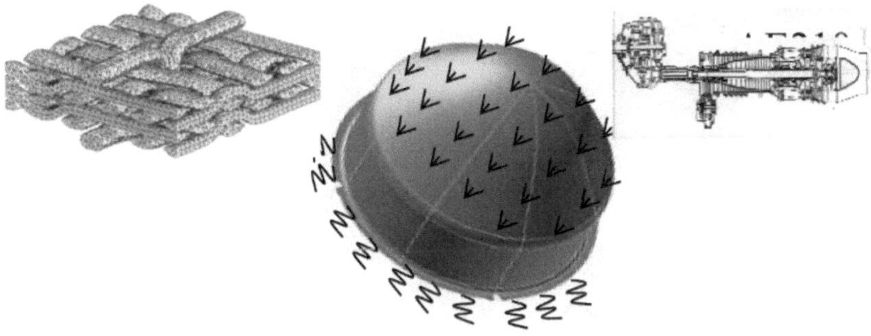

Fig. 5. Multiscale modeling of fatigue: microstructure (left), engine tailcone (middle), AE2100 engine (right).

respect to the number of load cycles can be approximated reasonably well. Consequently, the damage evolution can be modeled as a first order initial value problem with respect to the number of load cycles

$$\left.\frac{d\dot{\omega}^{(m)}}{dN}\right|_K \approx \int_t^{t+t_0} \dot{\omega}^{(m)} dt = \Delta\omega^{(m)}|_K = \omega^{(m)}|_K - \omega^{(m)}|_{K-1} \qquad (20)$$

where N denotes the number of load cycles; $\omega^{(m)}|_K$ is the phase damage (matrix of fiber) at the end of load cycle K. It is obtained by the incremental finite element analysis for this cycle with the initial damage of $\omega^{(m)}|_{K-1}$ and the corresponding initial strain/stress conditions. The block size can then be adaptively controlled using predictor-corrector time integrators. It is adaptively selected to keep the maximal phase damage increment sufficiently small.

The adaptive cycle block technique has been validated for various applications. One such problem is shown in Figure 5 where a composite tail cone is subjected to pressure on its upper surface resulting from noise vibration. The tailcone attachment is modeled using springs. Four cut-outs used for the attachment of the tailcone give rise to stress concentration and damage initiation. Figure 6 depicts 1/8th of the finite element model due to symmetry.

The distribution of the damage parameter is depicted in Figure 6. Elements which were totally damaged were deleted giving rise to fatigue crack growth. The damage initiated in the vicinity of the cut-out and propagated along the attachment. The number of load steps that were needed to perform the analysis using the adaptive block cycle scheme was about 10000 whereas a cycle-by-cycle simulation would have required about 75 million load steps corresponding to life prediction of 1.5 million load cycles.

Although the adaptive cycle block technique has been shown to be effective for simulation of the evolution of state variables, the governing equations are not guaranteed to be satisfied since not all the cycles are simulated. In [44,45] an alternative simulation based fatigue life prediction methodology that satisfies governing equations at each cycle has been developed.

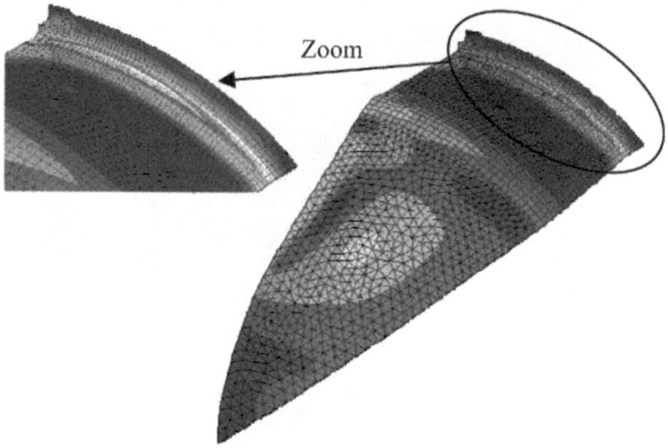

Fig. 6. Damage distribution in the tailcone.

It takes advantage of the fact that in a typical fatigue process, accumulation of damage, and the resulting macrocrack initiation and propagation is relatively slow, compared to rapid fluctuations of displacements within each load cycle. This suggests capturing the disparity between the two characteristic time scales using two time coordinates. By this approach a *macro-chronological* scale denoted by the intrinsic time coordinate, t, and a *micro-chronological* scale denoted by the fast time coordinate, τ are defined. These two scales are related through a scaling parameter

$$\tau = \frac{t}{\zeta}; \quad 0 < \zeta \ll 1 \tag{21}$$

in which ζ is defined by the ratio of the characteristic lengths in micro- and macro-chronological time. The response fields, ϕ, are assumed to depend on two temporal scales

$$\phi^{\zeta}(\mathbf{x}, t) = \phi(\mathbf{x}, t, \tau(t)) \tag{22}$$

where ϕ^{ζ} is a rapidly oscillating function in time; and $\mathbf{x}$ is a spatial coordinate. Time differentiation in the presence of multiple temporal scales is obtained using Eqs. (21) and (22) which gives

$$\dot{\phi}^{\zeta}(\mathbf{x}, t) = \dot{\phi}(\mathbf{x}, t, \tau(t)) = \phi_{,t}(\mathbf{x}, t, \tau(t)) + \frac{1}{\zeta}\phi_{,t}(\mathbf{x}, t, \tau(t)) \tag{23}$$

in which the comma followed by a subscript coordinate denotes partial derivative and a superposed dot is a total time derivative.

Consequently, the initial-boundary value problem can be decomposed into (i) the macro-chronological initial-boundary value problem loading for the long term response, and (ii) the micro-chronological initial-boundary value problem for the remaining (oscillatory) behavior.

In [45] the multiple temporal scale method has been generalized to account for non-periodicity in time domain and nonlocality in space. The non-periodicity is a

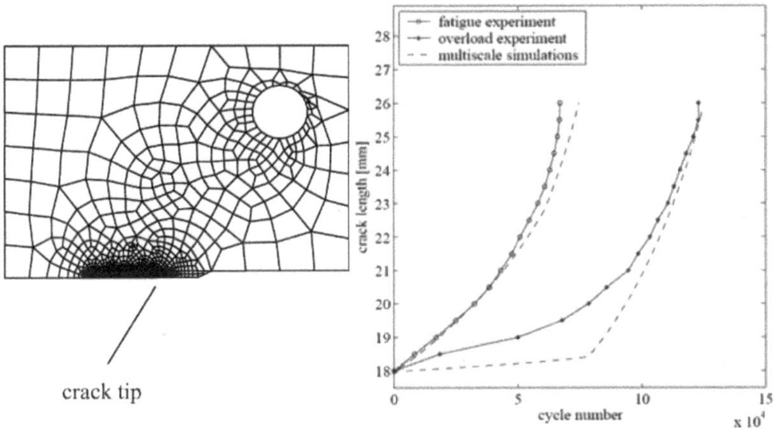

Fig. 7. Finite element mesh of the CT specimen (left) and crack growth (right).

byproduct of irreversible processes, such as damage accumulation, which naturally violate the condition of temporal periodicity. The non-periodic contribution has been assumed to be of an order $\varepsilon \ll 1$ perturbation to the periodic field.

The non-local model with two time scales [45] has been validated against experiments of Wheatley et. al. [53]. Half of the geometry was modeled due to symmetry. Figure 7 on the left displays the finite element mesh used to conduct the simulations. An initial void volume fraction, $f_0 = 5.0 \times 10^{-4}$ was assumed. The Tvergaard constants, q_1 and $q_2$were taken to be 1.5 and 1.0, respectively as suggested in [10]. Figure 7 on the right displays the fatigue crack growth curves simulated using the multiscale GTN model against the experimental results. It was found that a good agreement was observed when the mixed hardening parameter, b was taken to be 0.6. It is also observed that the simulations conducted after introducing of void coalescence and void nucleation functions overestimated the rate of crack growth.

References

1. J. Fish (Ed.), *Bridging the Scales in Science and Engineering*, Oxford University Press, 2009.
2. J. Fish, Multiscale computations: Boom or bust, *IACM Expressions*, 22, January 2008.
3. E. Sanchez-Palencia, Lecture Notes in Physics, Vol. 127. Springer-Verlag, Berlin, 1980.
4. J.M. Guedes and N. Kikuchi, *Comput. Methods Appl. Mech. Engng.*, 83:143–198, 1990.
5. K. Terada and N. Kikuchi, In S. Ghosh and M. Ostoja-Starzewski (Eds.), *Computational Methods in Micromechanics*, Vol. AMD-212/MD-62, pp. 1–16, ASME, 1995.
6. Z. Yuan and J. Fish, Towards realization of computational homogenization in practice, *J. Numer. Methods Engrg.*, 73(3):361–380, 2008.
7. M. Jardak and R. Ghanem, *Comput. Methods Appl. Mech. Engng.*, 193(6/8):429–447, 2004.

8. C. Oskay and J. Fish, On calibration and validation of eigendeformation-based multiscale models for failure analysis of heterogeneous systems, *Comput. Mech.*, 42(2):181–195, 2008.
9. J. Fish and A. Ghouli, Multiscale analytical sensitivity analysis for composite materials, *Int. J. Numer. Methods Engrg.*, 50:1501–1520, 2001.
10. F. Feyel and J.-L. Chaboche, *Comput. Methods Appl. Mech. Engrg.*, 183:309–330, 2000.
11. E. Giladi and H.B. Keller, *Numer. Math.*, 93(2), 2002.
12. H. Waisman and J. Fish, *Comput. Methods Appl. Mech. Engrg.*, 195(44/47):6542–6559, 2006.
13. S. Ghosh and S. Moorthy, *Comput. Methods Appl. Mech. Engrg.*, 121(1/4):373–409, 1995.
14. J. Aboudi, *J. Engrg. Sci.*, 20(55):605–621, 1982.
15. L.V. Berlyand and A.G. Kolpakov, *Arch. for Rat. Mech. Anal.*, 159:179–227, 2001.
16. H. Moulinec and P. Suquet, *C.R. Acad. Sci. Paris II*, 318:1417–1423, 1994.
17. H. Moulinec and P. Suquet, *Comput. Meth. Appl. Mech. Engrg.*, 157:69–94, 1998.
18. G.J. Dvorak, *Proc. R. Soc. Lond. A*, 437:311–327, 1992.
19. Y.A. Bahei-El-Din, A.M. Rajendran, and M.A. Zikry, *Int. J. Sol. Structures*, 41:2307–2330, 2004.
20. G.J. Dvorak and Y. Benveniste, *Proc. R. Soc. Lond. A*, 437:291–310, 1992.
21. N. Laws, On the thermostatics of composite materials, *J. Mech. Phys. Solids*, 21:9–17, 1973.
22. J. Willis. In C.S. Yih (Ed.), *Advances in Applied Mechanics*, Vol. 21, pp. 1–78, Academic Press, New York, 1981.
23. G.J. Dvorak, *Proc. R. Soc. Lond. A*, 431:89–110, 1990.
24. C. Oskay and J. Fish, Eigendeformation-based reduced order homogenization, *Comp. Methods Appl. Mech. Engrg.*, 196:1216–1243, 2007.
25. Z. Yuan and J. Fish, Multiple scale eigendeformation-based reduced order homogenization, *Comput. Methods Appl. Mech. Engrg.*, 198(21/26), 2016–2038, 2009.
26. C. Oskay and J. Fish, *Comput. Mech.*, 42(2):181–195, 2008.
27. J. Fish and Z. Yuan, N-scale model reduction theory. In *Bridging the Scales in Science and Engineering*, J. Fish (Ed.), Oxford University Press, 2009.
28. T.I. Zohdi, J.T. Oden, and G.J. Rodin, *Comput. Methods Appl. Mech. Engrg.*, 138:273–298, 1996.
29. S. Ghosh, K. Lee, and P. Raghavan, *Int. J. Solids Struct.*, 38:2335–2385, 2001.
30. M.G.D. Geers, V. Kouznetsova, and W.A.M. Brekelmans, *J. Phys. IV*, 11:145–152, 2001.
31. E. Weinan and B. Engquist, *Comm. Math. Sci.*, 1:87–132, 2003.
32. J. Fish and R. Fan, Mathematical homogenization of nonperiodic heterogeneous media subjected to large deformation transient loading, *Int. J. Numer. Methods Engrg.*, 76(7):1044–1064, 2008.
33. J. Fish, P. Nayak and M.H. Holmes, Microscale reduction error indicators and estimators for a periodic heterogeneous medium, *Comput. Mech.*, 14:1–16, 1994.
34. M. Ostoja-Starzewski, S.D. Boccara, and I. Jasiuk, Couple-stress moduli and characteristic length of a two-phase composite, *Mech. Res. Comm.*, 26(4):387–396, 1999.
35. O. van der Sluis, P.H.J. Vosbeek, P.J.G. Schreurs, and H.E.H. Meijer, Homogenization of heterogeneous polymers, *Int. J. Solids Struct.*, 36:3193–3214, 1999.
36. J. Fish and Z. Yuan, Multiscale enrichment based on the partition of Unity, *Int. J. Numer. Methods Engrg.*, 62(10):1341–1359, 2005.
37. J. Fish and Z. Yuan, Multiscale enrichment based on partition of unity for nonperiodic fields and nonlinear problems, *Comput. Mech.*, 40:249–259, 2007.

38. I. Babuska, G. Caloz, and J.E. Osborn, Special finite element methods for a class of second order elliptic problems with rough coefficients, *SIAM. J. Numer. Anal.*, 4:945–981, 1994.
39. I. Babuska and J.M. Melenk, The partition of unity finite element method: Basic theory and applications, *Comput. Methods Appl. Mech. Engrg.*, 139:289–314, 1996.
40. T. Belytschko and T. Black, Elastic crack growth in finite elements with minimal remeshing, *Int. J. Numer. Methods Engrg.*, 45(5):601–620, 1999.
41. N. Moës, J. Dolbow and T. Belytschko, A finite element method for crack growth without remeshing, *Int. J. Numer. Methods Engrg.*, 46(5):131–150, 1999.
42. M. Quaresimin and M. Ricotta, Life prediction of bonded joints in composite materials, *Int. J. Fatigue*, 28(10):1166–1176, October 2006.
43. J. Fish and Q. Yu, Computational mechanics of fatigue and life predictions for composite materials and structures, *Comput. Methods Appl. Mech. Engrg.*, 191:4827–4849, 2002.
44. C. Oskay and J. Fish, Multiscale modeling of fatigue for ductile materials, *Int. J. Multiscale Comput. Engrg.*, 2(3):1–25, 2004.
45. J. Fish and C. Oskay, Nonlocal multiscale fatigue model, *Mech. Adv. Mater. Struct.*, 12(6):485–500, 2005.
46. E. Gal, Z. Yuan, W. Wu, and J. Fish, A multiscale design system for fatigue life prediction, *Int. J. Multiscale Comput. Engrg.*, 5(6):435–446, 2007.
47. J. Fish and R. Fan, Mathematical homogenization of nonperiodic heterogeneous media subjected to large deformation transient loading, *Int. J. Numer. Methods Engrg.*, 76:1044–1064, 2008.
48. C. Oskay and J. Fish, Eigendeformation-based reduced order homogenization, *Comput. Methods Appl. Mech. Engrg.*, 196:1216–1243, 2007.
49. Simmetrix, Inc., SBIR Phase II Contract F33615-99-C-5021.
50. J. Fish and V. Belsky, Multi-grid method for periodic heterogeneous media, Part 2: Multiscale modeling and quality control in multidimensional case, *Comput. Methods Appl. Mech. Engrg.*, 126:17–38, 1995.
51. J. Fish and V. Belsky, Multigrid method for periodic heterogeneous media, Part 1: Convergence studies for one-dimensional case, *Comput. Methods Appl. Mech. Engrg.*, 126:1–16, 1995.
52. J. Fish, Bridging the scales in nano engineering and science, *J. Nanoparticle Res.*, 8:577–594, 2006.
53. G. Wheatley, R. Niefanger, Y. Estrin, and X.Z. Hu, Fatigue crack growth in 316l stainless steel. *Key Engrg. Mater.*, 145–149:631–636, 1998.

Multiscale Modelling of the Failure Behaviour of Fibre-Reinforced Laminates

M.V. Cid Alfaro[1], A.S.J. Suiker[1] and R. de Borst[2]

[1]Delft University of Technology, P.O. Box 5058, 2600 GB, Delft, The Netherlands;
a.s.j.suiker@tudelft.nl
[2]Eindhoven University of Technology, P.O. Box 513, 5600 MB, Eindhoven, The Netherlands

Abstract The static failure behaviour of fibre-reinforced laminates is examined at the mesoscale and microscale levels of observation using finite element simulations. The actual cracking and delamination processes are simulated with interface elements equipped with a mixed-mode damage model. The mesoscale simulations consider a fibre-metal laminate GLARE that is subjected to uniaxial tension. The effect of plasticity in the metal layers on the cracking and delamination processes in the laminate is analysed. The results for a brittle laminate are compared against a closed-form expression derived from energy considerations. In the microscale simulations the transverse failure response is studied of unidirectional fibre-epoxy systems subjected to uniaxial tension. The influence on the failure response by the relative strength of the fibre-epoxy interface and the epoxy matrix is demonstrated for single-fibre epoxy systems. For multiple-fibre epoxy systems the effect of the volume fraction on the failure response is assessed. Finally, the discrete, microscale fracture processes in thin fibre-epoxy layers are coupled to a mesoscale traction-separation law by means of a numerical homogenization approach. It is demonstrated how the effective traction-separation response and the corresponding microscale fracture patterns under mesoscale tensile conditions depend on the presence of microscale imperfections.

Key words: Delamination, fibre debonding, matrix cracking, cohesive zone model, discrete fracture

1 Introduction

The mesoscopic failure response of fibre-reinforced laminates is characterised by a combination of complex microscopic failure modes, such as fibre debonding, epoxy cracking, and fibre pull-out. For optimising the design and manufacturing of such materials it is important to understand these failure modes in detail. This requires the development of advanced numerical models that can simulate the microscale and mesoscale failure properties in a numerically robust and accurate fashion. In addition, a coupling between these scales needs to be established, so that it becomes clear how the microscopic fracture behaviour characterises the mesoscopic failure response.

R. de Borst and E. Ramm (eds.), *Multiscale Methods in Computational Mechanics*, Lecture Notes in Applied and Computational Mechanics 55, DOI 10.1007/978-90-481-9809-2_13,

The present chapter studies the fracture response of fibre-reinforced structures at the *mesoscopic* and *microscopic* scales of observation and establishes a coupling between those scales by means of a numerical homogenization framework. The numerical homogenization method focuses on connecting the fracture response of a thin fibre-reinforced layer, as characterised by microscopic failure phenomena such as fibre-debonding and matrix cracking, to a mesoscopic cohesive zone model. Accordingly, the organisation of this chapter is as follows. Section 2 starts with a review of a mixed-mode interface damage model used in both the microscale and mesoscale simulations. The formulation of the mode-mixity is based upon an energy criterion regularly applied in linear elastic fracture mechanics studies [1,2], using a derivation procedure similar to the one proposed recently by Turon et al. [3]. In addition, the kinetic law describing the evolution of the damage process is taken as rate-dependent. This is done to account for rate effects generated during interfacial delamination, and to avoid numerical convergence problems induced by crack bifurcations.

Subsequently, in Section 3 *mesoscale* simulations are discussed, in which the failure responses of GLARE specimens containing elastic aluminium layers and elasto-plastic aluminium layers are compared. The configuration studied is composed of two aluminium alloy sheets sandwiching a cross-plied, fibre-epoxy layer, where a horizontal initial crack is placed at the centre of the specimen, across the thickness of the aluminium sheets, i.e., a centre-cracked tensile specimen is studied. After imposing the tensile loading, the centre crack starts to tunnel in the aluminium layers and induces delamination at the interfaces between the aluminium and fibre-epoxy layers. The individual aluminium and fibre-epoxy layers of the laminate are modelled by solid-like shell elements [4–6]. These elements allow for a linear strain field in thickness direction, which avoids the effect of Poisson-thickness locking that appears in conventional volume elements with a high aspect ratio in spatial dimensions, i.e., 'slender' volume elements [7]. The isotropic, elasto-plastic behaviour of the aluminium layer is simulated using a J_2-plasticity model with an exponentially saturating hardening law. The cross-plied, fibre-epoxy layer is modelled as isotropically elastic. This simplification, which is reasonable if the elastic mismatch between fibres and matrix is moderate, allows part of the numerical results to be quantitatively compared to those presented in [8].

Section 4 focuses on the numerical simulation of the transverse failure behaviour of unidirectional fibre-epoxy systems. The *microscale* failure behaviour is characterised by a combination of fibre debonding and epoxy cracking, as typically observed in experiments on fibre-epoxy systems [9, 10]. In order to allow for the development of arbitrary, complex fracture patterns in the epoxy material, interface elements endowed with the interface damage model are introduced between all continuum elements that simulate the bulk response of the epoxy. This modelling approach was originally proposed by Xu and Needleman [11] somewhat more than a decade ago, and has since then been successfully applied in failure analyses of various materials, such as crazing in polymers [12], and quasi-brittle fracture in cementitious composites [13]. The analyses start with the examination of a single fibre embedded in an epoxy matrix, which is loaded under uniaxial tension. For this configuration the effect on the failure response by the relative strength of the fibre-epoxy interface and the epoxy matrix is assessed. Subsequently, in Section 5 numerical simulations of uniaxially loaded epoxy systems containing multiple fibres are studied. where the effect of the fibre volume fraction is considered.

In Section 6 a numerical homogenization method is discussed that connects the discrete *microscale* fracture processes in thin fibre-epoxy layers to a *mesoscale* traction-separation curve. Such a traction-separation law curve can, for example, be applied in material points of interface elements used for simulating mesoscopic fracture in fibre adhesives, under the condition that the thickness of the adhesive is negligible with respect to the length scales of the macroscopic structure considered (i.e., a separation of length scales can be warranted). The applicability of the numerical homogenization approach is demonstrated by studying fibre-epoxy samples loaded under uniaxial tension. Samples of different size are examined in order to check the convergence behaviour of the response to that of an RVE. Further, the effect of imperfections on the traction-separation response is analysed. Finally, in Section 7 some concluding remarks are given.

2 Review of the Interface Damage Model

The mesoscopic and microscopic failure analyses in this communication are carried out with the finite element method, where the fracture processes are modelled with interface elements endowed with a rate-dependent damage model. The interface damage model has been previously explained in detail in [14, 15], in which it has been applied to study the three-dimensional laminate failure mechanisms reported in [8, 16]. The damage model is phenomenological, and has a bi-linear traction-separation curve to model a quasi-brittle failure response generated by two separating crack faces, see Figure 1. For completeness, only the main equations of the interface damage model in [14, 15] are summarised below. The constitutive equation relating the tractions to the relative displacements across an interface is given by

$$t_i = (1 - d)K\delta_{ij}v_j - dC_{ij}\delta_{1j}\langle -v_1\rangle \qquad \text{where} \qquad i, j \in \{1, 2, 3\}, \tag{1}$$

with the normal and the two tangential components of the tractions t_i and relative displacements v_i designated by the indices '1', '2' and '3' respectively. Furthermore, K is the elastic stiffness and δ_{ij} is the Kronecker delta symbol. The damage parameter d is bounded as $0 \leq d \leq 1$, such that an undamaged integration point is represented by $d = 0$ and a completely damaged integration point is reflected by $d = 1$. Penetration of two opposite crack faces is prevented by the second term in the right-hand side of Eq. (1), where the Macauley brackets $\langle x\rangle = \frac{1}{2}(x + |x|)$ ensure that crack faces interact elastically when the normal relative displacement v_1 becomes negative. The evolution of damage is assumed to occur in a rate-dependent fashion, with the rate of damage formulated as

$$\dot{d} = \begin{cases} \dfrac{\hat{F}(\lambda, \kappa)}{\eta} & \text{for} \quad \lambda \geq \kappa \quad \text{and } v^0 \leq \kappa < v^f , \\ 0 & \text{for} \quad 0 \leq \lambda < \kappa \text{ or } \quad \kappa = v^f . \end{cases} \tag{2}$$

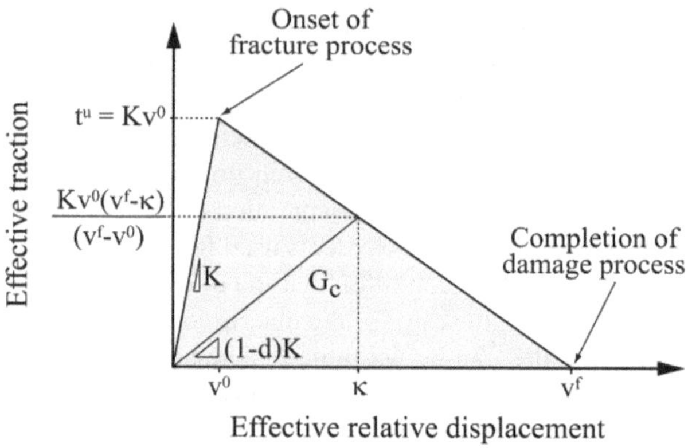

Fig. 1. Traction-separation law.

Here, η is a relaxation parameter (with dimension of time) and $\hat{F}(\lambda, \kappa)$ is the damage loading function, which is dependent upon a deformation measure λ and a history variable κ. The deformation measure λ is taken as the Euclidean norm of the vector of relative crack face displacements, $\lambda = \|\boldsymbol{v}\| = \sqrt{v_1^2 + v_{sh}^2}$, with v_{sh} the total shear displacement $v_{sh} = \sqrt{v_2^2 + v_3^2}$. The upper expression in Eq. (2) defines the rate of damage when the effective deformation λ exceeds the value of the history variable κ, whereas the lower expression sets the rate of damage equal to zero when (i) the threshold value for the initiation of damage has not (yet) been reached, (ii) the interfacial material point is in a state of unloading, or (iii) the damage process has been completed. Note from Eq. (2) that the history parameter is bounded as $v^0 \leq \kappa \leq v^f$, with v^0 and v^f the relative crack face displacements at damage initiation ($d = 0$) and damage completion ($d = 1$), respectively.

The actual form of the damage loading function $\hat{F}(\lambda, \kappa)$ is connected to the shape of the softening curve in the traction-separation relation, see [15] for more details. As illustrated in Figure 1, the damage process is characterised by a linear softening curve, which is in correspondence with the following damage loading function:

$$\hat{F}(\lambda, \kappa) = \hat{f}(\lambda) - \hat{d}(\kappa) = \frac{v^f(\lambda - v^0)}{\lambda(v^f - v^0)} - \frac{v^f(\kappa - v^0)}{\kappa(v^f - v^0)} . \tag{3}$$

It can be observed from Eq. (2) that in the limit case of the relaxation parameter going to zero, $\eta \to 0$, the rate-independent loading condition is recovered: $\hat{F}(\lambda, \kappa) = 0$, which, as shown by Eq. (3), corresponds to $\lambda = \kappa$.

In order to simulate fracture processes of arbitrary mode-mixity, the relative displacements at damage initiation, v^0, and damage completion, v^f, are made dependent of a mode-mixity parameter β, see also [3], i.e.,

$$\beta = \frac{v_{sh}}{v_{sh} + \langle v_1 \rangle} . \tag{4}$$

In correspondence with this expression, under tensile loading conditions $v_{sh} = 0$ and thus $\beta = 0$, whereas under shear loading conditions $v_1 = 0$ and thus $\beta = 1$. The functions $v^0 = \hat{v}^0(\beta)$ and $v^f = \hat{v}^f(\beta)$ can be computed adopting an energy-based, mixed-mode failure criterion from linear elastic fracture mechanics, see [3, 15]. For this purpose, a fracture criterion is taken that is often used to characterize mixed-mode toughness data for brittle interfacial fracture [1, 2]:

$$\frac{G_I}{G_{I,c}} + \frac{G_{II}}{G_{II,c}} + \frac{G_{III}}{G_{III,c}} = 1\,, \tag{5}$$

with G_I, G_{II} and G_{III} the mode I, mode II and mode III energy release rates, and $G_{I,c}$, $G_{II,c}$ and $G_{III,c}$ the toughnesses under pure mode I, pure mode II and pure mode III loading conditions respectively. From Eqs. (4) and (5), the effective relative displacements at damage initiation and damage completion can be, respectively, computed as

$$v^0 = \hat{v}^0(\beta) = v_1^0\, v_{sh}^0 \sqrt{\frac{1 + 2\beta^2 - 2\beta}{\left(\beta v_1^0\right)^2 + \left((1-\beta) v_{sh}^0\right)^2}}\,, \tag{6}$$

and

$$v^f = \hat{v}^f(\beta) = \frac{2(1 + 2\beta^2 - 2\beta)}{K\, v^0}\left[\left(\frac{(1-\beta)^2}{G_{I,c}}\right) + \left(\frac{\beta^2}{G_{II,c} + G_{III,c}}\right)\right]^{-1}, \tag{7}$$

where $v_1^0 = t_1^u/K$ and $v_{sh}^0 = t_{sh}^u/K$ are the relative displacements at which damage initiates under pure tension and pure shear loading conditions, respectively, and t_1^u and $t_{sh}^u = \sqrt{(t_2^u)^2 + (t_3^u)^2}$ are the ultimate tractions under tension and shear conditions, respectively. More details about the above interface damage model and its numerical discretization can be found in [15].

3 Mesoscale Simulations of a Centre-Cracked 2/1 GLARE Laminate

The interface damage model presented in the previous section is used to study mode I fracture and interfacial delamination in a centre-cracked 2/1 laminate subjected to uniaxial tension, see Figure 2. The 2/1 lay-up refers to a laminate composed of two aluminium layers sandwiching a single fibre-epoxy layer. Laminates reinforced by glass fibres are commercially known as GLARE. For this material the aluminium and fibre-epoxy layers commonly have thicknesses of 0.3 and 0.25 mm, respectively (see for example, [17]), which leads to a total specimen thickness of 0.85 mm.

3.1 Geometry and Boundary Conditions

The specimen length is taken as $L = 60$ mm and its width equals $W = 40$ mm. The initial pre-crack at the centre of the specimen is placed across the complete thickness

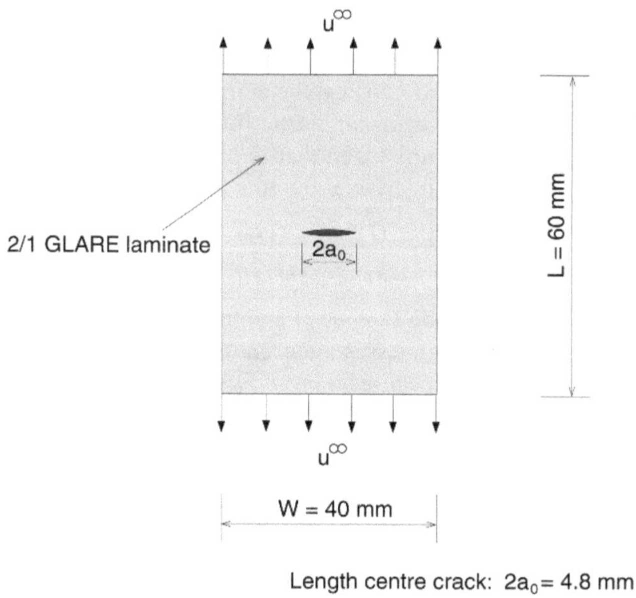

Fig. 2. Dimensions of a centre-cracked tensile specimen made of 2/1 GLARE.

of the two aluminium layers, and has a length of $2a_0 = 0.12\,W = 4.8$ mm. The tensile loading is imposed by prescribing a remote normal displacement u^∞ at the top and bottom edges of the laminate, which induces a uniform remote tensile stress σ^∞ in the aluminium layer at short distance from the top and bottom edges. Quasi-static loading conditions are warranted by applying a relatively small nominal strain rate $\dot{u}^\infty/L = 10^{-5}\ s^{-1}$. As demonstrated by experimental and modelling studies on the static and fatigue failure behaviour of centre-cracked laminates [8, 16–20], under the applied remote tensile loading the initial pre-crack will start to tunnel as a mode I crack through the aluminium sheet, thereby initiating delamination along the interface between the aluminium (material #1) and fibre-epoxy (material #2) layers. The complete fracture mechanism has been sketched in Figure 3 (left), and can be denoted as a *tunneling, doubly-deflected crack* [8, 16].

In order to examine how plasticity in the aluminium layer affects the failure characteristics of the specimen, simulations in which the aluminium layer behaves isotropically elastic are compared to simulations where the aluminium is modelled as isotropically elasto-plastic. In both cases the fibre-epoxy layer is modelled as isotropically elastic. This is an acceptable simplification if the internal structure of the fibre-epoxy layer is cross-plied, i.e., made of two sublayers of 0.125 mm thickness with the uni-directional fibres oriented under 0° and 90°, respectively, see Figure 3 (left), and the elastic mismatch between the fibres and the epoxy is moderate. The assumption of an isotropic fibre-epoxy layer increases the degree of symmetry of the problem, such that only one octant of the actual geometry has to be modelled, see Figure 3 (right). Trivially, this significantly reduces the computational time of the

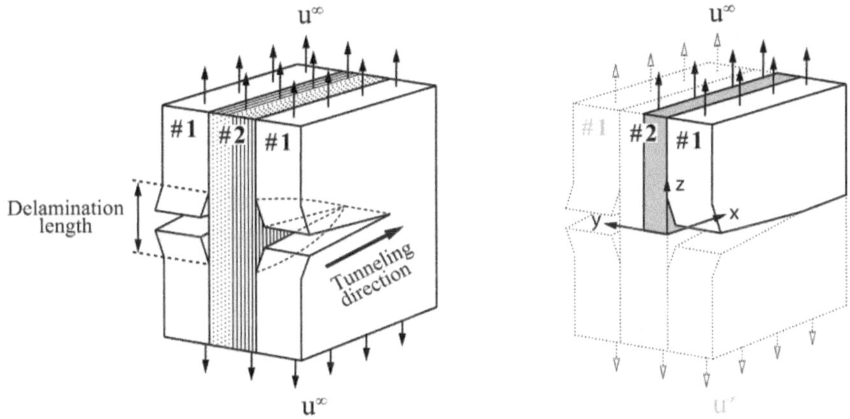

Fig. 3. Left: Doubly-deflected crack in the two outer layers of a 2/1 GLARE laminate subjected to uniaxial tension (imposed by a remote displacement u^{∞}). The laminate is composed of two aluminium (material #1) layers sandwiching a single, cross-plied, fibre-epoxy (material #2) layer. Right: One symmetrical octant of the centre-cracked tensile specimen, used in the numerical simulations.

analysis. Furthermore, this simplification allows the present computational results to be compared to results reported by Suiker and Fleck [8], which relate to fracture and delamination processes in lay-ups composed of brittle, isotropic elastic layers. The symmetry of the configuration depicted in Figure 3 (right) is warranted by fixed and roller supports, and by using a customized interface element that relates the interface tractions of the mode I crack in the aluminium (propagating along the horizontal x–y symmetry plane half-way the specimen) to *half* of the relative crack face displacements across the interface.

The numerical analyses are performed within a large-displacement, small-strain framework, where the aluminium and fibre-epoxy layers are meshed with sixteen-node, iso-parametric solid-like shell elements, with a $2 \times 2 \times 2$ Gauss quadrature [4–6]. For more details on the finite element discretisation, see [15]. The elasto-plastic response of the aluminium layers in the GLARE specimen is simulated using a standard J_2-flow theory (Von Mises plasticity), where the yield strength σ_y evolves in accordance with an exponentially-saturating hardening law. The dummy elastic stiffness K of the interface elements modelling the mode I crack and the delaminating crack is taken relatively high, such that in the elastic regime the interfacial deformations remain negligibly small. Values for the ultimate normal traction of the mode I crack in the aluminium, t_1^u, have not been widely reported in the literature. Correspondingly, in the present analysis the ultimate traction is varied, considering the following three values: $t_1^u = 1.4\sigma^*$, $2.0\sigma^*$ and $3.0\sigma^*$, with the reference stress taken as $\sigma^* = 305$ MPa (which equals the initial yield strength σ_y^0 in the elasto-plastic model for the aluminium). The ultimate tractions for the delaminating crack, t_1^u, t_2^u and t_3^u, are taken equal to the values reported in [4], which were determined from shear and tensile tests on a prepreg material typically used in GLARE. The relaxation parameter η for the mode I crack is taken very small such that under quasi-static loading conditions the crack advancement occurs virtually rate-independently. For the delaminating crack, the relaxation parameter η is chosen slightly larger in order to avoid convergence problems due to crack bifurcations; additional simulations not presented here have shown that in the limit of a truly rate-independent fracture behaviour (i.e., $\eta \rightarrow 0$) the convergence of the present numerical simulations can not be warranted. The actual values of the material parameters used in the simulations can be found in [15].

3.2 Results for a 2/1 Lay-up with Elastic Aluminium Layers

For the case where the aluminium layer behaves elastically, Figure 4a depicts the remote stress in the aluminium layer (evaluated at the top of the laminate, near the horizontal edge at which the loading is applied) plotted against the normal opening of the mode I crack in the aluminium (measured at the tip position of the initial pre-crack) for three different values of the ultimate normal traction, t_1^u, of the mode I crack. The characters 'a-f', which indicate different deformation stages, correspond to the delamination profiles in Figures 4b and c for the cases where the ultimate traction of the mode I crack is $t_1^u = 1.4\sigma^*$ and $t_1^u = 2.0\sigma^*$, respectively. Since the delamination evolution for the case with $t_1^u = 3.0\sigma^*$ looks similar to that of the case with $t_1^u = 2.0\sigma^*$ (with the only difference being a somewhat stronger delamination development during the initial loading stage), the former case has not been visualized. From Figure 4a it can be observed that the remote stress initially increases, subsequently reaches a maximum value, then decreases (thus representing an unstable, softening response) and finally asymptotes to a (residual) stress level that remains approximately constant under continuing crack opening. Clearly, the value of the maximum remote stress is higher for a higher value of the fracture strength t_1^u of the aluminium, i.e., $\sigma_{max}^{\infty} = 720$, 760 and 913 MPa for $t_1^u = 1.4\sigma^*$, $2.0\sigma^*$ and $3.0\sigma^*$, respectively. The fact that the local ultimate traction influences the remote stress indicates that the size of the fracture process zone of the tunneling crack can *not* be ignored with respect to its length and/or the total specimen width. If this would be the case, the crack would be of the Griffith type, where the remote stress only depends on the local toughness characteristics of the crack [1]. However, due to the crack bridging behaviour by the fibre-epoxy layer, the size of the process zone of the tunneling crack is significant, see [15] for more details.

The delamination profiles in Figures 4b and c are obtained by graphically connecting the integration points in which the damage d has just exceeded the value of 0.99 (which thus is close to the maximum damage value $d = 1$). The x- and z-coordinates used for denoting the location of the profiles are in correspondence with a coordinate system that has its origin taken at the centre of the specimen. It can be observed that the differences in delamination growth for $t_1^u = 1.4\sigma^*$ and $t_1^u = 2.0\sigma^*$ are small. In both cases the delamination profile initially develops with increasing length of the mode I crack in the aluminium (stages 'a' and 'b'), where the maximum delamination takes place at the specimen centre. When the mode I crack

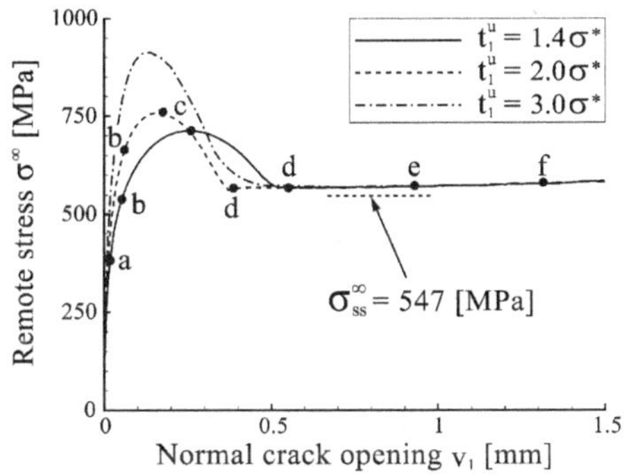

(a) Normal opening v_1 of the mode I crack in the aluminium (measured at the tip position of the initial pre-crack) versus the remote stress σ^{∞} in the aluminium layer, for three different values of the ultimate normal traction t_1^u of the mode I crack in the aluminium. The horizontal, dashed line indicates the steady-state remote stress σ_{SS}^{∞} = 547 MPa computed with Eq. (8), and the characters 'a-f' indicate different deformation stages during the loading process.

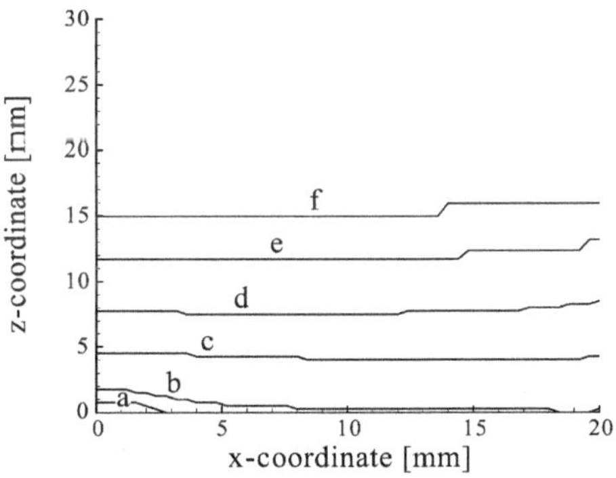

(b) Delamination profiles for deformation stages 'a-f' (see Figure 4a), for the case where the ultimate traction of the mode I crack in the aluminium is $t_1^u = 1.4\sigma^*$ (= 430 MPa). The x- and z-coordinates relate to a coordinate system that has its origin at the centre of the specimen.

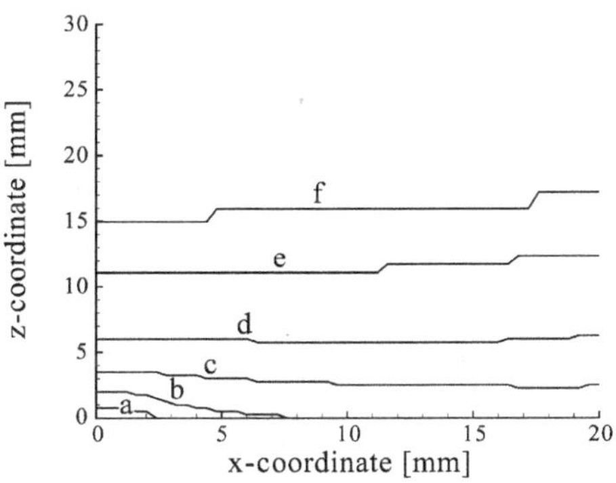

(c) Delamination profiles for deformation stages 'a-f' (see Figure 4a), for the case where the ultimate traction of the mode I crack in the aluminium is $t_1^u = 2.0\sigma^*$ (= 610 MPa). The x- and z-coordinates relate to a coordinate system that has its origin at the centre of the specimen.

Fig. 4. Failure response for a 2/1 lay-up with *elastic* aluminium layers.

reaches the vertical specimen edge, the delamination starts to develop mainly in the vertical direction. After stage 'd' the delamination profile has become almost uniform across the specimen width (only near the free edge the delamination is slightly larger), where the growth process closely approximates the condition of steady-state delamination. In [8] a closed-form expression has been presented for the remote failure stress for steady-state delamination that propagates along the two outermost plies of a uniaxially-loaded, brittle laminate. This expression is derived from the energy difference upstream and downstream of the delamination front, and reads

$$\sigma_{ss}^{\infty} = \sqrt{\frac{4\bar{E}_1\, G_c\left((n-2)w_1\bar{E}_1 + (n-1)w_2\bar{E}_2\right)}{2w_1\left(n\, w_1\bar{E}_1 + (n-1)w_2\bar{E}_2\right)}}\,. \tag{8}$$

Here, σ_{ss}^{∞} is the remote stress in the aluminium layer, n is the number of material #1 layers, G_c is the delamination toughness, and w_1 and w_2 are the thicknesses of the material #1 and #2 layers. Further, the stiffness $\bar{E}_i$ is defined by

$$\bar{E}_i = \begin{cases} E_i/(1-\nu_i^2) & \text{for plane strain}\,, \\ E_i & \text{for plane stress}\,, \end{cases} \tag{9}$$

where E_i and ν_i are the Young's moduli and Poisson's ratios of materials $i = 1$ (aluminium) and 2 (fibre-epoxy). During uniform, steady-state delamination the laminate approaches a *plane-stress* condition, since the right edge of the specimen is traction-free, in accordance with the boundary conditions applied to the GLARE specimen. Substitution of the actual material parameters into Eq. (8), where (as in the numerical simulations) the delamination toughness equals $G_c = 4$ N/mm and the number of aluminium layers equals $n = 2$ (in correspondence with a 2/1 lay-up), for a plane-stress condition leads to a remote stress of $\sigma_{ss}^{\infty} = 547$ MPa. This stress value, which is represented in Figure 4a by a horizontal, dashed line, is in close agreement with the residual strength following from the numerical analyses.

As illustrated by Eq. (8), the residual strength during steady-state delamination mainly depends on the delamination toughness G_c, and is independent of the aluminium fracture strength t_1^u. The latter aspect can also be observed from Figures 4a, b and c, which illustrate that at stages 'e' and 'f' both the $\sigma^{\infty} - v_1$ response and the delamination profiles are virtually identical for the different fracture strengths t_1^u. The brittle, almost rate-independent behaviour of the delamination process is further confirmed by the appearance of a relatively small process zone at the delamination tip, which actually can be ignored with respect to the specimen length L.

During the final phase of the crack tunneling process (i.e., between stages 'c' and 'd'), significant delamination has developed, see Figures 4b and c, where the effective response of the laminate has become unstable, see Figure 4a. In the study of Suiker and Fleck [8] a similar, unstable failure mechanism has been identified. Although the study [8] relates to an 'ideally brittle' laminate of infinite size, where the size of the process zone of the tunneling crack is equal to zero, the typical unstable propagation characteristics of this failure mechanism are thus also observed for the present 'quasi-brittle' laminate where the tunneling crack is characterised by a process zone of substantial length.

3.3 Results for a 2/1 Lay-up with Elasto-Plastic Aluminium Layers

The results reported in the above section for a GLARE laminate with elastic aluminium layers are now compared to those for a GLARE laminate with elasto-plastic aluminium layers. In Figure 5a the remote stress in the aluminium layers is plotted versus the normal opening of the mode I crack in the elasto-plastic aluminium layers. It can be observed that a higher fracture strength t_1^u of the aluminium leads to a higher remote stress, i.e., at a crack opening $v_1 = 1.9$ mm the remote stress in the aluminium equals $\sigma^\infty = 557$, 611 and 642 MPa for $t_1^u = 1.4\sigma^*$, $2.0\sigma^*$ and $3.0\sigma^*$, respectively. For the case with the highest aluminium fracture strength, $t_1^u = 3.0\sigma^*$, at $v_1 = 1.9$ mm the continuum elements close to the crack tip have become strongly distorted, which is due to the generation of substantial plastic deformations in these elements. Consequently, at this stage the numerical procedure fails to converge. Although it is expected that this problem can be overcome by further decreasing the size of the elements near the crack tip, this strategy has not been explored in more detail due to the already large computational times of the present simulations (i.e., the solid-like shell elements used in the numerical model are computationally expensive). Note that the curves depicted in Figure 5a do not show a softening branch, which is in contrast to the effective response of a lay-up with elastic aluminium layers, see Figure 4a. Hence, it may be concluded that the dissipation in the plastic zone ahead of the crack in the aluminium stabilizes the effective failure response of the laminate. For a larger fracture strength t_1^u of the aluminium, the plastic zone, and thus the dissipation in this zone, becomes larger, as a result of which the failure response becomes more stable, see Figure 4a.

The delamination profiles for the deformation stages 'a-f' indicated in Figure 5a are plotted in Figures 5b ($t_1^u = 1.4\sigma^*$) and 5c ($t_1^u = 2.0\sigma^*$). Apparently, during crack tunneling the delamination front for a specimen with elasto-plastic aluminium layers is much steeper than for a specimen with elastic aluminium layers, see Figures 4b,c. In addition, plastic deformations in the aluminium layer induce substantial free-edge delamination along the vertical specimen edge. Essentially, near the free edge the plastic deformations in the aluminium layer increase the mismatch in deformation with the fibre-epoxy layer that wants to contract laterally. This gives rise to the development of stress concentrations at the interface between the two layers, which are relaxed through the appearance of free-edge delamination.

At the level of uniform delamination (stage 'f'), the normal crack opening v_1 and the remote failure strength σ^∞ are larger for a higher fracture strength t_1^u of the aluminium, see Figure 5a. In other words, the failure strength of the panel is dependent on the plastic deformation history. Recall that this effect has *not* been observed for the 2/1 laminate with elastic aluminium layers; the strength and deformation characteristics at uniform delamination are independent of the fracture strength of elastic aluminium layers, see stages 'e' and 'f' in Figures 4a, b and c and Eq. (8). It is further interesting to notice that during uniform delamination the remote strength for a laminate containing elasto-plastic aluminium layers may be somewhat higher than for a laminate containing elastic aluminium layers, especially for higher values of the aluminium fracture strength t_1^u, see Figures 5a and 4a. Hence, the plastic deform-

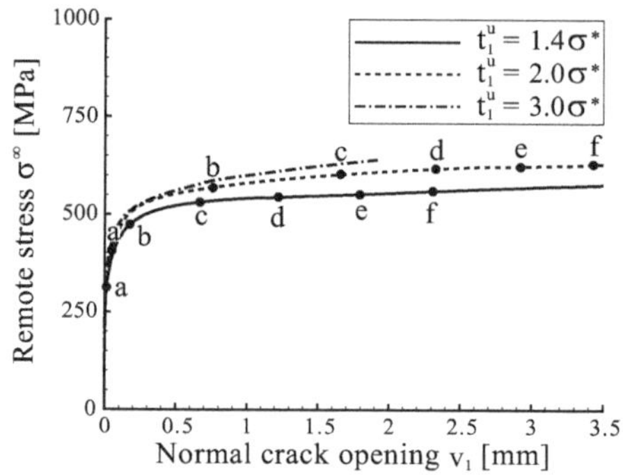

(a) Normal opening v_1 of the mode I crack in the aluminium (measured at at the tip position of the initial pre-crack) versus the remote stress σ^{∞} in the aluminium layer, for three different values of the ultimate normal traction t_1^u of the mode I crack in the aluminium. The characters 'a-f' indicate different deformation stages during the loading process.

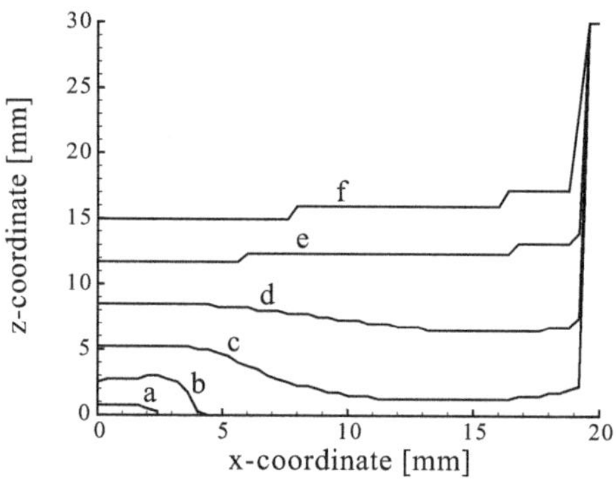

(b) Delamination profiles for deformation stages 'a-f' (see Figure 5a), for the case where the ultimate traction of the mode I crack in the aluminium is $t_1^u = 1.4\sigma^*$ $(= 1.4\sigma_y^0 = 430$ MPa). The x- and z-coordinates relate to a coordinate system that has its origin at the centre of the specimen.

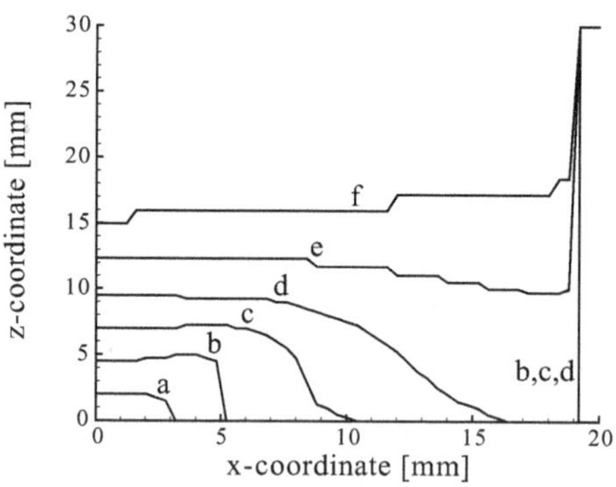

(c) Delamination profiles for deformation stages 'a-f' (see Figure 5a), for the case where the ultimate traction of the mode I crack in the aluminium is $t_1^u = 2.0\sigma^*$ $(= 2.0\sigma_y^0 = 610$ MPa). The x- and z-coordinates relate to a coordinate system that has its origin at the centre of the specimen.

Fig. 5. Failure response for a 2/1 lay-up with *elasto-plastic* aluminium layers.

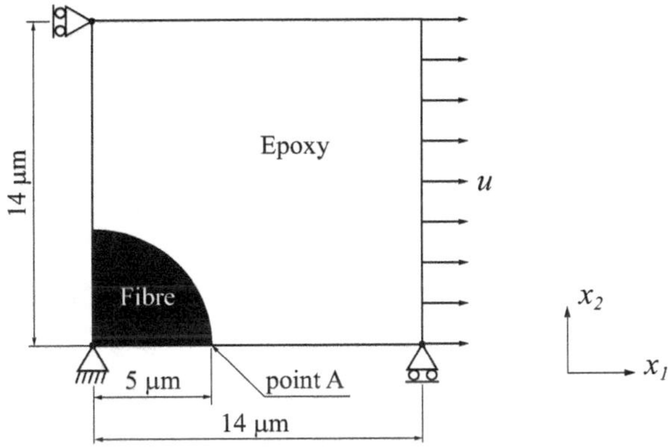

Fig. 6. Geometry and boundary conditions of a single fibre embedded in an epoxy matrix. Using symmetry boundary conditions only a quarter of the sample is simulated.

ation history developed during crack tunneling then provides a positive contribution to the final residual strength of the laminate specimen. A more extensive analysis of the present GLARE laminate can be found in [14, 15].

4 Microscale Simulations of Single-Fibre Epoxy Systems

In this section the fracture behaviour of fibre-epoxy systems is analysed in detail by performing microscale analyses. As a start, the failure response of a uniaxially-loaded sample composed of a single S2 glass fibre embedded in a FM94 epoxy matrix is studied, which are the typical components of the prepreg layers used in GLARE [18]. The modelled sample is assumed to be under plane-strain conditions, which are representative of a fibre with a relatively large (actually infinite) length. As a result of symmetry of the sample geometry, one quarter of the actual configuration may be modelled.

The geometry and boundary conditions of the model are illustrated in Figure 6. The symmetry is warranted by placing fixed and roller supports at the left and bottom edges of the sample. The diameter of the S2 glass fibre is 10 µm, which is representative of the glass fibres used in GLARE [18]. The specimen is square-shaped, with a side length of $L = 28$ µm, and has a fibre volume fraction $V_f = 0.1$. The tensile loading is imposed applying a uniform displacement u at the right edge of the specimen. Quasi-static loading conditions are warranted by prescribing a relatively small nominal strain rate of $\dot{u}/L = 4 \times 10^{-3}$ s^{-1}. The fibre and the epoxy matrix are meshed with 6-node triangular elements equipped with a 7-point Gauss quadrature. Debonding between fibre and matrix is simulated with the interface damage model outlined in Section 2, using 6-node interface elements equipped with a 3-point Newton–Cotes quadrature. Fracture processes within the epoxy material are also simulated with the interface damage model, placing 6-node interface elements furnished with a 3-point Newton–Cotes quadrature in between the triangular

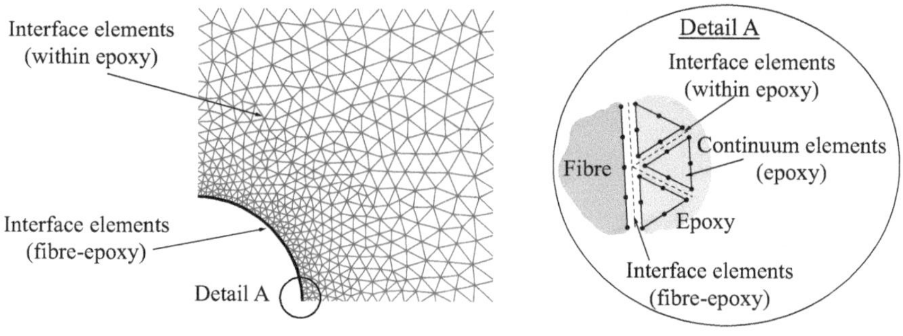

Fig. 7. Finite element mesh where interface elements are placed at the fibre-epoxy interface and within the epoxy.

continuum elements constructing the epoxy matrix, see Figure 7. This approach for the simulation of fracture was originally proposed by Xu and Needleman [11], and warrants that the crack initiation and crack propagation processes in the sample are naturally determined by the geometry and boundary conditions applied, and by the parameter values used in the interface traction-separation law. The influence of the finite element discretisation on the fracture response can be minimised by choosing a sufficiently fine, randomly oriented mesh. Nonetheless, the mesh fineness is bounded by a maximum in order to limit artificial response contributions related to the use of an interface 'dummy' stiffness K in the traction-separation law, see Eq. (1). In fact, the application of interface elements with an elastic dummy stiffness causes that the elastic response of the finite element model does not converge to that of an ideal continuum in the limit of an 'infinitely fine' mesh. More details on the influence of the numerical discretisation on the computed response can be found in [14, 21].

In the simulations, the elastic material parameters of the fibre and the epoxy matrix (i.e., their Young's modulus and Poisson's ratio) have been taken from [22]. Values for the ultimate strengths and fracture toughnesses of the epoxy material and the fibre-epoxy interface have not been reported in the literature, and therefore were estimated from the failure response of delamination tests on double spliced GLARE specimens [23]. For simplicity, both for the epoxy matrix and the fibre-epoxy interface the strength and toughness values were taken the same under tension (mode I) and shear (mode II) loading. The subscripts '*coh*' and '*adh*' that will be used in the denotation of these parameters refer to the *cohesive* strength (toughness) of the epoxy material and the *adhesive* strength (toughness) of the fibre-epoxy interface, respectively. The influence of these strength parameters on the effective failure response of the sample is studied in detail by means of parameter variation studies. The actual values of the material parameters used can be found in [14, 21].

4.1 Fibre-Epoxy Interfacial Strength versus Epoxy Strength

The influence of the relative strength of the fibre-epoxy interface and the epoxy matrix on the effective failure response of the single-fibre composite is studied consid-

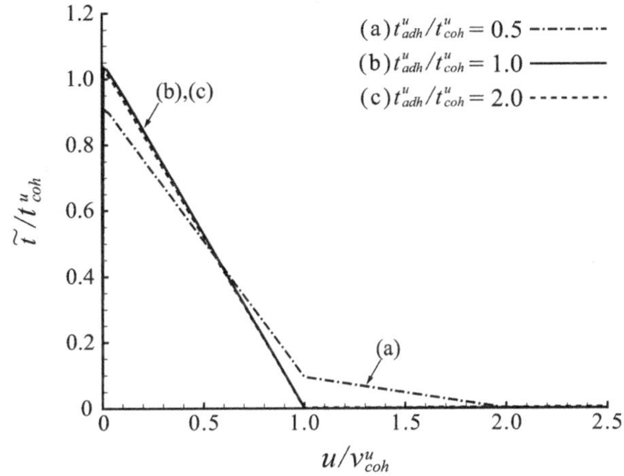

Fig. 8. Effective sample response for three different strengths ratios $t^u_{adh}/t^u_{coh} = 0.5$, 1.0 and 2.0. The traction $\tilde{t}$ represents the average value at the right sample boundary at which the displacement u is prescribed, see Figure 6.

ering three different strength ratios: $t^u_{adh}/t^u_{coh} = 0.5$, 1.0 and 2.0, where the ultimate strength of the epoxy is kept constant, $t^u_{coh} = 50$ MPa.

The effective sample response for the three different strength ratios is depicted in Figure 8. It can be observed that for the low strength ratio, $t^u_{adh}/t^u_{coh} = 0.5$, the softening curve has a bilinear shape. Because the sample consists for 90% of epoxy material, the effective tensile strength of the sample (= 46 N/mm^2) is close to the tensile strength of the epoxy (= 50 N/mm^2), i.e. $\tilde{t}_{max}/t^u_{coh} = 0.92$. The kink in the softening curve arises because the interface models for fibre-epoxy interfacial cracking and epoxy cracking have different ultimate relative displacements v^u. Since the values of the ultimate strength and fracture toughness were chosen equal for mode I and mode II loading conditions, and the softening branch of the applied traction-separation law is linear (see Figure 1), the ultimate relative displacement for mode I and II can be straightforwardly estimated as $v^u \approx 2G_c/t^u$. With the actual material data used this corresponds to $v^u_{coh} \approx 0.02$ mm for the interface damage model of the epoxy (which is reached in Figure 8 at the kink of the bilinear softening branch, when $u = v^u_{coh}$) and $v^u_{adh} \approx 0.04$ mm for the interface damage model of the fibre-epoxy interface (which is reached in Figure 8 at the end of the softening branch, when $u = v^u_{adh} = 2v^u_{coh}$). Hence, the crack through the epoxy has completely opened during the deformation stage associated with the second (and final) part of the bilinear softening branch, and the remaining load is transfered solely via the fibre-epoxy interface. The corresponding fracture pattern has been depicted in Figure 9(a). The dominant failure crack nucleates at the intersection between the fibre-epoxy interface and the horizontal symmetry axis of the fibre. Observe that, after some crack growth along the fibre-epoxy interface, the stress locally reaches the tensile strength of the epoxy, where the crack deflects into the epoxy material.

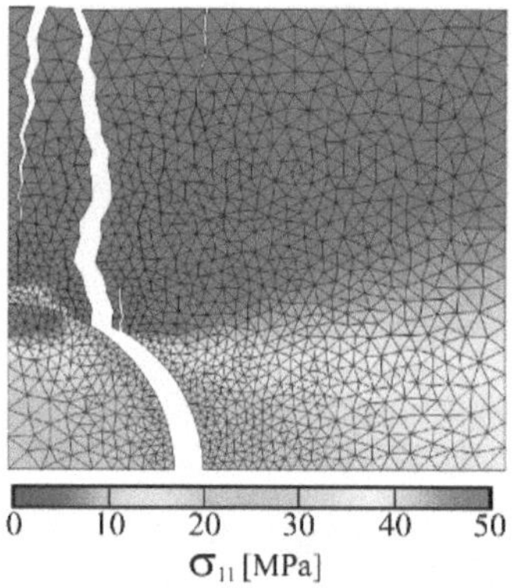

(a) Response for the case $t^u_{adh} < t^u_{coh}$.

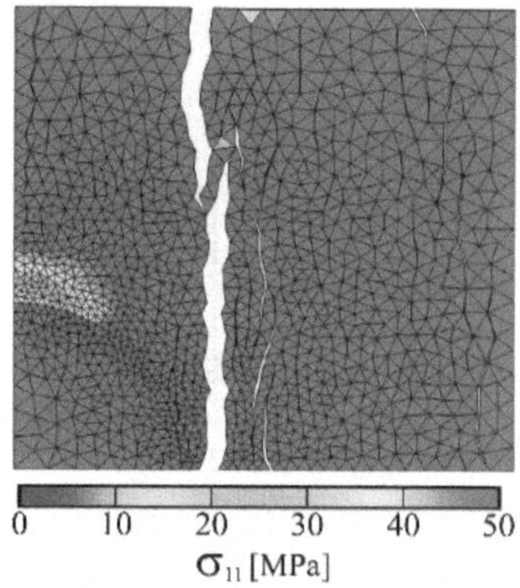

(b) Response for the case $t^u_{adh} > t^u_{coh}$.

Fig. 9. Axial normal stress σ_{11} depicted in the deformed configuration (true scale). The response states are taken after the effective peak strength is passed, at a prescribed displacement of $u = 0.05\, v^u_{coh}$, see Figure 8.

In contrast to this behaviour, for the intermediate and high strength ratios, $t^u_{adh}/t^u_{coh} = 1.0$ and 2.0, the response is characterised by a linear softening branch, which can be explained as follows. When the adhesive strength of the fibre-epoxy interface is larger than the cohesive strength of the epoxy, $t^u_{adh} > t^u_{coh}$, the failure response of the sample is determined by the fracture properties of the epoxy only and, as such, the strength of the fibre-epoxy interface becomes irrelevant. Correspondingly, the ultimate failure crack runs fully through the epoxy material, see Figure 9(b). The fact that this crack is virtually straight indicates that the epoxy matrix fails in mode I. The effective failure response in Figure 8 thus closely approximates the mode I traction-separation law of the epoxy material (which is characterised by a linear softening branch, see Figure 1), where the maximum effective traction equals $\tilde{t}_{max} = t^u_{coh}$ and the displacement at complete failure equals $u = v^u_{coh}$.

When the adhesive strength of the fibre-epoxy interface is equal to the cohesive strength of the epoxy, $t^u_{adh} = t^u_{coh}$, both fracture patterns depicted in Figure 9 may occur. Under these circumstances the actual fracture pattern becomes determined by the chosen finite element discretisation and the numerical round-off errors in the solution procedure. For the present finite element discretisation this has led to a failure response characterised by cracking through the epoxy only, in a similar fashion as depicted in Figure 9(b) for the case $t^u_{adh} > t^u_{coh}$.

In the examples above use had been made of symmetry boundary conditions. However, it should be mentioned that, depending on the actual finite element discretisation chosen, the symmetry of the failure pattern can be substantially broken, under which circumstances the effective traction-separation curve of the sample becomes rather different than that of the symmetric specimen. More details on this aspect can be found in [14, 21].

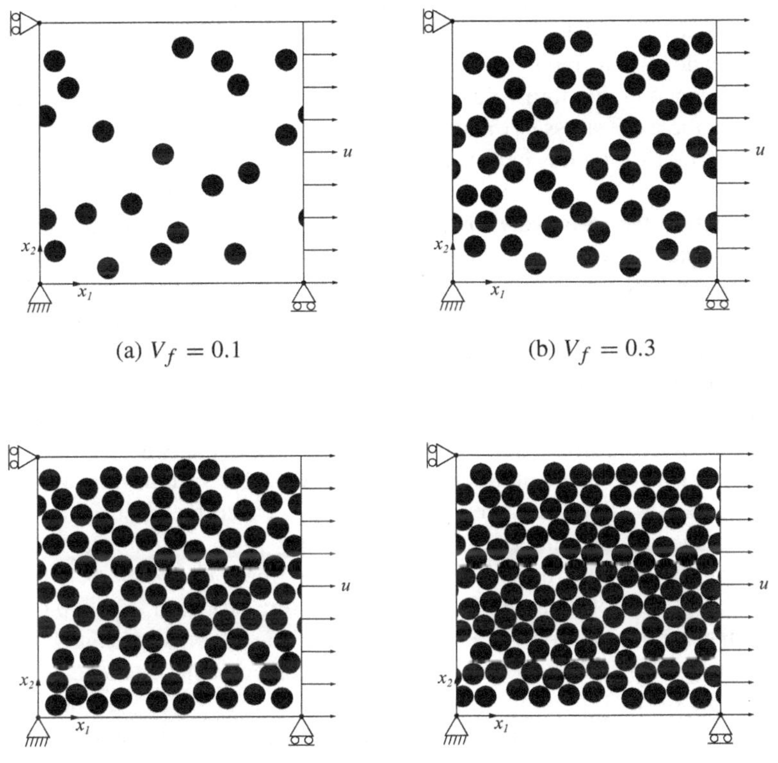

(a) $V_f = 0.1$ (b) $V_f = 0.3$

(c) $V_f = 0.5$ (d) $V_f = 0.6$

Fig. 10. Uniaxially-loaded samples with different fibre volume fractions V_f.

5 Microscale Simulations of Multiple-Fibre Epoxy Systems

The transverse failure response now is analysed for epoxy samples containing multiple, randomly placed fibres. The sample is square-shaped, and has a side length of 125 μm. The diameter of the fibres is 10 μm. The sample width and fibre diameter are representative of fibre-epoxy layers typically used in the fibre-metal laminate GLARE [18]. In the tensile direction (i.e., the x_1−direction) the internal material structure of the sample is modelled as periodic.

5.1 Influence of the Fibre Volume Fraction

To study the influence of the fibre volume fraction on the failure response of the sample, four different fibre volume fractions are considered, i.e., $V_f = 0.1, 0.3, 0.5$ and 0.6. The fibre distributions in the samples and the boundary conditions applied are sketched in Figure 10.

For the present simulations the relative strength of the fibre-epoxy interface and the epoxy matrix is taken as $t^u_{adh}/t^u_{coh} = 0.5$. For the analysis of higher strength ratios the reader is referred to [14,21].

The influence of the fibre volume fraction on the response is illustrated in Figure 11, showing the axial stress σ_{11} at three deformed states for the case where the fibre-epoxy interfacial strength is half of the epoxy strength, $t^u_{adh}/t^u_{coh} = 0.5$. Irrespective of the value of the fibre volume fraction, cracking in the sample starts when at a specific location the (relatively low) tensile strength of 25 MPa of the fibre-epoxy interface is reached. When the loading is further increased up to the tensile strength of the epoxy matrix (= 50 MPa), the interfacial cracks start deflecting in the epoxy matrix. Upon further deformation, the coalescence of these cracks leads to the development of a dominant crack across the full width of the specimen, preluding overall failure of the sample. It can be observed from Figure 11 that the position of the dominant failure crack varies for the different fibre volume fractions, and is determined by the actual fibre distribution in the sample (i.e., the sample heterogeneity).

Figure 12 illustrates the effective sample response for various fibre volume fractions. As a result of the relatively low fibre-epoxy interfacial strength, the peak strength of the sample increases with decreasing fibre volume fraction. Essentially, the peak strength for the smallest volume fraction, $V_f = 0.1$, is a factor $0.9/0.67 = 1.34$ larger than that for the largest volume fraction, $V_f = 0.6$. Furthermore, for the specimens with $V_f = 0.1, 0.3$ and 0.5 the effective strength has reached a value close to zero when the prescribed displacement u has become equal to the ultimate separation of the fibre-epoxy interface, $u = v^u_{adh} = 2v^u_{coh}$. The strength of the specimen with $V_f = 0.6$ at this deformation stage is still somewhat larger, i.e., $\tilde{t} \approx 0.1\ t^u_{coh}$. This is caused by the fact that the crack faces of the dominant failure crack are locally connected by parts of epoxy, see also Figure 11d, as a result of which transfer of loading across the crack faces remains possible at larger deformations, $2v^u_{coh} \leq u \leq 3v^u_{coh}$.

6 Coupling between Microscale and Mesoscale Crack Modelling

As the examples in the previous sections have indicated, the effective failure behaviour of FMLs may strongly depend on small-scale fracture events occurring within individual fibre-epoxy (prepreg) layers, such as fibre decohesion and matrix cracking. The present section focuses on the derivation of the effective mesoscale failure response of a thin fibre-epoxy layer from its complex microscale fracture behaviour. The mesoscale failure response is represented by a traction-separation curve constructed from numerically homogenizing the fracture response of a periodic fibre-epoxy microstructure loaded under uniaxial tension. The traction-separation curve can be applied in material points of interface elements that are used for simulating mesoscopic fracture in macroscopic laminate failure problems. This modelling approach is an appealing and acceptable alternative to the (costly) direct simulation of microscale fracture in macroscopic problems if the fluctuations of the microscale crack trajectories in the thickness direction of the laminate are much smaller than

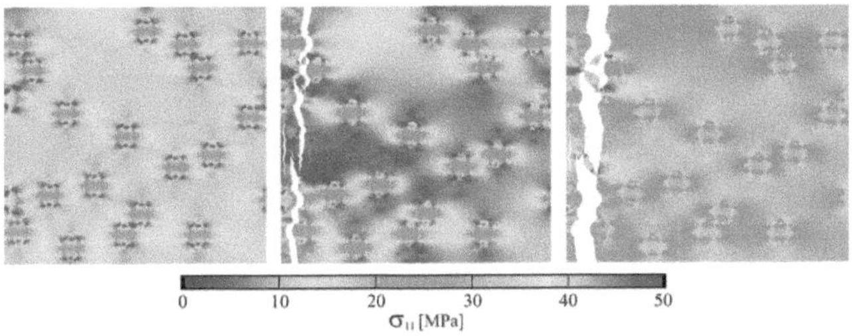

(a) $V_f = 0.1$. Axial normal stress σ_{11} taken at the deformed states $u = 0.06\ v^u_{coh}$, $0.23\ v^u_{coh}$ and $0.50\ v^u_{coh}$.

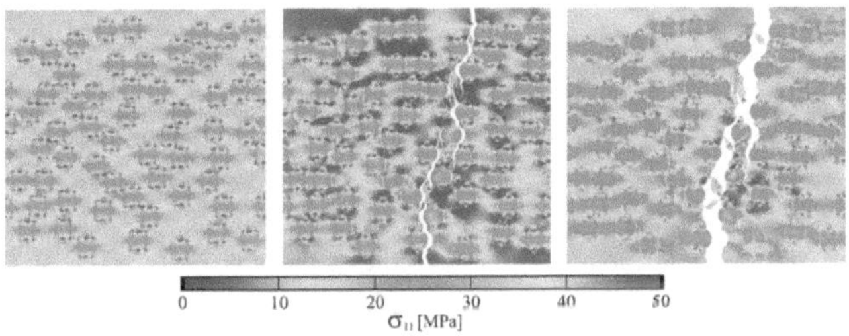

(b) $V_f = 0.3$. Axial normal stress σ_{11} plotted at the deformed states $u = 0.05\ v^u_{coh}$, $0.18\ v^u_{coh}$ and $0.44\ v^u_{coh}$.

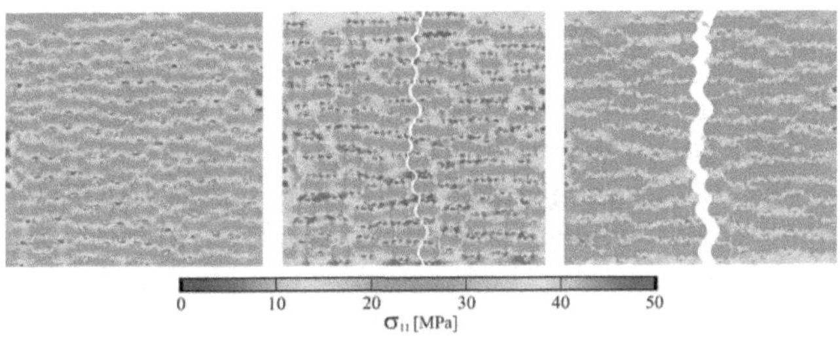

(c) $V_f = 0.5$. Axial normal stress σ_{11} plotted at the deformed states $u = 0.03\ v^u_{coh}$, $0.13\ v^u_{coh}$ and $0.37\ v^u_{coh}$.

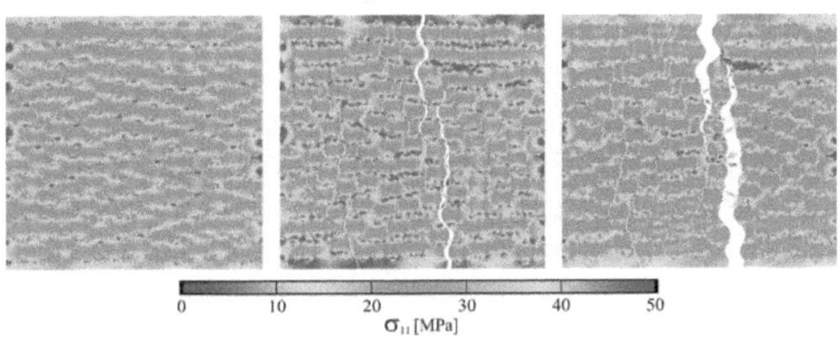

(d) $V_f = 0.6$. Axial normal stress σ_{11} plotted at the deformed states $u = 0.02\ v^u_{coh}$, $0.18\ v^u_{coh}$ and $0.43\ v^u_{coh}$.

Fig. 11. Axial stress σ_{11} at three deformed states, for different fibre volume fractions, $V_f = 0.1, 0.3, 0.5$ and 0.6.

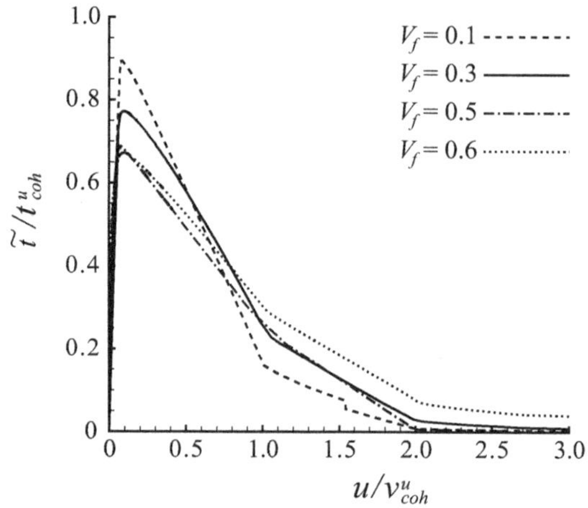

Fig. 12. Effective sample response for different fibre volume fractions. The average traction $\tilde{t}$ is measured at the right sample boundary at which the displacement u is prescribed. The ratio between the fibre-epoxy interfacial strength and the epoxy strength is $t^u_{adh}/t^u_{coh} = 0.5$.

the laminate thickness itself (i.e., a separation of length scales can be warranted). The attention is directed here to the numerical homogenization of the response of a microstructural RVE to a mesoscopic traction-separation curve, thereby considering the influence of microstructural sample size and local imperfections. Although the numerical examples treated focus upon uniaxial tension, the homogenization framework used is applicable to arbitrary loading conditions.

Consider a mesoscopic domain $\Omega^M \subset \mathbb{R}^2$ with an external boundary Γ^M, see Figure 13. The external boundary is subjected to displacements $\boldsymbol{u}^M$ and tractions $\boldsymbol{t}^M$ at Γ^M_u and Γ^M_t, respectively. The mesoscopic domain is crossed by an adhesive layer Γ^M_{coh}. The response in a material point of the adhesive layer is connected to the lower-scale response of a heterogeneous, microscopic domain $\Omega^m \subset \mathbb{R}^2$. The microscopic domain is represented by a fibre-epoxy sample of width b. Fracture in the microscopic domain occurs along the cracking path Γ^m_{coh}. The boundary conditions at the outer edges Γ^m of the microscopic domain are assumed to be periodic. For establishing an expression for the mesoscopic traction, t^M_i, in a material point at the mesoscale cohesive interface Γ^M_{coh}, the computational homogenization scheme presented in [14,24] is applied. This scheme uses the averaging principle of Hill [25], which states that the spatial average of the virtual work at the microscale, $\overline{\delta W^m}$, needs to be equal to the virtual work in a local material point of the mesoscale cohesive interface, δw^M_{coh}. Accordingly, the following expression for the mesoscopic traction is obtained [14,24]

$$t^M_i = \frac{1}{b} \int_{\Gamma^m_T} t^T_i \, d\Gamma^m_T, \tag{10}$$

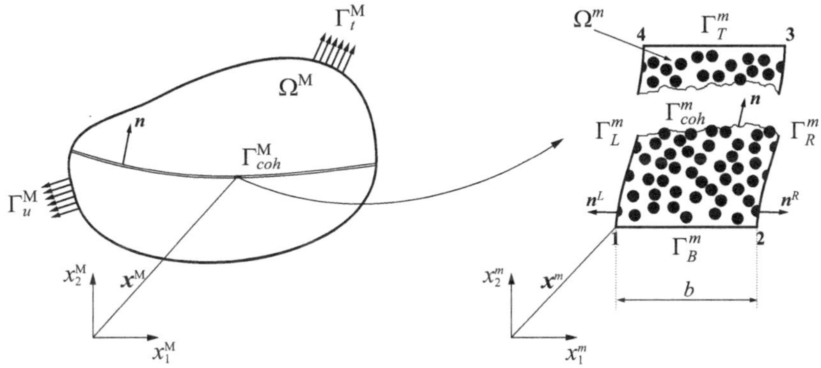

Fig. 13. Mesoscopic domain Ω^M (left) and microscopic domain Ω^m (right).

where t_i^T is the microscopic traction at the top surface Γ_T^m of the microscopic domain Ω^m, and b is the width of the microscopic domain, see Figure 13 (left). Within a finite element discretisation the integral term in the right-hand side of Eq. (10) is approximated by the summation of the nodal forces at the top edge of the RVE.

6.1 Fibre-Epoxy Specimen Subjected to Uniaxial Tension

Although Eq. (10) can be applied to samples subjected to arbitrary combinations of tensile and shear loading, for simplicity the examples studied in this section focus upon uniaxial tension. The interface damage model presented in Section 2 is used to simulate matrix cracking and debonding between fibres and matrix. In order to check whether the sample response converges to that of a representative volume element, in the numerical analyses two different specimen widths are considered, see Figure 14. The samples have a volume fraction of 0.3, and are subjected to uniaxial tension by prescribing the vertical displacement, $\hat{u}$, at the top edge of the sample. Periodic boundary conditions are prescribed at the left and right edges of the sample. The displacement at the bottom edge of the sample is constrained in the vertical direction, and the displacement of the node at the bottom-left corner of the specimen is also constrained in the horizontal direction, see Figure 14.

The first sample studied is square-shaped, with a width (and height) equal to 0.125 mm. The second sample is rectangular-shaped, where, in comparison with the square-shaped sample, the width is chosen two times larger, i.e., 0.250 mm, while the height is kept the same. Essentially, the chosen sample height is representative of the thickness of fibre-epoxy layers used in the fibre-metal laminate GLARE [18]. The diameter of the S2 glass fibres is 10 μm. As illustrated in Figure 14, the internal material structure of the samples relates to a random fibre distribution that is geometrically periodic in the x_1-direction.

The finite element model used for the microscale computations is plane-strain and thus is representative of fibres with a relatively large (actually infinite) length.

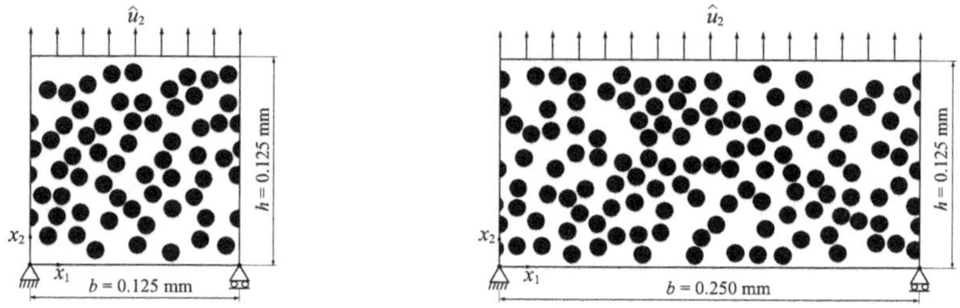

(a) Square-shaped specimen. (b) Rectangular-shaped specimen.

Fig. 14. Geometry and boundary conditions of square-shaped (left) and rectangular-shaped (right) fibre-epoxy samples with a fibre volume fraction of 0.3. The diameter of the glass fibres is 10 μm.

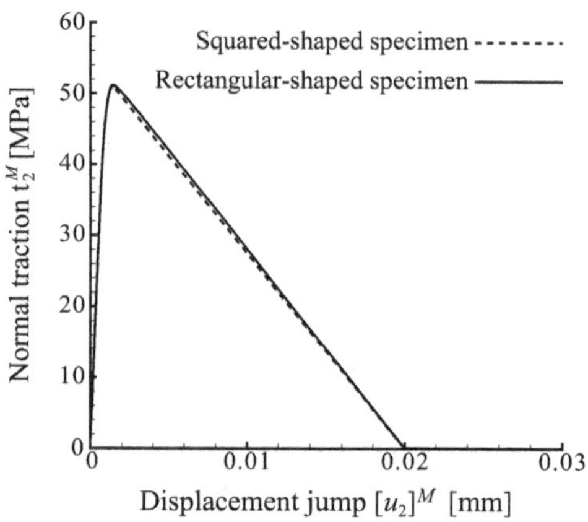

Fig. 15. Mesoscopic traction-separation response for the square-shaped and rectangular-shaped samples with a fibre volume fraction $V_f = 0.3$, shown in Figure 14.

The finite element meshes and material properties are similar to those used for the microscale simulations described in Sections 4 and 5. More details on these issues can be found in [14, 24]. It is further important to mention that in the present simulations the cohesive strength of the epoxy material is set equal to the adhesive strength of the fibre-epoxy interface, $t^u_{coh} = t^u_{adh} = 50$ MPa. Note that this strength ratio is different than for the microscale simulations presented in Section 5, where $t^u_{adh}/t^u_{coh} = 0.5$.

6.2 Influence of Sample Size

The failure responses of the samples with the two different widths are compared to assess the convergence of the numerical result upon an increasing microstructural sample volume. The mesoscopic traction-separation relation obtained after applying Eq. (10) to the numerical results is shown in Figure 15. It can be observed that the

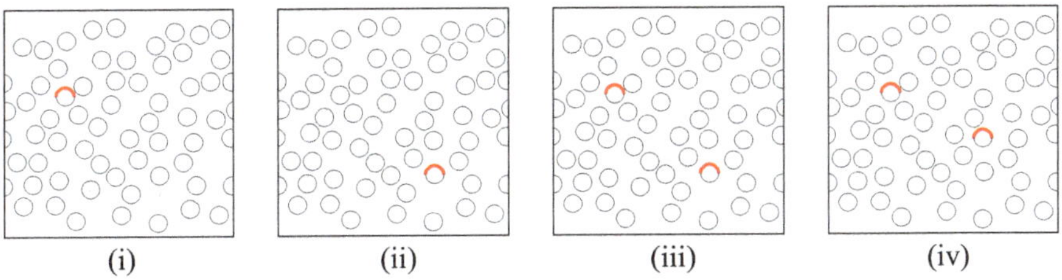

Fig. 16. Four fibre-epoxy samples with different locations and/or number of imperfections. The fibre volume fraction of the sample is 0.3 and imperfections are indicated by bold lines.

traction-separation responses of the square-shaped and rectangular-shaped samples are similar, indicating that the square-shaped sample is sufficiently large for being considered as an RVE. Essentially, the traction-separation responses closely follow the mode I traction-separation law of the epoxy material. Although not shown here, this is due to the fact that the final failure crack develops mainly through the epoxy material and is mode I dominated (i.e., the orientation of the crack is more or less perpendicular to the direction of the tensile loading).

Because the present analysis demonstrates that the square-shaped sample can be considered as a representative volume element, this sample will be used for the forthcoming computations in this chapter.

6.3 Influence of Imperfections

For studying how the homogenized traction-separation response is influenced by imperfections at fibre-epoxy interfaces, four different configurations are considered, see Figure 16. The fibre volume fraction of the sample is 0.3 and the imperfections are represented by a local absence of the adhesive bonding strength at specific fibre-epoxy interfaces. As illustrated in Figure 16, the four configurations considered are, respectively, characterised by (i) a single imperfection in the upper region of the sample (i.e., in the left-top quadrant), (ii) a single imperfection in the lower region of the sample (i.e., in the right-bottom quadrant), (iii) two imperfections in the upper and lower regions of the sample (i.e., a combination of configurations (i) and (ii)), and (iv) two imperfections in the upper and lower regions of the sample, where the upper imperfection coincides with that in configuration (i) and the lower imperfection is placed somewhat higher than that of configuration (ii).

The mesoscopic traction-separation response for the different configurations in Figure 16 is plotted in Figure 17. For comparison, the response of the specimen without imperfections, as plotted in Figure 15, has also been included in this figure. The corresponding fracture patterns are depicted in Figures 18(a) to 18(d). In all configurations the imperfections clearly act as nucleation sites for crack development. In addition, their location typically is included in the geometry of the dominant crack that develops upon complete failure of the fibre-epoxy specimen. An exception in this respect, however, is the imperfection located in the upper half of configuration

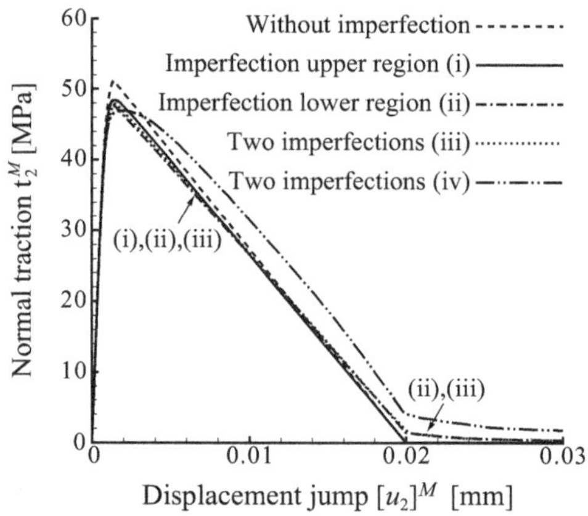

Fig. 17. Mesoscopic (mode I) traction-separation law corresponding to the square-shaped specimen with a fibre volume fraction of 0.3 (shown in Figure 14(a)), plotted for a different number and positions of imperfections, see Figure 16.

(iii), see Figure 18(c); this imperfection initially acts as a nucleation site for cracking but eventually is not included in the geometry of the dominant failure crack, due to a local unloading behaviour during the loading process. This is an important difference with configuration (iv), where *both* imperfections are included in the geometry of the dominant failure crack, see Figure 18(d). The net result of this feature is that the total crack length of configuration (iv) is substantially larger than that of configuration (iii), which thus requires the dissipation of more energy in order to fail the sample. Correspondingly, the effective fracture toughness of configuration (iv) is larger than that of configuration (iii), see Figure 17, and also is larger than that of the sample without imperfections. Hence, the presence of imperfections in the fibre-epoxy sample may have a positive effect on its overall failure characteristics.

7 Concluding Remarks

The mesoscale and microscale fracture behaviour of fibre-reinforced laminates has been modelled in a robust and accurate fashion. The interface damage model used in the simulations has proven to adequately simulate the rather complex fracture processes, which are characterised by numerous events of crack coalescence and crack bifurcation. The mesoscale failure mechanism computed for a fibre-metal laminate GLARE subjected to uniaxial tension is characterised by a mode I crack tunneling through the metal layers with continuous delamination developing along the interfaces with the adjacent fibre-epoxy layer. This mechanism is in good correspondence with experimental observations [17–20]. The microscale mechanisms of fibre debonding and epoxy cracking that are computed for fibre-epoxy systems loaded under uniaxial tension are also in good agreement with experiments on fibre-epoxy

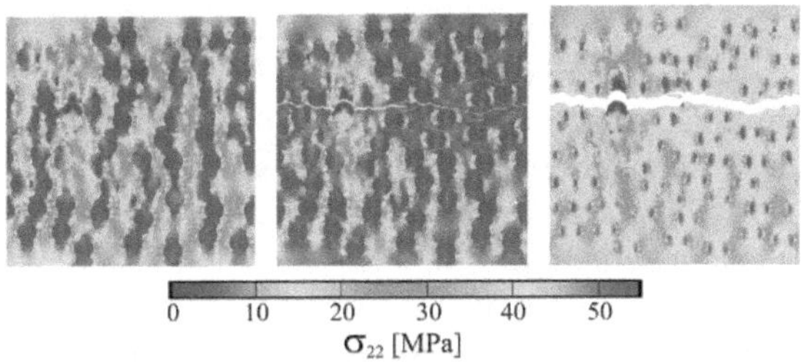

(a) One imperfection according to case (i) in Figure 16. The deformed states relate to $[\![u_2]\!]^M = 0.821$, 1.94 and 5.99 μm, respectively.

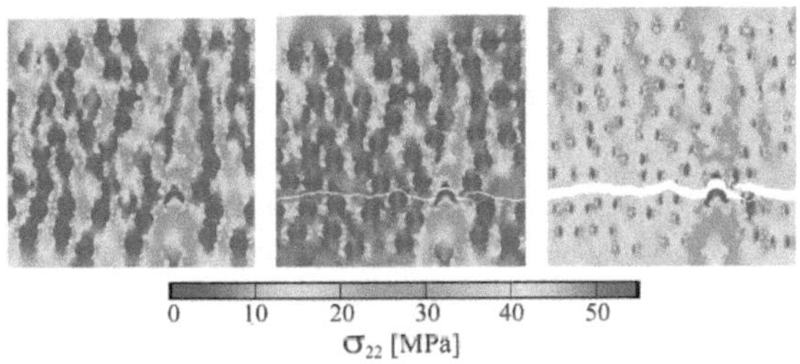

(b) One imperfection according to case (ii) in Figure 16. The deformed states relate to $[\![u_2]\!]_M = 0.820$, 1.85 and 5.86 μm, respectively.

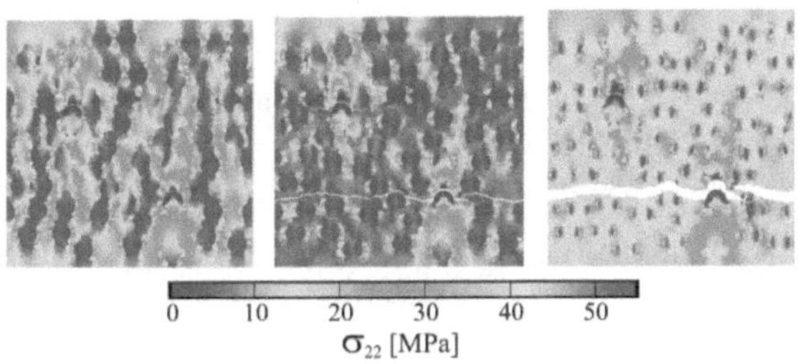

(c) Two imperfections according to case (iii) in Figure 16. The deformed states relate to $[\![u_2]\!]^M = 0.821$, 1.99 and 5.89 μm, respectively.

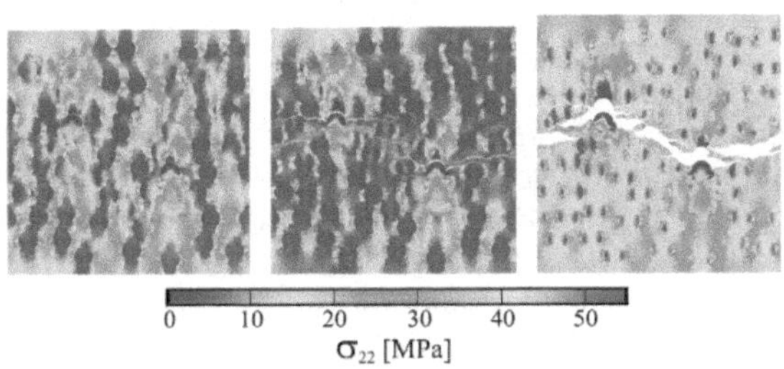

(d) Two imperfections according to case (iv) in Figure 16. The deformed states relate to $[\![u_2]\!]^M = 0.822$, 1.86 and 8.04 μm, respectively.

Fig. 18. Axial normal stress σ_{22} at three deformed, cracked states.

systems [9, 10]. The applicability of the numerical homogenization approach that connects the microscale fracture response of thin fibre-epoxy layers to a mesoscale interfacial traction-separation curve has been demonstrated by means of fibre-epoxy samples subjected to uniaxial tension. Nonetheless, it is emphasised that the final expression for the effective traction can be applied to *arbitrary* mixed-mode (i.e., tensile and shear) loading conditions. It is expected that under mixed-mode loading conditions the difference between the effective failure response of the fibre-epoxy layer and the microscopic failure response of the epoxy matrix will be larger than under the mode I loading conditions studied in this chapter, since under increasing mode-mixity the fracture patterns in fibre-epoxy systems typically become more distributive (or less localised), see [26].

Acknowledgements

The following persons are gratefully acknowledged for the helpful discussions on numerical implementation issues within the finite element framework JEM/JIVE: Dr. Erik-Jan Lingen and Dr. Martijn Stroeven from the software company Habanera, Delft, the Netherlands, Dr. Clemens Verhoosel and Dr. Joris Remmers of the Eindhoven University of Technology.

References

1. J.W. Hutchinson and Z. Suo. Mixed mode cracking in layered materials. *Advances in Applied Mechanics*, 29:63–191, 1992.
2. H.M. Jensen. Mixed mode interface fracture criteria. *Acta Metallurgica et Materialia*, 38:2637–2644, 1990.
3. A. Turon, P.P. Camanho, J. Costa, and C.G. Davila. A damage model for the simulation of delamination in advanced composites under variable-mode loading. *Mechanics of Materials*, 38:1072–1089, 2006.
4. F. Hashagen. *Numerical analysis of failure mechanisms in fibre metal laminates*. PhD thesis, Delft University of Technology, 1998.
5. J.J.C. Remmers, G.N. Wells, and R. de Borst. A solid-like shell element allowing for arbitrary delaminations. *International Journal for Numerical Methods in Engineering*, 58:2013–2040, 2003.
6. H. Parisch. A continuum-based shell theory for non-linear applications. *International Journal for Numerical Methods in Engineering*, 38(11):1855–1883, 1995.
7. M. Bischoff and E. Ramm. Shear deformable shell elements for large strains and rotations. *International Journal for Numerical Methods in Engineering*, 40:4427–4449, 1997.
8. A.S.J. Suiker and N.A. Fleck. Crack tunneling and plane-strain delamination in layered solids. *International Journal of Fracture*, 125:1–32, 2004.
9. E.K. Gamstedt and B.A. Sjogren. Micromechanisms in tension-compression fatigue of composite laminates containing transverse plies. *Composites Science and Technology*, 59:167–178, 1999.
10. T. Hobbiebrunken, M. Hojo, T. Adachi, C. De Jong, and B. Fiedler. Evaluation of interfacial strength in CF/epoxies using FEM and in-situ experiments. *Composites Part A - Applied Science and Manufacturing*, 37:2248–2256, 2006.

11. X.P. Xu and A. Needleman. Numerical simulations of fast crack growth in brittle solids. *Journal of the Mechanics and Physics of Solids*, 42:1397–1407, 1994.
12. M.G.A. Tijssens, E. van der Giessen, and L.J. Sluys. Simulation of mode I crack growth in polymers by crazing. *International Journal of Solids and Structures*, 37:7307–7327, 2000.
13. M.G.A. Tijssens, L.J. Sluys, and E. van der Giessen. Simulation of fracture of cementitious composites with explicit modeling of microstructural features. *Engineering Fracture Mechanics*, 68:1245–1263, 2001.
14. M.V. Cid Alfaro. *Multiscale Analyses of Fibre Metal Laminates*. PhD thesis, Delft University of Technology, Delft, 2008.
15. M.V. Cid Alfaro, A.S.J. Suiker, R. de Borst, and J.J.C. Remmers. Analysis of fracture and delamination in laminates using 3D numerical modelling. *Engineering Fracture Mechanics*, 76:761–780, 2009.
16. A.S.J. Suiker and N.A. Fleck. Modelling of fatigue crack tunneling and delamination in layered composites. *Composites Part A: Applied Science and Manufacturing*, 37:1722–1733, 2006.
17. T.J. de Vries. *Blunt and Sharp Notch Behaviour of GLARE laminates*. PhD thesis, Delft University of Technology, The Netherlands, 2001.
18. A. Vlot and J.W. (eds.) Gunnink. *Fibre Metal Laminates - An Introduction*. Kluwer Academic Publishers, Dordrecht, 2001.
19. T. Takamatsu, T. Matsumura, N. Ogura, T. Shimokawa, and Y. Kakatu. Fatigue crack growth properties of a GLARE-3-5/4 fiber/metal laminate. *Engineering Fracture Mechanics*, 63:253–272, 1999.
20. D.J. Shim, R.C. Alderliesten, S.M. Spearing, and D.A. Burianek. Fatigue crack growth in GLARE hybrid laminates. *Composites Science and Technology*, 63:1759–1767, 2003.
21. M.V. Cid Alfaro, A.S.J. Suiker, and R. de Borst. Transverse failure behaviour of fibre-epoxy systems. *Journal of Composite Materials*, 2010. Accepted for publication.
22. M. Hagenbeek. *Characterization of Fibre Metal Laminates under Thermo-mechanical loadings*. PhD thesis, Delft University of Technology, Delft, 2005.
23. F. Hashagen and R. de Borst. Numerical assessment of delamination in fibre metal laminates. *Computer Methods in Applied Mechanics and Engineering*, 185:141–159, 2000.
24. M.V. Cid Alfaro, A.S.J. Suiker, C.V. Verhoosel, and R. de Borst. Numerical homogenization of cracking processes in thin fibre-epoxy layers. *European Journal of Mechanics A/Solids*, 29:119–131, 2010.
25. R. Hill. The essential structure of constitutive laws for metal composites and polycrystals. *Journal of the Mechanics and Physics of Solids*, 15:79–95, 1967.
26. M.G. Kulkarni, P.H. Geubelle, and K. Matouš. Multi-scale modelling of heterogeneous adhesives: Effect of particle decohesion. *Mechanics of Materials*, 41:573–583, 2009.

Improved Multiscale Computational Strategies for Delamination

O. Allix, P. Gosselet and P. Kerfriden

LMT-Cachan (ENS-Cachan/CNRS/UPMC/Pres UniverSud Paris), 61 avenue du Président Wilson, 94235 Cachan, France; {allix,kerfriden,gosselet}@lmt.ens-cachan.fr

Abstract We propose a three-scale computational strategy for the simulation of laminated composite parts modelled at the meso-scale. Two nested domain decompositions are used: a LaTIn method is employed in the inner substructuring so that the debounding behaviour is bore by the interfaces between subdomains (first scale) while the outer decomposition permits to solve in parallel the LaTIn macro (second scale) problem which grants the method its scalability, a super-macro problem (third scale) is introduced to accelerate the transmission of largest wavelength numerical information. The strategy thus teams up various levels of parallelism, which makes it well suited to modern hardware architectures.

Key words: Delamination, domain decomposition, LaTIn method, three-scale approach, parallel processing

1 Introduction

The virtual testing of delamination is a goal shared by many practitioners, especially in the aeronautical field. In order to reach such an objective, two research topics which have undergone drastic changes over the last twenty years must be linked: the relevant modeling of composites and the efficient analysis of structures.

Indeed, there have been many advances toward a better understanding of the mechanics of laminated composites and the mechanisms of damage. The validity of two types of models, microscale models and mesoscale models, has been proven. Microscale models are closely connected to the physics of the material and, thus, provide a reliable framework for simulation. They take into account many damage processes, such as diffuse intralaminar degradations percolating into transverse cracking, diffuse interface degradations leading to distributed delamination, chemically- or thermally-induced degradations, or fiber breakage [8, 12]. On the microscale, simulations can combine continuous (damage) and discrete (fracture) degradation models [18]. Unfortunately, the analysis of models defined on the microscale requires such a refined discretization that only small test specimens can be simulated. Industrial-size structural calculations are beyond the reach of even recent computers. Mesomodels [1, 3, 13] are defined on a scale which makes both

R. de Borst and E. Ramm (eds.), *Multiscale Methods in Computational Mechanics*, Lecture Notes in Applied and Computational Mechanics 55, DOI 10.1007/978-90-481-9809-2_14,

the introduction of physics-based components and the simulation of small industrial structures possible. Very often these models rely on the definition of two mesoconstituents, the ply (a three-dimensional entity) and the interface (a two-dimensional entity), which are modeled using continuum (damage) mechanics and behavior derived from the homogenization of micromodels [16, 17]. Nevertheless, in order to achieve reliable simulations, refined discretizations are still required for the correct representation of the stress gradients induced by edge effects, which are responsible for the initiation of many degradations. Therefore, the resulting problems remain very large (in terms of the number of degrees of freedom) and highly nonlinear, which creates potential instabilities.

In a first approach to the reliable simulation of delamination in composite structures, we chose to neglect the effect of damage within plies and concentrate on the degradations at the interfaces. Thus, we adopted the mesomodel presented in [1], in which the debonding phenomenon is localized at the interfaces and handled through cohesive behavior. A similar approach with a different interface behavior (degradations based on plasticity) was applied in [22].

In order to handle the large nonlinear systems associated with this modeling approach, one can consider using one of the several multiscale [4, 5, 14, 19] and enrichment [6, 10, 20, 21] techniques developed recently. We based our strategy on the mixed domain decomposition method described in [14], which places special emphasis on the interfaces between substructures. Consequently, the reference problem resulting from the mesomodel chosen is substructured by nature, and the cohesive interfaces of the model are handled within the interfaces of the domain decomposition method. This idea is developed in Section 2. Furthermore, the resolution of the substructured problem by a LATIN iterative solver has very advantageous numerical properties: the nonlinearities are dealt with through local problems, and very little matrix reassembling is required. The incremental micro-macro LATIN algorithm as a resolution strategy for delamination problems is presented in Section 2.2. As shown in Section 2.3, the direct application of this method leads to a number of numerical difficulties. A first issue occurs when setting the parameters of the method: in Section 3, we present the indispensable tuning of the search directions according to the interface's status. Subsequently, the main remaining difficulties concern the treatment of the macroscale of the problem. In this paper, the emphasis is on the adaptation of our strategy in order to deal with large macroproblems. (Important remarks on how to make the macroproblem more relevant can be found in [11].) In Section 4, we present the parallelization of the resolution, which was inspired by Mandel [19]. In order to do that, we introduce a third level of discretization after the first and most refined level (the finite element) and the second level (the substructure): interconnected substructures are combined into "super-substructures" (which fill up the memory capacity of processors) connected to one another through "super-interfaces" using Message Passing Interface (MPI). The method is validated in Section 5 using a complex test case. The handling of the instabilities is not described in this paper. The interested reader may refer to [11] where the adaptation of an arc-length algorithm with local control is presented.

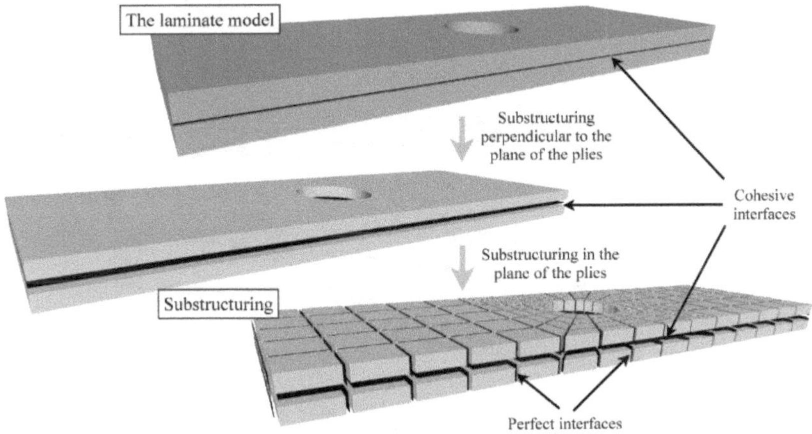

Fig. 1. Decomposition of the laminated composite structure into substructures.

2 Application of the Two-Scale Domain Decomposition Strategy to Delamination Analysis

2.1 The Substructured Delamination Problem

Let us consider a laminated structure **E** defined in a domain Ω bounded by $\partial\Omega$ and consisting of N_P adjacent plies P, each defined in a domain Ω_P, with $\Omega = \bigcup_P \Omega_P$. Adjacent plies P and P' are joined by cohesive interfaces $I_{PP'}$. An external traction field $\mathbf{F}_d$ is prescribed over a part $\partial\Omega_f$ of Ω, and a displacement field $\mathbf{U}_d$ is prescribed over the complementary part $\partial\Omega_u$ of Ω. Let $\mathbf{n}_P$ denote the outer normal to the boundary $\partial\Omega_P$ of Ply P, $\mathbf{f}_d$ the volume force, σ the Cauchy stress tensor and ϵ the symmetric part of the displacement gradient. The simulation is performed using a classical incremental scheme, assuming small perturbations and quasi-static isothermal evolution over time.

The laminated structure **E** is decomposed into substructures and interfaces as shown in Figure 1. Each of these mechanical entities has its own kinematic and static unknown fields as well as its own constitutive law. The substructuring pattern is defined in such a way that the domain decomposition interfaces coincide with the material's cohesive interfaces, so that each substructure belongs to a unique ply P and has a constant linear constitutive law. A substructure E defined in domain Ω_E is connected to an adjacent substructure E' through an interface $\Gamma_{EE'} = \partial\Omega_E \cap \partial\Omega_{E'}$ (Figure 2). The surface entity $\Gamma_{EE'}$ applies force distributions $\mathbf{F}_E$, $\mathbf{F}_{E'}$ and displacement distributions $\mathbf{W}_E$, $\mathbf{W}_{E'}$ to E and E'. Let $\Gamma_E = \bigcup_{E' \in \mathbf{E}} \Gamma_{EE'}$. Over a substructure E such that $\Gamma_E \cap \partial\Omega \neq \emptyset$, the boundary condition $(\mathbf{U}_d, \mathbf{F}_d)$ is applied through a boundary interface Γ_{E_d}.

At each step of the incremental time resolution algorithm, the substructured quasi-static problem consists in finding $s = (s_E)_{E\in\mathbf{E}}$ (where $s_E = (\mathbf{u}_E, \mathbf{W}_E, \sigma_E, \mathbf{F}_E)$) which is a solution of the following equations:

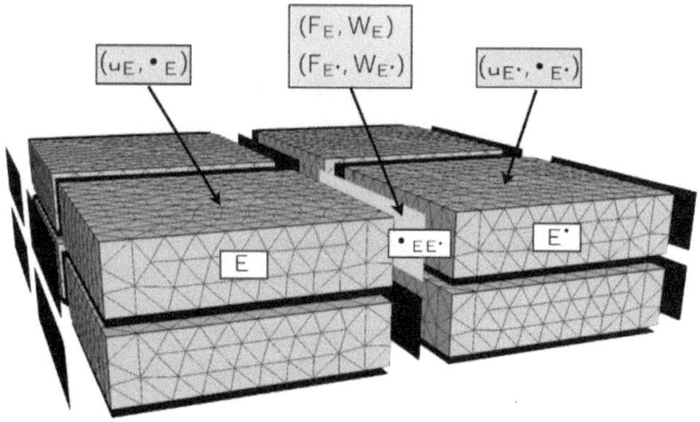

Fig. 2. Decomposition of the laminated composite structure into substructures: mixed unknown fields.

- kinematic admissibility of substructure E:

$$\text{at each point of } \Gamma_E, \quad \mathbf{u}_E = \mathbf{W}_E \tag{1}$$

- static admissibility of substructure E:

$$\begin{aligned} &\forall(\mathbf{u}_E{}^\star, \mathbf{W}_E{}^\star) \in \mathcal{U}_E \times \mathcal{W}_E \; / \; \mathbf{u}_E{}^\star{}_{|\partial\Omega_E} = \mathbf{W}_E{}^\star, \\ &\int_{\Omega_E} Tr\left(\sigma_E \, \epsilon(\mathbf{u}_E{}^\star)\right) \, d\,\Omega = \int_{\Omega_E} \mathbf{f}_d.\mathbf{u}_E{}^\star \, d\Omega \\ &\qquad\qquad + \int_{\partial\Gamma_E} \mathbf{F}_E.\mathbf{W}_E{}^\star \, d\Gamma \end{aligned} \tag{2}$$

- linear orthotropic constitutive law of substructure E:

$$\text{at each point of } \Omega_E, \quad \sigma_E = \mathbf{K}\,\epsilon(\mathbf{u}_E) \tag{3}$$

- behavior of the interfaces $\Gamma_{EE'} \in \Gamma_E$:

$$\begin{aligned} &\text{at each point of } \Gamma_{EE'} \in \Gamma_E, \\ &R_{EE'}(\mathbf{W}_E, \mathbf{W}_{E'}, \mathbf{F}_E, \mathbf{F}_{E'}) = 0 \end{aligned} \tag{4}$$

- behavior of the interfaces at the boundary $\partial\Omega \cap \Gamma_E$:

$$\begin{aligned} &\text{at each point of } \Gamma_{E_d}, \quad R_{Ed}(\mathbf{W}_E, \mathbf{F}_E) = 0 \\ &(\mathbf{W}_E = \mathbf{u}_d \text{ on } \partial\Omega_u \text{ and } \mathbf{F}_E = \mathbf{F}_d \text{ on } \partial\Omega_f) \end{aligned} \tag{5}$$

We make the formal relation $R_{EE'} = 0$ explicit in the two cases we will be considering:

- perfect interface:

$$\begin{cases} \mathbf{F}_E + \mathbf{F}_{E'} = 0 \\ \mathbf{W}_E - \mathbf{W}_{E'} = 0 \end{cases} \tag{6}$$

- cohesive interface:

$$\begin{cases} \mathbf{F}_E = \mathsf{K}_P([\mathbf{W}]_{EE'}(\tau < t)).[\mathbf{W}]_{EE'} \\ \mathbf{F}_E + \mathbf{F}_{E'} = 0 \end{cases} \tag{7}$$

where $[\mathbf{W}]_{EE'} = \mathbf{W}_{E'} - \mathbf{W}_E$. The stiffness operator K_P can be expressed in the $(\mathbf{N}_1, \mathbf{N}_2, \mathbf{n}_P)$ basis as (Figure 3):

$$\begin{pmatrix} (1-d_1)k_1{}^0 & 0 & 0 \\ 0 & (1-d_2)k_2{}^0 & 0 \\ 0 & 0 & (1-h_+[\mathbf{W}]_{EE'}\, d_3)\, k_3{}^0 \end{pmatrix} \tag{8}$$

where h_+ is the positive indicator function.

The cohesive constitutive law of an interface joining two adjacent plies is described classically through continuum damage mechanics. Local damage variables $(d_i)_{i\in[1\ 3]}$, with values ranging from 0 (healthy interface) to 1 (completely damaged interface), are introduced into the interface model in order to simulate its progressive softening. The parameters $(d_i)_{i\in[1\ 3]}$ are related to the local energy release rates $(Y_i)_{i\in[1\ 3]}$ of the interface's degradation modes (traction along direction $\mathbf{N}_3$ and shear along directions $\mathbf{N}_1$ and $\mathbf{N}_2$).

$$Y_i = -\frac{\partial e_d}{\partial d_i} \qquad \text{where} \qquad \begin{cases} Y_1 = \dfrac{1}{2}k_1\,([\mathbf{W}]_{EE'}.\mathbf{N}_1)^2 \\ Y_2 = \dfrac{1}{2}k_2\,([\mathbf{W}]_{EE'}.\mathbf{N}_2)^2 \\ Y_3 = \dfrac{1}{2}k_{3+}\,([\mathbf{W}]_{EE'}.\mathbf{N}_3)_+^2 \end{cases} \tag{9}$$

The damage variables are assumed to be functions of a single quantity: the maximum over time $Y_{|t}$ of a combination of the energy release rates $(Y_{i\,|\tau})_{i\in[1\ 3],\ \tau\leqslant t}$:

$$Y_{|t} = sup_{(\tau\leq t)}\left(Y_3{}^{\alpha}_{|\tau} + \gamma_1 Y_1{}^{\alpha}_{|\tau} + \gamma_2 Y_2{}^{\alpha}_{|\tau}\right)^{\frac{1}{\alpha}} \tag{10}$$

The evolution laws express that:

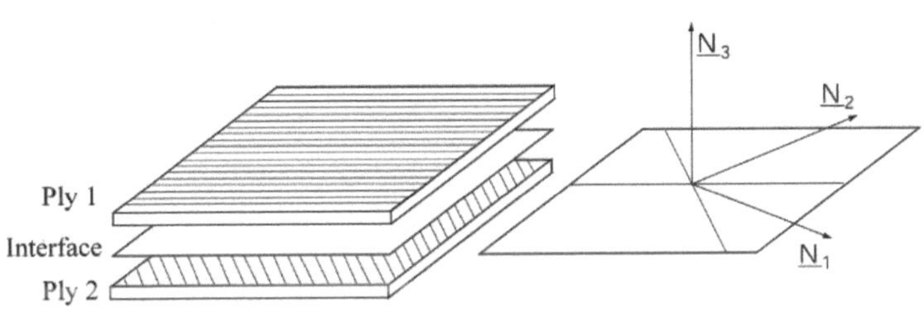

Fig. 3. The components of the mesomodel.

$$d_1 = d_2 = d_3 = w(Y) \quad \text{where, in general,} \quad w(Y) = \frac{n}{n+1}\left(\frac{Y}{Y_c}\right)^n \tag{11}$$

n and α being scalar parameters of the model. When Parameters γ_1 and γ_2 are set to identified physical values such that $\gamma_1 \neq \gamma_2 \neq 1$, the energies dissipated during the propagation of the crack are different for the three modes. Details on the identification procedure for such a model can be found in [2].

After a cohesive interface has become fully damaged, it is converted into a (frictionless) contact interface.

2.2 Two-Scale Iterative Resolution of the Substructured Problem

Introduction of the Macroscopic Scale

In the end, the substructured problem defined in the previous section will be solved using an iterative LATIN algorithm, which will be described in the next section. In order to ensure the scalability of the strategy, a coarse global problem, associated with the equilibrium and continuity of what one calls the "macro" force and displacement fields at the interfaces, must be solved at each iteration.

Over each interface $\Gamma_{EE'}$ such that $(E, E') \in \mathbf{E}^2$, the interface fields are divided into a macro part (superscript M) and a micro part (superscript m). The macro part belongs to a small subspace (9 macro degrees of freedom per plane interface for a 3D problem).

$$\begin{aligned} \mathbf{F}_E &= \mathbf{F}_E^M + \mathbf{F}_E^m \\ \mathbf{W}_E &= \mathbf{W}_E^M + \mathbf{W}_E^m \end{aligned} \tag{12}$$

The macro and micro data are uncoupled with respect to the interface's virtual work:

$$\begin{aligned} &\forall (\mathbf{F}_E, \mathbf{W}_E) \in \mathcal{F}_E \times \mathcal{W}_E, \\ &\int_{\Gamma_{EE'}} \mathbf{F}_E.\mathbf{W}_E \, d\Gamma = \int_{\Gamma_{EE'}} \mathbf{F}_E^M.\mathbf{W}_E^M \, d\Gamma + \int_{\Gamma_{EE'}} \mathbf{F}_E^m.\mathbf{W}_E^m \, d\Gamma \end{aligned} \tag{13}$$

Each macrospace is defined by one's choice of its basis. Numerical tests have shown that the use of a linear macro basis gives the method good scalability properties. Indeed, the corresponding macrospace includes the part of the interface fields with the largest wavelength. Consequently, according to Saint-Venant's principle, the micro complement resulting from the iterative resolution of the local problems has only a local influence.

The Iterative Algorithm

Here, the iterative LATIN algorithm for the resolution of nonlinear problems is applied to the resolution of the substructured reference problem, the nonlinearities being localized in the (cohesive) interfaces.

The equations of the problem can be divided into two groups:

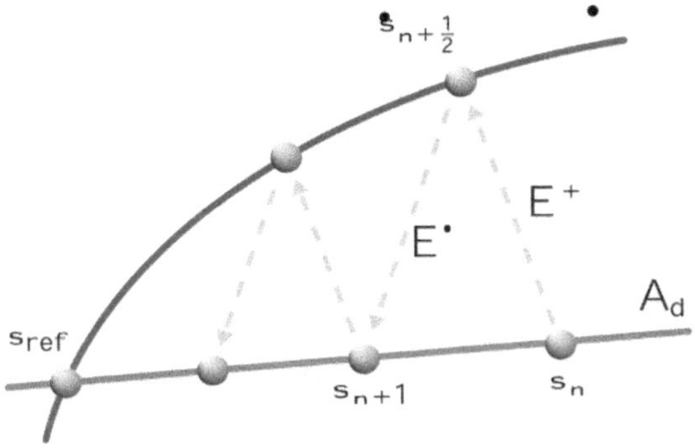

Fig. 4. Illustration of the LATIN iterative algorithm.

1. linear equations in substructure variables and interface macroscopic variables:
   - static admissibility of the substructures
   - kinematic admissibility of the substructures
   - linear constitutive law of the substructures
   - linear equilibrium of the macro interface forces
2. local equations in interface variables:
   - behavior of the interfaces

The solutions $s = (s_E)_{E\in\mathbf{E}} = (\mathbf{u}_E, \mathbf{W}_E, \sigma_E, \mathbf{F}_E)_{E\in\mathbf{E}}$ of the first set of equations belong to Space $\mathbf{A_d}$ and the solutions $\widehat{s} = (\widehat{s}_E)_{E\in\mathbf{E}} = (\widehat{\mathbf{u}}_E, \widehat{\mathbf{W}}_E, \widehat{\sigma}_E, \widehat{\mathbf{F}}_E)_{E\in\mathbf{E}}$ of the second set of equations belong to $\mathbf{\Gamma}$. The converged solution s_{ref} is such that

$$s_{ref} \in \mathbf{A_d} \bigcap \mathbf{\Gamma} \tag{14}$$

The resolution process consists in seeking the solution s_{ref} alternatively in these two spaces: first, a solution s_n is found in $\mathbf{A_d}$, then a solution $\widehat{s}_{n+\frac{1}{2}}$ is found in $\mathbf{\Gamma}$. In order for the two problems to be well-posed, one introduces two search directions, $\mathbf{E}^+$ and $\mathbf{E}^-$, linking the solutions s and $\widehat{s}$ through the iterative process (see Figure 4). Hence, an iteration of the LATIN algorithm consists of two stages:

- a local stage:

$$\textit{Find}\ \widehat{s}_{n+\frac{1}{2}} \in \mathbf{\Gamma}\ \textit{such that}\ \left(\widehat{s}_{n+\frac{1}{2}} - s_n\right) \in \mathbf{E}^+ \tag{15}$$

- and a linear stage:

$$\textit{Find}\ s_{n+1} \in \mathbf{A_d}\ \textit{such that}\ \left(s_{n+1} - \widehat{s}_{n+\frac{1}{2}}\right) \in \mathbf{E}^- \tag{16}$$

In the following sections, the subscript n will be omitted.

The local stage

During the local stage, uncoupled problems are solved at each point of the interfaces $(\Gamma_{EE'})_{|(E,E')\in\mathbf{E}^2}$ (as well as $(\Gamma_{E_d})_{E\in\mathbf{E}}$ for the interfaces which belong to the boundary $\partial\Omega$):

$$\text{Find } (\widehat{\mathbf{F}}_E, \widehat{\mathbf{W}}_E, \widehat{\mathbf{F}}_{E'}, \widehat{\mathbf{W}}_{E'}) \text{ such that:}$$

$$\begin{cases} \mathcal{R}_{EE'}(\widehat{\mathbf{F}}_E, \widehat{\mathbf{W}}_E, \widehat{\mathbf{F}}_{E'}, \widehat{\mathbf{W}}_{E'}) = 0 \\ (\widehat{\mathbf{F}}_E - \mathbf{F}_E) - k^+(\widehat{\mathbf{W}}_E - \mathbf{W}_E) = 0 \\ (\widehat{\mathbf{F}}_{E'} - \mathbf{F}_{E'}) - k^+(\widehat{\mathbf{W}}_{E'} - \mathbf{W}_{E'}) = 0 \end{cases} \tag{17}$$

The last two equations of this system define the search direction E^+ (k^+ and k^- are scalar search direction which, physically, are analogous to "stiffnesses"). In the case of a cohesive interface, problem (17) is nonlinear and its solution is obtained through a Newton–Raphson scheme.

The linear stage

The linear stage consists in the resolution of a series of linear systems within the substructures under the constraint of macroscopic equilibrium of the interface forces.

$$\text{at Interface } \Gamma_{EE'|(E,E')\in\mathbf{E}^2}, \quad \mathbf{F}_E^M + \mathbf{F}_{E'}^M = 0 \tag{18}$$

In order to verify the macroscopic condition exactly and the search direction E^- defined in (16) as well as possible, we use a Lagrangian formulation, which leads to:

$$\begin{aligned} \forall \mathbf{W}_E^\star \in \mathcal{W}_E, \quad & \int_{\Gamma_E} \left(\mathbf{F}_E - \widehat{\mathbf{F}}_E\right).\mathbf{W}_E{}^\star \, d\Gamma \\ & + \int_{\Gamma_E} \left(k^- \left(\mathbf{W}_E - \widehat{\mathbf{W}}_E\right) - k^- \widetilde{\mathbf{W}}^M\right).\mathbf{W}_E{}^\star \, d\Gamma = 0 \end{aligned} \tag{19}$$

which can be viewed as a modified search direction, the Lagrange multiplier $\widetilde{\mathbf{W}}^M$ becoming an additional unknown of Interface $\Gamma_{EE'}$.

The problem which needs to be solved for each substructure E is obtained by substituting (19) into (2):

$$\begin{aligned} & \forall (\mathbf{u}_E^\star, \mathbf{W}_E^\star) \in \mathcal{U}_E \times \mathcal{W}_E, \\ & \int_{\Omega_E} Tr(\epsilon(\mathbf{u}_E) K \epsilon(\mathbf{u}_E{}^\star)) \, d\Omega + \int_{\Gamma_E} k^- \mathbf{W}_E.\mathbf{W}_E{}^\star \, d\Gamma \\ & = \int_{\Omega_E} \mathbf{f}_d.\mathbf{u}_E{}^\star \, d\Omega + \int_{\Gamma_E} (\widehat{\mathbf{F}}_E + k^- \widehat{\mathbf{W}}_E + k^- \widetilde{\mathbf{W}}^M).\mathbf{W}_E{}^\star \, d\Gamma \end{aligned} \tag{20}$$

The condensation of this equation onto the macro degrees of freedom leads to a relation between $\mathbf{F}_E^M$ and $\widetilde{\mathbf{W}}_E^M$ which can be introduced into the macro equilibrium equation (18). Finally, one obtains a small linear system defined in the macro degrees of freedom. All the subdomains contribute to that "global" system through homogenized (condensed) flexibilities $\mathbb{L}_E^M$ calculated explicitly.

$$
\begin{aligned}
\forall\, \widetilde{\mathbf{W}}^{M\star} \in \mathcal{W}^M, \quad & \sum_E \int_{\Gamma_E} \mathbb{L}_E^M\, \widetilde{\mathbf{W}}^M . \widetilde{\mathbf{W}}^{M\star}\, d\Gamma \\
&= \sum_E \int_{\partial\Omega_f} \mathbf{F}_d . \widetilde{\mathbf{W}}^{M\star}\, d\Gamma - \sum_E \int_{\Gamma_E} \widetilde{\mathbf{F}}_E . \widetilde{\mathbf{W}}^{M\star}\, d\Gamma
\end{aligned} \tag{21}
$$

The macroscopic problem is discrete by nature and is expressed in matrix form as $\mathbf{L}^M\, \widetilde{W}^M = F^M$, where $\widetilde{W}^M$ is the vector of the components of the Lagrange multiplier in the macro basis.

The right-hand side of Equation (21) can be viewed as a macroscopic static residual from the calculation of a single-scale linear stage. In order to derive this term, problem (20) must be solved independently within each substructure. The resolution of the macroscopic problem (21) leads globally to the Lagrange multiplier $\widetilde{\mathbf{W}}^M$, which is finally used as a prescribed displacement for the resolution of the substructure-independent problems (20).

In order to carry out the resolutions of (20) in substructure variables, one uses the finite element method. Since the constitutive law of the substructures is linear, the stiffness operator of each substructure can be factorized once at the beginning of the calculation and reused without modifications throughout the analysis, which makes the method numerically advantageous.

Algorithm 1 summarizes the iterative procedure described in this section.

Algorithm 1: The two-scale domain decomposition solver

Construction of each substructure's operators
Calculation of each substructure's macro homogenized behavior $\mathbb{L}_E^M$
Global assembly of the macroscopic operator
Initialization $s_0 \in \mathbf{\Gamma}$ **for** $n = 0, \ldots, N$ **do**
  Linear stage: calculation of $s_n \in \mathbf{A_d}$
    □ Calculation of the macroscopic right-hand term $\widetilde{F}_E$ for each substructure
    □ Global assembly of the macroscopic right-hand term
    □ Resolution of the macro problem
    □ Resolution of the micro problems
  Local stage: calculation of $s_{n+\frac{1}{2}} \in \mathbf{\Gamma}$
    □ Resolution of the local problems at the boundary interfaces $(\Gamma_{cl})_{E\in\mathbf{E}}$
    □ Resolution of the local problems at the interfaces $(\Gamma_{EE'})_{(E,E')\in\mathbf{E}^2}$
  Calculation of a global error indicator
end

2.3 First Example of a Delamination Analysis

A first example of quasi-static delamination analysis is shown in Figure 5. The structure is a $[0\,90]_s$ double cantilever beam (DCB). The loading leading to Mode-I quasi-static propagation of the crack is increased linearly through ten time steps, the first

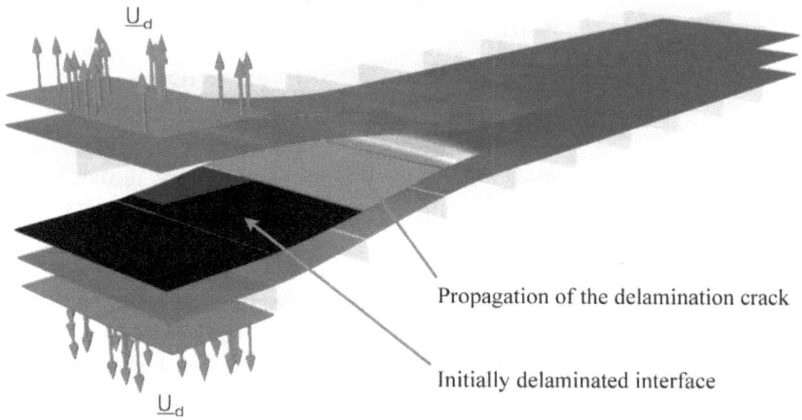

Fig. 5. The four-ply DCB test example.

two corresponding to the initiation of the delamination and the remainder to the crack's propagation.

The calculations were performed using a C++ implementation of the mixed domain decomposition method capable of handling the quasi-static analysis of 3D nonlinear problems. In this code, the parallel computations use the MPI library to exchange data among several processors. Each processor is assigned a set of connected substructures and their interfaces; then it calculates the associated operators and solves the local problems (Figure 8). The allocation of the substructures among the CPUs is handled by a METIS routine. The resolution of the macroproblem does not take full advantage of the parallelism because the substructures send their contributions to the macro problem to a separate processor in which the matrix is assembled and factorized and the substitutions are performed.

The direct use of the multiscale domain decomposition strategy to simulate the DCB case led to a number of numerical difficulties:

- The convergence rate of the LATIN-based strategy is highly dependent on the residual stiffnesses of the cohesive interfaces as well as the values of the search direction parameters. The iterative solver is even likely to stall when using low values of the search direction parameters. In the next section, we will briefly describe a practical tuning algorithm for the strategy which guarantees convergence.
- The method looses its numerical scalability when the crack's tip propagates. This phenomenon appears clearly in Figure 6 under the label "No subresolution". When the delamination process propagates (time steps 3 to 10), the number of LATIN iterations required for convergence becomes very large. A solution to this problem, described in [11], enabled us to recover the scalability of the method for our test case (under the label "Subresolutions"). This approach, which still needs to be generalized, is based on a filtering of the long-range effects of the crack's tip achieved through the resolution, at each global LATIN iteration, of

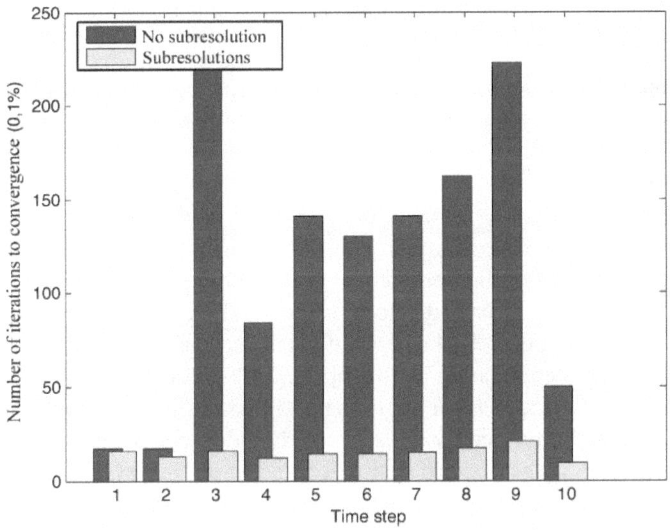

Fig. 6. Subiterations near the crack's tip.

local nonlinear problems in a box surrounding the front (where the main sources of nonlinearities are located).

- In this case, the ratio of the number of microscopic DOFs to the number of macroscopic DOFs is relatively small (40). The direct resolution of the macroscopic problem would become an issue if one were addressing the simulation of a realistic composite structure. A solution to this problem is discussed in Section 4.

3 Analysis of the Parameters of the Iterative Algorithm

A necessary condition for the algorithm to converge is for the search direction parameters k^+ and k^- to be positive definite, symmetrical operators. Previous studies have shown that there exists an optimum set of these operators. However, the optimum values are known to be difficult to interpret when the interface constitutive laws are complex, and even in simplified cases (perfect interfaces) are expensive to calculate. Therefore, our objective was to derive an efficient scalar approximation of these search direction operators for debonding analysis purposes.

As explained in [11], the non-monotonic relation between the interface stresses and the displacement gap due to damage imposes restrictions on the choice of parameter k^+. Concerning k^-, optimum values for a given damage map can be found after a micro/macro decomposition of this operator. Since the status of the interfaces changes with the evolution of the delamination (from elastic to damaged, then from damaged to ruined), the search direction parameters need to be updated often in order to remain optimum. Therefore, parameters whose efficiency range when they are not optimum is broad enough to require less frequent updating are preferred. An effective practical choice is the following:

- Parameter k^+: in order to avoid stalling or divergence of the algorithm, this parameter is set to a very high value (*i.e.* the search direction E^+ is quasi-infinitely stiff).
- Parameter k^-:
  - Perfect interfaces: k^- is set to the classically recommended value E/L [14], where E is the Young's modulus of the adjacent substructure and L a characteristic length of the interface.
  - Interfaces with prescribed forces (respectively displacements): k^- is set to a very small (respectively large) value in order to enforce the boundary condition through penalization in the adjacent substructure.
  - Cohesive interfaces: k^- is set to the stiffness of the undamaged interface.
  - Delaminated interfaces: k^- is set to zero in the shear direction. In the normal direction, k^- is set to zero in traction and to the initial stiffness in compression. Therefore, the status of the interface must be checked regularly (e.g. every 10 iterations).

The use of an infinitely stiff search direction E^+ and of the initial cohesive interface stiffness as the search direction parameter E^- brings us back to a well-known situation. The algorithm can be viewed as a secant Newton algorithm in which the solutions of the prediction steps are in equilibrium only in the macroscopic space, the equilibrium of the microscopic quantities being achieved at convergence.

4 The Three-Scale Domain Decomposition Strategy

The decomposition into substructures described in Section 2 leads to a very large macro problem and an unnecessarily refined macroscopic solution. In order to solve large problems such as that represented in Figure 7, one must place the emphasis on the parallel resolution of the macroproblem and on the selection and transmission of the large-wavelength part of the macroscopic solution.

These two features can be introduced into the method by using any Schur-complement-based domain decomposition technique [7]. We chose to solve the macroproblem using the BDD method [15, 19].

4.1 Resolution of the Macroproblem through the Balancing Domain Decomposition Method

Partitioning of the Macroproblem

The substructures of the initial partitioned problem are grouped into super-substructures ($\bar{E}$) separated by super-interfaces $\Gamma_{\bar{E}\bar{E}'}$ (Figure 8). The algebraic problem to be solved within each of these super-substructures (dropping superscript M) is

(a) Stresses and displacements in the substructures

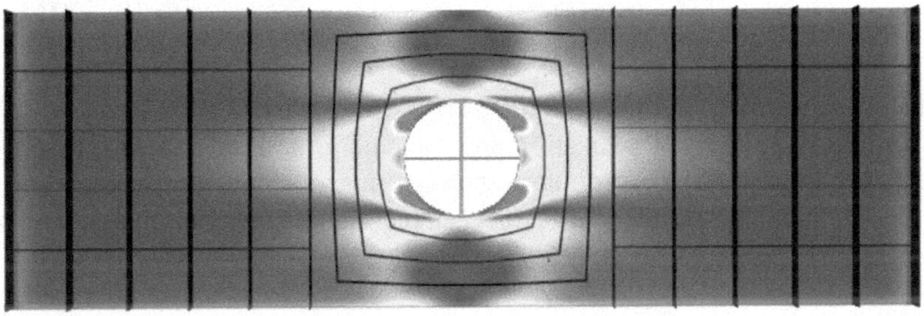

(b) Damage over the upper interface

Fig. 7. The four-ply perforated plate problem (3.4 MDOFs).

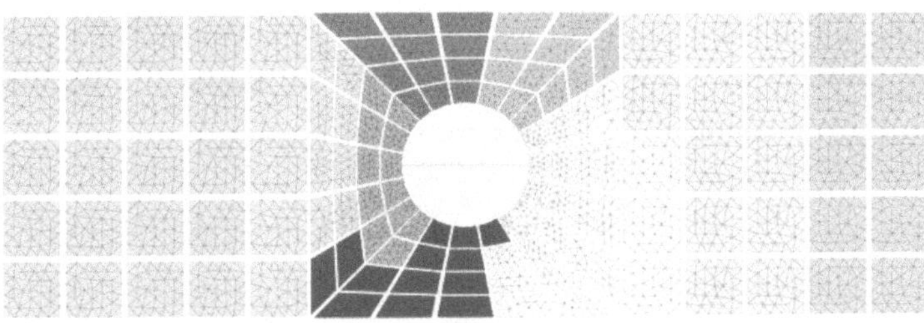

Fig. 8. Three-level substructuring: assignment of substructures to processors.

$$\left\{\begin{array}{l}\begin{pmatrix}\mathbf{L}_{ii}^{(\bar{E})} & \mathbf{L}_{ib}^{(\bar{E})}\\ \mathbf{L}_{bi}^{(\bar{E})} & \mathbf{L}_{bb}^{(\bar{E})}\end{pmatrix}\begin{pmatrix}\widetilde{W}_i^{(\bar{E})}\\ \widetilde{W}_b^{(\bar{E})}\end{pmatrix}=\begin{pmatrix}F_i^{(\bar{E})}\\ F_b^{(\bar{E})}+\widetilde{\lambda}_b^{(\bar{E})}\end{pmatrix}\\ \qquad\qquad \widetilde{W}_b^{(\bar{E})}=\mathbf{A}^{(\bar{E})^T}\widetilde{\mathbf{W}}_b\\ \qquad\qquad \sum_{\bar{E}}\mathbf{A}^{(\bar{E})}\widetilde{\lambda}_b^{(\bar{E})}=0\end{array}\right. \tag{22}$$

where subscripts b and i refer respectively to the super-interface quantities and to the internal quantities of the super-substructures. $\mathbf{A}^{(\bar{E})}$ is a Boolean operator which localizes data in such a way that the second equation of system (22) expresses the continuity of the kinematic unknowns (deduced from a single unknown $\widetilde{\mathbf{W}}_b$), while the third equation expresses the equilibrium of the nodal reactions at the super-interfaces.

First, the local equilibrium is condensed onto the super-interfaces by introducing the Schur complement $\mathbf{S}^{(\bar{E})}$ and the condensed force $\mathbf{F}_c^{(\bar{E})}$. The assembled condensed problem becomes:

$$\mathbf{S}\,\widetilde{\mathbf{W}}_b = \mathbf{F}_c \tag{23}$$

$$\text{where} \quad \left\{ \begin{array}{l|l} \mathbf{S} = \sum_{\bar{E}} \mathbf{A}^{(\bar{E})}\mathbf{S}^{(\bar{E})}\mathbf{A}^{(\bar{E})^T}\widetilde{\mathbf{W}}_b & \mathbf{S}^{(\bar{E})} = \mathbf{L}_{bb}^{(\bar{E})} - \mathbf{L}_{bi}^{(\bar{E})}\,\mathbf{L}_{ii}^{(\bar{E})^{-1}}\,\mathbf{L}_{ib}^{(\bar{E})} \\ F_c = \sum_{\bar{E}} \mathbf{A}^{(\bar{E})}\mathbf{F}_c^{(\bar{E})} & F_c^{(\bar{E})} = F_b^{(\bar{E})} - \mathbf{L}_{bi}^{(\bar{E})}\,\mathbf{L}_{ii}^{(\bar{E})^{-1}}\,F_i^{(\bar{E})} \end{array} \right.$$

This substructuring technique can be used exactly as in [9] in order to bind a domain which is prone to localization and damage to an undamaged region. Nevertheless, for large interface problems such as those encountered in our case, the condensed problem is much too large to be solved directly, and iterative solvers must be used.

Resolution of the Super-Interface Problem

The condensed macroproblem is solved iteratively using a conjugate gradient algorithm. Classically, this resolution involves only matrix-vector products and dot products, which are compatible with parallel computation. The recommended Neumann–Neumann preconditioner $\widetilde{\mathbf{S}}^{-1}$ involves the use of the pseudo-inverses $\mathbf{S}^{(\bar{E})^+}$ of the Schur complements of the super-substructures:

$$\widetilde{\mathbf{S}}^{-1} = \sum_{\bar{E}} \mathbf{A}^{(\bar{E})}\mathbf{S}^{(\bar{E})^+}\mathbf{A}^{(\bar{E})^T} \tag{24}$$

The use of this preconditioner means that the inverse of the global super-macro operator is approximated by the assembly of the inverses of the local Schur complements. Let us note that the description chosen for the interface macrofields precludes the existence of degrees of freedom belonging to more than two substructures; consequently, no scaling is required in the preconditioner (at least as long as the interfaces are not excessively heterogeneous). The use of pseudo-inverses is associated with an optimality condition which ensures that rigid body motions are not solicited (self-equilibrium of the floating super-substructures). This condition is verified thanks to a projector which makes the residual orthogonal to the kernels of the super-substructures (and possibly to other given subspaces) at each iteration of the conjugate gradient.

4.2 Results

Figure 9 shows the convergence rate of the LATIN algorithm when the conjugate gradient algorithm for the condensed macroproblem is stopped after a fixed number

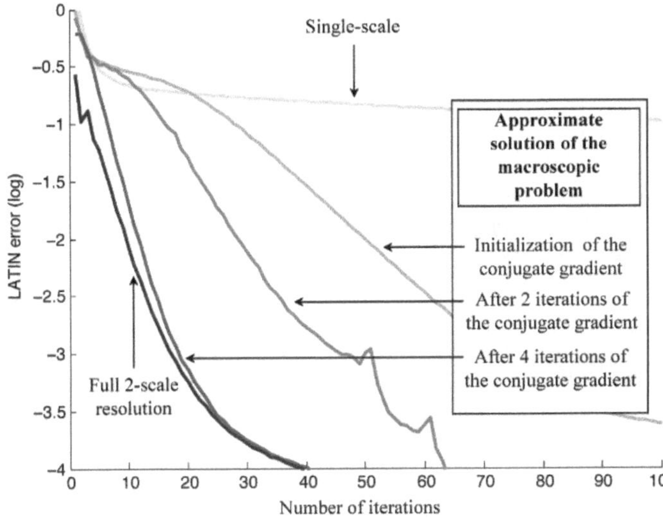

Fig. 9. The LATIN convergence curves (error criterion vs. the number of iterations) for several numbers of macroiterations.

of iterations. The test case was the perforated plate under traction represented in Figure 7 with the decomposition into super-substructures of Figure 8. It is clear that a rough approximation of the Lagrange multiplier, obtained after very few iterations of the conjugate gradient, is sufficient to reach the convergence rate of the multiscale LATIN algorithm. Typically, the algorithm is stopped when the residual error (normalized by the initial error) falls below 10^{-1}. Thus, the third-level enforcement of the admissibility of the macroforces (through the projection) appears to be sufficient for the determination of the large-wavelength part of the solution to be transmitted through the structure at each iteration of the resolution.

5 Efficiency of the Strategy: Study of a Complex Test Case

In this section, we illustrate the efficiency of the three-scale domain decomposition strategy through the simulation of the evolution of debonding in the bolted composite joint shown in Figure 10. Each composite plate interacts with the adjacent plates and with the two steel bolts through contact interfaces. The structure is subjected to prescribed displacements along the edges of the plates.

The discretization and the decomposition into substructures for this test case are illustrated in Figure 11. The total number of DOFs involved was 12×10^6, distributed among $10,600$ substructures. The number of macroscopic DOFs was 3×10^5, which would have made the direct resolution on a standard computer very inefficient. 29 processors each with 4 Gigabytes memory were used for this calculation, which led

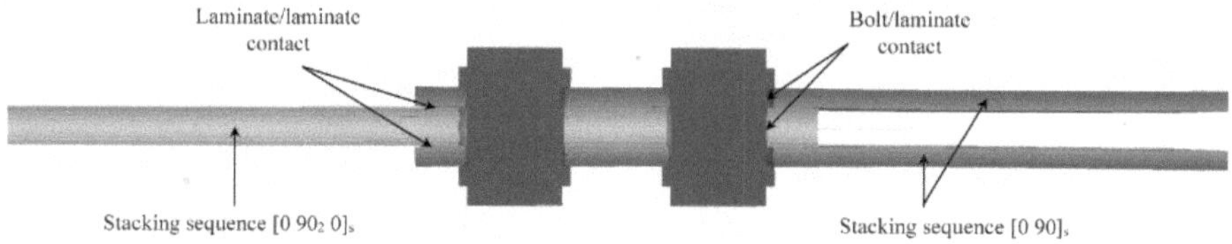

Fig. 10. Composite bolted joint made of 16 0.125 mm-thick plies. The in-plane dimensions of the structure are 30 × 5 mm. Prescribed displacements are applied along the left-hand side of the $[0\ 90_2\ 0]_s$ composite plate and along the right-hand sides of the $[0\ 90]_s$ composite plates.

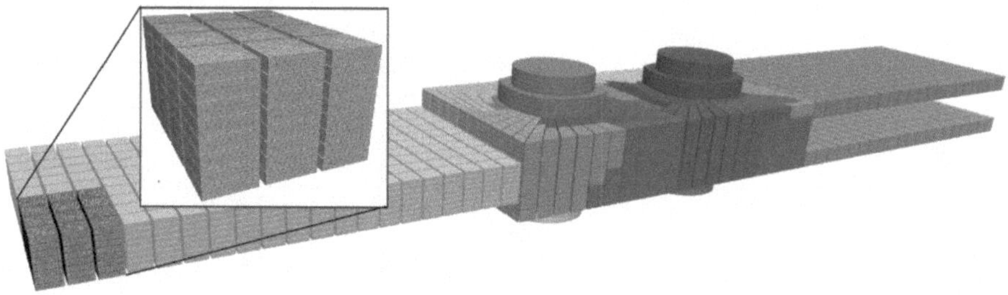

Fig. 11. Discretization of the composite bolted joint (12×10^6 DOFs), decomposition into substructures (10, 600 substructures) and assignment to processors (29 CPUs).

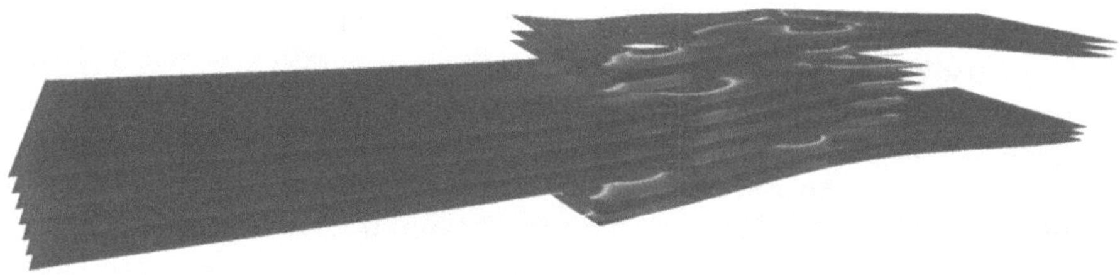

Fig. 12. Damage map of the cohesive interfaces of the composite bolted joint after the 70th time step of the quasi-static incremental analysis procedure.

to a super-coarse grid problem of dimension 150 (6 unknowns per floating super-substructure).

Figure 12 shows the damage map in the composite bolted joint after 70 time steps. The nonlinear calculation corresponding to each time step was carried out until the LATIN error criterion got below 10^{-2}, which occurred after an average of 80 LATIN iterations per time step. The average CPU time required for the calculation of each time step was 30 minutes, which is reasonable considering the small number of processors used.

However, the number of global LATIN iterations was quite high compared to what could have been obtained using the relocalization strategy of [11] in the vicinity of the crack's front (see the results of Figure 6). Figure 12 is a clear illustration of the difficulties which may arise in the use of this dedicated technique in a general case in which multiple crack front propagations may be involved. The front has a complex shape, which raises the difficult issue of the choice of the number of relocalization zones and their sizes. In addition, this test case is very unstable. These instabilities were handled globally using an arc-length algorithm along with the three-scale resolution strategy, but local instabilities might also appear within the region extracted for the relocalization calculations, a situation which has not yet been addressed at this stage in our development. Therefore, in the future, it might be necessary to generalize the relocalization strategy in order to improve the efficiency of the enhanced multiscale domain decomposition technique for complex laminated structures.

6 Conclusion

The accurate prediction of delamination in extended process zones of laminated composite structures requires refined models of the material's behavior, leading to the resolution of huge systems of equations. In order to solve such problems accurately, we used a two-scale domain decomposition strategy based on an iterative resolution algorithm. This method is particularly appropriate for laminated mesomodels, in which 3D and 2D entities are introduced separately.

This strategy was improved in order to enable it to handle very large delamination problems. A systematic analysis of the features of the method on the different scales was performed. First, we showed that in the high-gradient zones the classical scale separation was insufficient to ensure numerical scalability. Therefore, we developed a subresolution procedure which preserves the numerical scalability of the crack propagation analysis, but still needs to be automated for complex structures and regulated against local instabilities. We also proved that a third scale is required. Then, the problem on the intermediate scale was solved using a parallel iterative algorithm which enabled the rapid transmission of the very-large-wavelength part of the solution. Global instabilities were handled through a classical arc-length algorithm with local control (e.g. based on the maximum damage increment) and adjustment of the "time" steps during the calculation of the evolution of damage.

In future developments, 3D analysis in the process zone will be used in conjunction with plate analysis, which would be sufficient to describe the solution in the low-gradient zones. We will also, within the MAAXIMUS project, investigate the interaction between delamination and buckling for the simulation of components of aeronautical structures.

References

1. O. Allix and P. Ladevèze. Interlaminar interface modelling for the prediction of delamination. *Computers and structures*, 22:235–242, 1992.
2. O. Allix, D. Lévèque, and L. Perret. Identification and forecast of delamination in composite laminates by an interlaminar interface model. *Composites Science and Technology*, 58:671–678, 1998.

3. R. De Borst and J.C. Remmers. Computational modelling of delamination. *Composites Science and Technology*, 66:713–722, 2006.
4. F. Feyel and J.-L. Chaboche. Fe2 multiscale approach for modelling the elastoviscoplastic behaviour of long fibre sic/ti composite materials. *Computer Methods in Applied Mechanics and Engineering*, 183:309–330, 2000.
5. J. Fish, K. Shek, M. Pandheeradi, and M.S. Shephard. Computational plasticity for composite structures based on mathematical homogenization: Theory and practice. *Computer Methods in Applied Mechanics and Engineering*, 148:53–73, 1997.
6. S. Ghosh, K. Lee, and P. Raghavan. A multi-level computational model for multi-scale damage analysis in composite and porous materials. *International Journal of Solids and Structures*, 38:2335–2385, 2001.
7. P. Gosselet and C. Rey. Non-overlapping domain decomposition methods in structural mechanics. *Archives of Computational Methods in Engineering*, 13:515–572, 2006.
8. D. Guedra Degeorges and P. Ladevèze (Eds.). *Course on Emerging Techniques for Damage Prediction and Failure Analysis of Laminated Composite Stuctures*. Cepadues Editions, 2007.
9. T. Hettich, A. Hund, and E. Ramm. Modeling of failure in composites by X-FEM and level sets within a multiscale framework. *Computer Methods in Applied Mechanics and Engineering*, 197(5):414–424, 2008.
10. T.J.R. Hughes, G.R. Feijoo, L. Mazzei, and J.-B. Quincy. The variarional multiscale – A paradigm for computational mechanics. *Computer Methods in Applied Mechanics and Engineering*, 166:3–24, 1998.
11. P. Kerfriden, O. Allix, and P. Gosselet. A three-scale domain decomposition method for the 3d analysis of debonding in laminates. *Computational Mechanics*, 3(44):343–362, 2009.
12. P. Ladevèze. Multiscale computational damage modelling of laminate composites. In *Multiscale Modelling of Damage and Fracture Processes in Composite Materials*, T. Sadowski (Ed.). Springer-Verlag, 2005.
13. P. Ladevèze and G. Lubineau. An enhanced mesomodel for laminates based on micromechanics. *Composites Science and Technology*, 62(4):533–541, 2002.
14. P. Ladevèze and A. Nouy. On a multiscale computational strategy with time and space homogenization for structural mechanics. *Computer Methods in Applied Mechanics and Engineering*, 192:3061–3087, 2003.
15. P. Le Tallec. Domain decomposition methods in computational mechanics. In *Computational Mechanics Advances*, Volume 1. Elsevier, 1994.
16. G. Lubineau and P. Ladevèze. Construction of a micromechanics-based intralaminar mesomodel, and illustrations in Abaqus/standard. *Computational Materials Science*, 43(17/18):137–145, 2008.
17. G. Lubineau, P. Ladevèze, and D. Marsal. Towards a bridge between the micro- and mesomechanics of delamination for laminated composites. *Composites Science and Technology*, 66(6):698–712, 2007.
18. G. Lubineau, D. Violeau, and P. Ladevèze. Illustrations of a microdamage model for laminates under oxidizing thermal cycling. *Composites Science and Technology*, 69(1):3–9, 2009.
19. J. Mandel. Balancing domain decomposition. *Communications in Numerical Methods in Engineering*, 9:233–241, 1993.
20. J. Melenk and I. Babuška. The partition of unity finite element method: Basic theory and applications. *Computer Methods in Applied Mechanics and Engineering*, 39:289–314, 1996.

21. J.T. Oden, K. Vemaganti, and N. Moës. Hierarchical modeling of heterogeneous solids. *Computer Methods in Applied Mechanics and Engineering*, 172:3–25, 1999.
22. J.C.J. Schellekens and R. de Borst. Free edge delamination in carbon-epoxy laminates: A novel numerical/experimental approach. *Composite structures*, 28(4):357–373, 1994.

Damage Propagation in Composites – Multiscale Modeling and Optimization

Ekkehard Ramm[1], Andrea Erhart[2], Thomas Hettich[3], Ingrid Bruss[1], Frédéric Hilchenbach[1] and Junji Kato[1]

[1]Institute of Structural Mechanics, University of Stuttgart, 70550 Stuttgart, Germany; ramm@ibb.uni-stuttgart.de
[2]DYNAmore GmbH, Industriestr. 2, 70565 Stuttgart, Germany
[3]Mahle GmbH, Pragstr. 26–46, 70376 Stuttgart, Germany

Abstract The paper addresses general solution concepts for multiscale analyses and structural optimization of composite structures under damage evolution. Fiber Reinforced Composites, in particular Fiber Reinforced Concrete, are chosen as model material. First we describe undamaged and damaged material interfaces as well as matrix cracks utilizing the eXtended Finite Element Method (XFEM) and Level Set methods. The next part focuses on a solution strategy based on the Variational Multiscale Method (VMM) allowing to describe localized long range failure mechanisms for which standard homogenization is not applicable anymore. Either continuum damage or the discrete cohesive crack model applying XFEM is inserted into the VMM framework. Finally material optimization is performed to obtain an optimal fiber layout maximizing structural ductility.

Key words: Multiscale modeling, material optimization, composites, eXtended Finite Element Method (XFEM), Variational Multiscale Method (VMM)

1 Introduction

The present contribution describes mainly methodical concepts for nonlinear computations of composite structures. Without loss of generality it concentrates on a special class of composites where long carbon or glass fibers are embedded in a more or less soft matrix. Typical representatives of this class are Fiber Reinforced Polymers (FRP) with a ductile matrix. In this study we selected Fiber Reinforced Concrete (FRC) as model material, often also denoted as Textile Reinforced Concrete or Fiber Reinforced Cementitious Material [3, 16, 21]. This rather new material offers a particular challenge for design and simulation since both constituents fibers and mortar matrix are brittle and the toughness or ductility is supplied only via friction between both materials and within the rovings.

During loading of theses composites damage starts by local debonding and microcracking in matrix and fibers; this initially diffuse process usually ends in a failure mechanism with delamination and a few distinct macrocracks in the quasi-brittle

R. de Borst and E. Ramm (eds.), *Multiscale Methods in Computational Mechanics*, Lecture Notes in Applied and Computational Mechanics 55, DOI 10.1007/978-90-481-9809-2_15,

matrix. It is obvious that this macroscopic response is mainly governed by material properties on a smaller scale; for example for FRC the interface between both constituents plays a key role in this process. This in turn raises the questions of a suited description for weak and strong discontinuities and of an appropriate multiscale modeling.

Traditional multiscale approaches for heterogeneous materials use homogenization procedures based on the assumption that the separation of scales is guaranteed. This is usually the case when a certain periodicity or mean layout of the material microstructure exists. The point-wise scale transfer is managed via unit cell or RVE concepts in an uncoupled or coupled (concurrent) way. Extensions to higher order homogenization schemes have been proposed to capture a larger variation of the microstructure, e.g. [8,9], see also the review in [7]. In case of localization phenomena like shear banding and crack coalescense it remains for an RVE to zoom into the small scale zone [25].

In the present contribution we focus on problems where scale separation does not hold anymore. This may be either an a priori distinct variation of the material and structural layout, for example having a fiber reinforcement net, or a long range failure mechanisms evolving during loading, e.g. localized damage zones and cohesive cracks.

In the current study (Section 2) weak and strong discontinuities at material interfaces and cracks, respectively, are inserted in a structural mesh via the eXtended Finite Element Method (XFEM). A cohesive crack model is utilized, see for example [24, 29, 34]. The geometries of interfaces as well as of evolving cracks are parameterized by the Level Set Method.

Since homogenization schemes are not applicable we apply a volume coupled concurrent solution strategy based on the framework of the Variational Multiscale Method VMM [10], see Section 3. Here the displacement field is decomposed into a large scale and many small scale solutions each restricted to the domain of one large scale element; the compatibility between small scale domains is managed by applying Domain Decomposition (DD) schemes. The concept is first derived for a continuum damage model allowing to describe diffuse failure. Subsequently to this the discontinuum model using XFEM is inserted into the multiscale formulation so that material interfaces and discrete cracks can be analysed.

The delicate behavior of FRC suggest to search for an optimal layout of the fiber reinforcement in a structure maximizing the structural ductility [18–20]. This question is discussed in the last part (Section 4) applying a damage formulation where two optimization concepts are described, namely a multiphase material optimization with fiber fractions in design domains as design variables and shape optimization for the geometrical layout of the fibers. Even if these investigations are not directly related to the above multiscale solution the formulation of the material optimization has a clear multiscale aspect: The macroscopic structural response (ductility!) clearly depends on the material properties and layout on a microscale. It remains to further studies to optimize micromechanical material parameters and to embed the optimization into the multiscale formulation.

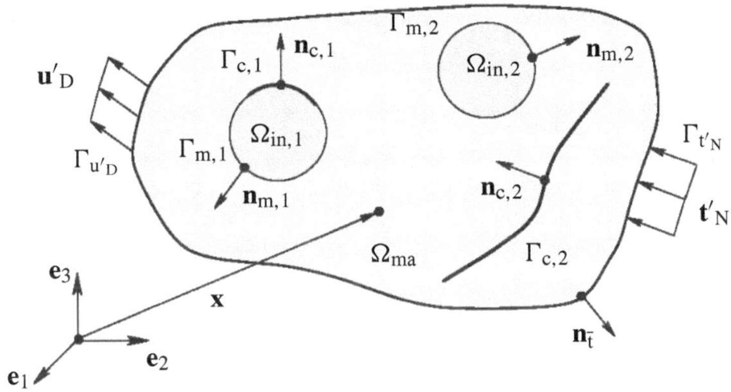

Fig. 1. Microstructure of a composite with interfacial and matrix crack [14].

2 Modeling of Discontinuities on Small Scale

2.1 Geometrical Description

First the geometry of the undamaged and damaged interfaces and matrix cracks have to be parameterized. Level Set methods have been proven as an appropriate geometrical description of surfaces since they provide also the position of the surface with respect to an arbitrary point. In contrast to the general case of a moving surface Γ solving a Hamilton–Jacobi differential equation the geometry of material interfaces is fixed and cracks move only at their tip. As an example for a two-dimensional case with circular inclusions the situation is depicted in Figure 1.

The composite consists of n_m inclusions with domains $\Omega_{in,i}$ in a matrix Ω_{ma}. The geometry of the union of individual undamaged material interfaces $\Gamma_m = \bigcup_i \Gamma_{m,i}$ is defined by the signed distance function

$$\phi_m(\mathbf{x}) = \min\{\underbrace{||\mathbf{x} - \mathbf{x}_{m,i}|| - r_{m,i}}_{\phi_{m,i}}\} \quad \text{for } i = 1, \dots, n_m \tag{1}$$

$\mathbf{x}_{m,i}$ and $r_{m,i}$ denote the position vector of the center and the radius of the ith inclusion, respectively. $\phi_{m,i}$ is the level set function which is zero per definition at the boundary between the two materials. The same signed distance function is used in 3D computations where the longitudinal axis of a cylindrical roving is aligned to one of the global axes. Correspondingly the geometry of the evolving interface cracks is also known a priori through the level set function.

For the growing matrix crack we apply the technique of Ventura et al. [32] in which a so-called compound object $\tilde{\phi}_{c,j}$ is introduced accounting for the location of a matrix crack

$$\tilde{\phi}_{c,j}(\mathbf{x}) = \{\Phi_{c,j}(\mathbf{x}), \chi_{c,j}(\Phi_{c,j})\} \quad \text{and} \quad \phi_{c,j} = \chi_{c,j}||\Phi_{c,j}|| \tag{2}$$

Vector $\Phi_{c,j}$ denotes the closest point projection from $\mathbf{x}$ onto the crack $\Gamma_{c,j}$; $\chi_{c,j}$ is a sign-function. Both objects allow calculating the signed distance function

$\phi_{c,j}$ at the nodes which is the basis for a finite element interpolation inside a cracked element.

2.2 Kinematic Description

The mechanics of undamaged and damaged interfaces and matrix cracks is described by the eXtended Finite Element Method (XFEM); see [4,31]. The displacement field is decomposed as follows:

$$\mathbf{u}'(\mathbf{x},t) = \underbrace{\hat{\mathbf{u}}(\mathbf{x},t)}_{cont.} + \sum_{i=1}^{n_\mathrm{m}} \underbrace{\chi_{\mathrm{m},i}(\mathbf{x})\,\tilde{\mathbf{u}}_i(\mathbf{x},t)}_{cont.} + \sum_{j=1}^{n_\mathrm{c}} \underbrace{\chi_{\mathrm{c},j}(\mathbf{x})\,\tilde{\tilde{\mathbf{u}}}_j(\mathbf{x},t)}_{discont.} \tag{3}$$

The upper index $\mathbf{u}'$ indicates already here that the field represents only the small scale solution in the multiscale formulation. $\hat{\mathbf{u}}$ is the usual continuous displacement field; $\chi_{\mathrm{m},i}$ and $\chi_{\mathrm{c},j}$ are enrichment functions acting on the continuous fields $\tilde{\mathbf{u}}_i$ and $\tilde{\tilde{\mathbf{u}}}_j$ and descibing the weak and strong discontinuities at the material interfaces and cracks, respectively. $\chi_{\mathrm{m},i}$ is a ridge function [23]. Using the level set representation interfacial and matrix cracks can be described through the sign-function $\phi_j := \phi_{\mathrm{m},j}$ and $\phi_j := \phi_{\mathrm{c},j}$; the function $\chi_{\mathrm{c},j}$ can be defined in terms of the Heaviside-function $\mathcal{H}_{\mathrm{c},j}$.

$$\chi_{\mathrm{c},j} = \mathrm{sign}(\phi_j) := 2\mathcal{H}_{\mathrm{c},j} - 1 \tag{4}$$

Assuming linear kinematics the strains are

$$\begin{aligned} \boldsymbol{\varepsilon} = \underbrace{\nabla^s \hat{\mathbf{u}}}_{cont.} &+ \sum_{i=1}^{n_\mathrm{m}} [\, \underbrace{(\tilde{\mathbf{u}}_i \otimes \nabla \chi_{\mathrm{m},i})^s}_{discont.} + \underbrace{\chi_{\mathrm{m},i} \nabla^s \tilde{\mathbf{u}}_i}_{cont.} \,] \\ &+ \sum_{j=1}^{n_\mathrm{c}} [\, \underbrace{2\,\delta_{\mathrm{c},j} (\tilde{\tilde{\mathbf{u}}}_j \otimes \mathbf{n}_{\mathrm{c},j})^s}_{unbounded} + \underbrace{\chi_{\mathrm{c},j} \nabla^s \tilde{\tilde{\mathbf{u}}}_j}_{discont.} \,] \end{aligned} \tag{5}$$

Now both extended fields contain discontinuous parts; the unbounded component is a consequence of the derivation of the Heaviside-function $\mathcal{H}_{\mathrm{c},j}$ leading to the Dirac-Delta distribution $\delta_{\mathrm{c},j}$. The strains are not defined at the crack $\Gamma_{c,j}$ and have to be replaced by the displacement jump as relevant kinematic variable:

$$[\![\mathbf{u}]\!]_{\mathrm{c},j} := \mathbf{u}^+ - \mathbf{u}^- = 2\,\tilde{\tilde{\mathbf{u}}}_j \tag{6}$$

The normal $\mathbf{n}_{\mathrm{c},i}$ of an interfacial crack is known a priori; for matrix cracks a criterion for the crack direction has to be introduced. It is based on the maximum principal nonlocal stress which is obtained through averaging within a predefined influence zone around the crack tip [34]. We assume that crack tips lie only on element edges avoiding extra crack tip displacements.

2.3 Cohesive Law

At a cohesive crack the traction vector $\mathbf{t} = \mathrm{t_n}\mathbf{n} + \mathbf{t}_T$ is decomposed into its normal and tangential components $\mathrm{t_n}$ and $\mathbf{t}_T$, respectively. Correspondingly also the total displacement jump $[\![\mathbf{u}]\!]$ is split into its normal and tangential components $[\![\mathrm{u}]\!]_\mathrm{n}$ and $[\![\mathbf{u}]\!]_T$.

The initiation of failure is controlled either by the Rankine criterion or the Brewer/Lagace criterion [1]:

$$\begin{array}{ll} \text{matrix failure} & \sigma_\mathrm{I} = f_t \\ \text{interface failure} & \sqrt{\langle t_\mathrm{n}\rangle^2 + \beta^{-2}\mathbf{t}_T^2} = c_\mathrm{n}, \end{array} \tag{7}$$

Thus a mode I failure is assumed for the matrix controlled by the tensile strength f_t. The response at the interface is described by a mixed mode model with β defining the ratio between the strengths in shear c_T and tension c_n. The Macauly bracket $\langle\cdot\rangle$ is introduced allowing to distinguish the different response in tension and compression. Accordingly an effective displacement jump η is introduced

$$\begin{array}{ll} \text{matrix failure} & \eta = [\![\mathrm{u}]\!]_\mathrm{n} \\ \text{interface failure} & \eta = \sqrt{\langle[\![\mathrm{u}]\!]_\mathrm{n}\rangle^2 + \beta^2[\![\mathbf{u}]\!]_T^2} \end{array} \tag{8}$$

The history variable is defined as the largest value of η attained during loading history.

$$\kappa = \max_{0 \le t < \tau} (\eta(t)) \tag{9}$$

The effective traction $\mathrm{t}^\star$ is obtained through an exponential softening law depending on the fracture energy G_f and on f_t and c_n, respectively.

$$\mathrm{t}^\star = \sigma \exp\left(-\frac{\sigma}{G_f}\kappa\right) \qquad \text{with} \qquad \sigma = f_t\ ,\ c_\mathrm{n} \tag{10a}$$

$\mathrm{t}^\star$ has been regularized for three dimensional interface failure to

$$\mathrm{t}^\star = c_\mathrm{n}\frac{\kappa}{\delta_\mathrm{c}}\exp\left(1 - \frac{\kappa}{\delta_\mathrm{c}}\right) \tag{10b}$$

with the critical crack opening δ_c.

The traction separation laws for the normal and tangential component $\mathrm{t_n}$ and $\mathbf{t}_T$ are for

$$\begin{array}{lll} \text{matrix failure} & \mathrm{t_n} = \mathrm{t}^\star/\kappa\, [\![\mathrm{u}]\!]_\mathrm{n} & \mathbf{t}_T = \mathrm{d}_T \exp(\mathrm{h}_T\kappa)\, [\![\mathbf{u}]\!]_T \\ \text{interface failure} & \begin{cases} \mathrm{t_n} = \mathrm{t}^\star/\kappa\, [\![\mathrm{u}]\!]_\mathrm{n} & \text{if } [\![\mathrm{u}]\!]_\mathrm{n} \ge 0 \\ \mathrm{t_n} = \mathrm{d_n}\, [\![\mathrm{u}]\!]_\mathrm{n} & \text{if } [\![\mathrm{u}]\!]_\mathrm{n} < 0 \end{cases} & \mathbf{t}_T = \beta^2\, \mathrm{t}^\star/\kappa\, [\![\mathbf{u}]\!]_T \end{array} \tag{11}$$

d_T is the initial shear stiffness whereas $\mathrm{d_n}$ plays the role of a penalty parameter avoiding penetration under compression. $\mathrm{h}_T \le 0$ controls the loss of shear stiffness and depends on κ.

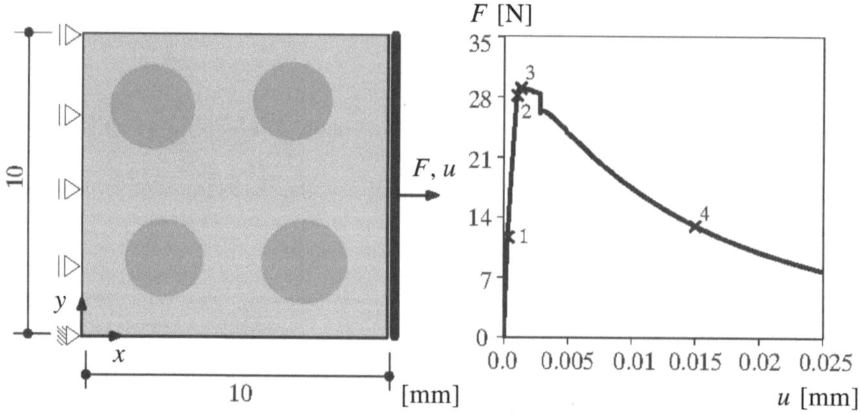

Fig. 2. Structure and load-displacement curve.

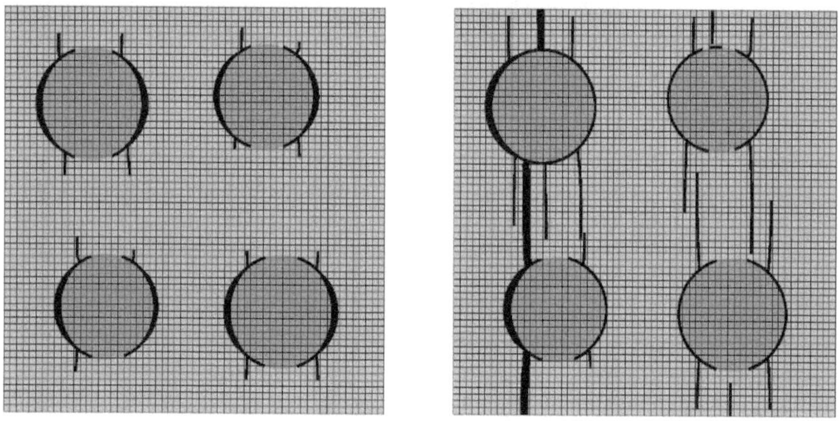

point 2

point 4

Fig. 3. Crack evolution.

2.4 Numerical Examples

Two-Phase Material under Tension [13]

The plane strain system of 1 mm thickness (Figure 2) represents a section through a fiber reinforced structure under displacement-controlled tension. The four circular fibers defined by level set functions derived from their center points and radii (2.7/2.6 – 1.3), (7.3/2.5 – 1.4), (6.9/7.8 – 1.3) and (2.3/7.6 – 1.4) have been assumed to be elastic with Young's modulus $E = 70000\,\mathrm{N/mm^2}$ and Poisson's ratio $\nu = 0.2$. The material data for the concrete matrix are $E = 35000\,\mathrm{N/mm^2}$, $\nu = 0.2$, $f_t = 4.0\,\mathrm{N/mm^2}$, $\mathrm{h}_T = 0.0/\mathrm{mm}$, $G_f = 0.05\,\mathrm{N/mm}$, $\mathrm{d}_T = 1.0\,\mathrm{N/mm^3}$; those for the interface are $c_n = 1.0\,\mathrm{N/mm^2}$, $\beta = 2.0$, $G_f = 0.05\,\mathrm{N/mm}$, $\mathrm{d}_n = 10^6\,\mathrm{N/mm^3}$.

The structure is discretized by a regular mesh of 60×60 bilinear displacement elements. The load displacement diagram is depicted in Figure 2; two snap-shots at points 2 (1000-times magnified) and 4 (15-times magnified) of the crack evolution

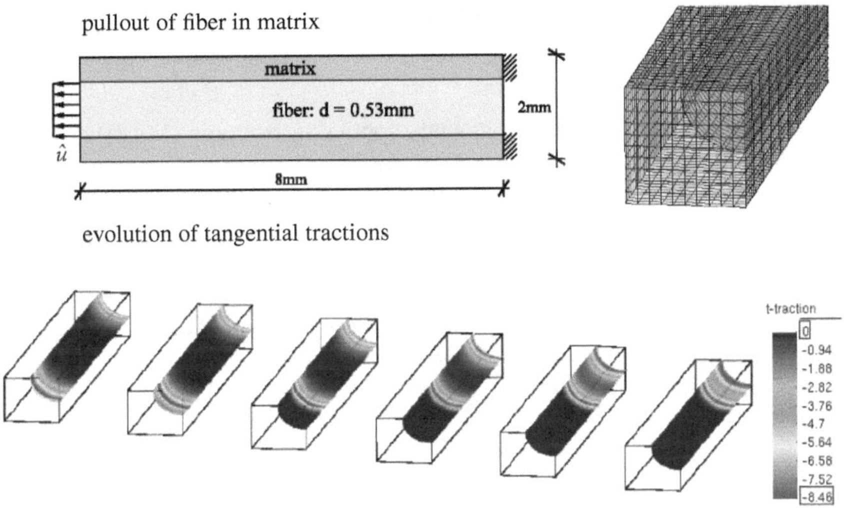

Fig. 4. Pull-out test. (For a color version of this image, see [2].)

are shown in Figure 3. They clearly show the debonding and the subsequent matrix cracks.

Pull-out of Fiber in Matrix

A cylindrical fiber embedded in an elastic concrete block is pulled out by a displacement controlled load (Figure 4). Matrix and fiber are assumed to be elastic with $E_c = 39000\,\mathrm{N/mm^2}$, $\nu_c = 0.2$ and $E_f = 72000\,\mathrm{N/mm^2}$, $\nu_f = 0.2$. The cohesive interface failure is described by the parameters $\delta_c = 0.0014$, $\beta = 1.8$ and $f_c = 4.7\,\mathrm{N/mm^2}$, see eqs. (10) and (11).

One quarter of the structure is discretized by 8-noded brick displacement elements. Figure 4 shows the evolution of the tangential tractions during loading clearly indicating the development of the axisymmetric process zone.

3 Multiscale Formulation [11, 12]

First we will introduce the basic concept of the multiscale formulation using a continuum damage model. Afterwards the above described discontinuous model will be inserted into the multiscale formulation allowing to model evolving localized long range failure mechanisms.

3.1 Formulation Using Continuum Damage

Material Model

Without loss of generality the present study applies an isotropic $(1-D)$ continuum damage formulation with de Vree's [33] definition of equivalent strains ε_v.

$$\varepsilon_v = \frac{k-1}{2k(1-2\nu)} I_1 + \frac{1}{2k}\sqrt{\frac{(k-1)^2}{(1-2\nu)^2} I_1^2 - \frac{12k}{(1+\nu)^2} J_2}. \tag{12}$$

I_1 is the first invariant of the strain tensor, J_2 the second invariant of the deviatoric strain tensor, ν is the Poisson ratio and k controls the sensitivity of the material for compression relative to tension. An exponential softening law is used for the damage evolution [22].

$$D = 1 - \frac{\kappa_0}{\kappa}\left(1 - \alpha + \alpha e^{-\beta(\kappa-\kappa_0)}\right) \quad \text{if} \quad \kappa \geq \kappa_0 \tag{13}$$

The material parameters α and β define the final softening stage and the rate of damage growth, respectively. The threshold variable κ_0 determines damage initiation; κ is the maximum memorized value of ε_v and controls the damage variable D which approaches 1 for $\kappa \to \infty$.

The formulation is regularized by a gradient enhancement where D is a function of a non-local equivalent strain $\tilde{\varepsilon}_v$ depending on the local strain ε_v within a radius of influence proportional to the square root of an internal length parameter c, see Peerlings et al. [27,28].

$$\tilde{\varepsilon}_v - c\nabla^2\tilde{\varepsilon}_v = \varepsilon_v(\varepsilon) \qquad \text{with} \quad \tilde{\varepsilon}_v \leq \kappa \tag{14}$$

For simplicity also a mesh adapted softening modulus may be used in which the ultimate strain κ_u is defined as function of fracture toughness or alternatively of the crackband width w, a normalized ultimate strain $\hat{\kappa}_u$ and the element size h.

Concept

As a frame for the formulation the Variational Multiscale Method (VMM) [10] is adopted decomposing the displacement field $\mathbf{u}$ into a large scale $\bar{\mathbf{u}}$ and a small scale solution $\mathbf{u}'$

$$\mathbf{u} = \bar{\mathbf{u}} + \mathbf{u}' \tag{15}$$

The small scale displacements $\mathbf{u}'$ represent the fluctuation of the mean field; they are due to an a-priori heterogeneous microstructure as in composites or evolve during loading, for example through microcracks and damage. Consequently $\mathbf{u}'$ can be restricted only to a subdomain Ω^d of the entire structure, for example the evolving failure zone in Figure 5. The remaining part Ω^{el} with $\mathbf{u}' = \mathbf{0}$ is still elastic.

Restricting the additive split to the displacement field eq. (14) the principle of virtual work leads to the three coupled nonlinear equations:

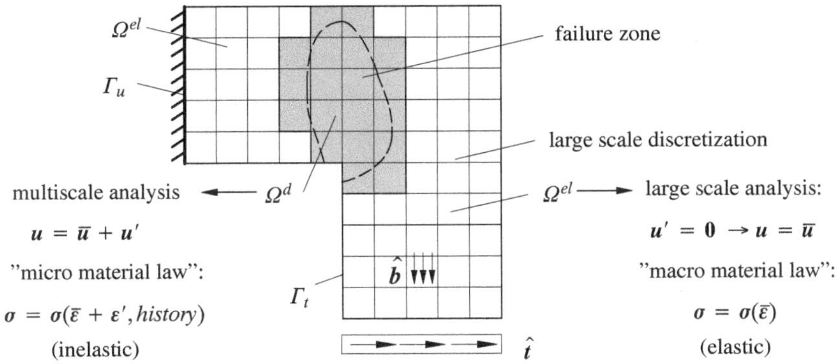

Fig. 5. Division of Ω into domains Ω^{el} and Ω^d.

$$\text{large-scale problem:} \quad \int_\Omega \delta\bar{\boldsymbol{\varepsilon}}^T \,\boldsymbol{\sigma}\, d\Omega = \int_\Omega \delta\bar{\mathbf{u}}^T \,\hat{\mathbf{b}}\, d\Omega + \int_{\Gamma_t} \delta\bar{\mathbf{u}}^T \,\hat{\mathbf{t}}\, d\Gamma \tag{16a}$$

$$\text{small-scale problem:} \quad \int_\Omega \delta\boldsymbol{\varepsilon}'^T \,\boldsymbol{\sigma}\, d\Omega = \int_\Omega \delta\mathbf{u}'^T \,\hat{\mathbf{b}}\, d\Omega + \int_{\Gamma_t} \delta\mathbf{u}'^T \,\hat{\mathbf{t}}\, d\Gamma \tag{16b}$$

$$\text{nonlocal strains:} \quad \int_\Omega c\,\delta\nabla\tilde{\varepsilon}_v^{\,l}\; \nabla\tilde{\varepsilon}_v\, d\Omega + \int_\Omega \delta\tilde{\varepsilon}_v\,\tilde{\varepsilon}_v\, d\Omega = \int_\Omega \delta\tilde{\varepsilon}_v\,\varepsilon_v\, d\Omega \tag{16c}$$

$\hat{\mathbf{b}}$ and $\hat{\mathbf{t}}$ are prescribed body and surface forces, respectively. All three independent variables $\bar{\mathbf{u}}$, $\mathbf{u}'$ and $\tilde{\varepsilon}_v$ are discretized by finite elements. Small-scale functions have to vanish at nodes of large-scale elements ensuring linear independence.

$$\text{large-scale elements:} \quad \bar{\mathbf{u}} \approx \bar{\mathbf{u}}^h = \bar{\mathbf{N}}\,\bar{\mathbf{d}} \tag{17a}$$

$$\text{small-scale elements:} \quad \mathbf{u}' \approx \mathbf{u}'^h = \mathbf{N}'\mathbf{d}' \tag{17b}$$

$$\tilde{\varepsilon}_v \approx \tilde{\varepsilon}_v^h = \mathbf{N}_e\mathbf{e} \tag{17c}$$

The nodal degrees of freedom $\bar{\mathbf{d}}$, $\mathbf{d}'$, $\mathbf{e}$ on element level become $\bar{\mathbf{D}}$, $\mathbf{D}'$, $\mathbf{E}$ on system level after assembly.

A key issue for the efficiency of the formulation is that we assume a local support for small-scale displacements $\mathbf{u}'$ and non-local strains within every large-scale element l (Figure 6). This assumption confines the small scale equations (16b) (16c) to the domain of every large-scale element and provides an a priori parallel solution algorithm for the small-scale problems.

This is a major difference to classical substructure methods where the degrees of freedom on the boundaries of large scale elements are not localized remaining in the global system of equations. This simplification has to be paid for with extra consideration of the kinematic compatibility between the boundaries Γ^d, see below.

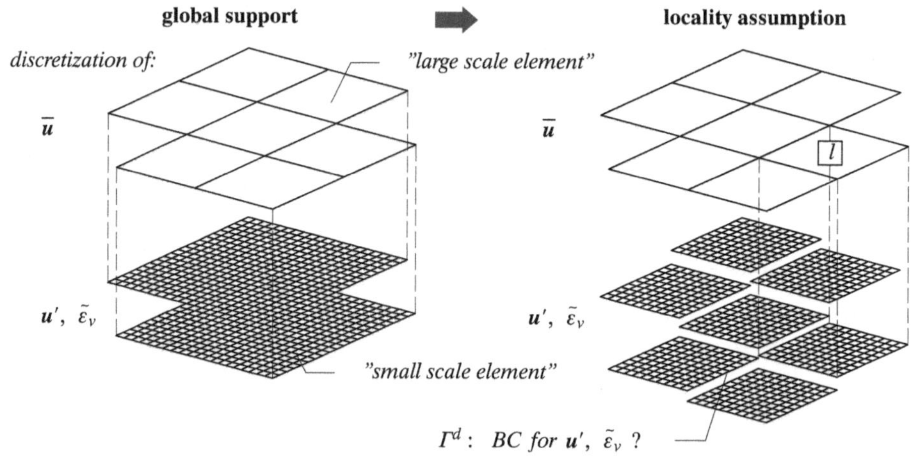

Fig. 6. Discretization and locality assumption.

$$\text{large-scale problem:} \quad \mathbf{K}_{\bar{D}\bar{D}}\ \Delta\bar{\mathbf{D}} + \sum_{l=1}^{ns}(\mathbf{K}^l_{\bar{d}D'}\ \Delta\mathbf{D}'^l + \mathbf{K}^l_{\bar{d}E}\ \Delta\mathbf{E}^l) \qquad = -\bar{\mathbf{Res}}$$

within every large-scale element l: update of $\Delta\mathbf{D}'^l$, $\Delta\mathbf{E}^l$ $\Downarrow$ $\Uparrow$ static condensation of $\Delta\mathbf{D}'^l$, $\Delta\mathbf{E}^l$

$$\text{small-scale problem:} \quad \mathbf{K}_{D'\bar{d}}\ \Delta\bar{\mathbf{d}}^l + \mathbf{K}^l_{D'D'}\ \Delta\mathbf{D}'^l + \mathbf{K}^l_{D'E}\ \Delta\mathbf{E}^l \qquad = -\mathbf{Res}'^l$$

$$\text{nonlocal strains:} \quad \mathbf{K}_{E\bar{d}}\ \Delta\bar{\mathbf{d}}^l + \mathbf{K}^l_{ED'}\ \Delta\mathbf{D}'^l + \mathbf{K}^l_{EE}\ \Delta\mathbf{E}^l \qquad = -\mathbf{Res}^l_E$$

$$\text{assumed BC for } \tilde{\varepsilon}_v : \tilde{\varepsilon}_{v,\mathbf{n}_\Gamma} = 0 \ + \ \text{additional constraints for } \mathbf{u}' \tag{18}$$

ns is the number of large scale elements in the system. The equations descibe the linearized system for one Newton–Raphson iteration within an arbitrary load step. The stiffness- and RHS- expressions are [11, 12]:

$$\mathbf{K}_{\bar{D}\bar{D}} = \int_\Omega \bar{\boldsymbol{\varepsilon}}^T_{,\bar{D}}\ \boldsymbol{\sigma}_{,\bar{D}}\ d\Omega, \quad \mathbf{K}^l_{\bar{d}D'} = \int_{\Omega^l} \bar{\boldsymbol{\varepsilon}}^T_{,\bar{D}}\ \boldsymbol{\sigma}_{,D'}\ d\Omega, \quad \mathbf{K}^l_{\bar{d}E} = \int_{\Omega^l} \bar{\boldsymbol{\varepsilon}}^T_{,\bar{D}}\ \boldsymbol{\sigma}_{,E}\ d\Omega$$

$$\bar{\mathbf{Res}} = \int_\Omega \bar{\boldsymbol{\varepsilon}}^T_{,\bar{D}}\ \boldsymbol{\sigma}\ d\Omega - \int_\Omega \bar{\mathbf{u}}^T_{,\bar{D}}\ \hat{\mathbf{b}}\ d\Omega - \int_{\Gamma^t} \bar{\mathbf{u}}^T_{,\bar{D}}\ \hat{\mathbf{t}}\ d\Gamma$$

$$\mathbf{K}^l_{D'\bar{d}} = \int_{\Omega^l} \boldsymbol{\varepsilon}'^T_{,D'}\ \boldsymbol{\sigma}_{,\bar{D}}\ d\Omega, \ \mathbf{K}^l_{D'D'} = \int_{\Omega^l} \boldsymbol{\varepsilon}'^T_{,D'}\ \boldsymbol{\sigma}_{,D'}\ d\Omega, \ \mathbf{K}^l_{D'E} = \int_{\Omega^l} \boldsymbol{\varepsilon}'^T_{,D'}\ \boldsymbol{\sigma}_{,E}\ d\Omega$$

$$\mathbf{Res}'^l = \int_{\Omega^l} \boldsymbol{\varepsilon}'^T_{,D'}\ \boldsymbol{\sigma}\ d\Omega - \int_\Omega \mathbf{u}'^T_{,D'}\ \hat{\mathbf{b}}\ d\Omega - \int_{\Gamma^t} \mathbf{u}'^T_{,D'}\ \hat{\mathbf{t}}\ d\Gamma$$

$$\mathbf{K}^l_{E\bar{d}} = \int_{\Omega^l} -\tilde{\varepsilon}^T_{v,E}\, \varepsilon_{v,\bar{D}}\, d\Omega, \quad \mathbf{K}^l_{ED'} = \int_{\Omega^l} -\tilde{\varepsilon}^T_{v,E}\, \varepsilon_{v,D'}\, d\Omega$$

$$\mathbf{K}^l_{EE} = \int_{\Omega^l} (c\nabla\tilde{\varepsilon}^T_{v,E}\, \nabla\tilde{\varepsilon}_{v,E} + \tilde{\varepsilon}^T_{v,E}\, \tilde{\varepsilon}_{v,E})\, d\Omega \tag{19}$$

$$\mathbf{Res}^l_E = \int_{\Omega^l} (c\nabla\tilde{\varepsilon}^T_{v,E}\, \nabla\tilde{\varepsilon}_v + \tilde{\varepsilon}^T_{v,E}\, \tilde{\varepsilon}_v - \tilde{\varepsilon}^T_{v,E}\, \varepsilon_v)\, d\Omega$$

In contrast to a classical homogenization procedure with a pointwise coupling the present algorithm can be denoted as a concurrent volume coupled method. Although not carried out in this study the decoupling of the small-scale solution also allows using different constitutive models on both scales.

The important question remains how the boundary conditions for $\mathbf{u}'$ are defined at the large-scale boundaries Γ_d. We may adopt the easiest and most efficient assumption $\mathbf{u}' = \mathbf{0}$, an assumption occasionally used for the application of the VMM in CFD; however it is obvious that this is only realistic for physically localized phenomena. Thus we have to enforce kinematic compatibility between the small-scale patches m and s $[\![\mathbf{u}']\!] = \mathbf{u}'^m - \mathbf{u}'^s = \mathbf{0}$. This in turn can be regarded as a parallel solution strategy similar to domain decomposition (DD) methods. Thus a direct information exchange between adjacent small-scale domains is provided which is an important feature for problems where the assumption of scale separation does not hold.

From DD methods the following versions have been investigated leading to additional variational terms $\delta\Pi_{constr}$ representing the constraint condition in the equilibrium equation.

Penalty Method $$\delta\Pi_{\text{constr}} = \int_{\Gamma_{int}} \epsilon\, \delta\mathbf{u}' \cdot [\![\mathbf{u}']\!]\, \mathrm{d}\Gamma \tag{20}$$

Lagrange Method $$\delta\Pi_{\text{constr}} = \int_{\Gamma_{int}} \delta\boldsymbol{\lambda} \cdot [\![\mathbf{u}']\!]\, \mathrm{d}\Gamma + \int_{\Gamma_{int}} \boldsymbol{\lambda} \cdot [\![\delta\mathbf{u}']\!]\, \mathrm{d}\Gamma \tag{21}$$

ϵ is a penalty parameter and λ a Lagrange parameter.

The Lagrange method follows the Finite-Element Tearing and Interconecting (FETI) procedure [5, 6] and can be exploited either in an integral or a pointwise format. It needs an extra solution effort because the Lagrange parameters turn out to be global variables. This can be avoided applying the Penalty or an Augmented Lagrange method. For details it is referred to [11,12], where also the underlying iteration concept and acceleration procedures are discussed.

Numerical Examples [11]

• L-shaped plate

The material parameters for the damage of the plate under plane strain conditions shown in Figure 7 are $E = 10000\,\text{kN/m}^2$, $\nu = 0.2$, $\text{k} = 10.0$, $\kappa_0 = 0.0004$, $\hat{\kappa}_\text{u} =$

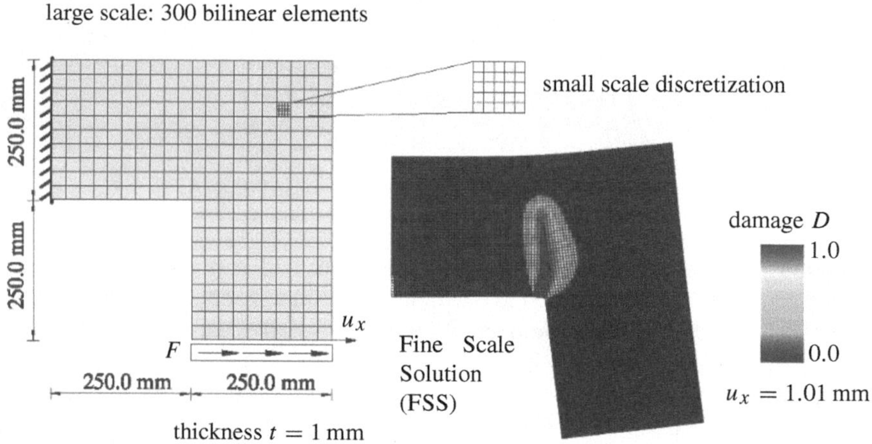

Fig. 7. L-shaped plate: structure and deformation with damage distribution for $u_x = 1.01$ mm (magnification 50). (For a color version of this image, see [11].)

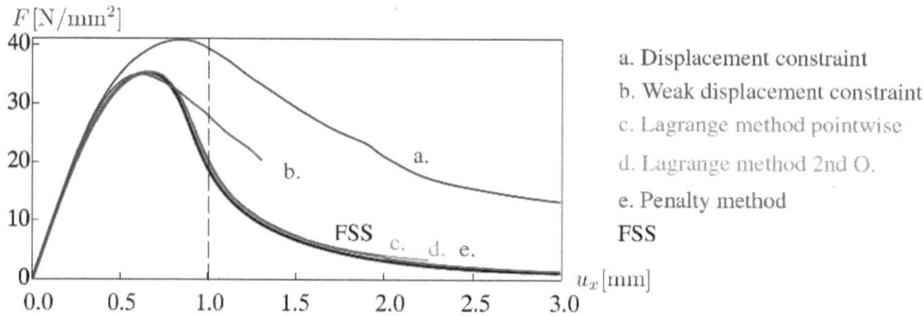

Fig. 8. L-shaped plate: load-displacement curves.

0.05, $\alpha = 1.0, \beta = 2.0$ applying a modified power law for the damage evolution [12]. Zero displacements have been assumed on the left boundary of the plate representing a clamped condition. The structure is loaded by a uniform horizontal load at its lower edge. The nonlinear solution is obtained by controlling the horizontal component u_x of the lower right node.

Instead of a gradient enhancement as "regularization" a mesh adapted softening modulus with a crackband $w = 10.0$ based on a mesh size $h = 5.0$ has been adopted for simplicity. The two-scale formulation is only introduced in the neighbourhood of the localized zone, i.e. the size Ω_d is adapted during loading. A snapshot of the damage zone for a displacement $u_x = 1.01$ mm is added to Figure 7. As indicated in the load-displacement diagram Figure 8 five different approaches have been investigated to satisfy the constraints of the small scale displacements between adjacent large scale elements; the results are compared to a Fine Scale Solution (FSS) taken as

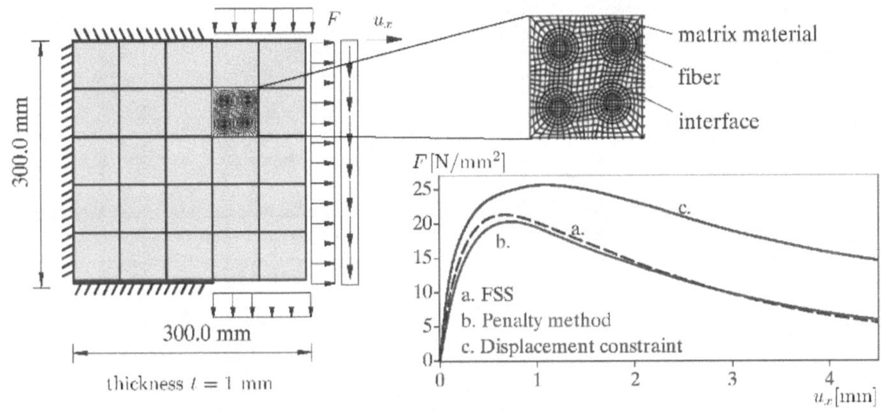

Fig. 9. Reinforced plate: system with discretization & load displacement curves.

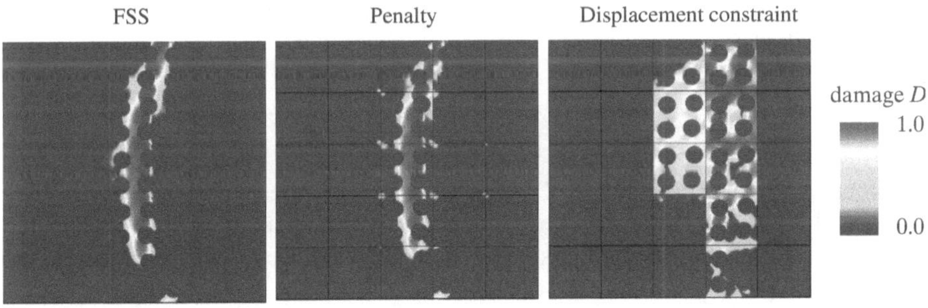

Fig. 10. Reinforced plate: damage distribution for $u_x = 3.73$ mm. (For a color version of this image, see [11].)

reference. As expected neither the displacement constraint $\mathbf{u}' = \mathbf{0}$ nor its weak counterpart can represent the response of the FSS whereas all three other versions are almost identical to the reference. This two-scale solution came along with a drastic reduction of the CPU-time, see [12].

• Reinforced plate

The plate (Figure 9) consists of elastic inclusions (section of fibers $E = 735000\,\mathrm{N/mm^2}$, $\nu = 0.18$) embedded in a soft matrix ($E = 32000\,\mathrm{N/mm^2}$, $\nu = 0.2$) with damage parameters eq. (13) k $= 3.0$, $\kappa_0 = 0.002$, $\alpha = 0.98$, $\beta = 5.0$; regularization is provided by gradient enhancement eq. (14) with $c = 2.0\,\mathrm{mm^2}$. The debonding (damage) of the fibers ($E = 7350\,\mathrm{N/mm^2}$, $\nu = 0.2$) is modeled by conventional interface elements with a thickness of $t = 0.01$ mm: $\kappa_0^n = 5 \cdot 10^{-5}$, $\kappa_0^t = 10^{-4}$, $\alpha^{n/t} = 1.0$, $\beta^{n/t} = 2.0$.

The displacements on the clamped boundaries are constraint. The remaining boundaries are loaded by horizontal and vertical uniform loads, respectively. The

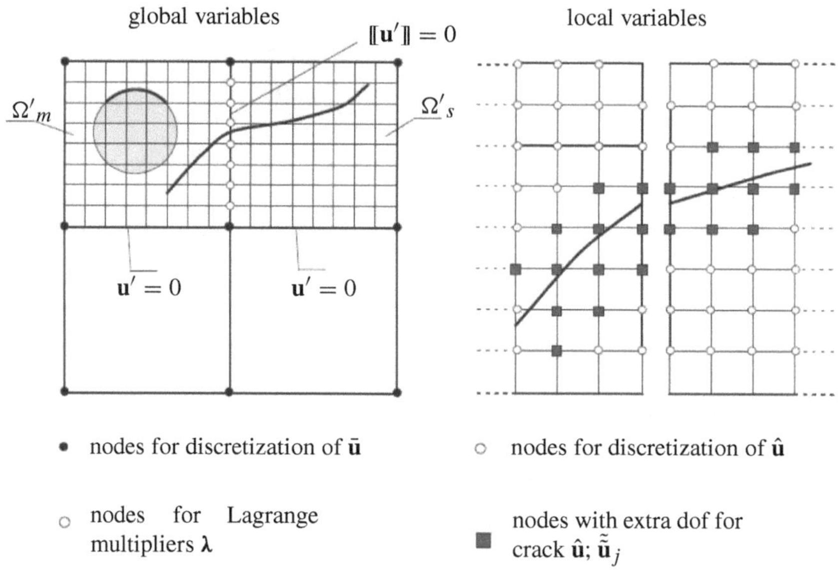

Fig. 11. Discontinuous model in VMM.

non-local strains are assumed to be zero on all boundaries for simplicity. In Figures 9 and 10 the results of the two-scale models for the penalty method and for the displacement constraint are compared to the FSS reference. They clearly show the quality of the penalty solution and the inadequate result for the $\mathbf{u}' = \mathbf{0}$ constraint.

3.2 Discontinuum Model in Multiscale Formulation

Concept

Following the additive decomposition of the VMM $\mathbf{u} = \bar{\mathbf{u}} + \mathbf{u}'$ the small-scale displacement field $\mathbf{u}'$ is now expressed by eq. (3)

$$\mathbf{u} = \begin{cases} \bar{\mathbf{u}} & \forall \mathbf{x} \in R/R' \\ \bar{\mathbf{u}} + \underbrace{\hat{\mathbf{u}} + \sum_{i=1}^{n_\mathrm{m}} \chi_{\mathrm{m},i}\, \tilde{\mathbf{u}}_i + \sum_{j=1}^{n_\mathrm{c}} \chi_{\mathrm{c},j}\, \tilde{\tilde{\mathbf{u}}}_j}_{\mathbf{u}'} & \forall \mathbf{x} \in R' \end{cases} \tag{22}$$

which is discretized in the same manner as described above, see also Figure 11.

Note that due to constraint equation $\mathbf{u}_m = \mathbf{u}_s$ and the compatibility of the large-scale solution $\bar{\mathbf{u}}_m = \bar{\mathbf{u}}_s$ an assumption with respect to the continuity of the small-scale solution has to be made $\mathbf{u}'_m = \mathbf{u}'_s$; due to the common Lagrange multiplier this leads for a matrix crack to

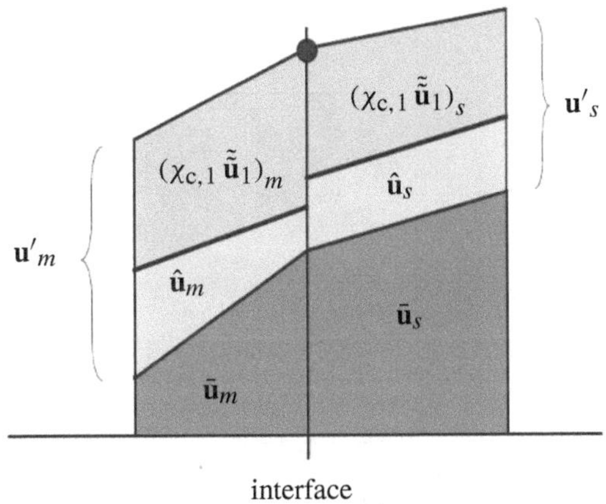

Fig. 12. Crack across interface (1D).

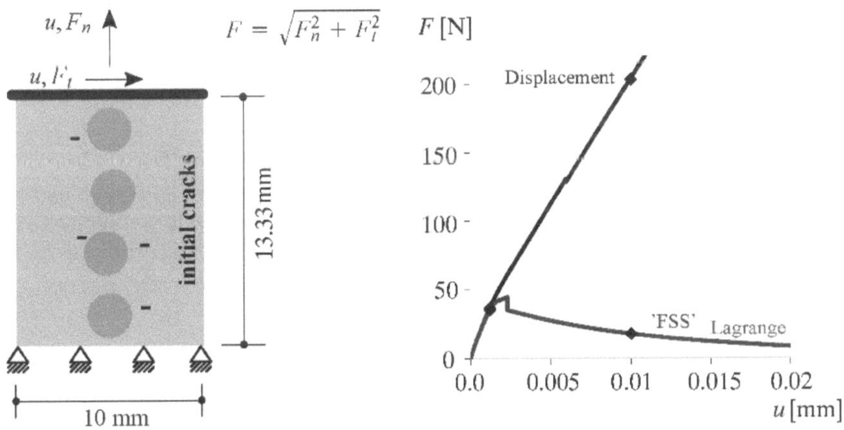

Fig. 13. Plate under tension and shear: system and load-displacement curves.

$$\left(\hat{\mathbf{u}} + \chi_{c,1}\,\tilde{\bar{\mathbf{u}}}_1\right)_m = \left(\hat{\mathbf{u}} + \chi_{c,1}\,\tilde{\bar{\mathbf{u}}}_1\right)_s \tag{23}$$

In other words the individual parts $\hat{\mathbf{u}}$ and $\chi_{c,1}\,\tilde{\bar{\mathbf{u}}}_1$ are in general not compatible (see Figure 12 for 1D case) unless additional multipliers are introduced.

Numerical Examples [13]

• Plate under tension and shear

The combination of XFEM and the multiscale concept is shown in example, Figure 13. Both constituents are assumed to be linear elastic: matrix ($E = 35000\,\mathrm{N/mm^2}$, $\nu = 0.2$) and fibers ($E = 70000\,\mathrm{N/mm^2}$, $\nu = 0.2$).

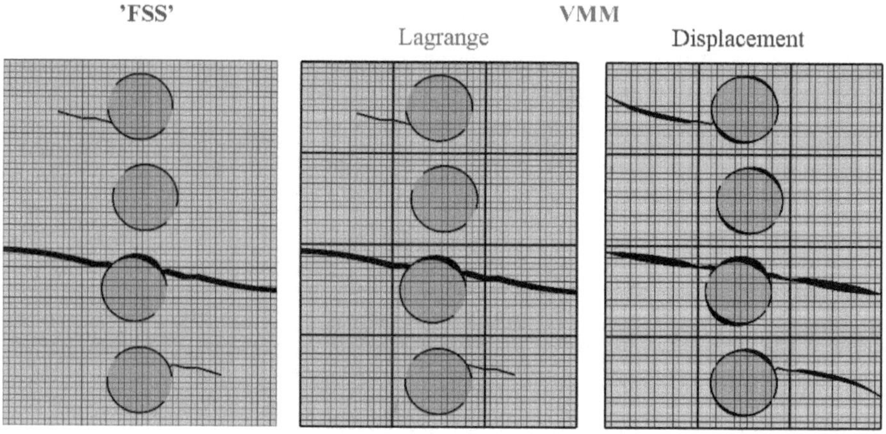

Fig. 14. Plate under tension and shear : crack pattern for $u = 0.0012$ mm.

The parameters for the cohesive crack models are: matrix $f_t = 4.0\,\text{N/mm}^2$, $\text{d}_T = 1.0\,\text{N/mm}^3$, $\text{h}_T = 0.01/\text{mm}$; interface $c_n = 1.0\text{N/mm}^2$, $\beta = 2.0$, $G_f = 0.05\,\text{N/mm}$.

The plane strain structure is discretized by 3×4 bilinear large scale elements, each subdivided into 15×15 smale scale elements. The load-displacement response and the crack pattern (Figure 14) of FSS and the multiscale solution using Lagrange multipliers can hardly be distinguished; again a displacement constraint $\mathbf{u}' = \mathbf{0}$ fails.

• **Deep beam with notch**

The notched plain concrete beam, Figure 15, has been analytically and experimentally investigated by John and Shah [17]. The chosen material parameters are $E = 30000\,\text{N/mm}^2$, $\nu = 0.2$, $f_t = 3.0\,\text{N/mm}^2$, $G_f = 0.13\,\text{N/mm}$, $\text{d}_T = 1.0\,\text{N/mm}^3$. Each large-scale element is discretized by 15×16 small-scale elements.

Results of FSS, Lagrange method and experiments [17] are shown in Figures 15 and 16.

4 Optimal Fiber Layout [18–20]

In this part a conceptually very general optimization strategy for fiber reinforced composites is described. As said in the introduction we again select Fiber Reinforced Concrete (FRC) with long carbon or AR-glass fibers and a fine grade concrete (mortar) matrix for this investigation.

Different from the previous studies we analyse plane beam- and wall-like structures where the layout of the long fibers is the target of optimization.

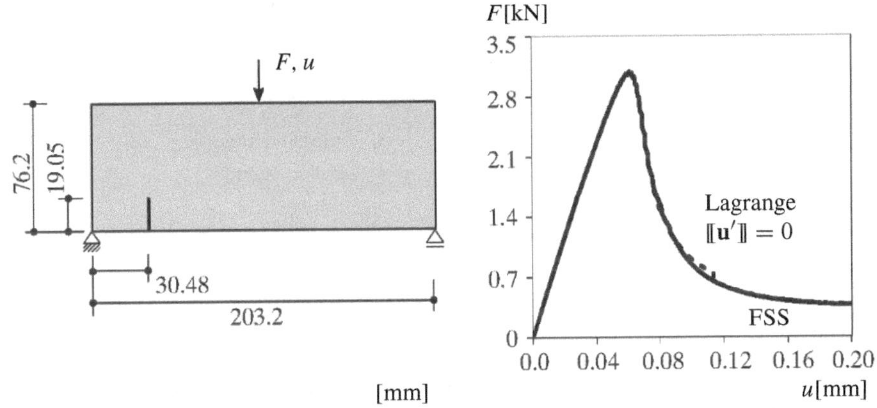

Fig. 15. Notched deep beam: system and load-displacement curve.

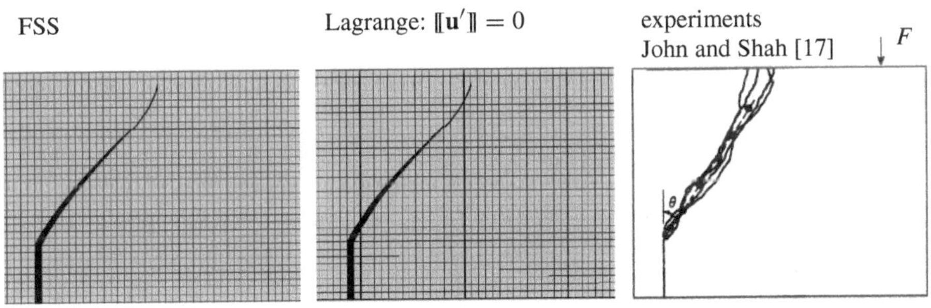

Fig. 16. Notched deep beam: Crack pattern.

4.1 Material Model

The underlying material model utilize a gradient enhanced continuum damage model for matrix and fibers, see Section 3.1, supplemented by an interface model [21] depicted in Figure 17 and defined by the bond stress–slip ($\sigma_L^i - u_L^i$) relation

$$\sigma_L^i = \tilde{w} \cdot \left\{ b + (1-b) \cdot \left(\frac{1}{1+\tilde{w}^R} \right)^{\frac{1}{R}} \right\} \cdot \sigma_0, \quad \text{for} \quad u_L^i \leq w_1 \tag{24}$$

$\tilde{w} = u_L^i / w_0$ is the normalized slip, u_L^i the slip length and R the curvature at w_1. $b = k_2/k_1$ and $\sigma_0 = k_1 \cdot w_0$ describe parameters to calculate the stress. The adhesion strength $\sigma_m = \sigma_{m,0}\,\psi$ and the friction bond strength $\sigma_f = \sigma_{f,0}\,\psi$ depend on $1 < \psi < 2$

$$\psi = 1 + \tanh\left[\alpha_r \frac{\sigma_R}{0.1 f_c} - \alpha_f\, \nu\, \varepsilon_s \left(1 - \frac{r^2}{(r+h)^2} \right)^{-1} \right] \tag{25}$$

$\sigma_{m,0}$ and $\sigma_{f,0}$ denote initial strengths; r is the fiber radius. α_r and α_f are constraints describing the lateral deformation of a fiber; f_c is the uniaxial compressive strength of

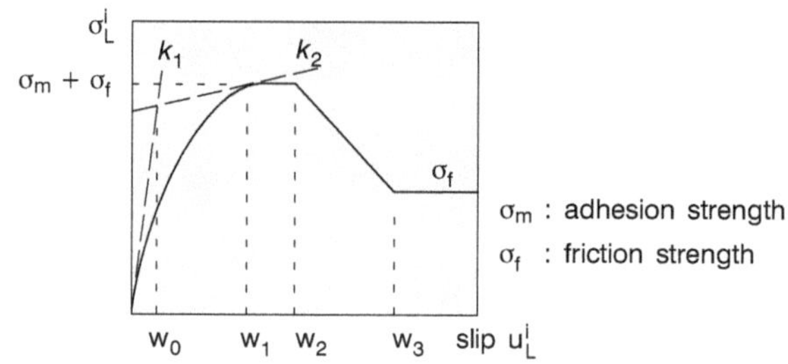

Fig. 17. Discrete bond model.

concrete, ε_s the uniaxial strain and σ_R the stress perpendicular to a fiber, see also [21]. The model derived for a fiber in 3D is modified to represent the fiber in 2D leading to the original interface surface.

4.2 Optimization Concept

As said in the beginning friction in the interface between the two rather brittle constituents is an important feature of FRC. For a given fiber volume $\hat{V}$ we minimize the negative value of structural ductility.

$$\text{minimize} \quad f(\mathbf{s}) = -\int_{\Omega}\int_{\hat{\varepsilon}} \sigma \, d\varepsilon \, d\Omega \tag{26a}$$

$$\text{subject to} \quad h(\mathbf{s}) = \int_{\Omega} \mathbf{s} \, d\Omega - \hat{V} = \mathbf{0} \tag{26b}$$

$$s_L \leq s_i \leq s_U \,, \quad i = 1, \ldots, n_s \tag{26c}$$

where s_L and s_U are lower and upper bounds of design variables $\mathbf{s}$. $\hat{\varepsilon}$ stands for the strain tensor after convergence at each time step. An optimality criteria method [26] and the method of the moving asymptotes [30] are applied to solve the optimization problem. For the sensitivity analysis a variational semi-analytical direct method is used.

Three concepts are followed, a multiphase material optimization and shape optimization of the fiber geometry as well as their combination.

4.3 Multiphase Material Optimization

In analogy to the *Solid Isotropic Microstructure with Penalization of intermediate densities* (SIMP) approach with two or more phases (e.g. void and one phase), where the intermediate stages of material parameters are defined for regularization purposes only, here the intermediate values represent a physically existing mixture of two or more materials, namely fibers and concrete matrix in a design area, e.g. one finite element or one fiber layer, see Figure 19. In the present study a material parameter ζ

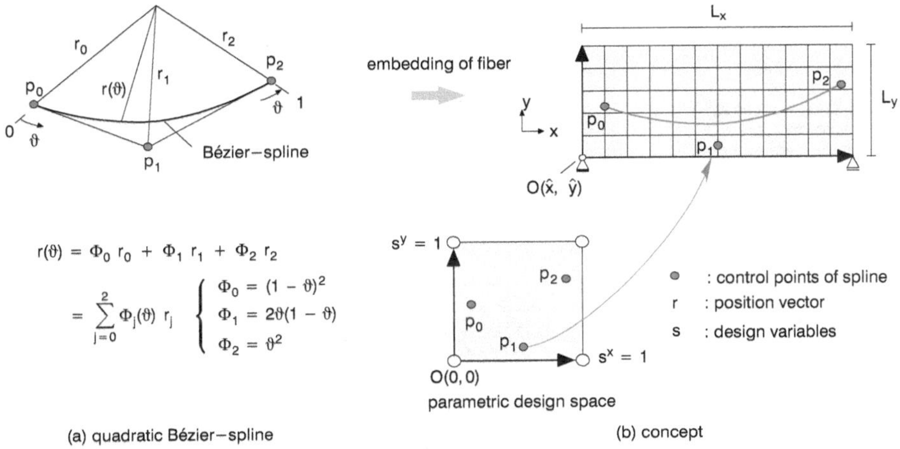

Fig. 18. Geometry of fibers.

which stands for Young's modulus E, κ_0, α and β, see eq. (13), are interpolated by power laws, for example for $\zeta = E$

$$\begin{aligned} \zeta &= \left(1 - s^{\hat{\eta}}\right)\zeta_1 + s^{\hat{\eta}}\zeta_2 && \text{for} \quad \zeta_1 \leq \zeta_2 \\ \zeta &= (1-s)^{\hat{\eta}}\,\zeta_1 + \left[1 - (1-s)^{\hat{\eta}}\right]\zeta_2 && \text{for} \quad \zeta_1 > \zeta_2 \end{aligned} \tag{27}$$

where $\zeta_1 = E_1$ and $\zeta_2 = E_2$ are material parameters of phase 1 (e.g. concrete) and phase 2 (e.g. fiber). The design variable s indicate a 'volume fraction' of the constituent materials in a design area. The exponent $\hat{\eta}$ is a fitting parameter. Similar interpolation formulas may be also derived by homogenization or by experiments.

Further details of the approach are described in [18, 19].

4.4 Shape Optimization of Fiber Geometry

The geometry of a continuous long fiber is defined by Bézier-splines in global coordinates, Figure 18. ϑ $(0 \leq \vartheta \leq 1)$ is a local coordinate of the spline. The entire structure is defined in a parametric space s $(0 \leq s \leq 1)$; the normalized coordinates of control points are chosen as design variables, see [18, 20]. The shape and multiphase material optimization can be combined allowing to adapt the fiber content in the design area.

4.5 Numerical Examples

Multiphase Material Optimization for a Beam

The beam, half of it shown in Figure 19 together with the material data, is decomposed into the concrete domain and four design layers which are the target of optimization. The fiber volume is set to 4.5% for the entire structure, i.e. 50% of the layers

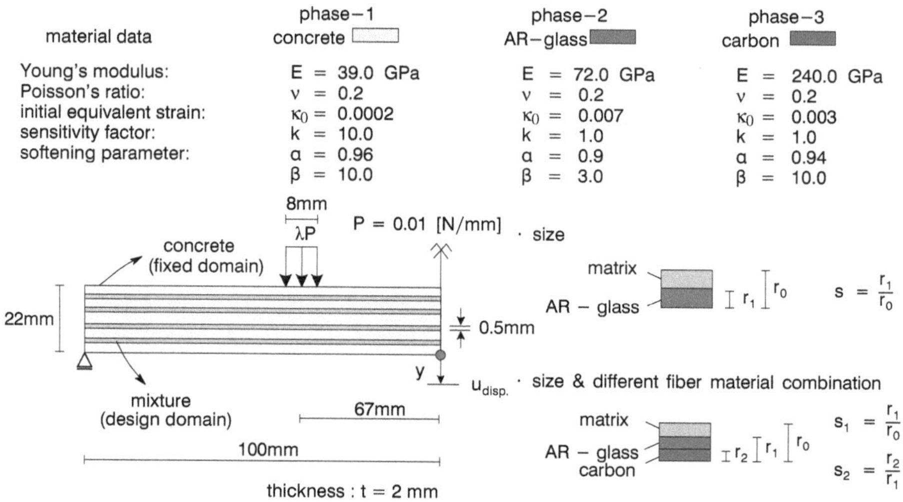

Fig. 19. Fiber reinforced beam.

are initially filled by fiber materials. Two studies have been performed as indicated in Figure 19. First only AR-glass fibers are used leading to one design variable s per layer.

In the second investigation we allow for glass and carbon fibers introducing two design variables s_1 and s_2 per layer. The results are given in Figure 20 where the original layout and that of the two optimized structures are shown.

In both cases the fiber material moves, as expected, to the lower part increasing the overall ductility by 278 and 410%, respectively. It is interesting to see that in the second case carbon is maintained increasing the stiffness whereas glass leads to an increase in ductility.

Multiphase and Shape Optimization for Deep Beam

Each of the four AR-glass fibers for the deep beam (Figure 21) with the same material parameters of the previous example are defined by the three geometrical design parameters s_1, s_2 and s_3.

The first study (Figure 22a) is based on pure shape optimization of fibers with a thickness of 0.4 mm corresponding to a total fiber content of 1.4%. The fourth fiber moves to the upper edge and does not contribute to the increase in ductility of 37%.

This value could be further increased to 45% if shape and multiphase optimization are combined allowing also variable thicknesses of fibers. Starting with an initial value $r_1 = 0.4$ mm the fiber thicknesses are allowed to be within $0\,\text{mm} < r_1 < 0.8$ mm, eventually leading to 0.5, 0.7, and 0.8 mm and a vanishing fourth fiber.

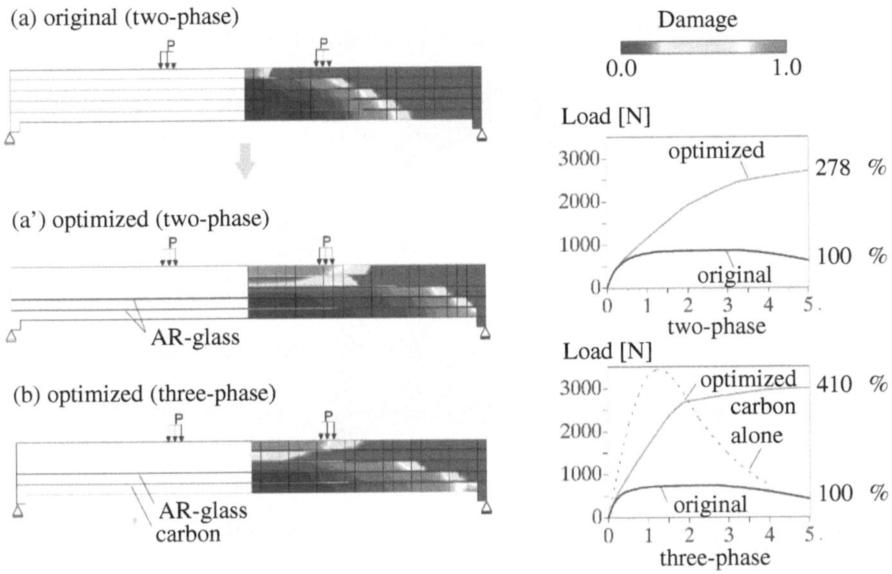

Fig. 20. Fiber reinforced beam based on damage formulation. (For a color version of this image, see [18].)

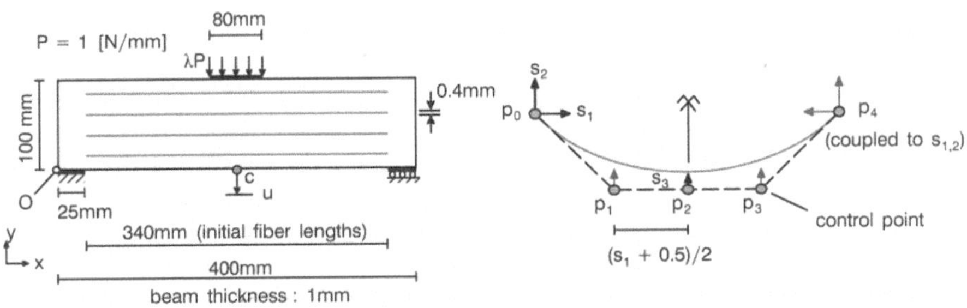

Fig. 21. Deep beam: structure & fiber geomety by quadratic Bézier-splines.

5 Conclusions

The present multiscale model turned out to be a promising and efficient scheme for long range failure mechanisms where traditional homogenization approaches are not applicable. The concept is currently transferred to three-dimensional problems. Applying different constitutive models on large- and small-scale levels is another option to be investigated.

Structures composed of several material phases like composites are ideally suited for material optimization based on nonlinear constitutive behaviour. We applied multiphase material and shape optimization to Fiber Reinforced Concrete maximizing the overall structural ductility. An extension to material design on a small scale with a spatial variation of the microstructure is a further possibility.

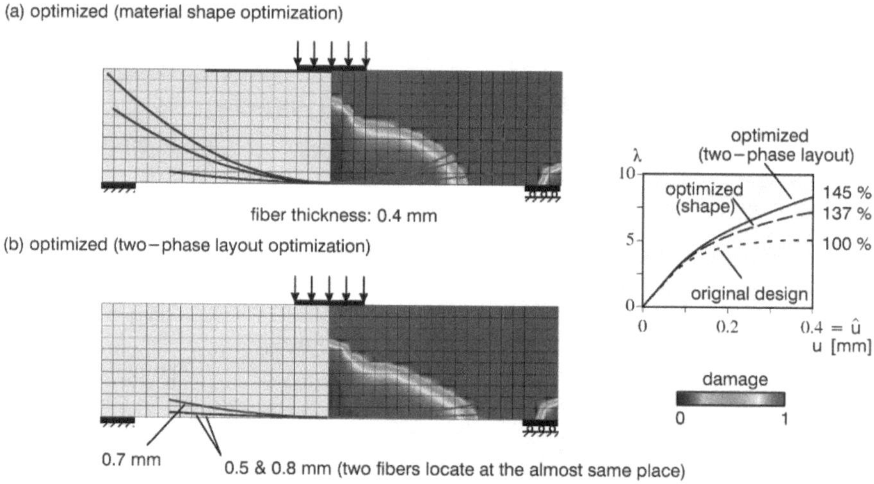

Fig. 22. Deep beam: optimization results. (For a color version of this image, see [18].)

Acknowledgement

The present study is supported by a grant of the foundation 'Deutsche Forschungsgemeinschaft' (DFG) within the German-Dutch Research Unit FOG 509. This support is gratefully acknowledged.

References

1. J.C. Brewer and P.A. Lagace, Quadratic stress criterion for initiation of delamination. *Journal of Composite Materials*, 22:1141–1155, 1988.
2. I. Bruss and E. Ramm, Modeling of continuous curved crack surfaces in three dimensional composite structures. In *Proc. of the Int. Conf. on Extended Finite Element Methods – Recent Developments and Applications*, pp. 23–26, http://www.ibb.uni-stuttgart.de/forschung/xfem/index.html, 2009.
3. M. Curbach and F. Jesse, Eigenschaften und Anwendung von Textilbeton. *Beton- und Stahlbetonbau*, 104:9–16, 2009.
4. C. Daux, N. Moës, J. Dolbow, N. Sukumar, and T. Belytschko, Arbitrary branched and intersecting cracks with the extended finite element method. *Int. J. Numer. Meth. Engrg.*, 48:1741–1760, 2000.
5. C. Farhat, A method of finite element tearing and interconnecting and its parallel solution algorithm. *Int. J. Numer. Meth. Engrg.*, 32:1205–1227, 1991.
6. C. Farhat, I. Harari, and U. Hetmaniuk, The discontinuous enrichment method for multiscale analysis. *Comp. Meth. Appl. Mech. Engrg.*, 192:3195–3209, 2003.
7. J. Fish and Z. Yuan, Multiscale enrichment based on partition of unity for nonperiodic fields and nonlinear problems. *Comput. Mech.*, 40:249–259, 2007.
8. J. Fish, Multiscale modeling and simulation of composite materials and structures. In: R. de Borst and E. Ramm (Eds.), *Multiscale Methods in Computational Mechanics*, Lecture Notes in Applied and Computational Mechanics. Springer, Dordrecht, 2010 (this volume).

9. V. Kouznetsova, M.G.D. Geers, and W.A.M. Brekelmans, Multi-scale constitutive modelling of heterogeneous materials with a gradient-enhanced computational homogenization scheme. *Int. J. Numer. Meth. Engrg.*, 54:1235–1260, 2002.
10. T.J.R. Hughes, G.R. Feijóo, L. Mazzei, and J.B. Quincy, The variational multiscale method – A paradigm for computational mechanics. *Comp. Meth. Appl. Mech. Engrg.*, 166:3–24, 1998.
11. A. Hund, Hierarchische Mehrskalenmodellierung des Versagens von Werkstoffen mit Mikrostruktur, PhD Dissertation, Institut für Baustatik und Baudynamik, Universität Stuttgart, http://elib.uni-stuttgart.de/opus/frontdoor.php?source_opus=3108, 2007.
12. A. Hund and E. Ramm, Locality constraints within multiscale model for non-linear material behaviour. *Int. J. Numer. Meth. Engrg.*, 70:1613–1632, 2007.
13. T. Hettich, Diskontinuierliche Modellierung zur Versagensanalyse von Verbundmaterialien. PhD Dissertation, Institut für Baustatik und Baudynamik, Universität Stuttgart, 2007.
14. T. Hettich, A. Hund, and E. Ramm, Modeling of failure in composites by X-FEM and level sets within a multiscale framework. *Comp. Meth. Appl. Mech. Engrg.*, 197:414–424, 2008.
15. T. Hettich and E. Ramm, Interface material failure modeled by the extended finite-element method and level sets. *Comp. Meth. Appl. Mech. Engrg.*, 195:4753–4767, 2006.
16. F. Jesse, N. Will, M. Curbach, and J. Hegger, Load-bearing behavior of textile-reinforced concrete. In: A. Dubey (Ed.), *Textile-Reinforced Concrete*, Proceedings of ACI Fall Convention, Kansas City, November 2005.
17. R. John and S.P. Shah, Mixed-mode fracture of concrete subjected to impact loading. *ASCE J. Struct. Engrg.*, 116:585–602, 1990.
18. J. Kato, Material optimization for fiber reinforced composites applying a damage formulation, PhD Dissertation, Institut für Baustatik und Baudynamik, Universität Stuttgart, http://elib.uni-stuttgart.de/opus/volltexte/2010/5162/, 2010.
19. J. Kato, A. Lipka, and E. Ramm, Multiphase material optimization for fiber reinforced composites with strain softening. *Struct. Multidisc. Optim.*, 39:63–81, 2009.
20. J. Kato and E. Ramm, Optimization of fiber geometry for fiber reinforced composites considering damage. In *Finite Elements in Analysis and Design*, in press, 2010.
21. M. Krüger, J. Ozbolt and H.W. Reinhardt, A discrete bond model for 3D analysis of textile reinforced and prestressed concrete elements. *Otto-Graf-J., University of Stuttgart*, 13:111–128, 2002.
22. J. Mazars and G. Pijaudier-Cabot, Continuum damage theory – Application to concrete. *J. Engrg. Mech. ASCE*, 115:345–365, 1989.
23. N. Moës, M. Cloirec, P. Cartraud, and J.-F. Remacle, A computational approach to handle complex microstructure geometries. *Comp. Meth. Appl. Mech. Engrg.*, 192:3163–3177, 2003.
24. N. Moës and T. Belytschko, Extended finite element method for cohesive crack growth. *Engrg. Fracture Mech.*, 69:813–833, 2002.
25. H.S. Park and W.K. Liu, An introduction and tutorial on multiple-scale analysis in solids. *Comp. Meth. Appl. Mech. Engrg.*, 193:1733–1772, 2004.
26. S.N. Patnaik, J.D. Guptill, and L. Berke, Merits and limitations of optimality criteria method for structural optimization. *Int. J. Numer. Meth. Engrg.*, 38:3087–3120, 1995.
27. R.H.J. Peerlings, Enhanced damage modelling for fracture and fatigue. PhD Thesis, Eindhoven University of Technology, the Netherlands, 1999.
28. R.H.J. Peerlings, M.G.D. Geers, R. de Borst, and W.A.M. Brekelmans, A critical comparison of nonlocal and gradient-enhanced softening continua, *Int. J. Solids & Structures*, 38:7723–7746, 2001.

29. J.J.C. Remmers, R. de Borst, and A. Needleman, A cohesive segments method for the simulation of crack growth. *Comput. Mech.*, 31:69–77, 2003.
30. K. Svanberg, The method of moving asymptotes – A new method for structural optimization. *Int. J. Numer. Meth. Engrg.*, 24:359–373, 1987.
31. N. Sukumar, D.L. Chopp, N. Moës, and T. Belytschko, Modeling holes and inclusions by level sets in the extended finite-element method. *Comp. Meth. Appl. Mech. Engrg.*, 190:6183–6200, 2001.
32. G. Ventura, E. Budyn, and T. Belytschko, Vector level sets for the description of propagating cracks in finite elements. *Int. J. Numer. Meth. Engrg.*, 58:1571–1592, 2003.
33. J.H.P. Vree, W.A.M. Brekelmans, and M.A.J. van Gils, Comparison of nonlocal approaches in continuum damage mechanics. *Computers & Structures*, 55:581–588, 1995.
34. G.N. Wells and L.J. Sluys, A new method for modelling cohesive cracks using finite elements. *Int. J. Numer. Meth. Engrg.*, 50:2667–2682, 2001.

Computational Multiscale Model for NATM Tunnels: Micromechanics-Supported Hybrid Analyses

Stefan Scheiner, Bernhard Pichler, Christian Hellmich and Herbert A. Mang

Institute for Mechanics of Materials and Structures, Vienna University of Technology, Karlsplatz 13/202, A-1040 Vienna, Austria; herbert.mang@tuwien.ac.at

Abstract Knowledge of the stress state in shotcrete tunnel shells is a necessary requirement to assess the safety of these structures. Estimation of these stresses from measured 3D tunnel shell displacement requires material models for shotcrete. Preferably such models should be able to deal with on-site adaptations of the water-cement and aggregate-cement ratios. Therefore, we recall the fundamentals of continuum micromechanics and its application to shotcrete in order to upscale elastic and strength properties from the micron scale (where unhydrated cement can be discerned from its reaction products with water, called hydrates), via the intermediate scale of cement paste, to the shotcrete scale. Comparison of model predictions with experimental data shows that the elasticity and strength evolutions of hydrating shotcrete can be predicted reasonably well from mixture- and hydration-independent elastic properties of aggregates, clinker, hydrates, water, and air, and from strength properties of hydrates. At the structural level, the micromechanics models are combined with 3D displacement measurements of a shotcrete tunnel shell. This hybrid approach provides insight into the load-carrying behavior of the tunnel, and it allows for a safety assessment of the shell's structural integrity.

Key words: Elasticity, hydration, micromechanics, New Austrian Tunneling Method, strength, tunneling

1 Introduction

If a high degree of flexibility is required during the tunnel excavation process (e.g. in difficult ground conditions or in urban areas), the New Austrian Tunneling Method (NATM), pioneered by Rabcewicz [63], is well suited. When driving tunnels according to the NATM, after excavation of a cross-section of a tunnel, shotcrete is applied onto the tunnel walls, constituting a thin and flexible shell. Accordingly, the excavation mode (type of cross section, sequence of excavation steps, time span between excavation and shotcreting) and the shotcrete mixture characteristics (such as water-cement ratio w/c and aggregate-cement ratio a/c) are key design parameters in NATM-tunneling: for example, too stiff shells or shells put up too early, trying to restrict stress redistributions in the ground caused by the excavation, may be destroyed, whereas too soft shells or shells put up too late may not be able to prevent

R. de Borst and E. Ramm (eds.), *Multiscale Methods in Computational Mechanics*, Lecture Notes in Applied and Computational Mechanics 55, DOI 10.1007/978-90-481-9809-2_16,

loss of stability of the opening [58]. The stability of the opening is often improved by reinforcing the surrounding ground by bolts. The success of the NATM is primarily based on the in situ realization of its concept by experienced engineers. This realization is supported by elaborate measurement systems on site (see, e.g., [64]). These systems have been continuously improved, and nowadays, optical monitoring of 3D displacements of the tunnel shell [72, 76] is the golden standard. From changes of characteristic trend parameters, extracted from 3D displacement measurements, it is possible to estimate changes in the geological structure of the soil or rock [73].

Besides the monitoring of variations in the geological conditions, knowledge on the stresses in the tunnel shell is of great importance, as to assess its safety against severe cracking or failure. Therefore, the shell forces can be derived from quasi-instantaneous deformations of holes cored out of the tunnel shell [10]. Since this semi-destructive method is restricted in applicability and versatility, it turned out to be more beneficial to back-calculate, on the basis of elaborate shotcrete material models, shotcrete stresses from shotcrete strains [71]. Preferentially, the strains are gained from the aforementioned 3D displacement vector measurements, as proposed by [66] on the basis of relative movements of pairs of measurement points. At a more elaborate state, the strain fields can be estimated on the basis of a hybrid method [23, 27], in which displacement vector fields are approximated from measured displacement vectors at discrete points of the tunnel shell and in which these fields are prescribed as boundary values for a three-dimensional Finite Element structural model of the tunnel shell. This type of method has also been extended towards segmented tunnel linings [41, 49], and use of approximations from thin shell theory [42, 51] turned out to be beneficial for day-to-day use of the hybrid method in engineering practice [9].

Underlying shotcrete models need to reasonably represent the creep behavior of the material [46, 55, 65, 71], and consideration of hydration-induced, thermal, and chemical strains further improves the reliability of shell force estimation [29, 30, 46, 74]. All aforementioned material models rely on material properties, which are shotcrete mixture-specific. Hence, any change in mixture (e.g., variations in water-cement ratio, as often encountered in situ) can only be considered if additional experiments (related to strength, creep, and shrinkage) are performed on samples with the modified shotcrete composition. Practically, this is often unfeasible so that the engineers on site usually agree on a ‘typical' shotcrete for a tunnel track. Clearly, this situation is unsatisfactory. As a remedy, the shotcrete composition needs to be incorporated into the shotcrete material models, within a micromechanical framework: This was recently shown for the case of elasticity [26] – following earlier work on concrete [7] and on bone [28]. While elasticity is an important factor for attracting stresses into the tunnel shell, strength is the key factor to understand whether these stresses significantly compromise the shell's safety. Also the strength evolution is influenced by the w/c and a/c ratios [57], and this here motivates us to present recent developments on a micromechanical model for shotcrete elasticity and strength, and to employ this new model for a hybrid analysis of an NATM-tunnel shell, see [60, 61].

In Section 2, we briefly summarize fundamentals of continuum micromechanics of microheterogeneous materials. Sections 3 and 4 are devoted to micromechanics-based modeling of cement paste and shotcrete, respectively. Section 5 deals with quantitative experimental validation of these models. As to provide further support for our strength model, we show, in Section 6, that the presented predictions of final shotcrete strength are consistent with predictions from the famous Feret's formula. In Section 7, we describe a hybrid analysis of the shotcrete tunnel shell of the Sieberg tunnel. To this end, we focus on a measurement cross-section of that tunnel and we combine measurements of 3D displacements of the shotcrete tunnel shell with a material model for shotcrete including the presented micromechanical developments. In Section 8, we draw conclusions related both to the presented topic of shotcrete homogenization and to the more general use of continuum micromechanics as a theoretical framework for modeling microheterogeneous materials.

2 Continuum Micromechanics of Microheterogeneous Materials

State-of-the-art modeling of microheterogeneous media by means of continuum micromechanics is based on groundbreaking works of Hill [33, 34] and Zaoui [83, 84], see also the collections of pioneering contributions edited by Suquet [78] and by Dormieux and Ulm [14]. In recent years, continuum micromechanics was successfully applied to many classes of microheterogeneous materials, including bone [19, 24, 28], polymers [48], cementitious materials [7, 11, 26, 60, 67, 68, 75], wood [35, 36], and ceramics [52, 70].

2.1 Representative Volume Element (Separation of Scales)

In continuum micromechanics [84], a material is understood as a macro-homogeneous but micro-heterogeneous body filling a representative volume element (RVE) with characteristic length ℓ, with $\ell \ll d$, where d stands for the characteristic length of inhomogeneities within the RVE, and with $\ell \ll \mathcal{L}$, where $\mathcal{L}$ stands for the characteristic lengths of geometry or loading of a structure built up by the material defined on the RVE. In general, the microstructure within an RVE is too complicated to describe it in full detail. Therefore, quasi-homogeneous subdomains with known physical quantities (such as volume fractions or elastic properties) are identified. They are called material phases. The homogenized mechanical behavior of the overall material, i.e., the relation between homogeneous deformations acting on the boundary of the RVE and resulting (average) stresses, or the ultimate stresses sustainable by the RVE, can then be estimated from the mechanical behavior of the aforementioned homogeneous phases (representing the inhomogeneities within the RVE), their dosages within the RVE, their characteristic shapes, and their interactions. If a single phase exhibits a heterogeneous microstructure itself, its mechanical behavior can be estimated by introduction of an RVE within this phase, with dimensions $\ell_2 \leq d$, comprising again smaller phases with characteristic length $d_2 \ll \ell_2$, and so on, leading to a multistep homogenization scheme.

2.2 Homogenization of Elasticity

Within the volume Ω of the RVE, we consider field equations of linear elasticity, i.e. a linear constitutive relation between stress tensors $\boldsymbol{\sigma}$ and strain tensors $\boldsymbol{\varepsilon}$, equilibrium conditions, and linear geometric equations reading as

$$\text{in } \Omega: \quad \boldsymbol{\sigma}(\mathbf{x}) = \mathbf{C}(\mathbf{x}) : \boldsymbol{\varepsilon}(\mathbf{x}), \quad \operatorname{div} \boldsymbol{\sigma}(\mathbf{x}) = 0, \quad \boldsymbol{\varepsilon}(\mathbf{x}) = \frac{1}{2}\left[\nabla \boldsymbol{\xi}(\mathbf{x}) + {}^{t}\nabla \boldsymbol{\xi}(\mathbf{x})\right], \tag{1}$$

where $\mathbf{x}$ denotes the position vector, $\mathbf{C}(\mathbf{x})$ the field of fourth-order elasticity tensors, and $\boldsymbol{\xi}(\mathbf{x})$ the displacement vector field. At the boundary $\partial\Omega$ of the RVE, we impose a uniform macroscopic strain state $\mathbf{E}$ in terms of displacement vectors $\boldsymbol{\xi}$ [22],

$$\text{on } \partial\Omega: \ \boldsymbol{\xi}(\mathbf{x}) = \mathbf{E} \cdot \mathbf{x} \, . \tag{2}$$

The geometric compatibility of the microscopic strain field $\boldsymbol{\varepsilon}(\mathbf{x})$ with boundary condition (2) implies the following strain average rule

$$\mathbf{E} = \frac{1}{\Omega} \int_{\Omega} \boldsymbol{\varepsilon}(\mathbf{x}) \, \mathrm{d}V = \sum_{p} f_p \, \boldsymbol{\varepsilon}_p \tag{3}$$

where p denotes an index extending over all phases of the considered RVE, f_p is the volume fraction of phase p, and $\boldsymbol{\varepsilon}_p$ stands for the first-order (spatial average) phase strains defined as

$$\boldsymbol{\varepsilon}_p = \frac{1}{\Omega_p} \int_{\Omega_p} \boldsymbol{\varepsilon}(\mathbf{x}) \, \mathrm{d}V \, . \tag{4}$$

In (4), Ω_p denotes the subvolume of the RVE occupied by phase p. Analogous to (3), macroscopic stresses $\boldsymbol{\Sigma}$ are set equal to the spatial average of the equilibrated local stresses $\boldsymbol{\sigma}(\mathbf{x})$ inside the RVE,

$$\boldsymbol{\Sigma} = \frac{1}{\Omega} \int_{\Omega} \boldsymbol{\sigma}(\mathbf{x}) \, \mathrm{d}V = \sum_{p} f_p \, \boldsymbol{\sigma}_p \, , \tag{5}$$

with $\boldsymbol{\sigma}_p$ as the first-order (spatial average) phase stresses.

Linearity of all field equations (1) implies a linear strain concentration rule

$$\boldsymbol{\varepsilon}_p = \mathbf{A}_p : \mathbf{E} \, , \tag{6}$$

with $\mathbf{A}_p$ as the fourth-order strain concentration tensor of phase p. Combination of the elastic constitutive law of phase p,

$$\boldsymbol{\sigma}_p = \mathbf{C}_p : \boldsymbol{\varepsilon}_p \, , \tag{7}$$

with the strain concentration rule (6), and insertion of the resulting expression for the phase stresses $\boldsymbol{\sigma}_p$ into the stress average rule (5), delivers a relation between macrostress $\boldsymbol{\Sigma}$ and macrostrain $\mathbf{E}$. Comparison of this relation with the macroscopic elastic law $\boldsymbol{\Sigma} = \mathbb{C}^{\text{hom}} : \mathbf{E}$ allows for identification of the homogenized elasticity tensor as

$$\mathbb{C}^{\text{hom}} = \sum_p f_p \, \mathbb{C}_p : \mathbb{A}_p \, . \tag{8}$$

Eq. (8) highlights that knowledge of phase strain concentration tensors $\mathbb{A}_p$ allows for homogenization (upscaling) of phase stiffnesses to the homogenized elasticity tensor. Unfortunately, the microstructure within an RVE is usually rather complicated, which renders direct access to phase strain concentration tensors $\mathbb{A}_p$ impossible. As a remedy, these tensors can be estimated by means of the classical matrix-inclusion problems of Eshelby [15] and Laws [45], for details see Zaoui [84] and Benveniste [6], as

$$\mathbb{A}_p = \left[\mathbb{I} + \mathbb{P}_p^0 : (\mathbb{C}_p - \mathbb{C}^0)\right]^{-1} : \left\{\sum_q f_q \left[\mathbb{I} + \mathbb{P}_q^0 : (\mathbb{C}_q - \mathbb{C}^0)\right]^{-1}\right\}^{-1} \tag{9}$$

where $\mathbb{P}_p^0$ denotes the fourth-order Hill tensor accounting for the characteristic shape of phase p embedded in a matrix with stiffness $\mathbb{C}^0$. Insertion of (9) into (8) delivers the sought estimate for the homogenized elastic stiffness tensor as

$$\mathbb{C}^{\text{hom}} = \sum_p f_p \mathbb{C}_p : \left[\mathbb{I} + \mathbb{P}_p^0 : (\mathbb{C}_p - \mathbb{C}^0)\right]^{-1} : \left\{\sum_q f_q \left[\mathbb{I} + \mathbb{P}_q^0 : (\mathbb{C}_q - \mathbb{C}^0)\right]^{-1}\right\}^{-1} . \tag{10}$$

In Eqs. (9) and (10), $\mathbb{P}_p^0$ denotes the fourth-order Hill tensor, accounting for the characteristic shape of phase p embedded in a matrix with stiffness $\mathbb{C}^0$. The choice of $\mathbb{C}^0$ describes the interactions between material phases. The choice $\mathbb{C}^0 = \mathbb{C}^{\text{hom}}$ refers to so-called self-consistent schemes and relates to mutual interaction between all phases (as in a polycrystal). The choice of $\mathbb{C}^0$ as one (matrix) phase, $\mathbb{C}^0 = \mathbb{C}_{\text{m}}$, refers to so-called Mori-Tanaka schemes and relates to a morphology where one or several inclusion phases are embedded within one (contiguous) matrix phase (as is the case for a composite material).

2.3 Homogenization of Strength

In recent years, it was shown that brittle failure modes of cementitious materials [60,61], of wood [37], and of biomaterials [18] can be modeled within the framework of elastic limit analysis. In this case, microscopic failure of one material phase results in the failure of the entire RVE. However, microscopic phase failure is governed by strain peaks within this phase, rather than by average phase strains defined in (4). To overcome this problem, strain levels corresponding to volumetric and deviatoric strain peaks can be estimated by corresponding quadratic strain averages [12]. They

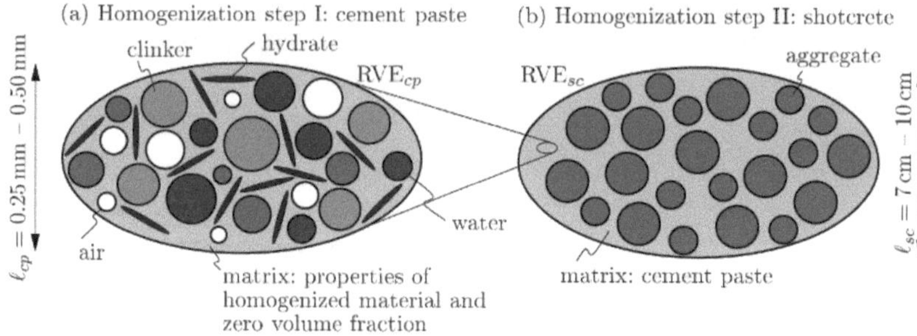

Fig. 1. Micromechanical representation of shotcrete microstructure through a two-step homogenization scheme: 2D sketches of 3D representative volume elements (RVEs) – (a) the polycrystalline RVE of 'cement paste', RVE$_{\text{cp}}$ is built up of clinker, water, needle-shaped hydrates, and air; (b) the RVE of the matrix-inclusion composite 'shotcrete', RVE$_{\text{sc}}$ is composed of a cement paste matrix with aggregate inclusions.

can be directly derived from the elastic energy stored in the RVE, provided that the phases within the RVE exhibit isotropic elastic behavior, characterized by

$$\mathbf{C}_p = 3k_p\mathbf{J} + 2\mu_p\mathbf{K} \tag{11}$$

where $\mathbf{K}$ is the deviatoric part of the fourth-order unity tensor, being defined as $\mathbf{K} = \mathbf{I} - \mathbf{J}$, with $\mathbf{I}$ as the symmetric fourth-order unity tensor with components $I_{ijrs} = 1/2(\delta_{ir}\delta_{js} + \delta_{is}\delta_{jr})$, and with $\mathbf{J} = 1/3(\mathbf{1} \otimes \mathbf{1})$ as the volumetric part of the fourth-order unity tensor, where $\mathbf{1}$ denotes the second-order unity tensor with components δ_{ij} (Kronecker delta), $\delta_{ij} = 1$ for $i = j$, and $\delta_{ij} = 0$ otherwise. In more detail, the quadratic averages of the volumetric strain $\operatorname{tr}\boldsymbol{\varepsilon}(\mathbf{x})$ and of the deviatoric strain $\boldsymbol{\varepsilon}^{\text{dev}}(\mathbf{x}) = \boldsymbol{\varepsilon}(\mathbf{x}) - [\operatorname{tr}\boldsymbol{\varepsilon}(\mathbf{x})/3]\mathbf{1}$ over phase p read as [12]

$$\begin{aligned} \overline{\overline{\varepsilon_p^{\text{vol}}}} &= \sqrt{\frac{1}{\Omega_p}\int_{\Omega_p}\left[\operatorname{tr}\boldsymbol{\varepsilon}(\mathbf{x})\right]^2 \mathrm{d}V} = \sqrt{\frac{1}{f_p}\mathbf{E} : \frac{\partial \mathbf{C}^{\text{hom}}}{\partial k_p} : \mathbf{E}}\,, \\ \overline{\overline{\varepsilon_p^{\text{dev}}}} &= \sqrt{\frac{1}{\Omega_p}\int_{\Omega_p}\frac{1}{2}\,\boldsymbol{\varepsilon}^{\text{dev}}(\mathbf{x}) : \boldsymbol{\varepsilon}^{\text{dev}}(\mathbf{x})\,\mathrm{d}V} = \sqrt{\frac{1}{4f_p}\mathbf{E} : \frac{\partial \mathbf{C}^{\text{hom}}}{\partial \mu_p} : \mathbf{E}}\,. \end{aligned} \tag{12}$$

3 Micromechanics at the Cement Paste Level

3.1 Micromechanical Representation

We consider an RVE$_{\text{cp}}$ of cement paste with characteristic length $\ell_{\text{cp}} = 0.25$–0.50 mm, see Figure 1a, consisting of four different material phases with characteristic dimensions $d_{\text{cp}} = 1$–50 microns (see also Figure 2): (i) a spherical clinker

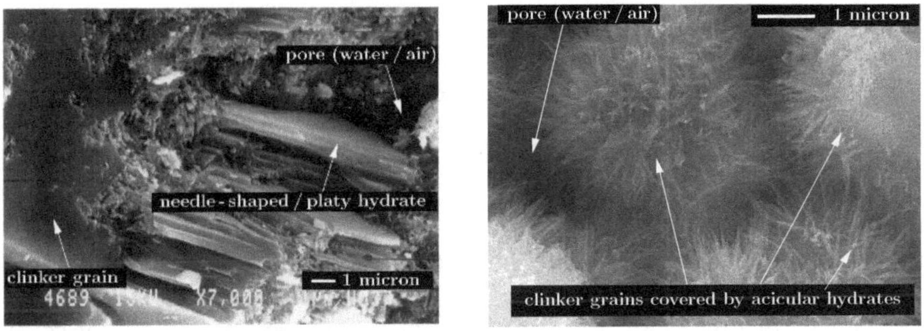

(a) (b)

Fig. 2. Images of cement paste obtained by Scanning Electron Microscopy showing non-spherical hydrates: (a) from Baroghel–Bouny [3], (b) from Tritthart and Häußler [79].

Table 1. Intrinsic mechanical properties of microstructural constituents of shotcrete.

Phase	bulk modulus k [GPa]	shear modulus μ [GPa]	Reference
Clinker	$k_{clin} = 116.7$	$\mu_{clin} = 53.8$	[1]
Water (drained RVE)	$k_{H_2O} = 0.0$	$\mu_{H_2O} = 0.0$	
Water (sealed RVE)	$k_{H_2O} = 2.3$	$\mu_{H_2O} = 0.0$	
Hydrates	$k_{hyd} = 14.1$	$\mu_{hyd} = 8.9$	[80]
Air	$k_{air} = 0.0$	$\mu_{air} = 0.0$	
Aggregate	$k_{agg} = 41.7$	$\mu_{agg} = 19.2$	[54, 82]

phase, (ii) needle-shaped hydrate phases with isotropically distributed spatial orientations, (iii) a spherical water phase, and (iv) a spherical air phase. Thus, the general strain averaging law, Eq. (3), must be adapted to the governing phase geometries,

$$\mathbf{E}_{cp} = \sum_p f_p \, \boldsymbol{\varepsilon}_p + f_{hyd} \int_{\varphi=0}^{2\pi} \int_{\vartheta=0}^{\pi} \boldsymbol{\varepsilon}_{hyd;\varphi,\vartheta} \frac{\sin\vartheta}{4\pi} \, d\vartheta \, d\varphi \, . \tag{13}$$

with $p = $ clin, H_2O, air, with $\boldsymbol{\varepsilon}_{hyd;\varphi,\vartheta}$ as the average phase strains in the hydrate needles, and with an orientation of the hydrate needles defined through spherical coordinates φ and ϑ, see [61] for details.

3.2 Constitutive Behavior of Clinker, Water, Hydrates, and Air

As linear elastic behavior and isotropy is assigned to all phases, Eqs. (7) and (11) describe their constitutive behavior. The phase stiffnesses $\mathbf{C}_p$ are available from experiments, in terms of bulk modulus k_p and shear modulus μ_p (see Table 1): Hydrate stiffnesses follow from nanoindentation experiments [1, 11]; we here consider the elastic properties of low-density calcium silicate hydrates (C-S-H) as representative for all hydrates, see also [26] for details. Clinker properties are taken from [1]. If, upon loading, no water can escape from the RVE_{cp} (sealed or undrained conditions), we consider an elastic water phase with negligible shear stiffness, being a very good approximation of 'undrained conditions in the sense of a fully *poro*-micromechanical theory' [31]; while a zero water stiffness relates to drained conditions. The zero-stiffness air phase is also in a 'drained' state.

Uniaxial compression tests on cement paste (or shotcrete) samples reveal linear elastic behavior until, in the vicinity of the compressive strength, axial strains increase overlinearly with increasing axial stresses. Once the peak load is reached, the material exhibits quasi-brittle failure. Hence, the strength of cement paste (or of shotcrete) can be estimated by means of an elastic limit analysis [60]. In this context, it is assumed that the elastic limits of hydrates govern the elastic limits of cement paste and of shotcrete. In more detail, we assume that each hydrate behaves linear elastically as long as microscopic deviatoric stress peaks remain below a specific critical value. If, due to (compressible uniaxial) macroscopic load increase, this critical value is reached in the most heavily stressed region of the hydrate phase, the elastic limit on the microscale is reached, which, in turn, corresponds to the macroscopic elastic limit of cement paste (or shotcrete), associated with failure of the material (under macroscopic uniaxial compression). In more detail, load bearing capacities of the hydrates are being bounded according to a *von Mises*-type elastic limit criterion [60],

$$\max_{\mathbf{x}\in\Omega_{\text{hyd}}} \sigma^{\text{dev}}(\mathbf{x}) = \max_{\mathbf{x}\in\Omega_{\text{hyd}}} \left[\left(\frac{1}{2} \boldsymbol{\sigma}^{\text{dev}}(\mathbf{x}) : \boldsymbol{\sigma}^{\text{dev}}(\mathbf{x}) \right)^{\frac{1}{2}} \right] \leq \sigma^{\text{dev}}_{\text{crit}} , \tag{14}$$

where $\sigma^{\text{dev}}(\mathbf{x})$ is the norm of stress deviator $\boldsymbol{\sigma}^{\text{dev}}(\mathbf{x})$. The choice of hydrates as weakest locations inside the RVE_{cp} of cement paste was corroborated in [60] where it was shown that the single strength-type value for $\sigma^{\text{dev}}_{\text{crit}} = 26$ MPa allows for prediction of uniaxial cube compressive strength of different cement pastes with w/c ranging from 0.35–0.60 [8, 77], at different degrees of hydration ξ.

3.3 Homogenized Elasticity of Cement Paste

As long as the material phases behave elastically, according to (11), the macroscopic elastic law of an RVE of cement paste reads as

$$\boldsymbol{\Sigma}_{\text{cp}} = \mathbf{C}_{\text{cp}} : \mathbf{E}_{\text{cp}} , \tag{15}$$

with the 'macroscopic' homogenized stiffness tensor of cement paste, $\mathbf{C}_{\text{cp}} = 3k_{\text{cp}}\mathbf{J} + 2\mu_{\text{cp}}\mathbf{K}$, and with bulk modulus k_{cp} and shear modulus μ_{cp}. Following the traditional approach in continuum micromechanics [84], the dependence of $\mathbf{C}_{\text{cp}}$ on the phase stiffness properties (Table 1) is established on the basis of Eshelby-Laws-type matrix-inclusion problems which are formulated separately for each phase p = clin, H_2O, air, and for infinitely many cylindrical hydrate phases oriented in all space directions. Specifying the concentration tensor $\mathbf{A}_p$ [Eq. (9)] for all of these different phases and choosing $\mathbf{C}^0 = \mathbf{C}_{\text{cp}}$ in Eq. (9), yields, when also considering

(8) and (10), a self-consistent scheme [32,40] for determination of the homogenized stiffness tensor of cement paste, $\mathbf{C}_{cp}$, reading as

$$\mathbf{C}_{cp} = \left\{ \sum_p f_p \mathbf{C}_p : \left[\mathbf{I} + \mathbf{P}^{cp}_{sph} : (\mathbf{C}_p - \mathbf{C}_{cp})\right]^{-1} + f_{hyd}\mathbf{C}_{hyd} : \right.$$
$$: \int_{\varphi=0}^{2\pi} \int_{\vartheta=0}^{\pi} \left[\mathbf{I} + \mathbf{P}^{cp}_{cyl}(\varphi,\vartheta) : (\mathbf{C}_{hyd} - \mathbf{C}_{cp})\right]^{-1} \frac{\sin\vartheta}{4\pi}\, \mathrm{d}\vartheta\, \mathrm{d}\varphi \bigg\} :$$
$$: \left\{ \sum_p f_p \left[\mathbf{I} + \mathbf{P}^{cp}_{sph} : (\mathbf{C}_p - \mathbf{C}_{cp})\right]^{-1} + f_{hyd} \int_{\varphi=0}^{2\pi} \int_{\vartheta=0}^{\pi} \left[\mathbf{I} + \mathbf{P}^{cp}_{cyl}(\varphi,\vartheta) : \right.\right.$$
$$\left.\left. (\mathbf{C}_{hyd} - \mathbf{C}_{cp})\right]^{-1} \frac{\sin\vartheta}{4\pi}\, \mathrm{d}\vartheta\, \mathrm{d}\varphi \right\}^{-1}, \tag{16}$$

where p = clin, H_2O, air. For detailed explanations regarding the numerical evaluation of (16), as well as for computation of the Hill tensors for spherical material phases, $\mathbf{P}^{cp}_{sph}$, and for cylindrical phases, $\mathbf{P}^{cp}_{cyl}$, respectively, see [60, 61].

3.4 Homogenized Strength of Cement Paste

From loading $\mathbf{E}_{cp}$ (or $\mathbf{\Sigma}_{cp}$) of an RVE$_{cp}$ of cement paste, we are left with estimating stress peaks $\max \boldsymbol{\sigma}^{dev}_{hyd}(\mathbf{x})$ in the hydrates relevant for quasi-brittle failure criterion (14). We have resolved the hydrate phase Ω_{hyd} down to bundles $\Omega_{hyd;\varphi,\vartheta}$ around angles φ and ϑ. We may specify

$$\max_{\mathbf{x}\in\Omega_{hyd}} \sigma^{dev}(\mathbf{x}) = \max_{\substack{\mathbf{x}\in\Omega_{hyd;\varphi,\vartheta} \\ \varphi\in[0,2\pi] \\ \vartheta\in[0,\pi]}} \sigma^{dev}(\mathbf{x},\varphi,\vartheta)\,. \tag{17}$$

Still, in the maximally stressed bundle, not the (first-order) average phase stresses $\boldsymbol{\sigma}_{hyd;\varphi,\vartheta}$, but stress *peaks* govern the failure of the (φ,ϑ)-oriented hydrate phase [60]. Such peaks can be appropriately estimated through quadratic stress averages over suitably chosen subdomains of the RVE$_{cp}$, such as 3D subdomains (bulk phases) [4,37,47] or 2D interfaces [13,17]. Herein, we introduce quadratic (or second-order) averages over (φ,ϑ)-oriented hydrates [12,38,39],

$$\overline{\overline{\sigma^{dev}_{hyd;\varphi,\vartheta}}} = \lim_{\Delta\varphi,\Delta\vartheta\to 0} \left(-\frac{\mu^2_{hyd}}{f_{hyd;\varphi,\vartheta}} \mathbf{\Sigma}_{cp} : \frac{\partial \mathbf{C}^{-1}_{cp}}{\partial \mu_{hyd;\varphi,\vartheta}} : \mathbf{\Sigma}_{cp} \right)^{\frac{1}{2}}. \tag{18}$$

Setting, as experimentally identified in [60], $\sigma^{dev}_{crit} = 26$ MPa, elastic limits of hydrates are related, according to (14) and (18), to the quasi-brittle failure of shotcrete, which may be represented by a uniaxial compressive strength $\Sigma^{comp,ult}_{cp,11}$: The latter follows from insertion of $\sigma_{crit}^{dev} = 26\,\text{MPa}$ into (18), and evaluating the latter for $\boldsymbol{\Sigma}_{cp} = -\left|\Sigma_{cp,11}^{comp,ult}\right| (\mathbf{e}_z \otimes \mathbf{e}_z)$,

$$\Sigma_{cp,11}^{comp,ult} = \left\{ \max_{\varphi,\vartheta} \left[\lim_{\Delta\varphi,\Delta\vartheta\to 0} \left(-\frac{\mu_{hyd}^2}{f_{hyd;\varphi,\vartheta}} (\mathbf{e}_z \otimes \mathbf{e}_z) : \right.\right.\right.$$
$$\left.\left.\left. : \frac{\partial \mathbf{C}_{cp}^{-1}}{\partial \mu_{hyd;\varphi,\vartheta}} : (\mathbf{e}_z \otimes \mathbf{e}_z) \right)^{\frac{1}{2}} \right] \right\}^{-1} \times \sigma_{crit}^{dev} . \quad (19)$$

An algorithm for computation of the aforementioned limit, $\lim_{\Delta\varphi,\Delta\vartheta\to 0}$, is presented in [60].

4 Micromechanics at the Shotcrete Level

We consider an RVE_{sc} of shotcrete with characteristic length $\ell_{sc} = 7$–10 cm, see Figure 1b, consisting of material phases with characteristic dimensions $d_{sc} = 1$–15 mm: (i) cement paste with volume fraction $\bar{f}_{cp}$ and stiffness according to (16), and (ii) aggregates with volume fraction $\bar{f}_{agg} = 1 - \bar{f}_{cp}$ and stiffness according to direct experiments on limestone aggregate (see Table 1). As for elasticity homogenization, resulting in a macroscopic elastic law of the form

$$\boldsymbol{\Sigma}_{sc} = \mathbf{C}_{sc} : \mathbf{E}_{sc} , \quad (20)$$

we proceed in a similar way as described by (2)–(16) for cement paste, with one basic difference: From a morphological viewpoint, the aggregates constitute *inclusions* in a cement paste *matrix*, so that the Mori-Tanaka scheme [6, 56, 84] is chosen for homogenization of shotcrete elasticity:

$$\mathbf{C}_{sc} = \left\{ \bar{f}_{cp}\,\mathbf{C}_{cp} + \bar{f}_{agg}\,\mathbf{C}_{agg} : \left[\mathbf{I} + \mathbf{P}_{sph}^{cp} : (\mathbf{C}_{agg} - \mathbf{C}_{cp})\right]^{-1} \right\} :$$
$$: \left\{ \bar{f}_{cp}\,\mathbf{I} + \bar{f}_{agg} \left[\mathbf{I} + \mathbf{P}_{sph}^{cp} : (\mathbf{C}_{agg} - \mathbf{C}_{cp})\right]^{-1} \right\}^{-1} . \quad (21)$$

The homogenized strength of shotcrete is obtained analogously to that of cement paste. Thus, σ_{crit}^{dev} is related to the uniaxial compressive strength of shotcrete, $\Sigma_{sc,11}^{comp,ult}$, by the homogenized compliance of shotcrete, $\mathbf{C}_{sc}^{-1}$, obtained from (21),

$$\Sigma_{sc,11}^{comp,ult} = \left\{ \max_{\varphi,\vartheta} \left[\lim_{\Delta\varphi,\Delta\vartheta\to 0} \left(-\frac{\mu_{hyd}^2}{f_{hyd;\varphi,\vartheta}} (\mathbf{e}_z \otimes \mathbf{e}_z) : \right.\right.\right.$$
$$\left.\left.\left. \frac{\partial \mathbf{C}_{sc}^{-1}}{\partial \mu_{hyd;\varphi,\vartheta}} : (\mathbf{e}_z \otimes \mathbf{e}_z) \right)^{\frac{1}{2}} \right] \right\}^{-1} \times \sigma_{crit}^{dev} . \quad (22)$$

An algorithm for computation of the aforementioned limit, $\lim_{\Delta\varphi,\Delta\vartheta\to 0}$, is presented in [61].

5 Experimental Validation of Micromechanics-Based Material Models

The micromechanics model presented in Sections 3 and 4, based on the universal phase properties of Table 1, will be fed with shotcrete mixture- and hydration kinetics-specific input data concerning material composition, i.e. with experiment-based values for the volume fractions of air, water, clinker, hydrates, cement paste, and aggregate: f_{air}, f_{H_2O}, f_{clin}, f_{hyd}, $\bar{f}_{cp}$ and $\bar{f}_{agg}$. For these input data, the model delivers predictions for mixture- and hydration-specific shotcrete stiffnesses and strengths. Comparison of these predictions to corresponding experimentally derived values allows for assessing the predictive capabilities of the model.

5.1 Mixture-Dependent Shotcrete Composition

The volume fractions inside an RVE_{cp} of cement paste depend on the degree of hydration ξ, which can be defined as mass of formed hydrates over the mass of hydrates formed if the entire clinker reacts with water, hence $0 \leq \xi \leq 1$. The volume fractions f_{air}, f_{H_2O}, f_{clin}, and f_{hyd} [1,60] are governed, besides ξ, by the w/c ratio, as well as by the mass densities of clinker, water, and hydrates, ρ_{clin}, ρ_{H_2O}, and ρ_{hyd}. In contrast to the volume fractions within the RVE_{cp}, the volume fractions of cement paste and aggregates do not change during hydration of the material. They can be determined from the water-cement ratio w/c, the aggregate-cement ratio a/c, and the mass densities of aggregates, water, and clinker, ρ_{agg}, ρ_{H_2O}, and ρ_{clin} [1,60,61]. The evolution of the degree of hydration is determined by means of adiabatic tests where the measured accumulated hydration heat is considered as an indicator for the hydration progress of the investigated shotcrete specimen [23,81].

5.2 Experimental Validation on Shotcrete Level

For the sake of experimental model validation, the model-predicted Young's modulus of shotcrete is compared to corresponding experimental values of Lafarge [44], obtained from resonant frequency tests on a shotcrete characterized by $w/c = 0.48$ and $a/c = 5.3$, see Figure 3a. The agreement between model predictions and experiments is excellent for sealed conditions (underlined by a mean relative error of 1.0% and a corresponding standard deviation of 6.5%), confirming the findings of Hellmich and Mang [26], where it is also concluded that the tests of Lafarge [44] are rather characterized by sealed conditions than by drained conditions. Model predictions related to sealed conditions correlate to experimental results by $r^2 = 98.8\%$, see Figure 3a.

Experimental validation of micromechanics-based homogenization of shotcrete strength is carried out according to the experiments of Lafarge [43] and Pillar [62]: The applied methodology of experimental strength determination (Hilti gun and penetrometer tests) suggests that model predictions referring to drained conditions are closer to the experimental findings than the ones referring to sealed conditions [61].

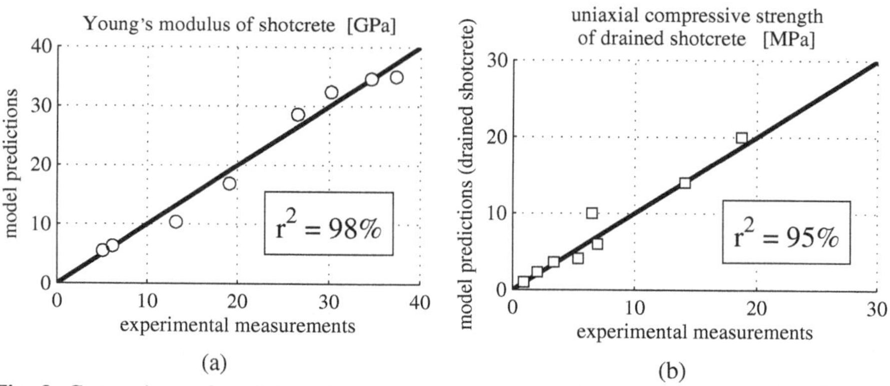

Fig. 3. Comparison of model predictions with experimental data from material testing: (a) Young's modulus, test data from [44]; (b) uniaxial compressive strength, test data from [43] and [62].

The agreement between model predictions (related to drained conditions) and experimental data, characterized by $w/c = 0.5$, $a/c = 3.8$ [62], and by $w/c = 0.4$, $a/c = 3.94$ [43], is quantified by a mean relative prediction error of -5.0%, and by a related standard deviation of 19.4%. The relative errors constitute upper bounds [61], suggesting the correlation between model predictions and experimental values of $r^2 = 95\%$ to be of satisfactory accuracy, see Figure 3b.

6 Micromechanics-Based Studies: Influence of Water-Cement and Aggregate-Cement Ratios on Evolutions of Elasticity and Strength of Shotcrete

The experimentally validated micromechanics model (see Sections 3 to 5) allows us to predict hydration degree-dependent evolutions of Young's modulus, Poisson's ratio, and the uniaxial compressive strength of different shotcrete mixtures, as functions of their w/c and a/c ratios.

The loss of final strength (at completed hydration) with increasing water-cement ratio almost perfectly follows Feret's famous empirical relationship [16], stating that the final strength [$\Sigma^{ult}(t = \infty)$] is proportional to the square of a ratio between the volumes of cement, water, and air contained in a material volume of concrete,

$$\Sigma^{\text{ult}}(t = \infty) = P \left[\frac{f_{\text{clin}}(\xi = 0)}{f_{clin}(\xi = 0) + f_{\text{H}_2\text{O}}(\xi = 0) + f_{\text{air}}(\xi = 0)} \right]^2 , \quad (23)$$

where P is a factor of proportionality. This relationship has proven remarkable usefulness, and it is widely used for mix designs in the cement and concrete industry: The match between this relationship (with $P = 197.2$ MPa) and the model-predicted strength at completed hydration, see Figure 4, further corroborates the relevance of our model, in addition to the experimental evidence given in Section 5 and in earlier publications [60, 61].

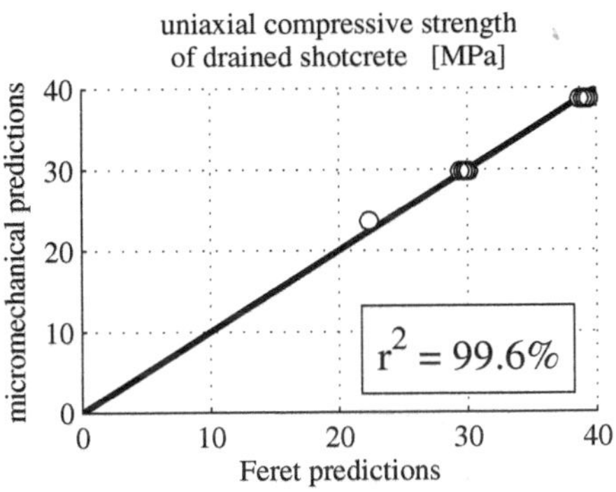

Fig. 4. Comparison of micromechanics-based model predictions with predictions of Feret's formula (23) where the factor of proportionality amounts to $P = 197.2$ MPa.

Remarkably, Feret's formula (23) does not include the aggregate volume, while our model does directly account for the influence of the aggregate-cement ratio. However, increase of the a/c ratio from 3.5 to 5.0 results in an increase of the Young's modulus of only up to 10%, while Poisson's ratio and uniaxial compressive strength are virtually unaffected by such variations of the shotcrete mixture. Conclusively, for typical shotcretes used in NATM tunneling, the a/c ratio plays a minor role in determination of the overall mechanical properties – and this has beneficial effects for the reliability of structural computations, as will be detailed in the next subsection.

7 Continuum Micromechanics-Based Safety Assessment of NATM Tunnel Shells

Finally, the continuum micromechanics-based, hydration degree-dependent evolutions of Young's modulus $E(\xi)$, Poisson's ratio $\nu(\xi)$, and uniaxial compressive strength $f_c(\xi)$ of shotcrete, given in Section 6, serve as input for the assessment of the degree of utilization of shotcrete tunnel shells by means of the hybrid method, according to [27]. This will elucidate the dependence of the structural safety of shotcrete tunnel shells on the shotcrete mixture, governed by the w/c and a/c ratios. Moreover, we examine the role of water with respect to the degree of utilization, by considering both sealed and drained conditions (see Table 1). All other thermochemomechanical phenomena in shotcrete, especially concerning autogeneous shrinkage and creep, are considered macroscopically. In detail, two types of computations are performed: (a) a thermochemical analysis for computation of hydration-induced temperature fields within the tunnel shell, and (b) a chemomechanical analysis for determination of the stress fields within the tunnel shell resulting from a given loading (in terms of prescribed displacement fields). In the course of the thermochemical analysis [29], the following governing equations are employed:

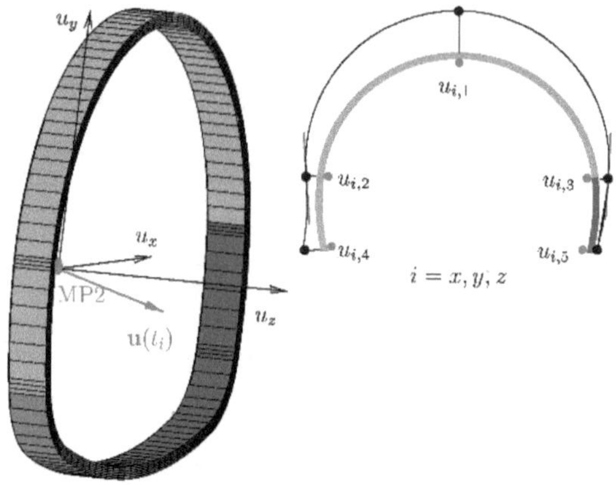

Fig. 5. Hybrid method for determination of the level of loading from prescription of measured displacements on a three-dimensional Finite Element model of the tunnel shell

- the first law of thermodynamics,

$$C\dot{T} + L^{\text{hyd}}\dot{\xi} = -\text{div}\,\mathbf{q}\,, \tag{24}$$

  relates the external heat supply $-\text{div}\,\mathbf{q}$ ($\mathbf{q}$ is the heat flux vector) to the internal heat change $C\dot{T}$ (C is the heat capacity of the considered shotcrete, and $\dot{T}$ is the rate of absolute temperature), and to the latent heat production due to hydration $L^{\text{hyd}}\dot{\xi}$ (L^{hyd} is the heat produced at complete hydration);
- Fourier's heat conduction law,

$$\mathbf{q} = -k\,\text{grad}T\,, \tag{25}$$

  with the shotcrete's conductivity k; and
- a macroscopic hydration kinetics law for shotcrete, describing the evolution of the degree of hydration ξ,

$$\dot{\xi} = \tilde{A}(\xi)\exp\left(-\frac{E_a}{RT}\right), \tag{26}$$

  with $\tilde{A}(\xi)$ as the normalized chemical affinity [30, 81], E_a as the activation energy, and R as the universal constant.

Based on shotcrete-specific values for C, k, L^{hyd}, E_a, and $\tilde{A}(\xi)$ (see [26, 46, 50] for the employed values, and for the heat characteristics of the surrounding soil), the thermochemical analysis delivers time-dependent fields of the temperature and the degree of hydration, see Figure 6. The maximum temperature of 45° is reached around 30 hours after shotcreting in a depth of around 10 cm. Subsequently, a chemomechanical analysis is performed, combining kinematic relations and force equilibrium with a chemoplastic material model including creep [74],

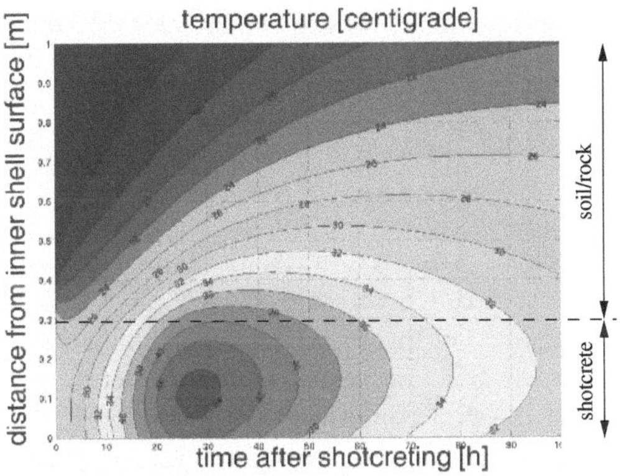

Fig. 6. Temperature evolution, in centigrade, in and around the tunnel shell, as function of time (in hours) since shotcreting and of distance (in m) from inner shell surface (shell thickness equals to 30 cm).

$$d\boldsymbol{\sigma} = \mathbb{C}_{sc}(\xi) : \left[d\boldsymbol{\varepsilon} - d\boldsymbol{\varepsilon}^{\mathrm{p}} - d\boldsymbol{\varepsilon}^{\mathrm{f}} - \alpha_T \mathbf{1} dT - \mathbf{1} d\varepsilon^{\mathrm{s}}(\xi) - d\boldsymbol{\varepsilon}^{\mathrm{v}} \right], \tag{27}$$

where $d\boldsymbol{\varepsilon}^{\mathrm{p}}$ is the increment of plastic strains (at the level of a representative volume element of shotcrete), $d\boldsymbol{\varepsilon}^{\mathrm{f}}$ is the increment of long-term creep strains, $d\boldsymbol{\varepsilon}^{\mathrm{v}}$ is the increment of short-term creep strains, α_T is the coefficient of thermal dilatation, $\mathbf{1}$ is the second-order unit tensor, and $d\varepsilon^{\mathrm{s}}(\xi)$ is the increment of shrinkage strains. The material functions for chemical affinity, creep, and shrinkage are given in [46], whereby short-term creep is considered according to [50].

For the hybrid method simulations, displacements measured at 156.990 km of the Sieberg tunnel are prescribed as boundary conditions on a three-dimensional Finite Element model of the tunnel shell, as indicated in Figure 5, compare [23,25,27]. Simulation results are illustrated in terms of the so-called level of loading $\mathscr{L}$ or degree of utilization. The latter is defined as the ratio of the loading (stress) over loading capacity (strength) of the shotcrete tunnel shell. More specifically, it is defined on the basis of a Drucker–Prager failure surface calibrated for uniaxial and biaxial compressive failure, $\Sigma_{\mathrm{sc},11}^{\mathrm{comp,ult}}$ from (22), and $\Sigma_{\mathrm{sc},11}^{\mathrm{Bicomp,ult}} = \Sigma_{\mathrm{sc},11}^{\mathrm{comp,ult}} \times \kappa$, $\kappa = 1.16 = \text{constant}$,

$$\mathscr{L} = \frac{\alpha \mathrm{tr}\, \boldsymbol{\Sigma}_{\mathrm{sc}} + \sqrt{\Sigma_{\mathrm{sc,ij}}^{\mathrm{dev}} \Sigma_{\mathrm{sc,ij}}^{\mathrm{dev}}}}{K}, \tag{28}$$

with

$$\alpha = \sqrt{\frac{2}{3}} \frac{\kappa - 1}{2\kappa - 1}, \quad K = \sqrt{\frac{2}{3}} \left(1 - \frac{\kappa - 1}{2\kappa - 1} \right) \Sigma_{\mathrm{sc},11}^{\mathrm{comp,ult}}. \tag{29}$$

We focus, in our present evaluation, on the average nature of $\mathscr{L}$ over the tunnel shell thickness h,

$$\overline{\mathscr{L}}(\varphi, t) = \frac{1}{h} \int_h \mathscr{L}(r, \varphi, t)\, dr\,, \tag{30}$$

with r and φ as polar coordinates defining (macroscopic) positions within a circular shell segment, and present the maximum value of $\overline{\mathscr{L}}(\varphi, t)$ in the tunnel shell, $\overline{\mathscr{L}}_{max}(t)$ as 'level of loading'.

7.1 Water-Cement Ratio-Dependence of Structural Safety

The simulation results show, presented in detail in [61, 69], that throughout the observed loading phase the resulting level of loading of the tunnel shell is considerably influenced by the w/c ratio. On the other hand, the deviations between the levels of loading predicted for sealed conditions in the shotcrete only negligibly differ from those predicted for drained conditions in the shotcrete. Still, we note that, in principle, sealed conditions result in slightly lower levels of loading as compared to drained conditions [61, 69].

7.2 Aggregate-Cement Ratio-Dependence of Structural Safety

During the shotcreting process, shotcrete constituents may detach from the sprayed material, which is referred to as rebound. Assuming rebound to concern aggregates only (m_{clin} = const. and m_{H_2O} = const.), the a/c ratio is decreased with respect to the targeted a/c ratio, denoted as $(a/c)_{target}$, $a/c = (a/c)_{target} - \Delta(a/c)_{rebound}$. Considering that the targeted mass of shotcrete, $m_{sc,target}$, is composed of the mass of clinker, water, and aggregates,

$$m_{sc,target} = m_{clin} + m_{H_2O} + m_{agg} = m_{clin}\left[1 + w/c + (a/c)_{target}\right] , \tag{31}$$

and that the shotcrete mass remaining on the tunnel wall, m_{sc}, is related to $m_{sc,target}$ by rebound R,

$$m_{sc} = m_{sc,target}(1 - R) = m_{clin}\,(1 + w/c + a/c) , \tag{32}$$

allows, through substitution of (31) into (32), for estimation of the actual a/c ratio as function of the rebound R, of the water-cement ratio w/c, and of the targeted aggregate-cement ratio $(a/c)_{target}$,

$$a/c = (a/c)_{target} - R\left[1 + w/c + (a/c)_{target}\right] . \tag{33}$$

For wet shotcreting, as has been used in the Sieberg tunnel, the rebound hardly exceeds 20% [21].[1] In order to elucidate the dependence of the structural safety on rebound-related variations of the a/c-ratio, eight different mixes are investigated (see Table 2 for corresponding rebounds as $(a/c)_{target} = 5.0$): $w/c = 0.40$ and $a/c = 3.5$ (mix IV), $w/c = 0.40$ and $a/c = 4.0$ (mix V), $w/c = 0.40$ and $a/c = 4.5$ (mix VI), $w/c = 0.40$ and $a/c = 5.0$ (mix VII), $w/c = 0.50$ and $a/c = 3.5$ (mix VIII), $w/c = 0.50$ and $a/c = 4.0$ (mix IX), $w/c = 0.50$ and $a/c = 4.5$ (mix X), as well as $w/c = 0.50$ and $a/c = 5.0$ (mix XI).

In contrast to the w/c ratio, the (effective) a/c ratio has no significant influence on the level of loading on the shotcrete tunnel shell, see Figure 7. This is because

[1] Dry shotcreting may lead to higher rebounds [2, 59].

Table 2. Shotcrete rebound R for given values of w/c and a/c; according to (33); for $(a/c)_{\text{target}} = 5.0$.

a/c	$w/c = 0.40$	$w/c = 0.50$
5.0	$R =$ 0.0%	$R =$ 0.0%
4.5	$R =$ 7.8%	$R =$ 7.7%
4.0	$R =$ 15.6%	$R =$ 15.4%
3.5	$R =$ 23.4%	$R =$ 23.1%

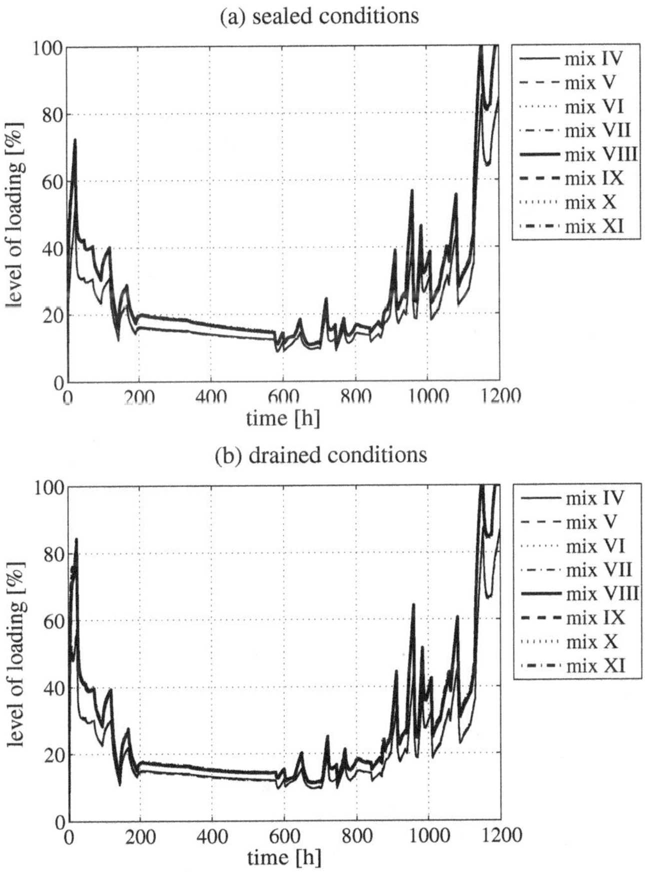

Fig. 7. Evolution of the level of loading as function of time after installation for (a) sealed, and for (b) drained conditions, determined through hybrid analyses.

the a/c ratio increase-related increase of Young's modulus, which lets expect higher utilization degrees in the tunnel shell, is compensated by stress redistributions in the tunnel shell. Hence, the level of loading shows no significant dependence on the (actual) a/c ratio for the investigated shotcrete mixtures. This structural behavior can be considered as beneficial: Our calculations suggest that such changes of the (actual)

a/c ratio because of shotcrete rebound do not compromise the structural safety of the tunnel shell. Since the investigated shotcrete mixes exhibit a maximum rebound of 23.4% (see Table 2), thus comprising common rebounds in wet shotcreting, the insensitivity of the loading level of the tunnel shell with respect to aggregate rebound is a robust feature of the NATM. It suggests that careful *in situ* monitoring of the w/c ratio, as compared to the a/c ratio, is much more critical.

8 Conclusions

We have developed a new micromechanics model which economically accounts for the impact of shotcrete composition onto the elasticity and strength properties of the material. In more detail, the principle of micromechanics, i.e. the formulation of a homogenization scheme in terms of composition-independent phase properties, phase shapes, and morphologies, allows to determine corresponding macroscopic material properties for a wide range of different material compositions. This suits exceptionally well to a material with a temporarily evolving microstructure, such as shotcrete due to ongoing hydration. On this basis, we have shown the potentially major role of the shotcrete composition (in particular of the water-cement ratio) on the forces induced in an NATM-tunnel shell. To further elucidate this role, it is highly desirable to further extend the micromechanical description of shotcrete towards consideration both of creep and of shrinkage. This is a topic of ongoing research. For latest developments on concrete creep, see [68]. Another open issue relates to the question whether high levels of loading (such as those encountered in Figure 7 at 1120 hours after shell installation) might sometimes be overestimations, as a result of enforcing $C1$-continuity of displacements between shell components installed at different time instants, see [25, 27] for details. This underlines that further improvement of data analysis during NATM-tunneling calls for even more refined applied mechanics tools, both at the material level (microstructural level = shotcrete) and at the (macro-) structure level (= tunnel shell).

Using homogenization of shotcrete as a vehicle, this contribution underlines that continuum micromechanics efficiently supports engineers to develop predictive material models which are as simple as possible and, at the same time, as complex as necessary. This is possible since the method allows for consideration of almost any type of knowledge on materials including information on morphological, mechanical, physical, and chemical properties. Related experimental insight can be introduced at the very level of hierarchical organization of the microstructure, at which it appears to be most appropriate. This is particularly important for materials whose microstructures undergo significant changes during their mechanical lifetime (such as early-age concrete or biological materials) whereas the role of multiscale material mechanics might be smaller for characterization of engineering structures built up of well-defined, chemically inert materials. Conclusively, continuum micromechanics provides a flexible and efficient framework for the mechanical description of microheterogeneous materials throughout their structural hierarchies. Thanks to the dynamic evolution of the field, a great variety of different types of material behavior may be considered, including not only elasticity and brittle strength, but also creep [68] and elasto-plasticity [20].

Acknowledgements

This work was part of the micromechanics-based activities within the integrated project 'Technology Innovation in Underground Construction – TUNCONSTRUCT' (http://www.tunconstruct.org), co-sponsored by the European Commission.

References

1. P. Acker. Micromechanical analysis of creep and shrinkage mechanisms. In F.-J. Ulm, Z.P. Bažant, and F.H. Wittmann (Eds.), *Creep, Shrinkage and Durability Mechanics of Concrete and Other Quasi-brittle Materials*, 6th International Conference CONCREEP@MIT, Amsterdam, pp. 15–26. Elsevier, 2001.
2. H.S. Armelin and N. Banthia. Development of a general model of aggregate rebound for dry-mix shotcrete (Part II). *Materials and Structures*, 31(207):195–202, 1998.
3. V. Baroghel-Bouny. Caractérisation des pâtes de ciment et des bétons-méthodes, analyse, interprétation [Characterization of cement pastes and concretes-methods, analysis, interpretations]. Technical Report, Laboratoire Central des Ponts et Chaussées, Paris, France, 1994 [in French].
4. J.-F. Barthélémy and L. Dormieux. Détermination du critère de rupture macroscopique d'un milieu poreux par homogénéisation non linèaire [Determination of the macroscopic strength criterion of a porous medium by nonlinear homogenization]. *Comptes Rendus Mécanique*, 331(4):271–276, 2003. In French.
5. G. Beer (Ed.), *Numerical Simulation in Tunneling*. Springer Verlag, Wien/New York, 2003.
6. Y. Benveniste. A new approach to the application of Mori–Tanaka's theory in composite materials. *Mechanics of Materials*, 6(2):147–157, 1987.
7. O. Bernard, F.-J. Ulm, and E. Lemarchand. A multiscale micromechanics-hydration model for the early-age elastic properties of cement-based materials. *Cement and Concrete Research*, 33(9):1293–1309, 2003.
8. A. Boumiz, C. Vernet, and F. Cohen-Tenoudji. Mechanical properties of cement pastes and mortars at early ages. *Advanced Cement Based Materials*, 3(3-4):94–106, 1996.
9. P. Brandtner, B. Moritz, and P. Schubert. On the challenge of evaluating stress in a shotcrete lining: Experiences gained in applying the hybrid analysis method. *Felsbau*, 25(5):93–98, 2007.
10. T. Celestino, M. de Mariano, A. Ferreira, and M. Guimaraes. Undercoring technique for stress measurement in shotcrete linings. In J. Golser, W. Hinkel, and W. Schubert (Eds.), *Tunnels for People*, Proceedings of the World Tunnel Congress 1997, Vienna, Austria, pp. 59–64. Balkema, Rotterdam, the Netherlands, 1997.
11. G. Constantinides and F.-J. Ulm. The effect of two types of C-S-H on the elasticity of cement-based materials: Results from nanoindentation and micromechanical modeling. *Cement and Concrete Research*, 34(1):67–80, 2004.
12. L. Dormieux, A. Molinari, and D. Kondo. Micromechanical approach to the behavior of poroelastic materials. *Journal of Mechanics and Physics of Solids*, 50(10):2203–2231, 2002.

13. L. Dormieux, J. Sanahuja, and Y. Maalej. Résistance d'un polycristal avec interfaces intergranulaires imparfaites [Strength of a polycrystal with imperfect intergranular interfaces]. *Comptes Rendus Mecanique*, 335(1):25–31, 2007 [in French].
14. L. Dormieux and F.-J. Ulm. *Applied Micromechanics of Porous Materials*, CISM Courses and Lectures, Vol. 480. Springer Verlag, Wien/New York, 2005.
15. J.D. Eshelby. The determination of the elastic field of an ellipsoidal inclusion, and related problems. *Proceedings of the Royal Society of London A*, 241:376–396, 1957. Reprinted in [53].
16. R. Feret. On the compactness of hydraulic mortars [Sur la compacitée des mortiers hydrauliques]. *Annales des Ponts et Chaussées*, 7:5–164, 1892 [in French].
17. A. Fritsch, L. Dormieux, C. Hellmich, and J. Sanahuja. Micromechanics of crystal interfaces in polycrystalline solid phases of porous media: Fundamentals and application to strength of hydroxyapatite biomaterials. *Journal of Materials Science*, 42(21):8824–8837, 2007.
18. A. Fritsch, L. Dormieux, C. Hellmich, and J. Sanahuja. Mechanical behavior of hydroxyapatite biomaterials: An experimentally validated micromechanical model for elasticity and strength. *Journal of Biomedical Materials Research Part A*, 88(1):149–161, 2009.
19. A. Fritsch and C. Hellmich. 'Universal' microstructural patterns in cortical and trabecular, extracellular and extravascular bone materials: Micromechanics-based prediction of anisotropic elasticity. *Journal of Theoretical Biology*, 244(4):597–620, 2007.
20. A. Fritsch, C. Hellmich, and L. Dormieux. Ductile sliding between mineral crystals followed by rupture of collagen crosslinks: Experimentally supported micromechanical explanation of bone strength. *Journal of Theoretical Biology*, 260(2):230–252, 2009.
21. I.D. Hague. Wet shotcreting – A total system approach, 2001. www.ugc.basf.com (last download: July 31, 2008).
22. Z. Hashin. Analysis of composite materials – A survey. *Journal of Applied Mechanics*, 50(3):481–505, 1983.
23. C. Hellmich. *Shotcrete as part of the New Austrian Tunneling Method: From thermochemomechanical material modeling to structural analysis and safety assessment of tunnels*. PhD Thesis, Vienna University of Technology, Vienna, Austria, 1999.
24. C. Hellmich, J.-F. Barthélémy, and L. Dormieux. Mineral-collagen interactions in elasticity of bone ultrastructure – A continuum micromechanics approach. *European Journal of Mechanics – A/Solids*, 23(5):783–810, 2004.
25. C. Hellmich, J. Macht, and H.A. Mang. Ein hybrides Verfahren zur Bestimmung der Auslastung von Spritzbetonschalen [A hybrid method for determination of the level of utilization of shotcrete shells]. *Felsbau*, 17(5):422–425, 1999. In German.
26. C. Hellmich and H.A. Mang. Shotcrete elasticity revisited in the framework of continuum micromechanics: From submicron to meter level. *Journal of Materials in Civil Engineering (ASCE)*, 17(3):246–256, 2005.
27. C. Hellmich, H.A. Mang, and F.-J. Ulm. Hybrid method for quantification of stress states in shotcrete tunnel shells: Combination of 3D in situ displacement measurements and thermochemoplastic material law. *Computers and Structures*, 79(22-25):2103–2115, 2001.
28. C. Hellmich and F.-J. Ulm. Micromechanical model for ultra-structural stiffness of mineralized tissues. *Journal of Engineering Mechanics (ASCE)*, 128(8):898–908, 2002.
29. C. Hellmich, F.-J. Ulm, and H.A. Mang. Consistent linearization in Finite Element analysis of coupled chemo-thermal problems with exo- or endothermal reactions. *Computational Mechanics*, 24(4):238–244, 1999.

30. C. Hellmich, F.-J. Ulm, and H.A. Mang. Multisurface chemoplasticity I: Material model for shotcrete. *Journal of Engineering Mechanics (ASCE)*, 125(6):692–701, 1999.
31. Ch. Hellmich and F.-J. Ulm. Drained and undrained poroelastic properties of healthy and pathological bone: A poro-micromechanical investigation. *Transport in Porous Media*, 58(3):243–268, 2005.
32. A.V. Hershey. The elasticity of an isotropic aggregate of anisotropic cubic crystals. *Journal of Applied Mechanics (ASME)*, 21:226–240, 1954.
33. R. Hill. Elastic properties of reinforced solids: Some theoretical principles. *Journal of Mechanics and Physics of Solids*, 11(5):357–372, 1963.
34. R. Hill. Continuum micro-mechanics of elastoplastic polycrystals. *Journal of Mechanics and Physics of Solids*, 13(2):89–101, 1965.
35. K. Hofstetter, C. Hellmich, and J. Eberhardsteiner. Development and experimental validation of a continuum micromechanics model for the elasticity of wood. *European Journal of Mechanics A/Solids*, 24(6):1030–1053, 2005.
36. K. Hofstetter, C. Hellmich, and J. Eberhardsteiner. Micromechanical modeling of solid-type and plate-type deformation patterns within softwood materials. A review and an improved approach. *European Journal of Mechanics A/Solids*, 61(4):343–351, 2007.
37. K. Hofstetter, Ch. Hellmich, and J. Eberhardsteiner. Predicting wood strength from composition and microstructure: development and experimental verification of a continuum micromechanics model. In L. Dormieux, D. Kondo, and K. Sab (Eds.), *Colloque en l'Honneur du Professeur Jean-Louis Auriault: Microstructure et Propriétés des Matériaux*, pp 217–222. Presses de l'Ecole Nationale des Ponts et Chaussées, 2005.
38. W. Kreher. Residual stresses and stored elastic energy of composites and polycrystals. *Journal of the Mechanics and Physics of Solids*, 38(1):115–128, 1990.
39. W. Kreher and A. Molinari. Residual stresses in polycrystals as influenced by grain shape and texture. *Journal of the Mechanics and Physics of Solids*, 41(12):1955–1977, 1993.
40. E. Kröner. Berechnung der elastischen Konstanten des Vielkristalls aus den Konstanten des Einkristalls [Computation of the elastic constants of a polycrystal based on the constants of the single crystal]. *Zeitschrift für Physik A Hadrons and Nuclei*, 151(4):504–518, 1958 [in German].
41. R. Lackner, J. Macht, C. Hellmich, and H.A. Mang. Hybrid method for analysis of segmented shotcrete tunnel linings. *Journal of Geotechnical and Geoenvironmental Engineering (ASCE)*, 128(4):298–308, 2002.
42. R. Lackner, J. Macht, and H.A. Mang. Hybrid analysis method for on-line quantification of stress states in tunnel shells. *Computer Methods in Applied Mechanics and Engineering*, 195(41-43):5361–5378, 2006.
43. Lafarge. Internal Report, Lafarge Centre Technique Europe Central, 1997. Commissioned by the Institute for Strength of Materials, Vienna University of Technology, Austria.
44. Lafarge. Internal Report, Lafarge Centre Technique Europe Central, 2002. Commissioned by the Institute for Strength of Materials, Vienna University of Technology, Austria.
45. N. Laws. The determination of stress and strain concentrations at an ellipsoidal inclusion in an anisotropic material. *Journal of Elasticity*, 7(1):91–97, 1977.
46. M. Lechner, C. Hellmich, and H.A. Mang. Short-term creep of shotcrete – Thermochemoplastic material modelling and nonlinear analysis of a laboratory test and of a NATM excavation by the Finite Element method. In P.A. Vermeer, S. Diebels, W. Ehlers, H.J. Herrmann, S. Luding, and E. Ramm (Eds.), *Proceedings of the International Symposium on Discontinuous Modelling of Cohesive-Frictional Materials*, Lecture Notes in Physics, pp. 47–62. Springer, Berlin, 2001.

47. E. Lemarchand, F.-J. Ulm, and L. Dormieux. Effect of inclusions on friction coefficient of highly filled composite materials. *Journal of Engineering Mechanics (ASCE)*, 128(8):876–884, 2002.
48. K. Li, X.-L. Gao, and A.K. Roy. Micromechanical modeling of viscoelastic properties of carbon nanotube – Reinforced polymer composites. *Mechanics of Advanced Materials and Structures*, 13(4):317–328, 2006.
49. J. Macht, C. Hellmich, R. Lackner, W. Schubert, and H.A. Mang. Assessment of a support system for squeezing rock conditions by means of a hybrid method. *Felsbau*, 18(6):9–15, 2000.
50. J. Macht, R. Lackner, C. Hellmich, and H.A. Mang. Shotcrete creep properties – Review of material tests and structural analyses of tunnels. In F.-J. Ulm, Z. Bažant, and F. Wittmann (Eds.), *Proceedings of the 6th International Conference on Creep, Shrinkage and Durability Mechanics of Concrete and Other Quasi-Brittle Materials*, pp. 285–300. Elsevier Science, Amsterdam, 2001.
51. J. Macht, R. Lackner, C. Hellmich, and H.A. Mang. Quantification of stress states in shotcrete shells. In G. Beer (Ed.), *Numerical Simulation in Tunneling*, pp. 225–248. Springer-Verlag, Wien/New York, 2003.
52. A. Malasoma, A. Fritsch, C. Kohlhauser, T. Brynk, C. Vitale-Brovarone, Z. Pakiela, J. Eberhardsteiner, and C. Hellmich. Micromechanics of bioresorbable porous cel2 glass ceramic scaffolds for bone tissue engineering. *Advances in Applied Ceramics*, 107(5):277–286, 2008.
53. X. Markenscoff and A. Gupta, *Collected Works of J.D.Eshelby – The Mechanics of Defects and Inhomogeneities*, Solid Mechanics and Its Applications, Vol. 133. Springer, 2006.
54. G. Mehlhorn. *Der Ingenieurbau: Grundwissen – Werkstoffe, Elastizitätstheorie [Basic Knowledge in Civil Engineering: Materials and Elasticity Theory]*, Vol. 4. Ernst & Sohn, Berlin, 1996 [in German].
55. G. Meschke. Consideration of aging of shotcrete in the context of a 3-D viscoplastic material model. *International Journal of Numerical Methods in Engineering*, 39(18):3123–3143, 1996.
56. T. Mori and K. Tanaka. Average stress in matrix and average elastic energy of materials with misfitting inclusions. *Acta Metallurgica*, 21:571–574, 1973.
57. A.M. Neville. *Properties of Concrete*, 3rd ed. Pitman Publishing Ltd., 1981.
58. F. Pacher. Deformationsmessungen im Versuchsstollen als Mittel zur Erforschung des Gebirgsverhaltens und zur Bemessung des Ausbaus [Measurements of deformations in experimental tunnels to explore the behavior of rock and to design the lining]. *Felsmechanik und Ingenieurgeologie*, I:149–161, 1964 [in German].
59. M. Pfeuffer and W. Kusterle. Rheology and rebound behaviour of dry-mix shotcrete. *Cement and Concrete Research*, 31(11):1619–1625, 2001.
60. B. Pichler, C. Hellmich, and J. Eberhardsteiner. Spherical and acicular representation of hydrates in a micromechanical model for cement paste – Prediction of early-age elasticity and strength. *Acta Mechanica*, 203(3/4):137–162, 2009.
61. B. Pichler, S. Scheiner, and C. Hellmich. From micron-sized needle-shaped hydrates to meter-sized shotcrete tunnel shells: micromechanical upscaling of stiffness and strength of hydrating shotcrete. *Acta Geotechnica*, 3(4):273–294, 2008.
62. N. Pillar. Determination of early age properties of fibre reinforced shotcrete to predict the cracking behavior. PhD Thesis, submitted to the University of New South Wales, Private Communication, Sydney, Australia, 2002.
63. L. Rabcewicz. Patentschrift. Österreichisches Patent Nr. 165573 [Patent specification, Austrian Patent No. 165573], 1948 [in German].

64. L. Rabcewicz, J. Golser, and E. Hackl. Die Bedeutung der Messung im Hohlraumbau, Teil I [The significance of measurement in tunneling]. *Der Bauingenieur*, 47(7):225–234, 1972 [in German].
65. R. Rokahr and K. Lux. Einfluß des rheologischen Verhaltens des Spritzbetons auf den Ausbauwiderstand [Influence of the rheological behavior of shotcrete on the excavation resistance]. *Felsbau*, 5:11–18, 1987.
66. R. Rokahr and R. Zachow. Ein neues Verfahren zur täglichen Kontrolle der Auslastung einer Spritzbetonschale [A new method for daily monitoring of the stress intensity of a sprayed concrete lining]. *Felsbau*, 15(6):430–434, 1997 [in German].
67. J. Sanahuja, L. Dormieux, and G. Chanvillard. Modelling elasticity of a hydrating cement paste. *Cement and Concrete Research*, 37(10):1427–1439, 2007.
68. S. Scheiner and C. Hellmich. Continuum microviscoelasticity model for aging basic creep of early-age concrete. *Journal of Engineering Mechanics (ASCE)*, 135(4):307–323, 2009.
69. S. Scheiner, B. Pichler, C. Hellmich, and H.A. Mang. Damage and disaster prevention in NATM tunnels during construction: Micromechanics-supported hybrid analyses. In A. Ibrahimbegovic and M. Zlatar (Eds.), *Proceedings of the NATO Advanced Research Workshop 'Damage assessment and reconstruction after natural disasters and previous military activities'*, pp. 145–171. Springer-Verlag, 2008.
70. S. Scheiner, R. Sinibaldi, B. Pichler, V. Komlev, C. Renghini, C. Vitale-Brovarone, F. Rustichelli, and C. Hellmich. Micromechanics of bone tissue-engineering scaffolds, based on resolution error-cleared computer tomography. *Biomaterials*, 30(12):2411–2419, 2009.
71. P. Schubert. Beitrag zum rheologischen Verhalten von Spritzbeton [Contribution to the rheological behavior of shotcrete]. *Felsbau*, 6:150–153, 1988.
72. W. Schubert and A. Steindorfer. Selective displacement monitoring during tunnel excavation. *Felsbau*, 14(2):93–97, 1996.
73. W. Schubert, A. Steindorfer, and E.A. Button. Displacement monitoring in tunnels – An overview. *Felsbau*, 20(2):7–15, 2002.
74. J. Sercombe, C. Hellmich, F.-J. Ulm, and H.A. Mang. Modeling of early-age creep of shotcrete I: Model and model parameters. *Journal of Engineering Mechanics (ASCE)*, 126(3):284–291, 2000.
75. L. Sorelli, G. Constantinides, F.-J. Ulm, and F. Toutlemonde. The nano-mechanical signature of Ultra High Performance Concrete by statistical nanoindentation techniques. *Cement and Concrete Research*, 38(12):1447–1456, 2008.
76. A. Steindorfer, W. Schubert, and K. Rabensteiner. Problemorientierte Auswertung geotechnischer Messungen [Advanced evaluation of geotechnical displacement monitoring data]. *Felsbau*, 13(6):386–390, 1995 [in German].
77. Z. Sun, G. Ye, and S. Shah. Microstructure and early-age properties of Portland cement paste – Effects of connectivity of solid phases. *ACI Materials Journal*, 102(1):122–129, 2005.
78. P.M. Suquet (Ed.). *Continuum Micromechanics*, CISM Courses and Lectures, Vol. 377. Springer Verlag, Wien/New York, 1997.
79. J. Tritthart and F. Häußler. Pore solution analysis of cement pastes and nanostructural investigations of hydrated C_3S. *Cement and Concrete Research*, 33(7):1063–1070, 2003.
80. F.-J. Ulm, G. Constantinides, and F.H. Heukamp. Is concrete a poromechanics material? – A multiscale investigation of poroelastic properties. *Materials and Structures / Materiaux et Constructions*, 37(1):43–58, 2004.
81. F.-J. Ulm and O. Coussy. Strength growth as chemo-plastic hardening in early age concrete. *Journal of Engineering Mechanics (ASCE)*, 122(12):1123–1132, 1996.
82. K. Wesche. *Baustoffe für tragende Bauteile [Building materials for structural components]*, 3rd ed. Bauverlag, Wiesbaden, Germany, 1974 [in German].

83. A. Zaoui. Structural morphology and constitutive behavior of microheterogeneous materials, P.M. Suquet (Ed.), *Continuum Micromechanics*, CISM Courses and Lectures, Vol. 377, chapter 6, pp. 291–347. Springer-Verlag, Wien New York, 1997.
84. A. Zaoui. Continuum micromechanics: Survey. *Journal of Engineering Mechanics (ASCE)*, 128(8):808–816, 2002.

Optimization of Corrugated Paperboard under Local and Global Buckling Constraints

Th. Flatscher, T. Daxner, D.H. Pahr and F.G. Rammerstorfer

Institute of Lightweight Design and Structural Biomechanics,
Vienna University of Technology, Gußhausstraße 27-29/E317, A-1040 Vienna, Austria;
ra@ilsb.tuwien.ac.at

Abstract An important design criterion for containers (boxes) made from corrugated paperboard is their resistance against buckling under compressive loads such as those arising from gravity as boxes are piled up. Since such boxes are typically means of packaging goods for transport, their weight should be as low as possible. These demands are taken into account in the presented optimization procedure for reducing the area-specific weight of corrugated paperboard under global, i.e., box wall buckling constraints and local buckling constraints pertaining to the buckling of flute and liner.

The critical load with respect to global buckling is correlated to the effective bending stiffness of the paperboard (obtained by homogenization). Local buckling is predicted by a unit cell approach in combination with the finite element method. The stiffness homogenization procedure as well as the unit cell approach for computing the buckling loads are embedded into an optimization process. The geometrical parameters describing the meso-scale geometry of the corrugated paperboard act as optimization parameters.

The presented approach is applied to a specific configuration of corrugated paper board, as it is used in packaging. Substantial weight saving could be achieved by the proposed optimization scheme.

A further consideration concerns the post-buckling behavior. Once the side walls of corrugated paperboard containers have buckled, they typically show the formation of folds. As demonstrated in non-linear finite element analyses, these folds are the result of the localization of the initially periodic local buckling pattern.

Key words: Corrugated paperboard, optimization, buckling, finite element method

1 Introduction

A high percentage of boxes used for packaging purposes and transportation is made of corrugated paperboard. This is because of its low costs, good properties regarding environmental aspects, and its excellent lightweight performance, i.e., its high weight related specific stiffness and specific strength.

Corrugated paperboard exists in different configurations with respect to the number and the arrangement of layers, see Figure 1. The wavy layers are called flutes, the flat layers are called liners.

R. de Borst and E. Ramm (eds.), *Multiscale Methods in Computational Mechanics*, Lecture Notes in Applied and Computational Mechanics 55, DOI 10.1007/978-90-481-9809-2_17,

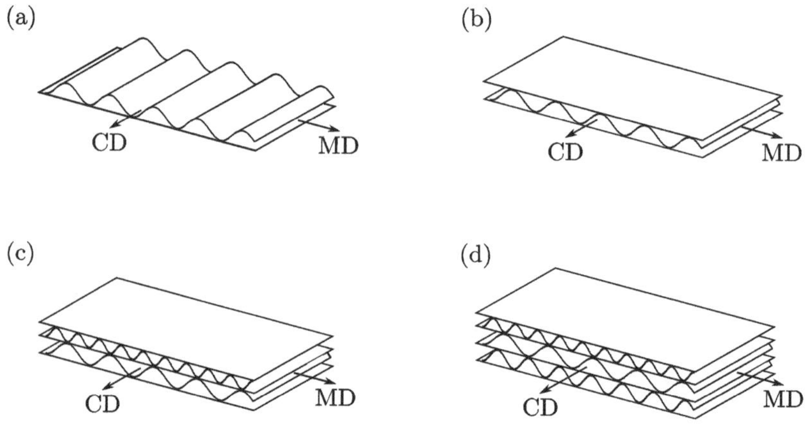

Fig. 1. Different configurations of paperboard; in this paper configuration (b), i.e., single wall corrugated paper is considered.

Boxes or containers made of corrugated paperboard are quite often stacked one over the other. Thus, side walls of such boxes are frequently loaded in-plane by substantial compression forces, typically in fluting direction, i.e., perpendicular to the waves of the flute. Under such loads corrugated paperboard is prone to buckling both on the global level of the box walls and on the local level of liner and/or flute.

There exist a number of papers dealing with buckling of corrugated paperboard; most of them are based on computational treatments. Full-scale finite element simulations were performed by Allanson and Svärd [3] considering a square corrugated paperboard under uniaxial loading, simply supported at all edges. Comparison with experimental results has shown very good agreement with respect to both predicted force-displacement relationships and the onset of failure. Patel et al. [4] have considered circular cylindrical shells made of corrugated paperboard under combined loading, namely axial compression, torque, and external pressure. Experimental failure envelopes have been derived under interacting global loads. In [5] Nyman has proposed a failure criterion for board liners, which has been used also in [6], taking into account both the strength of the material itself and its structural stability. Finite element simulations of whole corrugated paperboard boxes (including closure fins) using shell elements have been published in [7, 8]. In [7] it was shown by simulation results and by experimental data that the limit load under in-plane compression is typically approximately twice as high as the initial buckling load for the considered type of boxes. Global and, in particular, local buckling phenomena in conjunction with the investigation of creasing in corrugated paperboard are published in [9]. There, the results of detailed finite element simulations match well with experimental results in terms of load displacement curves during creasing.

In the present paper, which recalls and extends considerations published in [2], methods of improving corrugated paperboard designs are discussed by means of numerical optimization methods aiming at the reduction of the area-specific weight of the paperboard. The applied optimization constraints are intended to ensure a given required buckling strength both on the level of local buckling, i.e., buckling of liners and/or fluting and on the level of global buckling of whole plates made from corrugated paperboard (see also [10]).

Exclusively single wall corrugated paperboard, see Figure 1b, is considered. This type of corrugated paperboard is a sandwich construction consisting of one flute acting as the core and two liners acting as the faces of the sandwich. It is essential to mention that the raw paper material used in these layers exhibits a pronounced anisotropy resulting from the production process of the paper used. In Figure 1 the direction in which the raw material, i.e., the sheet paper moves during its production process is denoted by machine direction (MD). In this direction the sheet paper typically shows higher in-plane stiffness than in the cross direction (CD). The out-of-plane shear stiffness of this kind of sheet paper is very low [6].

The problem and the solution procedures treated in this paper belong to that class of multiscale analyses, which, in a more general form, are treated in [1].

2 Problem Description and Methods Used

In the following the principal features of the optimization of single wall corrugated paperboard aiming at minimizing the area-specific weight are described. This procedure requires in each optimization step the computation of the effective stiffness and the critical membrane force per unit width of the corrugated paperboard.

For the calculation of the effective stiffnesses homogenization techniques were applied in conjunction with unit cell finite element models with periodic boundary conditions. The unit cell approach for materials with periodic meso structures is described in more detail in [11, 12], or [13]. In the present contribution the unit cell approach is presented in Section 2.1 with focus on corrugated paperboard treated as meso structured plate.

Also for the calculation of the critical membrane force per unit width of the corrugated paperboard periodic unit cell models were employed. However, these unit cells models differ from the ones used for the homogenization process.

The optimization process is demonstrated for a specific reference configuration of the corrugated paperboard. This means that the main geometrical parameters (see Figure 2) were modified in order to minimize the area-specific weight without loosing the load carrying capacity of the reference configuration with respect to local and global buckling.

2.1 Unit Cell Approach

Corrugated paperboard represents a typical periodically meso structured "material". The cross section of the single wall corrugated paperboard considered in this paper is shown in Figure 2.

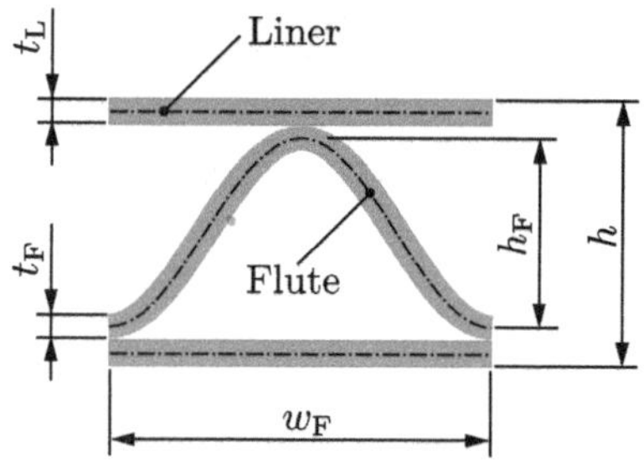

Fig. 2. Cross section (one period) and denotation of geometric parameters.

For the general case of plane periodic media in the form of a plate or shell configuration with in-plane periodicity of the meso structure and out-of-plane variations of the properties, the homogenization technique based on a unit cell approach is derived in [11, 12]. The macroscopic constitutive law in terms of classical lamination theory (see [14]) can be written as

$$\begin{pmatrix} \mathbf{N}^{\mathrm{h}} \\ \mathbf{M}^{\mathrm{h}} \end{pmatrix} = \begin{pmatrix} \mathbf{A}^{\mathrm{h}} & \mathbf{B}^{\mathrm{h}} \\ \mathbf{B}^{\mathrm{h}} & \mathbf{D}^{\mathrm{h}} \end{pmatrix} \begin{pmatrix} \bar{\boldsymbol{\varepsilon}}^{\mathrm{h}} \\ \bar{\boldsymbol{\chi}}^{\mathrm{h}} \end{pmatrix} \tag{1}$$

with the resultant membrane forces per unit width $\mathbf{N}^{\mathrm{h}} = (N_{11}\ N_{22}\ N_{12})^{\mathrm{T}}$, the resultant moments per unit width $\mathbf{M}^{\mathrm{h}} = (M_{11}\ M_{22}\ M_{12})^{\mathrm{T}}$, the strains in the reference surface (which is chosen to be the mid surface of the corrugated paperboard), $\bar{\boldsymbol{\varepsilon}}^{\mathrm{h}} = (\varepsilon_{11}\ \varepsilon_{22}\ 2\,\varepsilon_{12})^{\mathrm{T}}$, and the curvature of the reference surface, $\bar{\boldsymbol{\chi}}^{\mathrm{h}} = (-\chi_{11}\ -\chi_{22}\ -2\chi_{12})^{\mathrm{T}}$. Regarding orientations 1, 2, and 3 see Figure 3.

In the present contribution the homogenization problem is solved by using unit cells discretized by the finite element method (see Figure 4). Thus, these quantities have to be related to nodal forces and nodal displacements. The homogenized stiffness matrices $\mathbf{A}^{\mathrm{h}}$, $\mathbf{B}^{\mathrm{h}}$, and $\mathbf{D}^{\mathrm{h}}$ in Eqn. (1) can then be computed from results obtained by classical finite element analyses.

According to the periodicity of the corrugated paperboard's meso structure periodic boundary conditions (BCs) have to be applied at the boundaries of an appropriate unit cell. These boundaries are shown in Figure 3 for a general case of plane periodic media. For corrugated paperboard $d_2 = 0$ holds and, hence, l_1 and l_2 represent the periodicity of the meso structure. The BCs applied on the unit cell ensure that for each undeformed and deformed configuration the translated unit cells fit with each other, forming a complete macro structure. Considering the periodic BCs the macroscopic displacement field within the unit cell is represented by the displacements of characteristic points, the so called master nodes (A, A^+, B, B^+, 0, and 0^+ as shown in Figure 3). Accordingly, the distributed loads acting along the boundaries of the unit cell are condensed to concentrated nodal forces at the master nodes (e.g., at node A the force $\bar{\mathbf{F}}^A$ and the displacement $\mathbf{u}^A$ are shown in Figure 3).

The description of the relations between these nodal quantities and the quantities in the homogenized constitutive law (Eqn. (1)) start with the definition of the

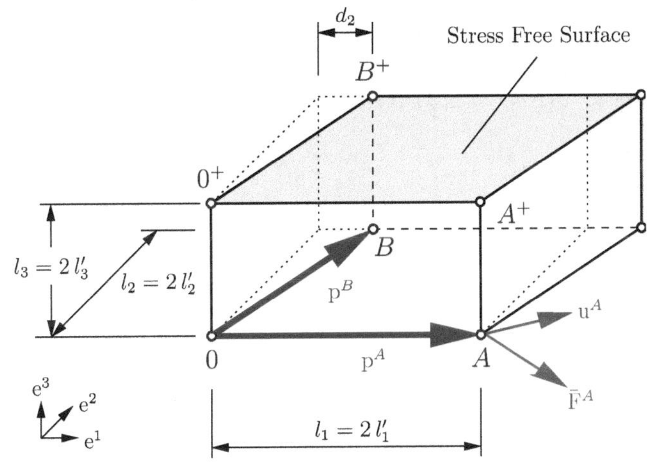

Fig. 3. Definition of the cubic unit cell for a plane periodic media.

characteristic rotations and displacements (for similar equations see [15], [16]). The rotations R and S as well as the twisting T of the unit cell are given by

$$
\begin{aligned}
R_1 &= -\frac{1}{2\,l_3'}\left((u_2^{A^+} - u_2^{A}) - (u_2^{0^+} - u_2^{0})\right) \\
R_2 &= \frac{1}{2\,l_3'}\left((u_1^{A^+} - u_1^{A}) - (u_1^{0^+} - u_1^{0})\right) \\
R_3 &= \frac{1}{2\,l_3'}\left((u_3^{A^+} - u_3^{A}) - (u_3^{0^+} - u_3^{0})\right) = 0 \\
S_1 &= -\frac{1}{2\,l_3'}\left((u_2^{B^+} - u_2^{B}) - (u_2^{0^+} - u_2^{0})\right) \\
S_2 &= \frac{1}{2\,l_3'}\left((u_1^{B^+} - u_1^{B}) - (u_1^{0^+} - u_1^{0})\right) \\
S_3 &= \frac{1}{2\,l_3'}\left((u_3^{B^+} - u_3^{B}) - (u_3^{0^+} - u_3^{0})\right) = 0 \\
T &= -\frac{1}{4\,l_1'\,l_3'}\left((u_2^{A^+} - u_2^{A}) - (u_2^{0^+} - u_2^{0})\right) .
\end{aligned}
\tag{2}
$$

In order to ensure that translated unit cells fit with each other forming the corrugated paperboard the relationship $R_3 = S_3 = 0$ has to hold.

The characteristic translations U and V are related to the master node displacements by

$$\begin{aligned}
U_1 &= \frac{1}{2}\left((u_1^{A^+} + u_1^A) - (u_1^{0^+} + u_1^0)\right) \\
U_2 &= \frac{1}{2}\left((u_2^{A^+} + u_2^A) - (u_2^{0^+} + u_2^0)\right) \\
U_3 &= \frac{1}{2}\left((u_3^{A^+} + u_3^A) - (u_3^{0^+} + u_3^0)\right) + \left(R_1 - R_2 \frac{d_2}{2l_2'}\right) l_2' \\
V_1 &= \frac{1}{2}\left((u_1^{B^+} + u_1^B) - (u_1^{0^+} + u_1^0)\right) \\
V_2 &= \frac{1}{2}\left((u_2^{B^+} + u_2^B) - (u_2^{0^+} + u_2^0)\right) \\
V_3 &= \frac{1}{2}\left((u_3^{B^+} + u_3^B) - (u_3^{0^+} + u_3^0)\right) - S_2 l_1' \,.
\end{aligned} \tag{3}$$

Thickness compatibility is ensured by the third and the sixth equation.

In [15] it is shown that with the definitions given by Eqn. (2) and Eqn. (3) the homogenized strains and curvature changes, respectively, follow from

$$\begin{aligned}
\varepsilon_{11} &= \frac{U_1}{2l_1'} & \chi_{11} &= -\frac{R_2}{2l_1'} \\
\varepsilon_{21} &= \frac{U_2}{2l_1'} & \chi_{21} &= T \\
\varepsilon_{12} &= \frac{V_1}{2l_2'} - \frac{U_1 d_2}{4l_1' l_2'} & \chi_{12} &= T \\
\varepsilon_{22} &= \frac{V_2}{2l_2'} - \frac{U_2 d_2}{4l_1' l_2'} & \chi_{22} &= \frac{S_1}{2l_2'} - \frac{T d_2}{2l_2'} \,.
\end{aligned} \tag{4}$$

These equations can be rewritten in matrix form as

$$\mathbf{U} = \boldsymbol{\varepsilon} \cdot \mathbf{P} \quad \Longleftrightarrow \quad U_{ij} = \varepsilon_{im} P_{mj} \,. \tag{5}$$

Here $\mathbf{U}$ is the matrix containing the master node displacements (it is also denoted as generalized master node displacement matrix), $\boldsymbol{\varepsilon}$ is the homogenized strain, and $\mathbf{P}$ is the generalized periodicity matrix, which contains the unit cell geometry l_1, l_2, and d_2.

The extended form of the matrix equation (5) reads

$$\begin{pmatrix} U_1 & V_1 \\ U_2 & V_2 \end{pmatrix} = \begin{pmatrix} \varepsilon_{11} & \varepsilon_{12} \\ \varepsilon_{21} & \varepsilon_{22} \end{pmatrix} \cdot \begin{pmatrix} p_1^A & p_1^B \\ p_2^A & p_2^B \end{pmatrix} \,. \tag{6}$$

In analogy to the displacements a relation is obtained for the curvature changes as

$$\mathbf{R} = \boldsymbol{\chi} \cdot \mathbf{P} \quad \Longleftrightarrow \quad R_{ij} = \chi_{im} P_{mj} \,. \tag{7}$$

Again, in extended form this equation reads

$$\begin{pmatrix} -R_2 & -S_2 \\ R_1 & S_1 \end{pmatrix} = \begin{pmatrix} \chi_{11} & \chi_{12} \\ \chi_{21} & \chi_{22} \end{pmatrix} \cdot \begin{pmatrix} p_1^A & p_1^B \\ p_2^A & p_2^B \end{pmatrix} \,. \tag{8}$$

The components of the vectors $\mathbf{N}^{\mathrm{h}}$ and $\mathbf{M}^{\mathrm{h}}$ in Eqn. (1), i.e., the resultant membrane forces N_{ij} and moments M_{ij}, respectively, are obtained by integration of the stresses and stress moments, respectively. The integrals read

$$N_{ij} = \frac{l_3}{V} \int_V \sigma_{ij}(\mathbf{x})\,\mathrm{d}V, \qquad M_{ij} = \frac{l_3}{V} \int_V \sigma_{ij}(\mathbf{x})\,x_3\,\mathrm{d}V\,. \tag{9}$$

Using Gauss's divergence theorem and by consideration of periodic stresses at the boundaries S_I the above equations leads to

$$N_{ij} = \sum_k \left(\frac{l_3}{2\,V} \int_{S_I^{k+}} (T_i^{k+}\,p_j^k + T_j^{k+}\,p_i^k)\,\mathrm{d}S_I^{k+} \right) \tag{10}$$

$$M_{ij} = \sum_k \left(\frac{l_3}{2\,V} \int_{S_I^{k+}} (T_i^{k+}\,p_j^k + T_j^{k+}\,p_i^k)\,x_3\,\mathrm{d}S_I^{k+} \right),$$

with $(k = A, B)$. There the tensors p_i^k are independent of location and can be extracted from the integral in Eqn. (10). Thus, the remaining integrals containing the traction vectors T_i can be rewritten as

$$\int_{S_I^{k+}} T_i^{k+}\mathrm{d}S_I^{k+} = (F_i^{k+} + F_i^k) = \bar{F}_i^k \tag{11}$$

$$\int_{S_I^{k+}} T_i^{k+}\,x_3\,\mathrm{d}S_I^{k+} = \frac{1}{2}(F_i^{k+} - F_i^k)\,l_3 = \Delta F_i^k\,l_3\,. \tag{12}$$

The introduction of a generalized master node force $\bar{\mathbf{F}}$ yields

$$\mathbf{N} = \frac{l_3}{2\,V} \left(\bar{\mathbf{F}} \cdot \mathbf{P}^{\mathrm{T}} + \mathbf{P} \cdot \bar{\mathbf{F}}^{\mathrm{T}} \right) = \begin{pmatrix} N_{11} & N_{12} \\ N_{21} & N_{22} \end{pmatrix}, \quad N_{21} = N_{12} \tag{13}$$

with

$$\bar{\mathbf{F}} = \begin{pmatrix} \bar{F}_1^A & \bar{F}_1^B \\ \bar{F}_2^A & \bar{F}_2^B \end{pmatrix} = \begin{pmatrix} F_1^{A+} + F_1^A & F_1^{B+} + F_1^B \\ F_2^{A+} + F_2^A & F_2^{B+} + F_2^B \end{pmatrix}, \tag{14}$$

where $\bar{\mathbf{F}}$ is the given/obtained generalized master node force matrix. That means, if $\boldsymbol{\mathcal{E}}$ is given (strain controlled loading), $\mathbf{U}$ follows from Eqn. (5), $\bar{\mathbf{F}}$ is computed by using FE-analyses (from master node forces), and $\mathbf{N}$ is computed from Eqn. (13).

Similarly, the matrix $\mathbf{M}$ of the resultant moments M_{ij} (see Eqn. (1)) can be found as

$$\mathbf{M} = \frac{l_3\,l_3}{2\,V} \left(\Delta\mathbf{F} \cdot \mathbf{P}^{\mathrm{T}} + \mathbf{P} \cdot \Delta\mathbf{F}^{\mathrm{T}} \right) \tag{15}$$

with

$$\Delta\mathbf{F} = \begin{pmatrix} \Delta F_1^A & \Delta F_1^B \\ \Delta F_2^A & \Delta F_2^B \end{pmatrix} = \frac{1}{2} \begin{pmatrix} F_1^{A+} - F_1^A & F_1^{B+} - F_1^B \\ F_2^{A+} - F_2^A & F_2^{B+} - F_2^B \end{pmatrix}, \tag{16}$$

where for strain controlled loading Eqns. (7) and (15) are used.

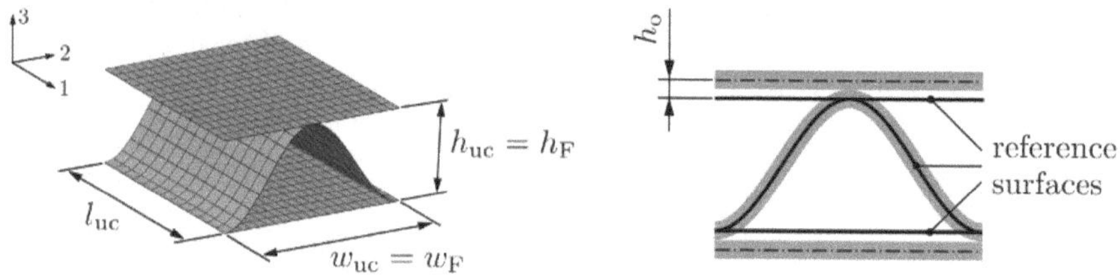

Fig. 4. FE model of the unit cell for stiffness homogenization.

For a given load case the elements of the matrices **N**, **M**, $\boldsymbol{\mathcal{E}}$, and χ can now be determined. Note that these elements are also the elements of the corresponding vectors $\mathbf{N}^{h}$, $\mathbf{M}^{h}$, $\bar{\boldsymbol{\varepsilon}}^{h}$, and $\bar{\chi}^{h}$ used in Eqn. (1). In order to compute the homogenized stiffness matrices $\mathbf{A}^{h}$, $\mathbf{B}^{h}$, and $\mathbf{D}^{h}$ of Eqn. (1) six independent load cases need to be considered.

If stress controlled loading is used, the equations look similar. There, the master nodal forces follow from given resultant forces by using the relations

$$\bar{\mathbf{F}} = \frac{V}{l_3} \mathbf{N} \cdot (\mathbf{P}^{\mathrm{T}})^{-1} \tag{17}$$

$$\Delta \mathbf{F} = \frac{V}{l_3\, l_3} \mathbf{M} \cdot (\mathbf{P}^{\mathrm{T}})^{-1} \,. \tag{18}$$

The relation between master node displacements and homogenized strains and curvature changes, respectively, follows directly from the inverse of Eqn. (5) and (7). These relations read

$$\boldsymbol{\mathcal{E}} = \frac{1}{2}\big(\mathbf{U} \cdot \mathbf{P}^{-1} + (\mathbf{P}^{-1})^{\mathrm{T}} \cdot \mathbf{U}^{T}\big) \tag{19}$$

and

$$\chi = \frac{1}{2}\big(\mathbf{R} \cdot \mathbf{P}^{-1} + (\mathbf{P}^{-1})^{\mathrm{T}} \cdot \mathbf{R}^{T}\big) \,. \tag{20}$$

Again, for the computation of the homogenized stiffness matrices in the above mentioned constitutive law, Eqn. (1), six independent load cases are needed.

The unit cell for the homogenization process of corrugated paperboard is shown in Figure 4. There h_{uc}, which is equal to h_{F} (see Figure 2), describes the height of the finite element model as the distance between the reference surfaces of the shell elements forming the liners. The parameter w_{uc} gives the width of the unit cell, which is equal to the wavelength w_{F} of the flute in the presented example (see Figure 2). The thickness of the flute is denoted as t_{F}, and the thickness of the liners as t_{L}. Finally, l_{uc} denotes the length of the unit cell in direction 1. The shape of the flute is represented by a sinus function.

The shell elements representing the liners are arranged with an offset h_{o} from the the mid-surfaces of the liners. This way the kinematic coupling between flute and liners was established by sharing the same FE nodes without changing the effective

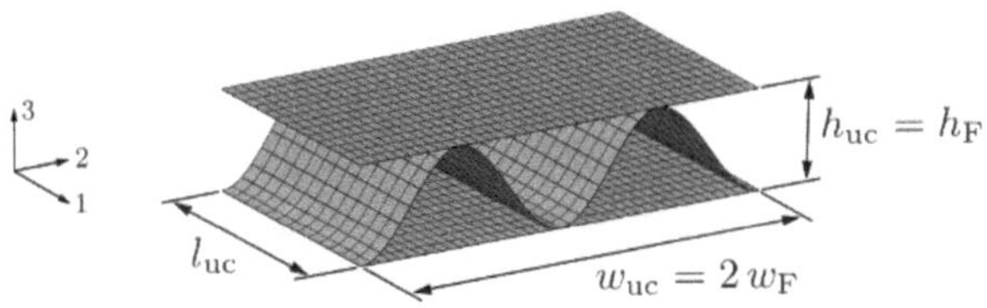

Fig. 5. Unit cell for calculation of local buckling loads and buckling modes. l_{uc} has to be determined.

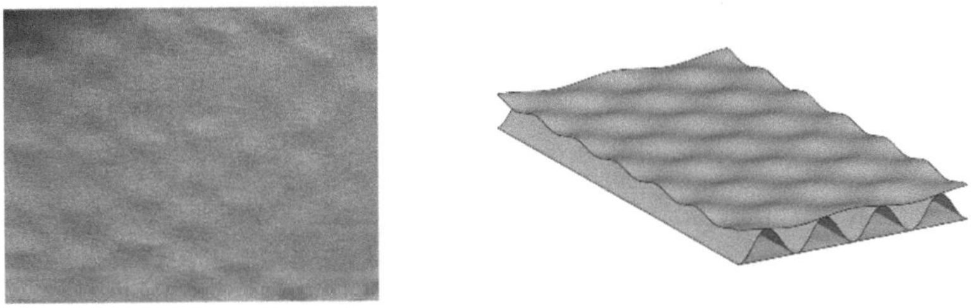

Fig. 6. Checkerboard buckling. Observed and computed buckling mode.

mechanical properties. Note that this modeling approach implies a perfect bonding between fluting and liners.

Due to the periodicity of the geometry of the cross section and the expected periodicity of the local buckling modes, periodic unit cell analyses appear to be appropriate computational approaches. However, the unit cell models used for buckling analysis differ from the ones used for stiffness homogenization. Different models are needed because the relevant local buckling mode does not necessarily need to have the same periodicity in the MD-direction (direction 2) as the geometry has and because these modes exhibit periodicity also in the CD-direction (direction 1).

In contrast to [2] the periodic unit cell for calculating local buckling contains two periods of the cross-section's geometry, see Figure 5. This is necessary in order to catch checkerboard buckling modes. Such patterns can be observed easily when the corrugated paperboard is carefully bended by hand, see Figure 6 left.

Furthermore, the length of the unit cell, l_{uc}, has to represent either the wave length of the first local buckling mode or an integer multiple of it. Certainly, this wave length is not known in advance. Hence, it has to be determined by minimizing the first local buckling eigenvalue with respect to l_{uc}. This optimization was performed using Brent's algorithm [17].

In the unit cells flute and liners were discretized with 8-noded, bi-quadratic finite shell elements with reduced integration taking also transverse, i.e., out-of-plane shear deformations into account. The latter was necessary because compared to the small dimensions at the meso level the paper used as raw material represents moderately thick plates or shells. In addition to this, the out-op-plane shear modulus of paper is very small.

Because each optimization procedure consisted of a large number of individual finite element analyses, the mesh density had to be a compromise between numerical accuracy and numerical effort. A convergence study has shown that the chosen mesh density gives reliable predictions.

2.2 Formulation of the Optimization Problem

The objective function, which should become minimum, is the corrugated paperboard's weight per unit area. For constant weight density of the paperboard this objective function can be expressed by the width-specific cross-sectional area $\bar{A}$ according to

$$\bar{A} = \frac{2w_F t_L + L_F t_F}{w_F} \,. \tag{21}$$

Therein L_F denotes the running length of flute paper for a full sinusoidal wave of the fluting, compare Figure 4. The geometrical parameters w_F, h_F, t_L, and t_F act as optimization parameters, and the constraints can be expressed by the following inequalities.

The optimized corrugated paperboard must have at least the same local buckling strength $|N_{11}^*|$ as the reference design. Here $|N_{11}^*|$ is that value of the resultant compressive membrane force per unit width of the corrugated paper board, $-N_{11}$, at which local buckling appears. Thus,

$$|N_{11}^*| \geq |N_{11,0}^*| \tag{22}$$

must hold. $|N_{11,0}^*|$ is the critical compressive membrane force for local buckling of the paperboard according to the reference design.

With respect to the global buckling strength of the board the constraint reads

$$D_{11} \geq D_{11,0} \,. \tag{23}$$

Here D_{11} is the homogenized bending stiffness of the corrugated paperboard with respect to direction 2, i.e., the (1,1)-element of the matrix $\mathbf{D}^{\mathrm{h}}$ in Eqn. (1).

It appears to be natural that all of the optimization parameters must fulfill

$$w_F, h_{uc}, t_L, t_F \geq \varepsilon \tag{24}$$

with a proper value for ε determined by restrictions coming from the production of the raw material.

The reference values used in the constraint conditions, $D_{11,0}$ and $|N_{11,0}^*|$, were calculated at the beginning of the optimization procedure. $|N_{11}^*|$ as well as D_{11} had to be determined in each step during the optimization.

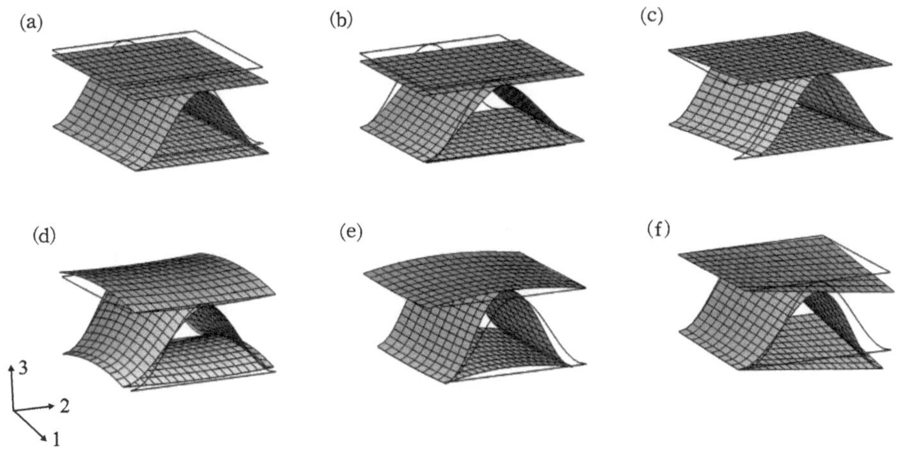

Fig. 7. Unit cell deformed under the action of six loading cases for stiffness homogenization.

2.3 Homogenization Process Using Unit Cell Models

For calculating effective properties of corrugated paperboard, such as effective stiffnesses, finite element unit cell models as shown in Figure 4 were used. The macroscopic deformation of the unit cell was super-imposed on a field of microscopic deformations that was periodic in directions 1 and 2, respectively, from one side of the unit cell to the other.

The six basic deformation modes, which are required in the application of the above described unit cell based homogenization procedure, are presented in Figure 7. Mode (a) corresponds to uni-axial tension in longitudinal direction 1, mode (b) is caused by uni-axial tension in direction 2, mode (c) is the result of in-plane shear forces. Modes (d) and (e) are predicted for bending about the direction 2 and 1, respectively. Mode (f) shows the effects of a macroscopic twisting moment.

As outlined above, from these six loading cases the stretching, coupling, and bending matrices $\mathbf{A}^{\mathrm{h}}$, $\mathbf{B}^{\mathrm{h}}$, and $\mathbf{D}^{\mathrm{h}}$ in Eqn. (1) can be determined.

As shown in [2], the value of D_{11} can be estimated by

$$D_{11} \approx E_{1,\mathrm{L}} \frac{1}{1-\nu_{12,\mathrm{L}}\,\nu_{21,\mathrm{L}}} \frac{2I_{22,\mathrm{L}}}{w_{\mathrm{F}}} + E_{1,\mathrm{F}} \frac{I_{22,\mathrm{F}}}{w_{\mathrm{F}}} . \tag{25}$$

Here E_1 is the Young's modulus in direction 1, ν_{21} and ν_{12} are the respective Poisson's ratios, and I_{22} is the moment of inertia per period of the cross section of the board with respect to the direction 2. As usual, the subscripts "F" and "L" denote properties of the flute and the liners, respectively.

2.4 Unit Cell Approach for Local Buckling Analysis

As mentioned in Section 2.1, for the local buckling analysis unit cells are required which differ from that ones used in the homogenization procedure. This is due to

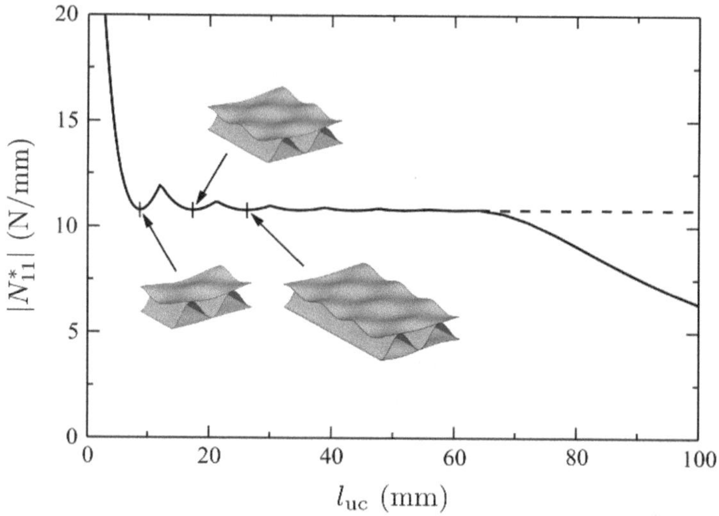

Fig. 8. Dependence of the critical compressive membrane force on the length of the unit cell, l_{uc}. The first minimum represents a proper l_{uc}-value.

the fact that buckling and post-buckling deformations may show a higher degree of periodicity as compared to the sub critical deformation on the meso level. Figure 5 shows a typical unit cell used in buckling analysis.

As also described above, the length of the unit cell, l_{uc}, requires special attention. The length l_{uc} has to represent either the wave length of the first local buckling mode or an integer multiple of it. Unit cells with l_{uc} not fulfilling this requirement lead to first eigenvalues of the buckling analysis which are higher than the correct ones. For too long unit cells the first buckling eigenvalue does no longer correspond to local but to global buckling. These facts are clearly demonstrated in Figure 8 and have to be considered during optimization.

2.5 Numerical Meso Structural Optimization Approach

The above described unit cell scheme for the determination of that membrane load in the longitudinal direction $|N^*_{11}|$, which is critical in terms of local buckling for a given set of governing geometrical parameters, was embedded as a constraint (cf. inequality (22)) into a numerical optimization scheme for minimizing the width specific cross-sectional area $\bar{A}$. The second constraint (cf. inequality (23)), which concerns the stiffness of the corrugated paperboard, is purely based on the optimization parameters and the elastic moduli of the raw material. The full numerical optimization scheme is shown schematically in Figure 9.

The structural optimization, which is represented by the outer loop in Figure 9, was performed by a numerical procedure in the form of the SciPy function "fmin_cobyla" [18]. This SciPy function is suitable for determining a local min-

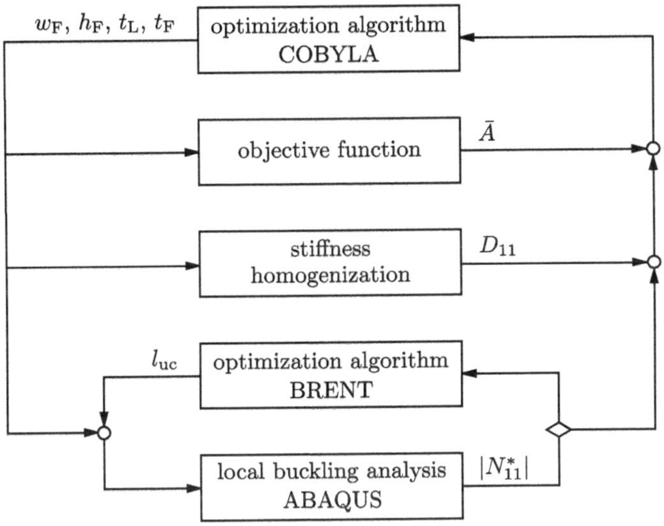

Fig. 9. Flow diagram of nested optimization loops.

imum for a given nonlinear objective function in the presence of nonlinear optimization constraints. "COBYLA" stands for "constrained optimization by linear approximation". The iterative algorithm allows for an optimization without calculation of derivatives. It was originally proposed by Powell [19].

As can be seen in Figure 9, within each structural optimization step (outer loop) a further optimization has to be performed (inner loop) in order to determine the actual length of the unit cell, l_{uc}. For this purpose Brent's algorithm [17] was applied in order to find the first minimum of the festoon shaped curve in Figure 8.

By using the described nested numerical optimization scheme it is possible to improve any given set of starting geometrical parameters in terms of minimizing the effective area-specific weight of corrugated paperboard while at the same time providing a design with the necessary strength for avoiding local and global buckling.

In the following section the results are presented for a given reference design.

3 Results of Optimization

The geometrical parameters of the reference configuration together with the material properties of the paper used for liner and flute, respectively, are given in Table 1.

The material data are taken from [4]. Here the l-direction coincides with the direction MD and the q-direction with the direction CD. The material orientation is sketched by example of one liner in Figure 10.

The results obtained by applying the above described optimization procedure depend on the start configuration. This is a consequence of the highly nonlinear de-

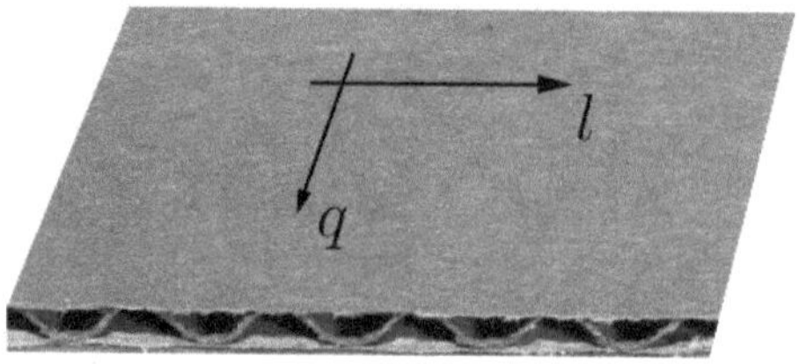

Fig. 10. Definition of the local material orientation.

Table 1. Geometrical parameters for the corrugated paperboard, which serves as the reference design in the optimization, along with orthotropic material data. All data correspond to an actual corrugated paperboard as published in [4].

Geometry
$w_F = 7.2$ mm
$h_F = 3.6$ mm
$h = 4.083$ mm
$t_L = 0.231$ mm
$t_F = 0.252$ mm

	Liner	Flute	
E_l	8220	5270	MPa
E_q	3010	2260	MPa
ν_{lq}	0.17	0.17	-
G_{lq}	1920	1340	MPa
G_{lt}	45	45	MPa
G_{qt}	45	45	MPa

pendence of the objective function and the constraint conditions on the respective optimization parameters. Due to the presence of a number of local minima the optimization procedure was performed 100 times with different start configurations. Thereby, the individual geometrical parameters of the start configurations were randomly selected within the intervals $0.5\, p_{\text{ref}} \leq p \leq 2\, p_{\text{ref}}$ with $p = w_F, h_{uc}, t_L, t_F$. The subscript "ref" denotes the respective dimension of the reference configuration. Finally, the minmum of all found local minima was searched and used as result of the optimization.

This procedure has lead to a result that is 18.7% lighter than the reference configuration while predicting the same critical membrane force $|N^*_{11}| = |N^*_{11,0}|$ regarding local buckling of the perfect meso structure and the same bending stiffness $D_{11} = D_{11,0}$ relevant for global buckling, respectively, as the reference configuration.

The corresponding geometrical parameters are given in Figure 11, where the "optimal" configuration is compared to the reference configuration.

This figure shows that the thickness of the paperboard, h, increases as a result of the optimization procedure while the wavelength, i.e., the length of the period of the flute, w_F, decreases. Hence, the flanks of the sinusoidal flute become steeper. Both the thickness of the fluting paper and the thickness of the liners are reduced. This is the primary reason for the weight decrease.

The optimized configuration of the corrugated paperboard is not only significantly lighter but exhibits also a different buckling mode, see Figure 12. The reference

Reference Configuration
$w_F = 7.20\,\text{mm}$
$h_F = 3.60\,\text{mm}$
$t_L = 0.231\,\text{mm}$
$t_F = 0.252\,\text{mm}$
$\bar{A} = 0.831\,\text{mm}^3/\text{mm}^2$
$\lvert N^*_{11} \rvert = 10.77\,\text{N/mm}$
$D_{11} = 7029\,\text{Nmm/mm}$

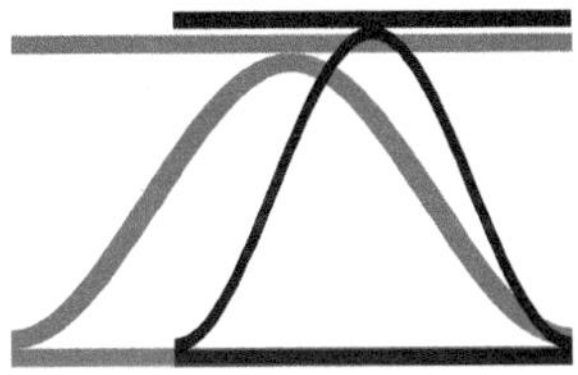

Optimized Configuration
$w_F = 5.10\,\text{mm}$
$h_F = 4.06\,\text{mm}$
$t_L = 0.205\,\text{mm}$
$t_F = 0.137\,\text{mm}$
$\bar{A} = 0.676\,\text{mm}^3/\text{mm}^2$
$\lvert N^*_{11} \rvert = 10.77\,\text{N/mm}$
$D_{11} = 7029\,\text{Nmm/mm}$

Fig. 11. Comparison between reference configuration (gray) and optimized configuration (dark).

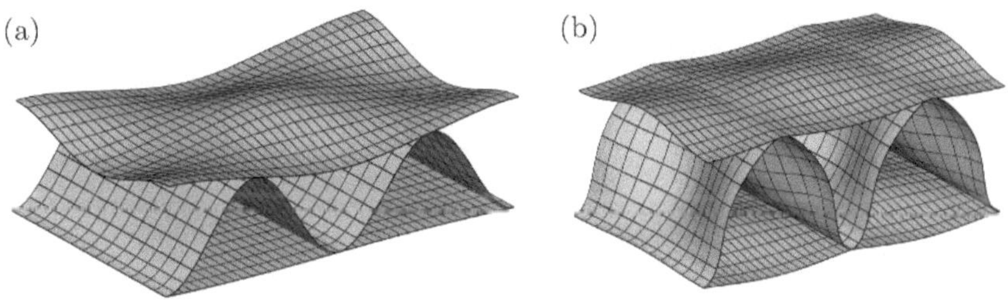

Fig. 12. Fundamental mode of local buckling, (a) for the reference configuration, (b) for the optimized configuration.

configuration shows local buckling of the liners in the checkerboard pattern, and the flute is not affected by buckling (left picture in Figure 12). In contrast to this behavior, in the optimal configuration simultaneous buckling of liner and flute appears, leading to a buckling pattern, which is symmetrical with respect to the apexes of the flute's wave, i.e., the bonding lines between fluting and liners.

As a drawback of this result it should be mentioned that optimizations leading to structures with simultaneous buckling of different components involves the danger of substantial increase of imperfection sensitivity, as outlined already in [20]. Accordingly, the optimized corrugated paperboard is more imperfection sensitive than the reference design, which does not show simultaneous buckling of liners and flute.

4 Fold Formation in the Post-Buckling Regime

In the handling of corrugated paperboard it is an everday experience that failure initiated by buckling ends up with the formation of sharp folds in the board, compare Figure 13.

The analysis of the post-buckling behavior of a strip of corrugated paperboard shows that these folds are the result of localization of the initially periodic local buckling pattern finally leading to global bending in the post-buckling regime, see Figure 14.

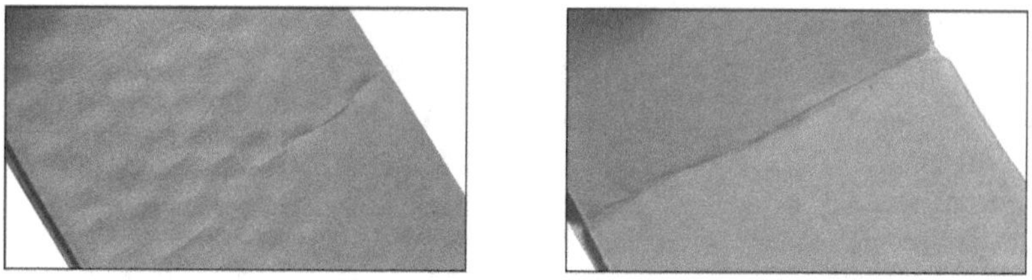

Fig. 13. Observation of formation of folds after local buckling.

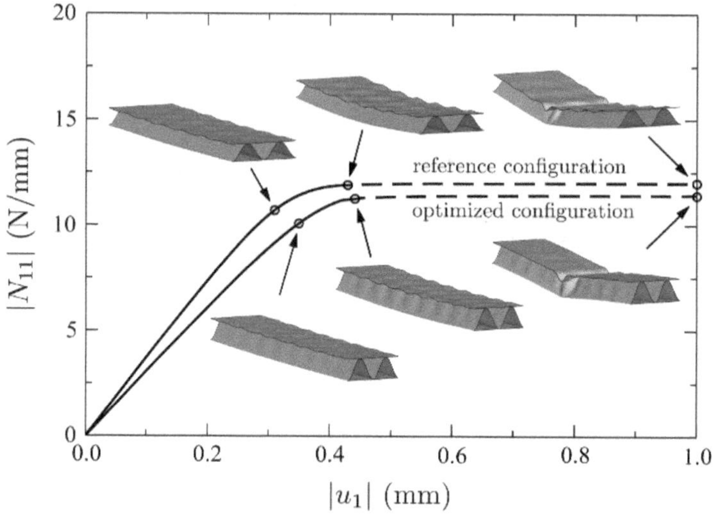

Fig. 14. Load displacement behavior of buckling and folding for the corrugated paperboard in reference configuration and in optimized configuration.

In this figure the behavior of a strip with a meso structure according to the reference configuration and the weight-optimized configuration, respectively, are considered. The load displacement paths are the result of implicit dynamic analyses under load control. The boundary conditions were chosen so that the small edges, where the in-plane compression load is applied, are assumed to be rigidly movable, and at the long edges symmetry conditions have been applied in order to simulate a strip with a large number of flute periods. In both considered configurations a small geometrical imperfection was introduced according to the respective first buckling mode. The imperfection amplitude was in both cases 10% of the liner's paper thickness.

As a consequence of the boundary conditions, which – for simulating a finite size plate – differ slightly from the unit cell's boundary conditions used in the optimization procedure, the optimized configuration starts buckling also with a checkerboard pattern as the reference configuration does. However, in contrast to the reference design, in the optimized meso structure the flute takes part on the post-buckling deformations from the beginning on.

Furthermore, the comparison between the two load displacement paths shows that the ultimate load of the corrugated paperboard with the optimized geometrical parameters is smaller than the one of the reference design. This supports what was stated in the discussion of the results of the optimization: The corrugated paperboard with the weight optimized geometrical parameters of the meso structure is prone to exhibit a higher imperfection sensitivity as compared to the reference design. The smaller in-plane stiffness of the weight-optimized paperboard is a direct consequence of the reduced cross-sectional area.

5 Conclusions

Based on unit cell approaches in combination with finite element methods in this paper procedures for homogenization of the stiffness behavior as well as for calculating critical loads and modes regarding local buckling of corrugated paperboard are presented. These procedures have been incorporated in an optimization strategy for reducing the area-specific weight of corrugated paperboard under strength constraints. Both a required minimum resistance of the board against local and against global elastic buckling must not be dropped by the design which is defined by the principal geometrical parameters of the corrugated paperboard on the meso level. Compared to a given reference configuration, which represents an actual corrugated paperboard, a 18.7% weight reduction was achieved. The design resulting from the optimization procedure shows simultaneous buckling of liners and fluting under compressive membrane loads in the longitudinal direction of the flute waves. Thus, the danger of increased imperfection sensitivity must be taken into account. Consequently, in real designs one should take the geometrical parameters close to but not identical with that of the "optimized" design in order to avoid simultaneous local buckling configurations.

The formation of folds as typical failure modes of corrugated paperboard is investigated and identified as localization of the initially periodical local buckling pattern in the deep post-buckling regime.

References

1. Daxner, T., Pahr, D.H., and Rammerstorfer, F.G., Micro- meso-instabilities in structured materials and sandwich structures. In: B.G. Falzon and M.H. Aliabadi (Eds.), *Buckling and Postbuckling Structures*. Imperial College Press, London, 2008.
2. Daxner, T., Flatscher, T., and Rammerstorfer, F.G., Optimum design of corrugated board under buckling constraints. In: B.M. Kwak (Ed.), *Proceedings 7th World Congress on Structures and Multidisciplinary Optimization*, BMD Co., Seoul, pp. 349–358, 2007.
3. Allansson, A. and Svärd. B., Stability and collapse of corrugated boards; numerical and experimental analysis. Master's Dissertation, Lund University, Lund, Sweden, 2001.

4. Patel, P., Nordstrand, T., and Carlsson, L.A., Local buckling and collapse of corrugated board under biaxial stress. *Compos. Struct.*, 39(1/2):93–110, 1997.
5. Nyman, U. and Gustafsson. P.J., Material and structural failure criterion of corrugated board facings. *Compos. Struct.*, 50:79–83, 2000.
6. Nyman, U., Continuum mechanis modelling of corrugated board. PhD Thesis, Lund University, Lund, Sweden, 2004.
7. Biancolini, M.E. and Brutti, C., Numerical and experimental investigations of the strength of corrugated board packages. *Pack. Tech. Sci.*, 16:47–60, 2003.
8. Nordstrand. T., Basic testing and strength design of corrugated board and containers. PhD Thesis, Lund University, Lund, Sweden, 2003.
9. Thakkar, B.K., Gooren, L.G.J., Peerlings, R.H.J., and Geers, M.G.D., Experimental and numerical investigation of creasing in corrugated paperboard. *Phil. Mag.*, 88(28/29):3299–3310, 2008.
10. Flatscher, T., Modellierung der Steifigkeit und Stabilität von Wellpappe. Diploma Thesis, Vienna University of Technology, Vienna, 2006.
11. Pahr, D.H., Experimental and Numerical Investigations of Perforated FRP-Laminates. Fortschritt-Berichte VDI Reihe 18 Nr. 284, VDI-Verlag, Düsseldorf, Germany, 2003.
12. Pahr, D.H. and Rammerstorfer, F.G., Buckling of honeycomb sandwiches: Periodic finite element considerations. *CMES – Comp. Model Eng.*, 12(3):229–242, 2006.
13. Rammerstorfer, F.G., Pahr, D.H., Daxner, T., and Vonach, W.K., Buckling in thin walled micro and meso structures of lightweight materials and material compounds. *Comput. Mech.*, 37(6):470–478, 2006.
14. Jones, R.M., *Mechanics of Composite Materials*, 2nd edn. Taylor and Francis, Philadelphia, USA, 1999.
15. Anthoine, A., Derivation of the in-plane elastic characteristics of masonry through homogenization theory. *Int. J. Solids Structures*, 32(2):137–163, 1995.
16. Hohe, J., A direct homogenisation approach for determination of the stiffness matrix for microheterogeneous plates with application to sandwich panels. *Composites Part B: Engineering*, 34:615–626, 2003.
17. Brent, R.P., *Algorithms for Minimization without Derivatives*. Prentice-Hall, Englewood Cliffs, NJ, chapters 3–4, 1973.
18. Jones, E., Oliphant, T., Peterson, P., et al (2001) SciPy: Open source scientific tools for Python. http://www.scipy.org/, 2001.
19. Powell, M.J.D., A direct search optimization method that models the objective and constraint functions by linear interpolation. In: S. Gomez and J.P. Hennart (Eds.), *Advances in Optimization and Numerical Analysis*. Kluwer Academic Publishers, Dordrecht, pp. 51–67, 1994.
20. Thompson, J.M.T. and Hunt, G.W., Dangers of structural optimization. *Eng. Opt.*, 1:99–110, 1974.

Framework for Multi-Level Optimization of Complex Systems

Albert de Wit[1,2] and Fred van Keulen[2]

[1]National Aerospace Laboratory – NLR, Anthony Fokkerweg 2, 1059 CM Amsterdam, The Netherlands; adewit@nlr.nl
[2]Department of Precision and Microsystems Engineering, Delft University of Technology, Mekelweg 2, 2628 CD Delft, The Netherlands; a.vankeulen@tudelft.nl

Abstract Complex systems, such as those encountered in aerospace engineering, can typically be considered as a hierarchy of individual elements. This hierarchy is reflected in the analysis techniques that are used to analyze the characteristics of the system as a whole. Consequently, a hierarchy of models is to be used, each accounting for different physical scales, components or disciplines. This chapter presents a framework for the multi-level optimization of such complex systems. The framework is demonstrated on maximizing the range of a supersonic business jet via multi-level optimization. Flexibility of the framework is shown considering two different decompositions of the model and two different coordination procedures. The latter, are illustrated using a simple two-bar truss structure. Furthermore, via adjusting algorithm settings a significant reduction of computational costs is shown at the cost of finding a design with marginally lower range.

Key words: Multi-level, multi-disciplinary, optimization, framework, decomposition, coordination

1 Introduction

Complex systems, such as those encountered in aerospace engineering, can typically be considered as a hierarchy of individual coupled elements. This hierarchy is often reflected in the analysis techniques that are used to analyze the physical characteristics of the system. Consequently, a hierarchy of coupled models is to be used, accounting for different physical scales, components and/or disciplines. Optimization of complex systems with embedded hierarchy is accomplished via multi-level optimization methods. Multi-level optimization methods utilize the hierarchical nature of complex systems to distribute the optimization process into smaller coupled, but less complex optimization problems located at the individual elements of the hierarchy.

In the literature a number of multi-level optimization methods have been documented for the optimization of complex systems, e.g., Analytical Target Cascading [6] and Bi-Level Integrated System Synthesis [9], among others. These methods differ in how coupled elements are temporarily decomposed and how the solution of

R. de Borst and E. Ramm (eds.), *Multiscale Methods in Computational Mechanics*, Lecture Notes in Applied and Computational Mechanics 55, DOI 10.1007/978-90-481-9809-2_18,

the individual elements is coordinated. Furthermore, each method defined in literature is formulated for specific problems. Therefore, there is a need for a generalized formulation that combines the generic aspects of multi-level optimization methods such that it can handle the majority of problems using a variety of optimization techniques.

This chapter aims at formulating a generic multi-level optimization framework that combines different aspects of multi-level optimization methods. Therefore, a notation is introduced that is similar for each individual element of the hierarchy. Furthermore, two decomposition formulations are distinguished as well as means to illustrate coupling in the optimization problem. Coordination is shown as part of the decision making process for which three different cases are distinguished. The framework is demonstrated on the optimization of a supersonic business jet via combining two different decomposition approaches and two different coordination methods.

This chapter is organized as follows. In Section 2 a notation for multi-level optimization problems is introduced. Section 3 describes the decomposition process via an illustrative coupling circle and problem matrix. Coordination of the individual coupled elements is discussed in Section 4. The structure of a software program that implements the generalized multi-level framework is discussed in Section 5. Numerical results for the multi-level optimization of a supersonic business jet problem are presented in Section 6. Finally, conclusions are presented in Section 7.

2 A Unifying Multi-Level Notation

A multi-level problem typically has a hierarchical structure of individual elements as is illustrated in Figure 1a. The top-level is denoted Level-0. At this level, the top-element or global design variables are distinguished, which are denoted ${}^{0}\mathbf{x}$. One level lower, i.e. at Level-1, the "children" of Level-0 ("parent") can be found. The elements at Level-1 are numbered via the left superscript, i.e. ${}^{\cdot\cdot}\mathbf{x}$. At the second level, Level-2, the children of the Level-1 elements are found. They have two numbers in the left superscript, i.e. ${}^{\cdot\cdot,\cdot\cdot}\mathbf{x}$. The first reflects the parent in the hierarchy and the second their place among the children of the parent element. In a similar fashion, more levels can be added to the hierarchy. Consequently, at every level the relative position can easily be seen from the superscript in front of a (design) variable. Similarly, the physical responses corresponding to the elements are identified. This is illustrated in Figure 1b.

Although for each element a set of physical responses can be identified, these responses are typically interacting with responses in neighboring elements, i.e. parent, brothers/sisters and children. In other words, there are couplings which need to be taken into account as is illustrated in Figure 2. For this purpose, the operators $\mathcal{H}$ are introduced. These operators map the response from one element, onto another. In front of the operator in the left superscript the origin of the information is indicated, i.e. ${}^{\cdot\cdot}\mathcal{H}$. In the left subscript, the destination of the information is specified, i.e. ${}_{\cdot\cdot}\mathcal{H}$.

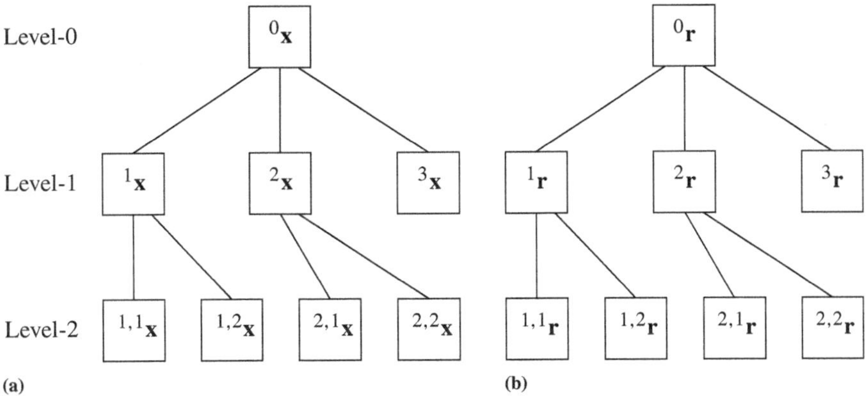

Fig. 1. Multi-level notation for a three-level hierarchy. (a) The design variables ($^{\cdot\cdot}\mathbf{x}$) corresponding to the individual elements are illustrated. (b) The physical responses ($^{\cdot\cdot}\mathbf{r}$) corresponding to each of the elements.

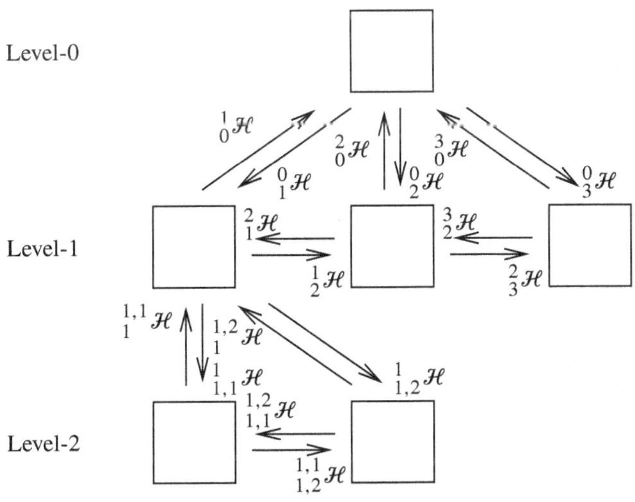

Fig. 2. Physical interactions between elements. The operators ${}^{\cdot\cdot}_{\cdot\cdot}\mathcal{H}$ map the physical responses of an element onto a neighboring element. This can be a mapping taking place at the same level, though it may also map between different levels.

Design variables can be present that are shared among multiple elements. These design variables are considered separately from the individual design variables $^{\cdot\cdot}\mathbf{x}$. The shared design variables are written as ${}^{\cdot\cdot}_{\cdot\cdot}\mathbf{z}$. The top-left index ($^{\cdot\cdot}$) represents the element to which the design variable is assigned and the bottom-left index ($_{\cdot\cdot}$) represents the element that shares that design variable. In case the same design variable is shared among the parent element and two or more child elements a dot (${}^{\cdot\cdot}_{\cdot\cdot\,.\,\cdot\cdot}\mathbf{z}$) separates the indices from each of the child elements. Shared design variables are shown

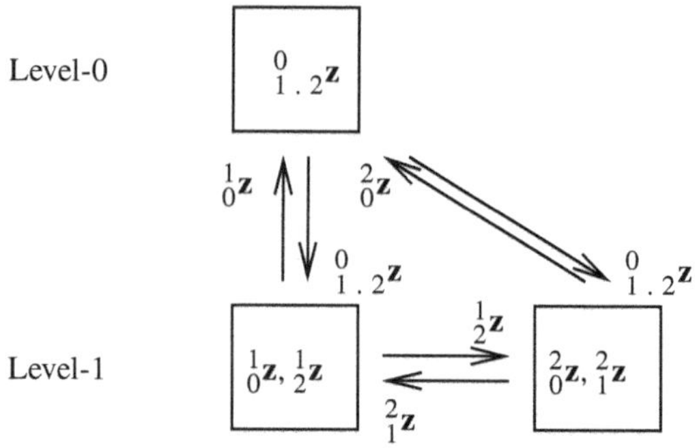

Fig. 3. Design variables that are shared among multiple elements. The superscript represents the element to which the design variable is associated and the subscript represents the neighboring element(s) that share the design variable.

in Figure 3, where a design variable is shared among three elements distributed over two levels of a hierarchy.

Responses that are mapped from one element onto another element are introduced via the operator $\mathcal{H}$. Once a response is mapped from one element onto another element it is written as ${}^{\cdot\cdot}_{\cdot\cdot}\mathbf{h}$, see Figure 5. This variable is called a coupling variable and is defined as: ${}^{\cdot\cdot}_{\cdot\cdot}\mathbf{h} = {}^{\cdot\cdot}_{\cdot\cdot}\mathcal{H}({}^{\cdot\cdot}\mathbf{r})$. This coupling variable can be used in one element as a desirable property of neighboring elements. Hence, instead of mapping the physical responses coming from a neighboring element onto the current element: ${}^{\cdot\cdot}_{\cdot\cdot}\mathcal{H}({}^{\cdot\cdot}\mathbf{r}) = {}^{\cdot\cdot}_{\cdot\cdot}\mathbf{h}$, the value of the coupling variable (${}^{\cdot\cdot}_{\cdot\cdot}\mathbf{h}$) could be chosen. The corresponding physical responses (${}^{\cdot\cdot}\mathbf{r}$) are determined via the neighboring element.

The abstract notation is illustrated via a simple mechanical problem shown in Figure 4. The two-bar truss problem is simple, yet captures the essence of large structural multi-level optimization problems. Analogous to Figure 1, two individual levels are distinguished.

The top level covering overall structural dimensions (Length L, wall mounts ${}^{0}x_1$ and ${}^{0}x_2$ and cross-sectional areas ${}^{1}_{0}h$ and ${}^{2}_{0}h$) that are required to compute overall physical responses (${}^{0}\mathbf{r}$), in the present example the top displacement of the structure. These overall physical responses are only visible when considering the entire system.

As a second level, two individual elements are distinguished. These individual elements are the bar columns for which a detailed cross-section representation (via ${}^{i}x_1, {}^{i}x_2$) can be distinguished. At this level, physical responses are evaluated, such as stresses (${}^{i}\mathbf{r}$), that are only visible when detailed modeling of the structure is conducted or in a decomposed setting a modeling of the subsystem is carried out.

Physical coupling is present between the levels via mapping of displacements (${}^{0}\mathbf{r}$), evaluated at Level-0, onto the individual elements that are located at Level-1. In the present example, displacements evaluated at Level-0 are mapped as internal forces onto the individual Level-1 elements. Likewise, stiffness parameters ${}^{i}_{0}h$ (cross-sectional areas) are mapped from Level-1 onto the Level-0 element.

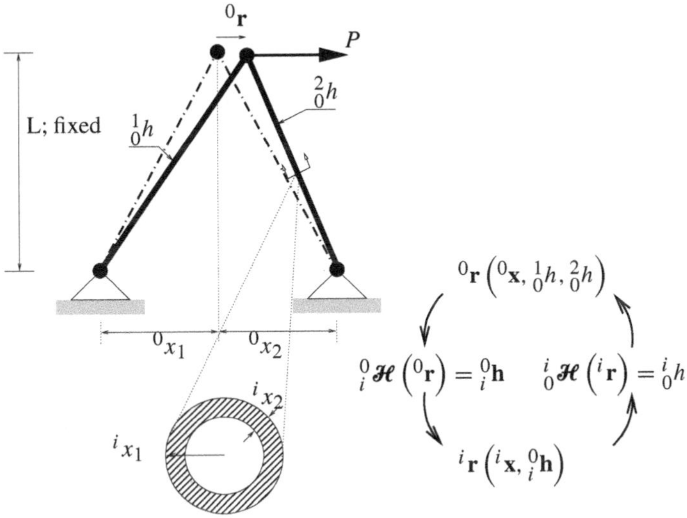

Fig. 4. Two-bar truss structure with embedded hierarchy. Level-0 responses are the displacement and the total mass of the structure. These responses depend on stiffness parameters of the two bar members. These stiffness parameters are determined via Level-1 responses.

Mapped responses in the present two-bar truss example (recall Figure 4) are the displacements from which internal forces are computed and the geometrical characteristics from which stiffness parameters (cross-sections) are computed. Hence, coupling parameters present in the two-bar truss example are the internal forces ${}^{0}_{i}\mathbf{h}$ and the cross-sectional areas ${}^{i}_{0}\mathbf{h}$, shown in Figure 4.

Consistency constraints are introduced in order to temporarily decouple the elements from elements to which they are coupled, such that each of them is solved without interacting with other elements. Decoupling is accomplished via requiring that mapped responses (${}^{\cdot\cdot}_{\cdot\cdot}\mathcal{H}({}^{\cdot\cdot}\mathbf{r})$) remain equal to the coupling variables (${}^{\cdot\cdot}_{\cdot\cdot}\mathbf{h}$) and that shared design variables (${}^{i}_{j}\mathbf{z}$ and ${}^{j}_{i}\mathbf{z}$) remain equal. Hence, these consistency constraints are mathematically expressed as: ${}^{\cdot\cdot}_{\cdot\cdot}\mathbf{c} = {}^{\cdot\cdot}_{\cdot\cdot}\mathcal{H}({}^{\cdot\cdot}\mathbf{r}) - {}^{\cdot\cdot}_{\cdot\cdot}\mathbf{h}$ and ${}^{i}_{j}\mathbf{c}_z = {}^{i}_{j}\mathbf{z} - {}^{j}_{i}\mathbf{z}$ $(i \neq j)$. A consistency constraint that is assigned to an element enables the element to change ${}^{\cdot\cdot}\mathbf{r}$ and/or ${}^{i}_{j}\mathbf{z}$ or when applicable ${}^{\cdot\cdot}_{\cdot\cdot}\mathbf{h}$ and/or ${}^{j}_{i}\mathbf{z}$ without communicating directly to the element to which it is coupled. These consistency constraints are written ${}^{\cdot\cdot}_{\cdot\cdot}\mathbf{c}$ and ${}^{\cdot\cdot}_{\cdot\cdot}\mathbf{c}_z$, where the upper left subscript ${}^{\cdot\cdot}\mathbf{c}$, ${}^{\cdot\cdot}\mathbf{c}_z$ indicates the origin of the mapped responses and/or shared design variables and the lower left subscript ${}_{\cdot\cdot}\mathbf{c}$, ${}_{\cdot\cdot}\mathbf{c}_z$ indicates the destination, see Figure 5. This notation is in accordance with the notation for coupling (hence operators $\mathcal{H}$) or shared design variables ($\mathbf{z}$) between two individual elements. Thus, decoupled hierarchical elements have been obtained via the introduction of consistency constraints and additional variables. Consequently, the consistency constraints are added to the optimization functions defined for the hierarchical elements.

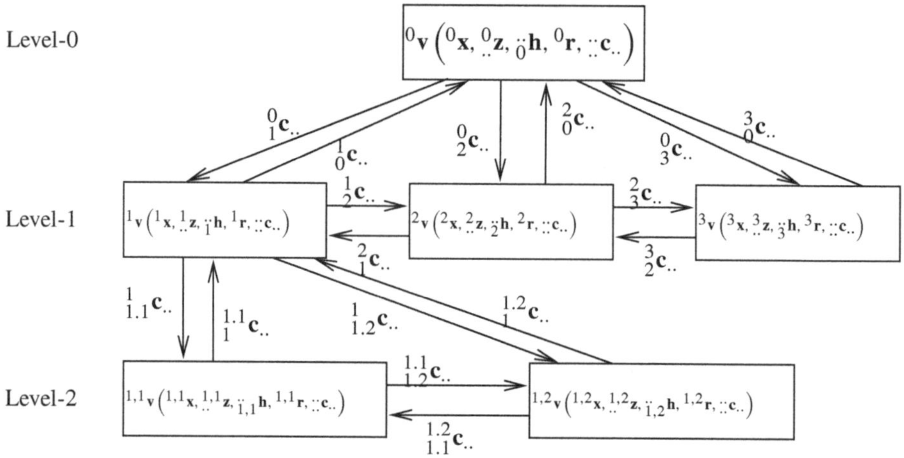

Fig. 5. Functions used in the optimization formulation of the hierarchy. These functions typically depend on design variables, the constraints and the physical responses of the corresponding element.

The final step in setting up an optimization problem is to formulate objective and constraint functions ($^{\cdot\cdot}\mathbf{v}$) that can be minimized (maximized) by the optimizer. These functions depend on local design variables ($^{\cdot\cdot}\mathbf{x}$), local responses ($^{\cdot\cdot}\mathbf{r}$) and consistency conditions ($^{\cdot\cdot}_{\cdot\cdot}\mathbf{c}$) as well as the coupling variables ($^{\cdot\cdot}_{\cdot\cdot}\mathbf{h}$ and $^{\cdot\cdot}_{\cdot\cdot}\mathbf{z}$) between individual elements. These design functions ($^{\cdot\cdot}v$) are illustrated in Figure 5.

Typical objective and constraint functions for structural optimization problems are minimal weight and maximum stiffness. In addition, typical constraints reflect buckling and stresses as well as manufacturing constraints. In the present two-bar truss example the Level-0 objective is total structural weight subjected to a maximum tip displacement at the top. One level lower constraints considering buckling and maximum stress are present.

The hierarchy described in this section is an abstract representation of a complex system. It can be interpreted in three different ways:

- The physical behavior of the system can be fully described via a huge amount of parameters. However, the size of the model poses problems on the computer infrastructure on which the model should be evaluated. Therefore, a hierarchy is defined that splits the optimization problem into smaller coupled problems that can be evaluated on the available computer infrastructure. Behavior of individual components that reflect the smaller coupled problems is computed via a relatively small number of parameters. Coupling is present between the individual components to translate properties of elements higher in the hierarchy into properties of elements lower in the hierarchy and *vice versa*. This coupling is covered via the coupling between levels.
- The physical behavior cannot be described via a single disciplinary model. Coupling is present, however there is no single model that can replace the individual

models. This type of coupling is present in the mapping of physical responses from one element onto a neighboring element. This coupling is covered via the coupling between elements that are present on the same level.

- The physical behavior covers multiple scales, the numerical computations of the underlying theory are too costly to accurately model the entire system using an all-in-one approach. Therefore, a mapping from a small-scale model onto a larger scale model is needed and *vice versa*. This coupling is covered via the coupling between levels.

The analysis of complex structures typically involves the above three categories. Optimization of a structure that embeds one or more of the above categories can be accomplished via an all-in-one optimization, provided enough resources to solve the problem are available. Hence, a single optimization problem is formulated where the responses are computed at the different elements of the hierarchy they occur. Consistency between individual elements is to be maintained between the physical analysis models throughout every iteration step of the optimization process. In contrast, a multi-level formulation takes the consistency constraints into account as part of the optimization problem formulation, such that part of the optimization process can be conducted at the individual elements.

3 Decomposition

Decomposition involves identifying (weak) links between elements that are coupled, allowing the elements to represent individual, though coupled, optimization problems. The origin of this coupling can be physically or can be prompted by the nature of the design optimization problem at hand. Generic properties of physical coupling between elements is captured via an illustrative coupling circle (Section 3.1) and the coupling within the optimization problem is shown via a problem matrix (Section 3.2).

3.1 Physical Coupling

Physical coupling between two elements is shown in Figure 6a. A top element (Level-0) produces a physical response ${}^{0}\mathbf{r}$ which is mapped into information ${}^{0}_{1}\mathbf{h}$ that is necessary to compute the physical responses at the bottom (Level-1) element. The bottom element computes a physical response which is mapped back to the top element.

Decoupling is accomplished via consistency constraints. There are two formulations that are used frequently in order to maintain consistency between two coupled elements. These formulations are shown in Figures 6b and 6c and can mathematically be expressed as:

1. strong (equality) constraints/conditions:

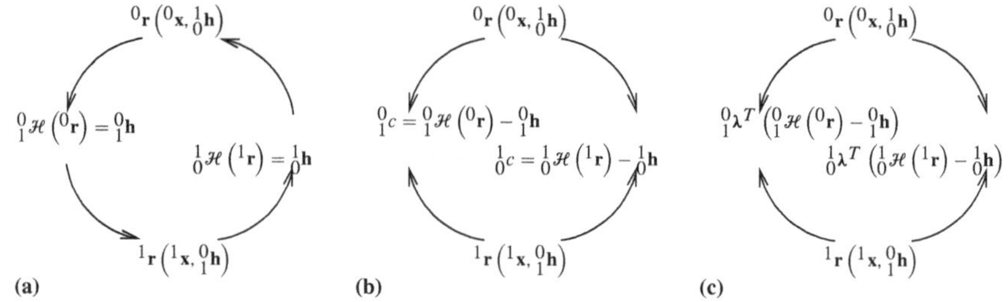

Fig. 6. (a) Physical response interaction between two elements. The responses ${}^{0}\mathbf{r}$ computed at the parent element are transformed by the operator ${}^{0}_{1}\mathcal{H}$ into information ${}^{0}_{1}\mathbf{h}$ influencing the responses of the child element ${}^{1}\mathbf{r}$. Likewise, the responses from the child element are transformed by the operator ${}^{1}_{0}\mathcal{H}$ into information ${}^{1}_{0}\mathbf{h}$ influencing the responses of the parent element. (b) Consistency between the coupled elements is maintained through interface compatibility or equilibrium. (c) Consistency is relaxed using Lagrange multipliers. Typically, these multipliers are controlled, such that the inconsistencies vanish at compatibility or equilibrium of the interface.

$$\begin{aligned} {}^{0}_{1}\mathbf{c} &= {}^{0}_{1}\mathcal{H}({}^{0}\mathbf{r}) - {}^{0}_{1}\mathbf{h} = \mathbf{0}; \\ {}^{1}_{0}\mathbf{c} &= {}^{1}_{0}\mathcal{H}({}^{1}\mathbf{r}) - {}^{1}_{0}\mathbf{h} = \mathbf{0}. \end{aligned} \tag{1}$$

2. weak (relaxation), e.g. using Lagrange multipliers:

$$\begin{aligned} {}^{0}_{1}\boldsymbol{\lambda}^{T}\,{}^{0}_{1}\mathbf{c} &= {}^{0}_{1}\boldsymbol{\lambda}^{T}\left({}^{0}_{1}\mathcal{H}({}^{0}\mathbf{r}) - {}^{0}_{1}\mathbf{h}\right); \\ {}^{1}_{0}\boldsymbol{\lambda}^{T}\,{}^{1}_{0}\mathbf{c} &= {}^{1}_{0}\boldsymbol{\lambda}^{T}\left({}^{1}_{0}\mathcal{H}({}^{1}\mathbf{r}) - {}^{1}_{0}\mathbf{h}\right). \end{aligned} \tag{2}$$

Other common relaxation formulations involve penalty functions or an Augmented Lagrangian function. Each decomposition formulation can be subdivided into:

1. hierarchic decomposition, which is subdivided into top-down or/and bottom-up formulations.
2. non-hierarchic decomposition, which treats all elements equal.

Hierarchic decomposition is the result of identifying elements in the hierarchy that dictate the output from other elements. This is shown in Figure 7a in case of a top-down decomposition with relaxed consistency constraints, e.g. Lagrange relaxation. In this figure, the top Level-0 element prescribes the necessary output from the Level-1 element, i.e. ${}^{1}_{0}\mathbf{h}$. Consistency constraints are assigned to the bottom element constraining the right part of the coupling circle. A similar but opposite approach is possible, where the Level-1 element prescribes the necessary output of the Level-0 element, i.e. ${}^{0}_{1}\mathbf{h}$. The latter approach is shown in Figure 7b and is called a bottom-up decomposition constraining the left part of the coupling circle.

Non-hierarchic decomposition is shown in Figure 7c and involves elements that do not prescribe output from each other. Instead, output (i.e. ${}^{i}_{j}\mathcal{H}\left({}^{i}\mathbf{r}\right)$) is estimated (i.e. ${}^{i}_{j}\mathbf{h}$) and after element solutions are found information is exchanged among the elements to update the input (i.e. ${}^{i}_{j}\mathcal{H}\left({}^{i}\mathbf{r}\right)$) parameters via a coordination method.

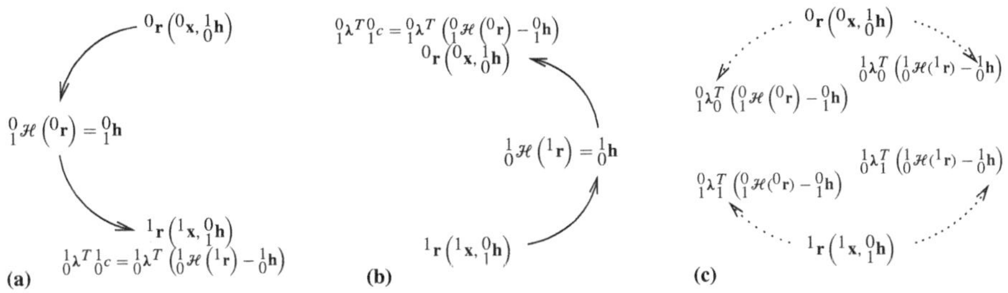

Fig. 7. (a) Top-down decomposition via relaxation of the consistency constraint and assigning the relaxation parameters to the Level-1 element. (b) Bottom-up decomposition via relaxation of the consistency constraint and assigning the relaxation parameters to the Level-0 element. (c) Non-hierarchic decomposition, none of the elements prescribes the output of a neighboring element. Consistency between two elements is relaxed and via coordination the relaxation parameters are updated. The relaxation parameters are not necessarily equal. Therefore, an additional index distinguishes the relaxation parameters, here $^{\cdot}_{\cdot}\boldsymbol{\lambda}_0$ and $^{\cdot}_{\cdot}\boldsymbol{\lambda}_1$

Non-hierarchic decomposition is shown in Figure 7c in case of decomposition through relaxed consistency constraints. Consistency constraints are assigned on both sides of the coupling circle.

Coupling that is present within the two-bar truss example is shown in Figure 8. The coupling circle shown on the right is an abstract representation of the structure shown on the left. On Level-0, physical responses ($^0\mathbf{r}$) that are evaluated involve displacements of the two-bar truss structure. These displacements depend on parameters $^0\mathbf{x}$ (here Level-0 design variables) and parameters ^1_0h, ^2_0h (here coupling variables). Displacements evaluated for the entire structure are mapped onto the individual column members via $^0_i\mathcal{H}\left(^0\mathbf{r}\right)$. The mapped displacements are used to evaluate internal forces ($^0_i\mathbf{h}$) for each member. The internal forces are used as boundary conditions on the individual columns. In the present example, a physical response that is of interest in Level-1 is stress ($^i\mathbf{r}$). Stress within an individual column depends on lay-out of the cross-section. This lay-out is adjusted via parameters $^i\mathbf{x}$ (here Level-1 design variables). The cross-section lay-out is a property that is required at Level-0 to evaluate the displacements. Hence, mapping from Level-1 onto Level-0 involves mapping of stiffness properties ($^i_0\mathbf{h}$) that are necessary to evaluate the displacements at Level-0.

3.2 Problem Matrix

The decomposition process of optimization problems involves identifying relationships between the design variables, physical responses and objectives/constraints that permit us to separate them into elements that are connected. These optimization problems for which no elements have been identified yet are defined as

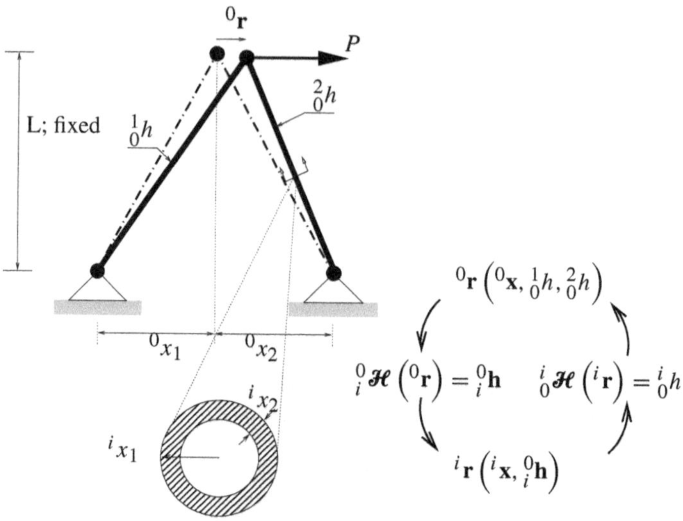

Fig. 8. Two-bar truss structure with embedded hierarchy. Level-0 responses are the displacement and the total mass of the structure. These responses depend on stiffness parameters of the two bar members. These stiffness parameters are determined via Level-1 responses.

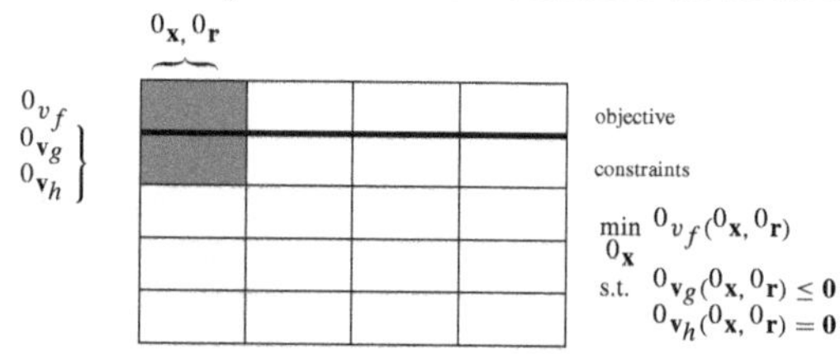

Fig. 9. Problem matrix, illustrating the relationship between variables and functions. A dark colored block indicates that specific functions (horizontally) depend on certain design variables and responses (vertically).

$$
\begin{aligned}
\min_{\mathbf{x}} \quad & v_f(\mathbf{x}, \mathbf{r}(\mathbf{x})) \\
\text{s.t.} \quad & \mathbf{v}_g(\mathbf{x}, \mathbf{r}(\mathbf{x})) \leq \mathbf{0} \\
& \mathbf{v}_h(\mathbf{x}, \mathbf{r}(\mathbf{x})) = \mathbf{0}
\end{aligned}
\tag{3}
$$

where $v_f(\ldots)$ is the objective function, $\mathbf{v}_g(\ldots)$ are inequality constraints and $\mathbf{v}_h(\ldots)$ are equality constraints. Equation (3) represents the so-called all-in-one optimization problem. The relationship between variables ($\mathbf{x}$), responses ($\mathbf{r}$) and functions (v_f, $\mathbf{v}_g$ and $\mathbf{v}_h$) is illustrated, amongst others by Barthelemy [2], with the *problem matrix* (also known as *Functional Dependence Table* [12]).

A single individual element depends on local design variables and local physical responses. Such an element is illustrated in Figure 9, where the dependence of $\mathbf{r}$ on the design variables $\mathbf{x}$ is dropped for brevity of notation. Above the thick line, the objective function ^0v_f is written and the block of variables $^0\mathbf{x}$ and responses $^0\mathbf{r}$ that this objective function depends on is dark. Likewise, the design constraints

${}^{0}\mathbf{v}_g$ (inequality) and ${}^{0}\mathbf{v}_h$ (equality) are listed below the thick line and the block of variables and responses corresponding to these functions is colored dark.

Four characteristic coupling patterns are shown in Figure 10. For compactness of Figure 10, shared design variables **z** are not explicitly shown, however these enter the problem matrix similar to the coupling variables. A dark colored block indicates that the functions (horizontally) depend on the design variables and responses (vertically). A light colored block indicates a weak dependence and a white colored block indicates dependence is absent.

In Figure 10, pattern (a) illustrates a problem matrix where the objective function only depends on the top-level element. Pattern (b) illustrates the case where a small number of coupling variables couples the objective and constraint functions of all the levels. Pattern (c) illustrates a problem where the design variables and/or physical responses are shared over multiple levels. Pattern (d) illustrates a problem where the objective function and the constraints depend on all the design variables and physical responses of all the elements.

A problem structure such as shown in Figures 10a and b is typically decomposed via an equality based decomposition procedure. Here, the number of parameters that are coupled are limited and a relatively small number of equality consistency constraints are added to the optimization problem. Because equality consistency constraints are used, the data that is exchanged and the intermediate designs are all related to the design problem. Contrary to relaxation approaches where the dimensionality of the problem is increased via relaxation parameters that become part of the solution.

A problem structure such as shown in Figures 10c and d is typically decomposed via a relaxation based method. Applying an equality based decomposition will introduce too many equality constraints into the individual optimization problems. The equality based constraints are difficult to satisfy via optimization algorithms. Therefore, relaxation of the consistency constraints is applied. The benefit is that intermediate steps (individual element optimizations) towards convergence of the entire hierarchy become less difficult as apposed to introducing equality constraints. The downside is that the relaxation parameters become part of the solution. Therefore, intermediate designs are solutions to optimization problems that have no straightforward translation into solutions that can be physically interpreted. In summary, the starting point of the multi-level optimization process and the final optimal solution are meaningful solutions. Intermediate solutions cannot be used since these intermediate solutions are inconsistent.

For the two-bar truss problem that is considered, typical optimization functions are mathematically expressed as

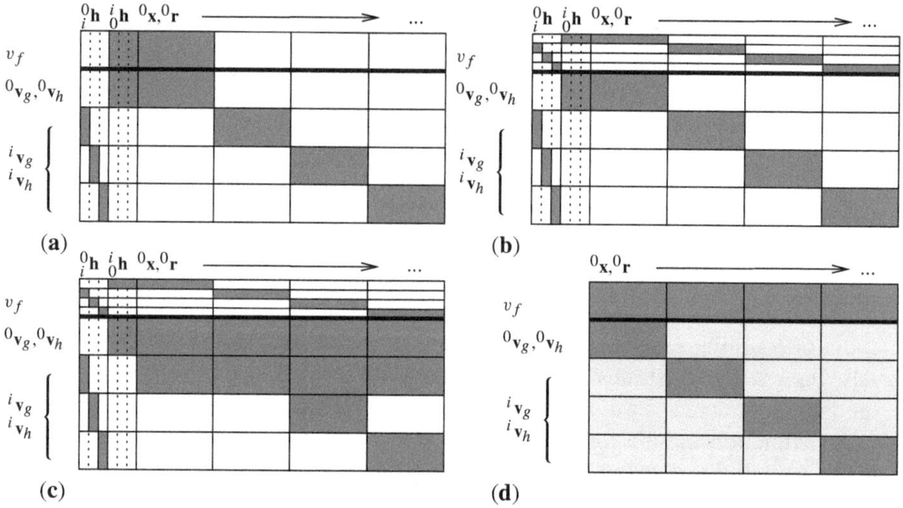

Fig. 10. Four typical patterns can be distinguished in the problem matrix. A dark colored block indicates that the functions (horizontally) depend on the design variables and responses (vertically). A light colored block indicates a weak dependence and a white colored block indicates dependence is absent. (a) A hierarchy where the objective function depends on the top-level design variables ${}^0\mathbf{x}$ and physical responses ${}^0\mathbf{r}$ and on the coupling variables ${}^i_0\mathbf{h}$. The lower level elements do not have an individual objective function. (b) A hierarchy where the objective function and constraint functions share a small number of coupling variables. Level-0 as well as elements present at Level-1 contribute to the minimization of the additively separable objective function v_f. (c) A hierarchy where the design variables and physical responses of different elements contribute to the constraint functions of neighboring elements. Some of the elements in lower levels depend on physical responses and/or design variables from neighboring elements. (d) No clear hierarchy can be distinguished. The objective function as well as the constraint functions depend on the physical responses and design variables from all the elements. A distinction between strong dependencies (dark) and weak dependencies (light) is made. Via Global Sensitivity Equations (GSE)[8] the elements are separated or separation is accomplished via relaxation of the consistency constraints between the elements.

$$
\begin{aligned}
{}^0v_f &= \frac{{}^0r_1({}^0\mathbf{x}, {}^1\mathbf{x}, {}^2\mathbf{x})}{{}^0r_{mmax}} \\
{}^0v_g &= \frac{{}^0r_2({}^0\mathbf{x}, {}^1\mathbf{x}, {}^2\mathbf{x})}{{}^0r_{umax}} - 1 \\
{}^1v_g &= \frac{{}^1r_1({}^1\mathbf{x}, {}^0\mathbf{r})}{0.9\,{}^1r_{cr}} - 1 \\
{}^2v_g &= \left(2\frac{{}^2r_1({}^2\mathbf{x}, {}^0\mathbf{r})}{{}^2r_{Euler}({}^0x_2, {}^2x_1)}\right)^2 - 1.
\end{aligned}
\tag{4}
$$

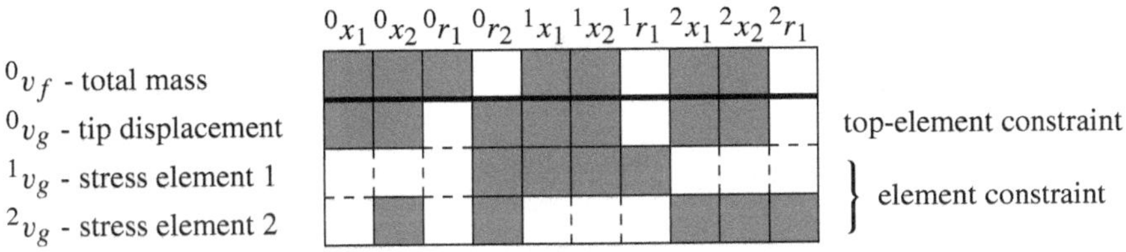

Fig. 11. Coupling matrix for the Two-bar truss optimization problem before decomposition. On the left, the optimization problem functions are listed and on the top the design variables and physical responses. A function that depends on a specific design variable or physical response is shaded.

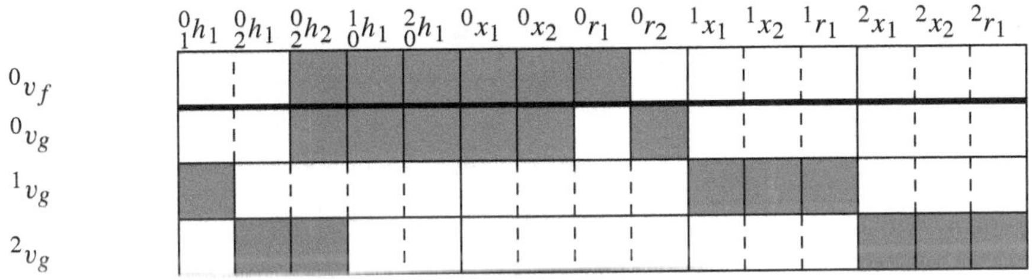

Fig. 12. Coupling parameters are introduced into the problem matrix, which illustrates the data that is required between the Level-0 element and the two Level-1 elements.

Hence, an objective function ${}^0v_f$ is present, e.g., total mass (${}^0r_1$). In addition, a constraint equation (${}^0v_g$) is present accounting for, e.g, maximum displacement (${}^0r_2$). The total mass depends on the cross-section of the members (calculated via lay-out parameters ${}^i\mathbf{x}$ associated with Level-1) and the length of each member (calculated via parameters ${}^0\mathbf{x}$ that are associated with Level-0). Likewise, the displacements depend on the cross-section of the members and the length of each member. Constraint equations (${}^1v_g$, ${}^2v_g$) are present accounting for, e.g., maximum allowable stress (${}^1r_1$) and Euler buckling (${}^2r_1$).

The problem matrix illustrating the dependencies of the optimization functions of Equation (4) on the design variables and the physical responses is shown in Figure 11. The objective function and constraint function of the Level-0 optimization problem depend on the design variables of Level-0 and Level-1 elements.

To illustrate coupling within the optimization problem, coupling variables are introduced, see Figure 12. The pattern of the coupling is similar to the pattern shown in Figure 10a. Hence, an equality based multi-level optimization technique is promising to solve the problem. Decomposing the optimization problem can be hierarchic or non-hierarchic. In case hierarchic decomposition is chosen, some of the coupling parameters are omitted in the coupling matrix.

In the present example a hierarchic top-down decomposition is considered for the two-bar truss coupling circle (Figure 4), see Figure 13. Hence, the physical properties ${}^i\mathbf{r}$ that are mapped from Level-1 onto Level-0 via ${}^i_0\mathcal{H}({}^i\mathbf{r})$ are expressed via coupling variables ${}^i_0h$. These mapped properties reflect stiffness parameters charac-

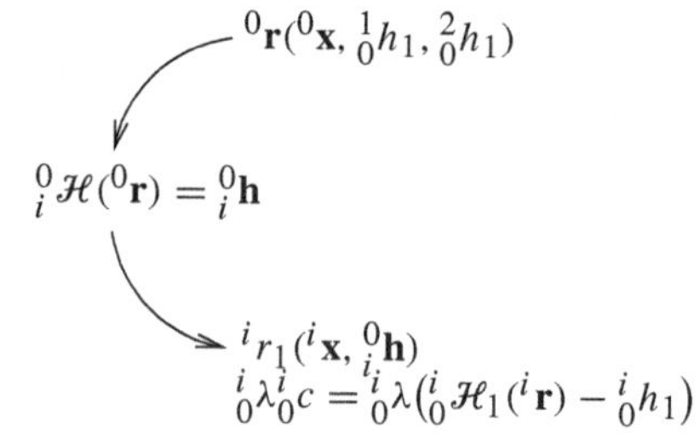

Fig. 13. Top-down hierarchical relaxation based decomposition via, e.g. relaxation of the consistency constraint with Lagrangian multipliers assigned to the Level-1 element.

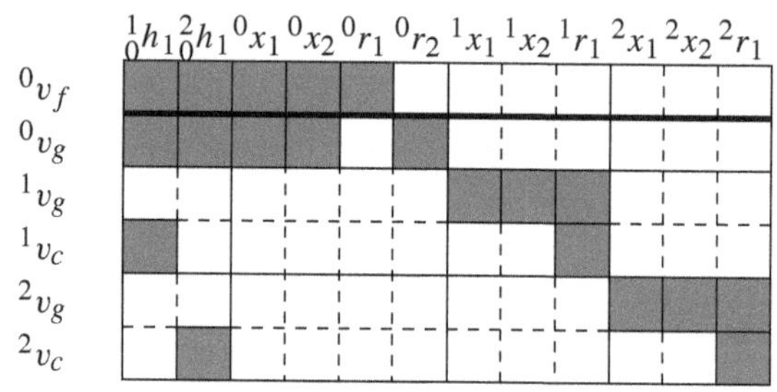

Fig. 14. Coupling matrix Two bar truss after decomposition. A top-down decomposition scheme is used, where the consistency constraints are added to the Level-1 elements.

terizing Element-1 and Element-2. In the present example these stiffness parameters are the cross-sectional areas.

Instead of mapping the physical properties from Element-1 onto Element-0, the cross-sectional area ${}^{1}_{0}h_1$ (and ${}^{2}_{0}h_1$) is introduced in the Element-0 problem as an additional design variable. Hence, Element-0 determines via, e.g. optimization, the value for ${}^{1}_{0}h_1$ (and ${}^{2}_{0}h_1$). The resulting problem matrix has an angular structure, see Figure 12.

The problem is now decomposed in a manner that Element-0 can be solved without prior knowledge of Element-1 (and Element-2). This is accomplished via introducing consistency constraints ${}^{i}v_c$, see Figure 14. After Element-0 is solved, the necessary parameters for the elements located at Level-1 are sent from Level-0 onto Level-1 to evaluate the Level-1 elements. In the Level-1 elements an individual objective function is absent. In case an equality based decomposition approach is applied, special measures are to be taken in order to formulate a multi-level optimization problem, i.e. constraint minimization. However, in the present example a relaxation based approach is adopted.

In case relaxation is applied to the consistency constraints via, e.g., Lagrange multipliers the relaxed consistency constraints read

$$
\begin{aligned}
{}^{1}_{0}\lambda^{1}_{0}c &= {}^{1}_{0}\lambda\left({}^{1}_{0}\mathcal{H}\left({}^{1}r\right) - {}^{1}_{0}h\right); \\
{}^{2}_{0}\lambda^{2}_{0}c &= {}^{2}_{0}\lambda\left({}^{2}_{0}\mathcal{H}\left({}^{2}r\right) - {}^{2}_{0}h\right).
\end{aligned}
\tag{5}
$$

Relaxation of the consistency constraints means that an inconsistency is present because the constraints are generally not exactly satisfied. An additional term is added to each individual optimization problem present at Level-1 that reflects in a weak manner the consistency constraint. A derivation of Lagrangian relaxation applied to multi-level optimization is beyond the scope of the present chapter. The interested reader is referred to the work of Kim *et al.*[5] for a derivation of the necessary mathematical expressions.

The optimization problem for the Level-0 element is

$$\begin{aligned} \min_{{}^0x_1, {}^0x_2, {}^1_0h_1, {}^2_0h_1} \quad {}^0v_f &= \frac{{}^0r_1\left({}^0\mathbf{x}, {}^1_0h_1, {}^2_0h_1\right)}{{}^0r_{mmax}} \\ \text{s.t.} \quad {}^0v_g &= \frac{{}^0r_2\left({}^0\mathbf{x}, {}^1_0h_1, {}^2_0h_1\right)}{{}^0r_{umax}} - 1 \leq 0 \\ & {}^0\underline{\mathbf{x}} \leq {}^0\mathbf{x} \leq {}^0\overline{\mathbf{x}} \end{aligned} \tag{6}$$

and for the Level-1 elements the individual optimization problems become:

Element 1:

$$\begin{aligned} \min_{{}^1x_1, {}^1x_2} \quad {}^1v_f &= {}^1v_c({}^1_0c) \\ \text{s.t.} \quad {}^1v_g &= \frac{{}^1r_1}{0.9\,{}^1r_{cr}} - 1 \quad \leq 0 \\ & {}^1\underline{\mathbf{x}} \leq {}^1\mathbf{x} \leq {}^1\overline{\mathbf{x}} \\ \text{where } {}^1v_c({}^1_0c) &= {}^1_0\lambda\left({}^1_0\mathcal{H}_1\left({}^1r_1\right) - {}^1_0h_1\right) \end{aligned} \tag{7}$$

Element 2:

$$\begin{aligned} \min_{{}^2x_1, {}^2x_2} \quad {}^2v_f &= {}^2v_c({}^2_0c) \\ \text{s.t.} \quad {}^2v_g &= \left(2\frac{{}^2r_1}{{}^2r_{Euler}}\right)^2 - 1 \quad \leq 0 \\ & {}^2\underline{\mathbf{x}} \leq {}^2\mathbf{x} \leq {}^2\overline{\mathbf{x}} \\ \text{where } {}^2v_c({}^2_0c) &= {}^2_0\lambda\left({}^2_0\mathcal{H}_1\left({}^2r_1\right) - {}^2_0h_1\right) \end{aligned} \tag{8}$$

Equations (6)–(8) represent three individual optimization problems that are coupled. Due to the coupling that is still present within the physical responses from Level-0 onto Level-1, Level-0 is optimized first and Level-1 afterwards. Optimal values for ${}^1_0h_1$ and ${}^2_0h_1$ are determined within Element-0 located at Level-0. These optimal values are send to the Level-1 elements and kept fixed during Level-1 element optimizations.

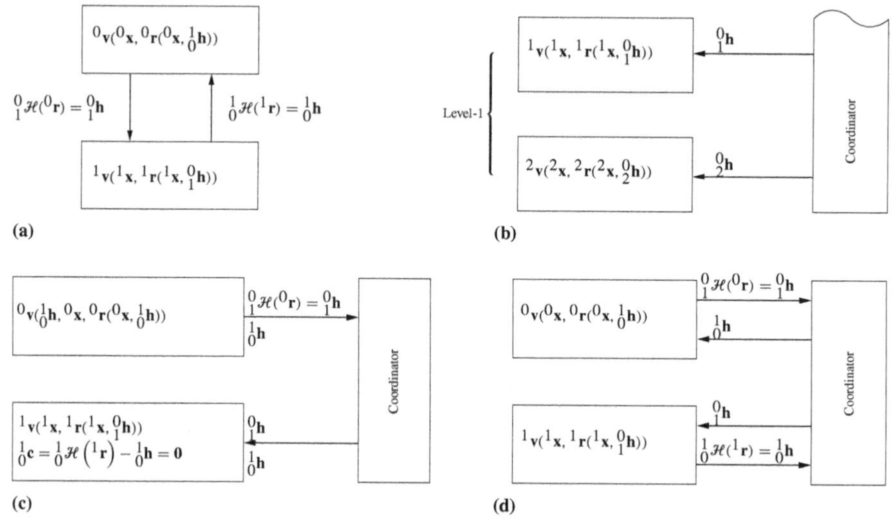

Fig. 15. (a) Information exchange between two individual though coupled optimization problems. Coordination of the data that is send between the two elements is necessary. (b) Separable decisions, the optimization problems are not coupled. Hence, the data for the first element at Level-1 does not change the outcome of the neighboring second element at Level-1. (c) Inseparable decisions, the optimization problems are coupled and data is send from one element to the neighboring element. (d) Inseparable coupled decisions require iteration. The solution of both elements depends on the solution of the neighboring element.

4 Coordination

Solutions of decomposed elements require coordination, such that the individual solutions combined lead to the solution for the entire hierarchy. Coordination of interaction between elements involves decision making. These decisions determine the steps in which the individual optimization problems are solved. In addition, decisions involve adding information to individual elements in order to steer the solution process.

Consider two individual element optimizations transferring data, see Figure 15a. The origin of this data can be, e.g., physical responses ($\mathbf{r}$) such as shown in Figure 15 or shared design variables ($\mathbf{z}$). This data cannot be arbitrarily send from one element to the neighboring element. Therefore, decisions on which data to send and in what order are necessary.

There are three types of decisions that can be distinguished:

1. separable decisions;
2. inseparable decisions;
3. inseparable and coupled decisions.

The first type of decisions, the separable decisions are shown in Figure 15b. These type of decisions occur in a multi-level hierarchy where the update of information of one element does not depend on the data that is send to neighboring elements. This is typically the case for elements on the same level that are not coupled to each other. Hence, after solving the parent element (Level-0), the updates of child elements (Level-1) are independent from one another.

The second type of decisions are inseparable decisions, as shown in Figure 15c. The output of one element is send to the input of the neighboring element. These decisions are done sequentially and are the result of top-down or bottom-up decomposition schemes. No operations are performed inside the coordinator block and no changes are made to the Level-0 optimization problem after it has finished. This type of coordination fails in cases where the Level-0 element prescribes physical responses (i.e. ${}^{1}_{0}\mathbf{h}$) or shared design variable vectors (i.e. ${}^{0}_{1}\mathbf{z} = {}^{1}_{0}\mathbf{z}$) that the Level-1 element is unable to meet. Either due to restrictions on design variables (i.e. lower bounds $\underline{\mathbf{x}}$, $\underline{\mathbf{z}}$ and/or upper bounds $\overline{\mathbf{x}}$, $\overline{\mathbf{z}}$) and/or physical restrictions on the Level-1 element. Therefore, the Level-0 optimization problem requires additional information via, e.g., Optimum Sensitivity Analysis [10] or a copy of the relaxed consistency function [6].

The third type of decisions are the inseparable and coupled decisions shown in Figure 15d. The output of one element is the input of the neighboring element and *vice versa*. These decisions are done sequentially via iterating or concurrently through optimization.

The two-bar truss multi-level optimization problem expressed in Equations (6)–(8) requires coordination of the individual element results such that the optimum of the entire hierarchy is found. The problem can be coordinated via first solving the Level-0 optimization problem and then mapping the requested optimal physical responses onto the Level-1 elements. In addition, the optimal coupling variables are send to the Level-1 optimization problems. This type of decisions are defined as inseparable decisions, see Figure 15c.

A problem occurs with coordination via inseparable decisions. In some cases the mapped responses coming from Level-0, together with the coupling variables determined at Level-0, correspond to Level-1 elements that are infeasible. Hence, a single (or multiple) individual element(s) located at Level-1 can be infeasible due to bounds on the element design variables or element constraint equations. To overcome the problem of infeasible Level-1 elements, the relaxed consistency constraints $({}^{i}_{0}\lambda({}^{i}_{0}\mathcal{H}_1({}^{i}r_1) - {}^{i}_{0}h_1))$ present at Level-1 are added to the Level-0 optimization problem. Thus, the modified Level-0 optimization problem is mathematically expressed as

$$
\begin{aligned}
\min_{{}^{0}x_1, {}^{0}x_2, {}^{1}_{0}h, {}^{2}_{0}h} \quad & {}^{0}v_f = \frac{{}^{0}r_1\left({}^{0}\mathbf{x}, {}^{1}_{0}h, {}^{2}_{0}h\right)}{{}^{0}r_{mmax}} + \sum_{i=1}^{2} {}^{i}_{0}\lambda\left({}^{i}_{0}\mathcal{H}({}^{i}\mathbf{r}) - {}^{i}_{0}h\right) \\
\text{s.t.} \quad & {}^{0}v_g = \frac{{}^{0}r_2\left({}^{0}\mathbf{x}, {}^{1}_{0}h, {}^{2}_{0}h\right)}{{}^{0}r_{umax}} - 1 \leq 0 \\
& {}^{0}\underline{\mathbf{x}} \leq {}^{0}\mathbf{x} \leq {}^{0}\overline{\mathbf{x}}
\end{aligned}
\qquad (9)
$$

The effect is that coordination becomes inseparable and coupled and that the coordinator balances via the Lagrange multipliers between the Level-0 optimization problem and the Level-1 optimization problem to maintain consistency between the elements. Infeasible Level-1 elements are avoided since the Level-0 optimization problem is now encouraged via the additional term in the objective function to remain consistency. The latter is accomplished via avoiding designs in Level-0 that result in mapped physical responses and coupling variables that produce infeasible Level-1 elements.

The coordinator problem involves solving the so-called dual problem[1] to determine optimal Lagrange multipliers. A derivation of the dual problem is beyond the scope of the present chapter and the interested reader is referred to [4] for an explanation of the duality method applied to a hierarchy of optimization problems.

5 Software Framework

The generalized formulation for decomposition and coordination allows for a single software framework that can handle the majority of multi-level optimization problems via various multi-level optimization methods. The framework is constructed based on two different types of individual optimization problem formulations. Via techniques from object oriented programming the expressions in the two formats are replaced or functionality is added according to the multi-level method that is used. The framework is build up as follows.

An individual element of a hierarchy consists of a local optimization problem, an analysis routine that computes responses and a coupling expression. Mathematically an individual element optimization problem is expressed as:

$$
\begin{array}{ll}
\text{Equality based:} & \text{Relaxation based:} \\
\begin{array}{ll} \min\limits_{\mathbf{x},\mathbf{h}} & v_f(\mathbf{h},\mathbf{x},\mathbf{r}(\mathbf{x},\mathbf{h})) \\
\text{s.t.} & \mathbf{v}_g(\mathbf{h},\mathbf{x},\mathbf{r}(\mathbf{x},\mathbf{h})) \leq 0 \\
& \mathbf{v}_h(\mathbf{h},\mathbf{x},\mathbf{r}(\mathbf{x},\mathbf{h})) = 0 \\
& \mathbf{v}_c(\mathbf{c}(\mathbf{r},\mathbf{h})) = 0 \\
& \mathbf{v}_a(\mathbf{h},\mathbf{x}) \leq 0
\end{array} &
\begin{array}{ll} \min\limits_{\mathbf{x},\mathbf{h}} & v_f(\mathbf{h},\mathbf{x},\mathbf{r}(\mathbf{x},\mathbf{h})) + \sum v_c(\mathbf{c}(\mathbf{r},\mathbf{h})) \\
\text{s.t.} & \mathbf{v}_g(\mathbf{h},\mathbf{x},\mathbf{r}(\mathbf{x},\mathbf{h})) \leq 0 \\
& \mathbf{v}_h(\mathbf{h},\mathbf{x},\mathbf{r}(\mathbf{x},\mathbf{h})) = 0
\end{array}
\end{array}
\tag{10}
$$

Equality based methods maintain consistency via equality constraints $\mathbf{v}_c(\mathbf{c})$ and gradient information is added via additional constraints $\mathbf{v}_a(\mathbf{h},\mathbf{x})$ that represent approximations of objective and/or constraint behavior of neighboring elements. On the right, relaxation based methods typically add a relaxation function to the objective.

The basic components of the framework are shown in Figure 16, which shows the component diagram of the framework. The individual optimization problem expressed by Equation (10) is computed in the "Element" component. The coupling part $(\mathbf{v}_c, \mathbf{v}_a)$ is constructed in the "Coupling" component and the normal objective

[1] The dual problem is defined in [3, propositions 5.1.1–5.1.6].

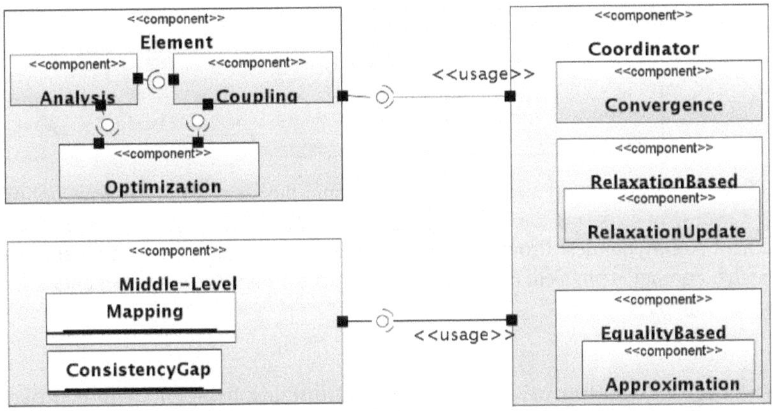

Fig. 16. Component diagram of the multi-level framework. There are three basic components. First, the Element component consisting of analysis, optimization and coupling. Second, the coordinator component consisting of convergence, equality based and relaxation based coordination methods. Third, the Middle-level component which consists of a mapping of element responses onto neighboring elements and procedures to compute inconsistencies.

(v_f) and constraint equations ($\mathbf{v}_g$, $\mathbf{v}_h$) are embedded inside the "Optimization" component. Analysis functions are evaluated inside the "Analysis" component.

Individual elements are coupled to neighboring elements. To enable exchange of data a "Middle-level" component is constructed which communicates the coupling between elements, hence the mapping: $\mathcal{H}(\mathbf{r}) = \mathbf{h}$. Furthermore, the size of inconsistencies $\mathbf{c} = \mathcal{H}(\mathbf{r}) - \mathbf{h}$ is computed. This information is required by the coordinator for the decision making process.

The "coordinator" component determines the order in which elements are solved (decision making). Furthermore, it updates coefficients for the approximations in case of equality based decomposition methods or updates relaxation parameters in case of relaxation based decomposition methods. Finally, the convergence of the hierarchy is determined inside the "Coordinator" component. Depending on the type of decision method, convergence of the hierarchy as a whole and between individual elements is required. This is accomplished via checking inconsistency ε_{v_c} from the "Middle-level" component. In addition, a measure of change of individual objective functions ε_{v_f} is present to determine when optimization of the individual members of the hierarchy has converged.

6 Supersonic Business Jet Optimization

In this section, the object oriented framework described in the previous chapter is applied to an example taken from the literature. A model of a supersonic business jet is optimized via multi-level optimization. Essential in the multi-level optimization

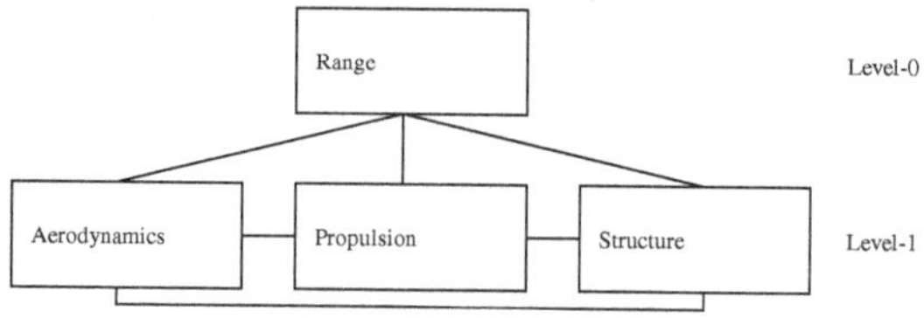

Fig. 17. Two levels are distinguished in the supersonic business jet multi-level optimization problem. Level-0 considering the range of the plane and at Level-1 3 individual elements are identified: Aerodynamics; Propulsion; and Structure. Each hierarchical element involves a discipline that computes physical characteristics required by neighboring elements to compute the physical responses.

process is the distribution of the all-in-one optimization problem into a hierarchy of smaller more tractable optimization problems. The supersonic business jet problem is suitable because it has coupling between a number of disciplines and can be described as a hierarchical and non-hierarchical problem. In addition, the mapping of physical responses involves mapping onto a single parameter. The latter simplifies coupling considerably as opposed to problems where the coupling does not map onto a single parameter, such as e.g. surfaces within finite element meshes.

The supersonic business jet was introduced by Agte et al. [1] as a benchmark problem for multi-disciplinary design optimization approaches. In the present work, for self-containment, a brief overview of the multi-disciplinary design optimization problem is presented. The interested reader is referred to Agte et al. [1] for a detailed derivation of the expressions used to model the supersonic business jet.

6.1 Multi-Level Optimization Problem

The supersonic business jet optimization hierarchy is illustrated in Figure 17 and the parameters determining the lay-out and performance of the supersonic business jet are listed in Table 1. There are 41 design variables distributed over four elements. Eight of these design variables are present in more than one element. The total structural weight is also listed twice, because this value is send to two different elements.

For this problem, the objective $({}^{0}v_f)$ is to maximize the range $({}^{0}r)$ of the aircraft. The supersonic business jet design problem is subjected to constraints on adverse pressure gradient[2] $({}^{1}v_{g_1})$ and constraints on the ratio between lift of the wing and lift of the horizontal tail wing (respectively ${}^{1}v_{g_2}$ and ${}^{1}v_{g_3}$). Furthermore, constraints on the maximum engine temperature $({}^{2}v_{g_1})$ and on the throttle setting $({}^{2}v_{g_2})$ are present. Finally, constraints on allowable stresses $({}^{3}v_{g_{1\dots60}})$ and geometrical constraints $({}^{3}v_{g_{61\dots72}})$ are considered in the design problem. Lower bounds ${}^{\cdot\cdot}\underline{\mathbf{x}}$, ${}^{\cdot\cdot}\underline{\mathbf{z}}$ and upper bounds ${}^{\cdot\cdot}\overline{\mathbf{x}}$, ${}^{\cdot\cdot}\overline{\mathbf{z}}$ are present on all the design variables. The numerical values for these upper and lower bounds are listed in Table 2.

[2] An adverse pressure gradient occurs when the static pressure increases along the surface in the direction of the flow. The adverse pressure gradient is used to predict separation of

Table 1. Definition of design variables, physical responses, physical interactions between subsystems and optimization parameters for the supersonic business jet.

Design	Description	Physical response	Description	Interaction	Description
${}^{0}_{1..2}z_1$	Mach number	${}^{0}r_1$	range		
${}^{0}_{1..2}z_2$	altitude				
${}^{1}_{0}z_1$	Mach number	${}^{1}r_1$	adverse pressure gradient	${}^{1}_{0}h$	lift to drag ratio
${}^{1}_{0}z_2$	altitude	${}^{1}r_2$	Lift coefficient	${}^{1}_{2}h$	drag
${}^{1}_{3}z_1$	wing aspect ratio	${}^{1}r_3$	Drag coefficient	${}^{1}_{3}h$	lift
${}^{1}_{3}z_2$	wing surface area				
${}^{1}_{3}z_3$	tail aspect ratio				
${}^{1}_{3}z_4$	tail surface area				
${}^{1}_{3}z_5$	wing sweep				
${}^{1}_{3}z_6$	thickness/chord				
${}^{1}x_1$	tail sweep				
${}^{1}x_2$	wing lift				
${}^{1}x_3$	tail wing lift				
${}^{2}_{0}z_1$	Mach number	${}^{2}r_1$	engine temperature	${}^{2}_{0}h$	spec. fuel cons.
${}^{2}_{0}z_2$	altitude	${}^{2}r_2$	throttle setting	${}^{2}_{1}h$	engine scale fact.
${}^{2}x_1$	throttle			${}^{2}_{3}h$	engine weight
${}^{3}\mathbf{x}_{1...9}$	skin thickness	${}^{3}\mathbf{r}_{1...6}$	σ at $1\ldots6$	${}^{3}_{0}h_1$	fuel weight
${}^{3}\mathbf{x}_{10...18}$	sandwich thickness	${}^{3}\mathbf{r}_{7...12}$	τ at $1\ldots7$	${}^{3}_{0}h_2$	total weight
${}^{3}x_{19}$	taper ratio	${}^{3}r_{13}$	mass wing	${}^{3}_{1}h_1$	total weight
${}^{3}_{1}z_1$	wing aspect ratio	${}^{3}r_{14}$	mass tail wing	${}^{3}_{1}h_2$	wing twist
${}^{3}_{1}z_2$	wing surface area	${}^{3}r_{15}$	wing deformation		
${}^{3}_{1}z_3$	tail aspect ratio				
${}^{3}_{1}z_4$	tail surface area				
${}^{3}_{1}z_5$	wing sweep				
${}^{3}_{1}z_6$	thickness/chord				

The problem matrix showing the dependencies of the optimization functions on the design variables and physical responses is shown in Figure 18. The objective function depends on mapped responses from the Level-1 elements. Likewise, the constraint equations of the Level-1 hierarchical elements depend on physical responses that are mapped between the levels and between the individual hierarchical elements. The coupling between hierarchical elements is shown via coupling variables in the problem matrix, see Figure 19.

Hence, from Figure 19 coupling is identified as coupling in the design variables and between elements over multiple constraints, see Figure 10c. This type of problems is solved via relaxation based methods. Therefore, a relaxation based method is applied to decompose the optimization problem.

---

flow. This separation has a serious effect on the lift and drag characteristics of the surface in the flow.

Table 2. Scaling parameters and lower and upper bounds for the supersonic business jet design variables and coupling variables.

Variable	Lower bound	Upper bound	Scaling variable	Variable	Lower bound	Upper bound	Scaling variable
${}^{0}_{1.2}z_1$	1.4	1.8	1.6	${}^{2}_{0}z_1$	1.4	1.8	1.6
${}^{0}_{1.2}z_2$	30000.0	60000.0	45000.0	${}^{2}_{0}z_2$	30000.0	60000.0	45000.0
${}^{1}_{0}h$	0.1	10.0	5.0	${}^{2}x_1$	0.1	1.0	0.5
${}^{2}_{0}h$	1.0	4.0	2.5	${}^{1}_{2}h$	100.0	70000.0	35000.0
${}^{3}_{0}h_1$	0.1	0.9	0.5				
${}^{3}_{0}h_2$	5000.0	100000.0	45000.0				
${}^{1}_{0}z_1$	1.4	1.8	1.6	${}^{3}\mathbf{x}_{1...9}$	0.1	4.0	2.0
${}^{1}_{0}z_2$	30000.0	60000.0	45000.0	${}^{3}\mathbf{x}_{10...18}$	0.1	9.0	4.5
${}^{1}_{3}z_1$	2.5	8.5	5.5	${}^{3}x_{19}$	0.1	0.4	0.25
${}^{1}_{3}z_2$	200.0	800.0	500.0	${}^{3}_{1}z_1$	2.5	8.5	5.5
${}^{1}_{3}z_3$	2.5	8.5	5.5	${}^{3}_{1}z_2$	200.0	800.0	500.0
${}^{1}_{3}z_4$	50.0	148.9	100.0	${}^{3}_{1}z_3$	2.5	8.5	5.5
${}^{1}_{3}z_5$	40.0	70.0	55.0	${}^{3}_{1}z_4$	50.0	148.9	100.0
${}^{1}_{3}z_6$	0.01	0.09	0.05	${}^{3}_{1}z_5$	40.0	70.0	55.0
${}^{1}x_1$	40.0	70.0	55.0	${}^{3}_{1}z_6$	0.01	0.09	0.05
${}^{1}x_2$	0.01	0.2	0.1	${}^{1}_{3}h$	5000.0	100000.0	45000.0
${}^{1}x_3$	1.0	3.5	2.5	${}^{2}_{3}h$	100.0	30000.0	15000.0
${}^{2}_{1}h$	1.0	4.0	2.5				
${}^{3}_{1}h_1$	5000.0	100000.0	45000.0				
${}^{3}_{1}h_2$	0.2	50.0	25.0				

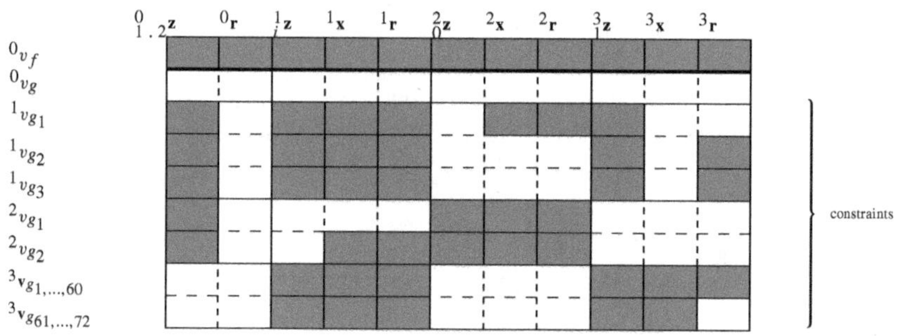

Fig. 18. Supersonic business jet problem matrix before decomposition. On the left the optimization problem functions are listed and on the top the design variables and physical responses. A function that depends on a specific design variable or physical response is represented via a dark colored block. No constraints are present in the Level-0 element, hence ${}^{0}v_g$ is left blank. Furthermore, design variables are present in multiple hierarchical elements that are shared. For brevity of the problem matrix, the design variables, responses and design optimization functions are combined into vectors.

Hierarchic Decomposition

A top-down hierarchic decomposition is considered, see Figure 20. Coupling variables (${}^{\cdot}_{\cdot\cdot}h$) are introduced at the elements following the notation introduced in Section 2. Consistency constraints are introduced between elements present on Level-1. These consistency constraints replace the mapping of physical responses between elements on Level-1 (${}^{\cdot}_{\cdot\cdot}\mathcal{H}({}^{\cdot\cdot}\mathbf{r}) = {}^{\cdot}_{\cdot\cdot}\mathbf{h}$) and design variables that are shared (${}^{i}_{j}\mathbf{z} =$

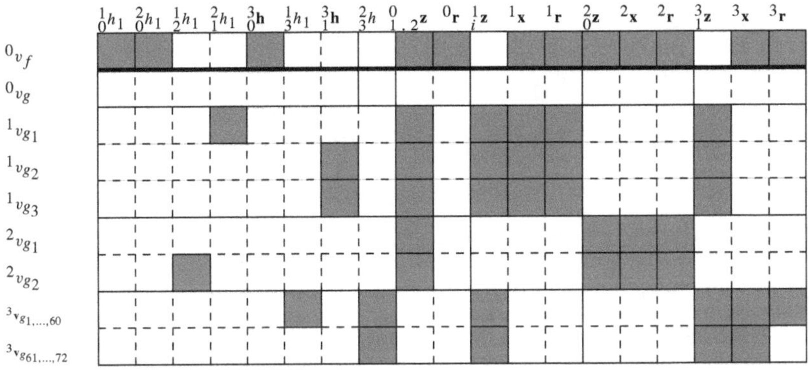

Fig. 19. Coupling parameters are introduced into the problem matrix, showing the data that is shared between the Level-0 element and the three Level-1 elements.

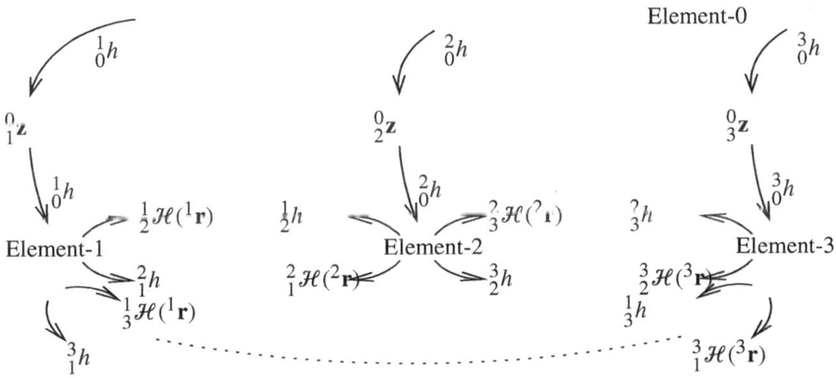

Fig. 20. Top-down hierarchic formulation of the supersonic business jet optimization problem. Physical responses that are mapped from Level-1 onto Level-0 are replaced within Level-0 via coupling variables. These coupling variables are determined via e.g. optimization in Element-0. Together with shared design variables the coupling variables are sent to the elements that are present at Level-1. In addition, coupling variables are introduced for elements located at Level-1 to decouple the interaction between elements on the same level.

${}^{j}_{i}\mathbf{z}$). Furthermore, consistency constraints are introduced on the physical coupling between elements that are located on different levels (${}^{\cdot\cdot}_{0}\mathbf{c} = {}^{\cdot\cdot}_{0}\mathcal{H}({}^{\cdot\cdot}\mathbf{r}) - {}^{\cdot\cdot}_{0}\mathbf{h}$).

Consistency constraints are introduced between the individual elements and these are relaxed via Augmented Lagrangian relaxation. Hence, consistency constraints ${}^{\cdot\cdot}_{\cdot\cdot}\mathbf{c}_{\cdot\cdot}(..) = \mathbf{0}$ are introduced and relaxed via ${}^{\cdot\cdot}_{\cdot\cdot}\boldsymbol{\lambda}^{T}_{\cdot\cdot}\,{}^{\cdot\cdot}_{\cdot\cdot}\mathbf{c}_{\cdot\cdot} + ||{}^{\cdot\cdot}_{\cdot\cdot}\mathbf{s}_{\cdot\cdot} \circ {}^{\cdot\cdot}_{\cdot\cdot}\mathbf{c}_{\cdot\cdot}||_2^2$. The expression $||..||_2$ is the l_2-norm and expression $\circ$ represents the Hadamard product.[3] Coupling variables (${}^{\cdot\cdot}_{0}h$) are added to the Level-0 optimization problem as design variables. These variables must match the mapped physical responses (${}^{\cdot\cdot}_{0}\mathcal{H}({}^{\cdot\cdot}r)$). The latter are evaluated during the Level-1 optimizations.

[3] The Hadamard product multiplies two vectors element by element.

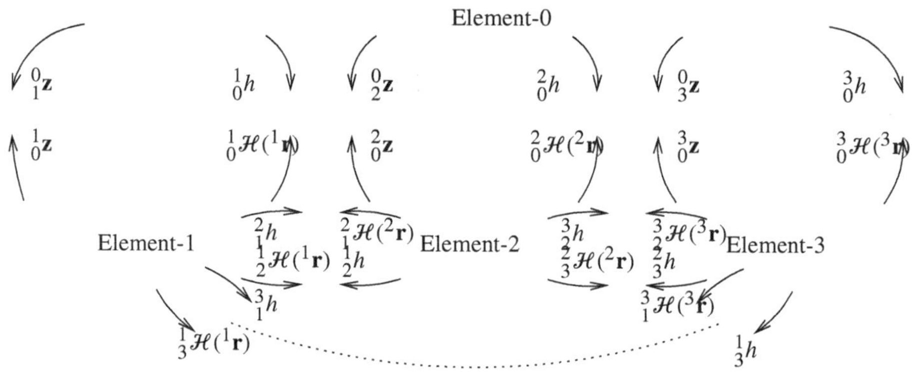

Fig. 21. Non-hierarchic formulation of the supersonic business jet optimization problem. Physical responses that are mapped from one element onto a neighboring element are replaced within the elements via coupling variables. These coupling variables are determined via e.g. optimization. Together with shared design variables the coupling variables are sent between the elements to exchange coupling data.

The Aerodynamics, Propulsion and Structures element are located on Level-1. Since these individual elements are coupled on the same level the consistency constraints are relaxed in both directions. Therefore, coupling variables $({}^{\cdot\cdot}_{\cdot\cdot}h)$ are added to the design variables of these elements. These variables must match the mapped physical responses $({}^{\cdot\cdot}_{\cdot\cdot}\mathcal{H}(r))$ that are computed in the neighboring element.

Non-Hierarchic Decomposition

Agte et al. [1] consider a non-hierarchic decomposition of the supersonic business jet model, see Figure 21. Hence, consistency constraints are introduced between Level-0 and Level-1 and these consistency constraints are relaxed in both directions (from Level-0 onto Level-1 and *vice versa*). Therefore, in addition to the consistency constraints already present in the top-down hierarchic decomposition, the mapping of design variables that are shared $({}^{0}_{\cdot\cdot}\mathbf{z}, {}^{\cdot\cdot}_{0}\mathbf{z})$ from Level-0 onto Level-1 is replaced via consistency constraints. These consistency constraints are relaxed via Augmented Lagrangian relaxation.

The non-hierarchic decomposition requires introducing additional coupling variables that are introduced as design variables within the individual elements. Therefore, the individual element optimization problems become more complex as compared to hierarchic top-down decomposition. An advantage of the non-hierarchic decomposition is that in case coupling is strong in both directions of the coupling circle the individual elements are updated with new information much quicker then a hierarchic top-down decomposition.

Coordination

The coordinator uses two criteria to determine convergence of the multi-level optimization algorithm, see Section 5.

1. The first criterion checks if the optimal value of the objective function of an individual element does not change after coupling data is exchanged and the individual optimizations are repeated. This criterion is mathematically expressed as:

$$\varepsilon_{v_f} \leq \max\{||{}^{i}v_f^{(t+1)} - {}^{i}v_f^{(t)}||_\infty\} \text{ where } i = 0, 1, 2, 3. \tag{11}$$

   If the change in individual objective function value is smaller then a predetermined value coupling data is exchanged and the relaxation parameters are updated. Else, coupling data is exchanged and the individual element optimizations are repeated.
   Due to the relaxation of the consistency constraints the coordinator has to update the relaxation parameters. For the present relaxation approach a method known as the method of multipliers is used [11]. The Lagrange multipliers (${}^{\cdot\cdot}_{\cdot\cdot}\boldsymbol{\lambda}_{\cdot\cdot}$) as well as the penalty weights (${}^{\cdot\cdot}_{\cdot\cdot}\mathbf{s}_{\cdot\cdot}$) are updated via the coordinator. The coordinator computes new relaxation parameters via:

$$\begin{aligned} {}^{\cdot\cdot}_{\cdot\cdot}\boldsymbol{\lambda}^{(t+1)}_{\cdot\cdot} &= {}^{\cdot\cdot}_{\cdot\cdot}\boldsymbol{\lambda}^{(t)}_{\cdot\cdot} + 2 \cdot {}^{\cdot\cdot}_{\cdot\cdot}\mathbf{s}^{(t)}_{\cdot\cdot} \circ {}^{\cdot\cdot}_{\cdot\cdot}\mathbf{s}^{(t)}_{\cdot\cdot} \circ {}^{\cdot\cdot}_{\cdot\cdot}\mathbf{c}^{(t)}_{\cdot\cdot} \\ {}^{\cdot\cdot}_{\cdot\cdot}\mathbf{s}^{(t+1)}_{\cdot\cdot} &= \begin{cases} {}^{\cdot\cdot}_{\cdot\cdot}\mathbf{s}^{(t)}_{\cdot\cdot} & \text{if } {}^{\cdot\cdot}_{\cdot\cdot}\mathbf{c}^{(t)}_{\cdot\cdot} \leq \gamma \cdot {}^{\cdot\cdot}_{\cdot\cdot}\mathbf{c}^{(t-1)}_{\cdot\cdot} \\ \beta \cdot {}^{\cdot\cdot}_{\cdot\cdot}\mathbf{s}^{(t)}_{\cdot\cdot} & \text{if } {}^{\cdot\cdot}_{\cdot\cdot}\mathbf{c}^{(t)}_{\cdot\cdot} > \gamma \cdot {}^{\cdot\cdot}_{\cdot\cdot}\mathbf{c}^{(t-1)}_{\cdot\cdot} \end{cases} \\ &\beta > 1 \text{ and } 0 < \gamma < 1 \end{aligned} \tag{12}$$

   where ${}^{\cdot\cdot}_{\cdot\cdot}\mathbf{c}^{(t)}_{\cdot\cdot} = {}^{\cdot\cdot}_{\cdot\cdot}\mathbf{c}^{(t)} = ({}^{\cdot\cdot}_{\cdot\cdot}\mathcal{H}({}^{\cdot\cdot}\mathbf{r}) - {}^{\cdot\cdot}_{\cdot\cdot}\mathbf{h})^{(t)}$ and/or ${}^{\cdot\cdot}_{\cdot\cdot}\mathbf{c}^{(t)}_{\cdot\cdot} = {}^{\cdot\cdot}_{\cdot\cdot}\mathbf{c}^{(t)}_z = ({}^{i}_{j}\mathbf{z} - {}^{j}_{i}\mathbf{z})^{(t)}$, the inconsistency at iteration ${}^{(t)}$.
2. The second criterion determines if the multi-level optimization has converged:

$$\varepsilon_{v_c} \leq \max\{||{}^{\cdot\cdot}v^{(t+1)}_{\cdot\cdot c} - {}^{\cdot\cdot}v^{(t)}_{\cdot\cdot c}||_\infty, ||{}^{\cdot\cdot}v^{(t+1)}_{\cdot\cdot c}||_\infty\}. \tag{13}$$

   Equation (13) compares the infinity norm of the inconsistencies between elements and the infinity norm of the rate of inconsistency change with respect to a predetermined tolerance.

Convergence settings ε_{v_c}, ε_{v_f} can be divided into three categories according to Tosserams et al. [11]. A method where a small value ε_{v_f} is chosen called Block Coordinate Descent (BCD). Data between individual elements is exchanged until the convergence tolerance (ε_{v_f}) is met. Once the convergence tolerance ε_{v_f} is met, coupling data is exchanged and the relaxation parameters are updated according to Equation (12). These three steps are repeated until ε_{v_c} is met. A method that chooses a 'mild' value for ε_{v_f} called Inexact (InE). Therefore, data between individual elements is exchanged lesser frequent then for the exact method before relaxation parameters are updated via Equation (12). A method that chooses infinity for ε_{v_f} called

Alternating Descent (AD). Multipliers are updated after each individual element optimization via Equation (12).

Exact, inexact and alternating descent prescribe convergence parameter settings that put computational effort on the individual elements (Exact), on the communication lines and the coordinator (Alternating Descent) or a combination of the two (Inexact). Via changes of β and γ updates of the relaxation parameters can be adjusted such that the coordination procedure converges.

Finally, for the coordination a distinction is made between sequential coordination and parallel coordination. Sequential coordination solves the top-element of the hierarchy first and then solves each individual element on Level-1 one at a time. This is similar to the traditional multi-level optimization solution approach. Parallel coordination solves each individual element at the same time. The coordinator waits until all individual elements are optimized and then redistributes the coupling data. The Augmented Lagrangian relaxation is parallelized via Diagonal Quadratic Scaling [7] and a damping parameter τ is introduced that damps exchange of the coupling data.

6.2 Numerical Results

Two different coordination methods are applied to the supersonic business jet multi-level optimization problem: sequential coordination (according to the procedure proposed by Tosserams et al. [11]) and parallel coordination (according to the procedure proposed by Li et al. [7]). The relaxation parameters are updated via the so-called method of multipliers [11].

Results

The optimal design variable values and coupling variable values for the best solution found via the multi-level optimization framework for the supersonic business jet multi-level optimization problem are listed in Table 3.

Results for different decomposition and coordination approaches applied to the supersonic business jet multi-level optimization are listed in Table 4. Two decomposition methods are applied: hierarchic top-down and non-hierarchic decomposition. Furthermore, results are compared for three different combinations of convergence criteria: Block Coordinate Descent (BCD) ($\epsilon_{v_c} = 1 \times 10^{-3}$ and $\epsilon_{v_f} = 1 \times 10^{-4}$), Inexact (InE) ($\epsilon_{v_c} = 1 \times 10^{-3}$ and $\epsilon_{v_f} = 1 \times 10^{-3}$) and Alternating Descent (AD) ($\epsilon_{v_c} = 1 \times 10^{-3}$ and $\epsilon_{v_f} = +\text{inf}$). Parameters β and γ are adjusted such that the coordination procedures converge. Finally, for settings that exhibited least computational cost via a sequential solution process, the sequential solution process is compared with parallel execution of the individual hierarchical elements.

The computational costs listed in Table 4 do not represent multi-level optimization runs that converged to the same optimum. The best optimum was obtained via non-hierarchic decomposition with convergence settings that correspond to Block Coordinate Descent (BCD) and a sequential solution process. The same optimal

Table 3. Optimal solution found via the implemented multi-level optimization procedure for the supersonic business jet multi-level optimization problem for an optimal range: ${}^{0}r_1 = 2626$. This optimum was found via a non-hierarchic decomposition, sequential coordination process, $\beta = 1.05$ and $\gamma = 0.95$ and convergence settings $\epsilon_{v_f} = +\inf$, $\epsilon_{v_c} = 1 \times 10^{-3}$.

Design	Value	Interaction	Value
${}^{0}_{1.2}z_1$	1.50		
${}^{0}_{1.2}z_2$	6.00×10^4		
${}^{1}_{0}z_1$	1.50	${}^{1}_{0}h$	6.24
${}^{1}_{0}z_2$	6.00×10^4	${}^{1}_{2}h$	7.97×10^3
${}^{1}_{3}z_1$	2.50	${}^{1}_{3}h$	5.04×10^4
${}^{1}_{3}z_2$	8.00×10^2		
${}^{1}_{3}z_3$	2.50		
${}^{1}_{3}z_4$	1.49×10^2		
${}^{1}_{3}z_5$	7.00×10^1		
${}^{1}_{3}z_6$	9.00×10^{-2}		
${}^{1}x_1$	7.00×10^1		
${}^{1}x_2$	1.00×10^{-2}		
${}^{1}x_3$	3.50		

Design	Value	Interaction	Value
${}^{2}_{0}z_1$	1.50	${}^{2}_{0}h$	1.00
${}^{2}_{0}z_2$	6.00×10^4	${}^{2}_{1}h$	1.43
${}^{2}x_1$	1.72×10^{-1}	${}^{2}_{3}h$	1.27×10^4
${}^{3}\mathbf{x}_{1...9} = [1.18, 0.59, 0.24, 4.00, 4.00, \ldots$ 1.11, 1.18, 0.59, 0.24]			
${}^{3}\mathbf{x}_{10...18} = [1.93, 1.30, 0.67, 4.40, 4.40, \ldots$ 1.22, 1.93, 1.30, 0.67]			
${}^{3}x_{19}$	1.00×10^{-1}	${}^{3}_{0}h_1$	3.87×10^{-1}
${}^{3}_{1}z_1$	2.50	${}^{3}_{0}h_2$	5.04×10^4
${}^{3}_{1}z_2$	8.00×10^2	${}^{3}_{1}h_1$	5.04×10^4
${}^{3}_{1}z_3$	2.50	${}^{3}_{1}h_2$	1.12×10^1
${}^{3}_{1}z_4$	1.49×10^2		
${}^{3}_{1}z_5$	7.00×10^1		
${}^{3}_{1}z_6$	9.00×10^{-2}		

range was also found via a parallel solution process with convergence settings that correspond to Inexact (InE). However, the sequential and parallel approach that converged to the best optimum were also the most expensive methods considering computational effort. An optimum value found close to the optimum value of 2626 was found via a sequential solution process and convergence settings that correspond to Alternating Descent (AD) (2489) at significantly lower computational effort. The worst optimum (628) was found via non-hierarchic decomposition with convergence settings that corresponded to the Inexact method.

In general, settings for γ and β are chosen such that the relaxation parameters are updated frequently and with small increments. If for the current example, combinations of γ and β are chosen such that large increments in the relaxation parameters occur, the multi-level optimization process diverges and no solution is found. Furthermore, as a consequence of large increments in the relaxation parameters individual optimization problems have difficulties to converge to an individually feasible solution.

A parallel solution process requires small increments in both relaxation parameters and coupling variable updates. No convergence was observed for Alternating Descent (AD) convergence settings together with a parallel solution process. However, inexact (InE) convergence settings together with a proper choice of β and γ were sufficient to update coupling variables such that the solution process was numerically stable.

In Figure 22 the history of the optimization process is plotted for three different convergence settings, (a) Alternating Descent (AD), (b) Block Coordinate Descent (BCD) and (c) Inexact (InE) convergence settings and a sequential solution process. In Figure 23 Inexact (InE) convergence settings and a parallel solution process are shown. During the shift from individual optimal to overall optimal design large jumps

Table 4. Computational cost of each solution strategy for solving the supersonic business jet. Number of function evaluations (func.eval.) represent the total amount of function evaluations. These function evaluations are required to compute sensitivities (finite differences) and design points for the optimization problems. Number of optimization iterations (opt.iter.) represent the number of optimization function evaluations that were evaluated during the optimization process. Number of hierarchical updates (hier.upd.) represent the number of times the complete hierarchy of individual elements is re-optimized. For function evaluations and number of optimization iterations the order of magnitude is presented. Settings that did not converge are indicated by nc. Elements present at Level-1 are solved in parallel for the hierarchic top-down sequential decomposition process. Therefore a damping parameter τ is necessary for updates of the coupling data. For the parallel coordination τ is also introduced to damp the updates of coupling data.

Decomp.	Coord.	Func.eval.	Opt.iter.	Hier.upd.	Conv. crit.
top-down	sequential				$\epsilon_{v_c} \leq 1 \times 10^{-3}$
BCD: $\beta = 2.2$, $\gamma = 0.4$, $\tau = 0.9$		9×10^5	4×10^4	480	$\epsilon_{v_f} \leq 1 \times 10^{-4}$
					$\epsilon_{v_c} \leq 1 \times 10^{-3}$
InE: $\beta = 1.1$, $\gamma = 0.9$, $\tau = 0.9$		6×10^5	3×10^4	1101	$\epsilon_{v_f} \leq 1 \times 10^{-3}$
					$\epsilon_{v_c} \leq 1 \times 10^{-3}$
AD: $\beta = 1.1$, $\gamma = 0.90$, $\tau = 0.85$		nc	nc	nc	$\epsilon_{v_f} \leq +inf$
top-down	parallel				$\epsilon_{v_c} \leq 1 \times 10^{-3}$
InE: $\beta = 1.05$, $\gamma = 0.95$, $\tau = 0.9$		nc	nc	nc	$\epsilon_{v_f} \leq 1 \times 10^{-3}$
non-hier.	sequential				$\epsilon_{v_c} \leq 1 \times 10^{-3}$
BCD: $\beta = 1.1$, $\gamma = 0.9$		3×10^6	3×10^5	4790	$\epsilon_{v_f} \leq 1 \times 10^{-4}$
					$\epsilon_{v_c} \leq 1 \times 10^{-3}$
InE: $\beta = 2.0$, $\gamma = 0.5$		5×10^5	3×10^4	560	$\epsilon_{v_f} \leq 1 \times 10^{-3}$
					$\epsilon_{v_c} \leq 1 \times 10^{-3}$
AD: $\beta = 1.05$, $\gamma = 0.95$		1×10^5	7×10^3	174	$\epsilon_{v_f} \leq +inf$
non-hier.	parallel				$\epsilon_{v_c} \leq 1 \times 10^{-3}$
InE: $\beta = 1.05$, $\gamma = 0.95$, $\tau = 0.9$		9×10^6	4×10^5	4450	$\epsilon_{v_f} \leq 1 \times 10^{-3}$

in the objective function values are observed. The computations that were conducted during this jump showed numerical difficulties with respect to convergence of the individual optimization problems and difficulties with finding feasible solutions.

The results for the supersonic business jet show that the framework is able to perform a multi-level optimization of a complex optimization problem involving many design constraints and a significant amount of coupling. However, the computational effort required to converge to an optimum is considerable and the optima found show a large difference between best optimum and worst optimum found. Because the computational effort is considerable it might not be feasible in a commercial environment to rerun these optimizations frequently enough to assure that one finds a good optimal point. The convergence rate still poses a challenge before multi-level optimization becomes an alternative in a commercial environment.

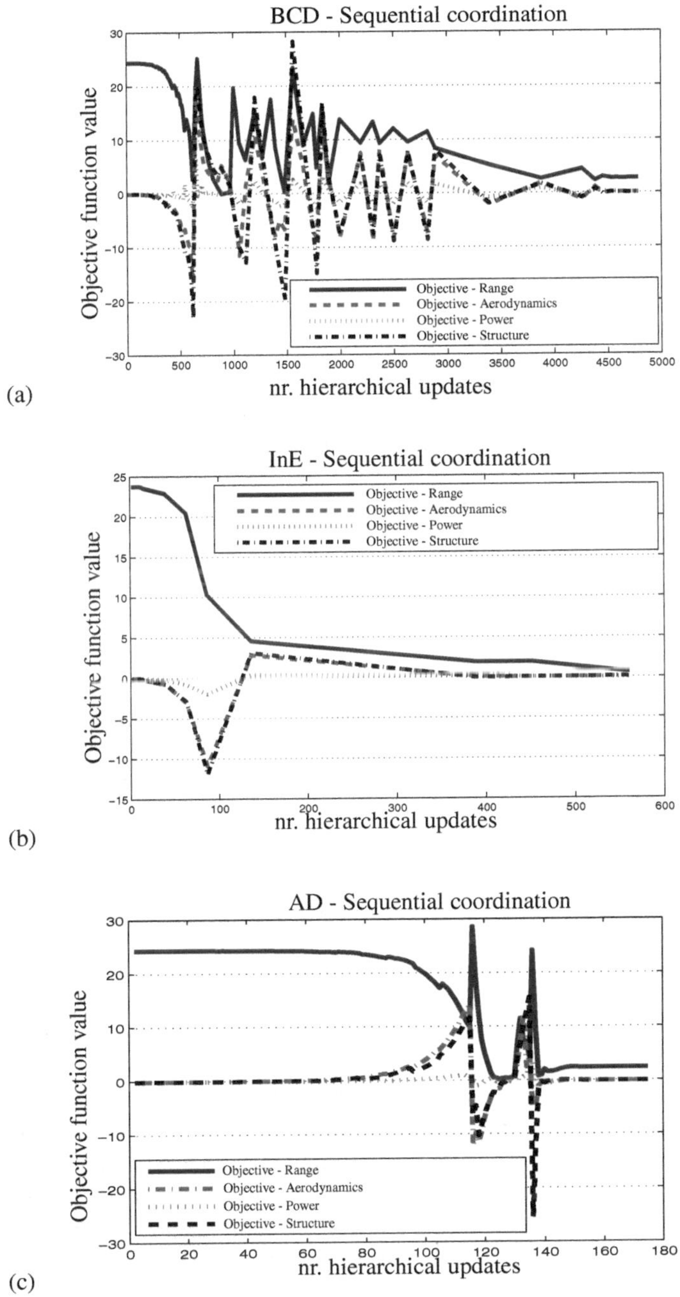

Fig. 22. Optimization history of the supersonic business jet using a non-hierarchic decomposition with a sequential coordination process and (a) exact (BCD) convergence criteria. (b) Inexact (InE) convergence criteria. (c) Alternating Descent (AD) convergence criteria. Note that the objective function value is scaled with 1×10^3 in all graphs.

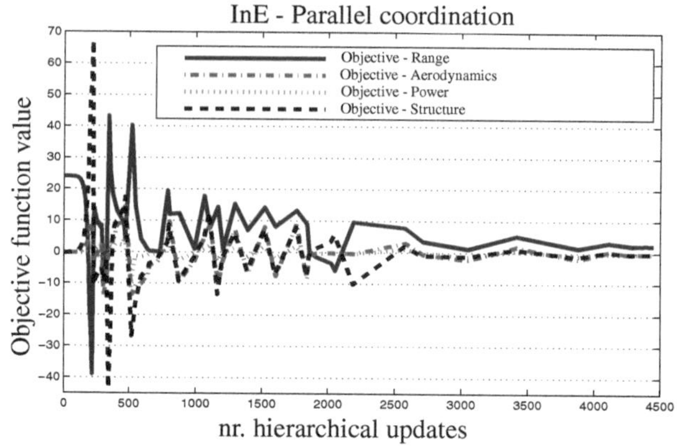

(d)

Fig. 23. Optimization history of the supersonic business jet using a non-hierarchic decomposition with a parallel coordination process and inexact convergence criteria. Note that the objective function value is scaled with 1×10^3.

7 Conclusions

The necessary steps for identifying a hierarchy and applying a decomposition formulation to a complex design problem were introduced. The benefit of this approach is that the developed steps are applicable to a wide range of complex systems with embedded hierarchy. Contrary to multi-level optimization methods that can be found in the literature that are developed for specific problems. The developed approach in this chapter allows for solving a wide range of complex problems with embedded hierarchy via different decomposition and coordination strategies. Via identifying coupling within the optimization problem individual optimization problems can be identified. Coordination of the individual optimization problems that are coupled requires decision making. Based on these decisions additional information is added to the individual optimization problems. Finally, solution strategies can be chosen that solve the decomposed problem sequentially or in parallel.

By capturing the generic steps of multi-level optimization the key components of a multi-level software program were defined. The flexibility of the framework was shown on a typical multi-level design problem for which different decomposition and coordination schemes can easily be exchanged. Furthermore, numerical results showed that via adjusting algorithm settings the computational costs can be reduced significantly at the cost of finding a marginally lower range for the supersonic business jet. It is expected that this framework will enable the development of multi-level methods that show better convergence characteristics then the methods that have been posted in literature.

Acknowledgments

This research is part of the larger project Multiscale Methods in Computational Mechanics which is supported by the Dutch Technology Foundation – STW, the Netherlands Organization for Scientific Research – NWO (DFG 5790) and the German Research Foundation – DFG (FOR 509).

References

1. J.S. Agte, J. Sobieszczanski-Sobieski, and R.R. Sandusky Jr. Supersonic business jet design through bi-level integrated system synthesis. In *Proceedings of the World Aviation Conference*, San Francisco, MCB Press, SAE paper 1999-01-5622, 1999.
2. J.-F.M. Barthelemy. Engineering design applications of multilevel optimization methods. In C.A. Brebbia and S. Hernandez (Eds.), *Computer Aided Optimum Design of Structures: Applications*. Springer-Verlag, 1989.
3. D.P. Bertsekas. *Nonlinear Programming*. Athena Scientific, Belmont, Massachusetts, 1995.
4. A.J. de Wit. A unified approach towards decomposition and coordination for multi-level optimization. PhD Thesis, Delft University of Technology, 2009.
5. H.M. Kim, W. Chen, and M.M. Wiecek. Lagrangian coordination for enhancing the convergence of analytical target cascading. *American Institute of Aeronautics and Astronautics Journal*, 44(10):2197–2207, 2006.
6. H.M. Kim, N.F. Michelena, P.Y. Papalambros, and T. Jiang. Target cascading in optimal system design. *Journal of Mechanical Design*, 125:474–480, September 2003.
7. Y. Li, Z. Lu, and J. Michalek. Diagonal Quadratic Approximation for Parallelization of Analytical Target Cascading. *ASME Journal of Mechanical Design*, 130(5), 2008.
8. J. Sobieszczanski-Sobieski. Sensitivity of complex, internally coupled systems. *AIAA Journal*, 28(1):153–160, 1990.
9. J. Sobieszczanski-Sobieski, T.D. Altus, M. Phillips, and R. Sandusky. Bilevel integrated system synthesis for concurrent and distributed processing. *American Institute of Aeronautics and Astronautics Journal*, 41(10):1996–2003, October 2003.
10. J. Sobieszczanski-Sobieski, J.F. Barthelemy, and K.M. Riley. Sensitivity of optimum solutions of problem parameters. *American Institute of Aeronautics and Astronautics Journal*, 20:1291–1299, 1982.
11. S. Tosserams, L.F.P. Etman, and J.E. Rooda. Augmented lagrangian coordination for distributed optimal design in MDO. *International Journal for Numerical Methods in Engineering*, 73:1885–1910, 2008.
12. T.C. Wagner. A general decomposition methodology for optimal system design. PhD Thesis, The University of Michigan, 1993.

PART 4

Coupled Problems and Porous Media

Multiscale/Multiphysics Model for Concrete

Bernhard A. Schrefler[1], Francesco Pesavento[1] and Dariusz Gawin[2]

[1]Department of Structural and Transportation Engineering, University of Padova, Via F. Marzolo 9, 34131 Padova, Italy; bas@caronte.dic.unipd.it
[2]Department of Building Physics and Building Materials, Technical University of Lodz, Al. Politechniki 6, 93-590 Lódz, Poland

Abstract In this paper a general model for the analysis of concrete as multiphase porous material, obtained from microscopic scale by applying the so-called Hybrid Mixture Theory, is presented. The final formulation of the governing equations at macro-level is obtained by upscaling their local form from the micro-scale. This procedure allows for taking into account both bulk phases and interfaces of the multiphase system, to define several quantities used in the model and to obtain some thermodynamic restrictions imposed on the evolution equations describing the material deterioration. Two specific forms of the general model adapted to the case of concrete structures under fire and to the case of concrete degradation due to the leaching process are shown. Some numerical simulations aimed at proving the validity of the approach adopted, are also presented and discussed.

Key words: Multiscale, multiphysics, concrete, leaching, high temperature

1 Introduction

In this work a general theoretical framework for the analysis of cementitious materials as multiphase porous media is presented. Nowadays, in material mechanics numerical multiscale procedures are often used for solving problems involving multiphysics aspects, but the thermodynamics principles are not automatically fulfilled. For instance in environmental engineering, where typically large scale problems are considered, a multi-scale approach is needed in the definition of the continuum mechanics. In the present approach the mathematical model is formulated by using different scales starting from micro level, i.e. from a local form of the governing equations at the pore scale. More precisely, the final form of the mathematical model is obtained by applying some averaging operators to the equations at micro-level, while the constitutive laws are defined directly at the upper scale, according to the so called Hybrid Mixture Theory. This approach allows for taking into account both bulk phases and interfaces of the multiphase system, assures that the Second Law of Thermodynamics is satisfied at macro-level, that no unwanted dissipations are generated and that the definition of the relevant quantities involved is thermodynamically correct. In particular, if the thermodynamically constrained theory (TCAT) is

R. de Borst and E. Ramm (eds.), *Multiscale Methods in Computational Mechanics*, Lecture Notes in Applied and Computational Mechanics 55, DOI 10.1007/978-90-481-9809-2_19,

used, also the satisfaction of the second law of thermodynamics for all constituents at micro-level is guaranteed. Within this last approach some stress measures are obtained and their form is described. The chosen procedure does not exclude however the use of a numerical multiscale approach in the formulation of the material properties. The numerical solution is obtained directly at macro-level by discretizing the governing equations in their final form.

Two specific applications of the general model adapted to the cases of interest are presented. The first form is applied to the case of concrete structures under fire for which surrounding high temperature and pressure are considered. In these conditions concrete structures experience spalling, which results in rapid loss of the surface layers of the concrete at temperature exceeding about 200–300°C. The second relevant application of the general model is the analysis of behaviour of concrete structures subject to leaching process. The latter chemical reaction is of importance during assessment of durability of concrete structures exposed to direct contact with deionised water. Usually thermodynamic equilibrium of the calcium ions in pore solution and the solid calcium in material skeleton, as well as purely diffusive calcium transport, are assumed in modeling of the process. Here we consider thermodynamic imbalance of the calcium in solid and liquid phases. Moreover, the leaching model is non isothermal. It allows for analyses of durability of concrete structure in various conditions, also those which were before impossible to be modelled, like for example leaching due to existing water pressure gradient and/or with thermal gradients.

Finally, some numerical simulations aimed to prove the validity of the approach adopted are presented and discussed.

2 General Mathematical Model

The balance equations are written by considering cementitious materials as a multiphase porous medium, which is assumed to be in hygral equilibrium state locally. More specifically, in the present case the solid skeleton voids are filled partly by liquid water (the wetting phase) and partly by a gas phase (the non-wetting phase). Below the critical temperature of water, T_{cr}, the liquid phase consists of physically bound water and capillary water, which appears when the degree of water saturation exceeds the upper limit of the hygroscopic region, S_{ssp}. Above the temperature T_{cr} the liquid phase consists of bound water only. In the whole temperature range the gas phase is a mixture of dry air and water vapour, which is a condensable gas constituent for temperatures $T < T_{cr}$.

In the following, subscripts mean physical quantities related to the whole volume of medium, and upper indices their intrinsic values related to the single phase or constituent only. By constituent we indicate a matter which is uniform throughout in chemical composition, while phase means here different physical state of a matter (solid, liquid or gaseous). Symbols s, w, g, gw and ga denote solid skeleton, pure liquid water, gas phase in general, vapour and dry air, respectively. The solid phase is assumed to be in contact with all fluids in the pores. The symbols l and Ca indicate the whole liquid phase (i.e. water + ions) and the calcium ions content in the liquid solution respectively, in the case leaching process is under consideration. In the latter case, only materials fully saturated with pore water are considered in the applications, but for the sake of generality, all the equations are developed for fully/partially saturated state.

The full development of the model equations, starting from the local, microscopic balance equations with successive volume averaging, and further transformations are presented in [1–5].

The final form of the governing equations, written at microscopic level in terms of the primary variables (p^g, p^c, T, $\mathbf{u}$, c_{Ca}) and taking into account the small deformation theory, is:

– *Dry air mass balance equation* (including the solid skeleton mass balance) takes into account both diffusive and advective air flow, as well as variations of porosity caused by hydration and leaching processes and deformations of the skeleton. It has the following form:

$$-n\frac{D^s S_w}{Dt} - \beta_s (1-n) S_g \frac{D^s T}{Dt} + S_g \mathrm{div}\, \mathbf{v}^s + \frac{S_g n}{\rho^{ga}} \frac{D^s \rho^{ga}}{Dt} + \frac{1}{\rho^{ga}} \mathrm{div}\, \mathbf{J}_g^{ga}$$
$$+ \frac{1}{\rho^{ga}} \mathrm{div}\left(n\, S_g \rho^{ga} \mathbf{v}^{gs}\right) - \frac{(1-n)\, S_g}{\rho^s} \frac{\partial \rho^s}{\partial \Gamma_{hydr}} \frac{D^s \Gamma_{hydr}}{Dt}$$
$$= \frac{\dot{m}_{hydr}}{\rho^s} S_g - \frac{\dot{m}_{diss}}{\rho^s} S_g \tag{1}$$

where S_w and S_g are the water and gas saturation level respectively, β_s is the thermal expansion coefficient of the solid phase, ρ^{ga} is the dry air density, Γ_{hydr} the hydration degree, $\mathbf{J}_g^{ga}$ the dry air diffusive flux and n the porosity. The terms at the right hand side represent the source terms related to the hydration ($\dot{m}_{hydr}$) and leaching process ($\dot{m}_{diss}$).

– *Water species (liquid+vapor) mass balance equation* (including the solid skeleton mass balance) considers diffusive and advective flow of water vapour, mass sources related to phase changes of vapour (evaporation-condensation, physical adsorption - desorption and dehydration), and variations of porosity caused by hydration/dehydration and leaching processes and deformations of the skeleton, resulting in the following equation:

$$n(\rho^w - \rho^{gw})\frac{D^s S_w}{Dt} + (\rho^w S_w + \rho^{gw} S_g)\alpha\, \mathrm{div}\, \mathbf{v}^s$$
$$- \beta^*_{swg} \frac{D^s T}{Dt} + S_g n \frac{D^s \rho^{gw}}{Dt} + \mathrm{div}\, \mathbf{J}_g^{gw} + div\,(n\, S_w \rho^w \mathbf{v}^{ls}) + \mathrm{div}\,(n\, S_g \rho^{gw} \mathbf{v}^{gs})$$
$$- (\rho^w S_w + \rho^{gw} S_g)\frac{(1-n)}{\rho^s}\left[\frac{\partial \rho^s}{\partial \Gamma_{hydr}} \frac{D^s \Gamma_{hydr}}{Dt} + \frac{D^s \rho^s}{D\Gamma_{leach}} \frac{D^s \Gamma_{leach}}{Dt}\right]$$
$$= -\frac{\dot{m}_{hydr}}{\rho^s}(\rho^w S_w + \rho^{gw} S_g - \rho^s) - \frac{\dot{m}_{diss}}{\rho^s}(\rho^{gw} S_g + \rho^w S_w) \tag{2}$$

where $\beta^*_{swg} = \beta_s(1-n)(S_g\rho^{gw} + \rho^w S_w) + n\,\beta_w\rho^w S_w$ is the thermal expansion coefficient of the multiphase system, ρ^{gw} is the water vapour density, Γ_{leach} the leaching degree and $\mathbf{J}_g^{gw}$ the water vapour diffusive flux.

– *Energy balance equation* (for the whole system) accounting for the conductive and convective heat flow, heat effects of phase changes and hydration/dehydration process, can be written as follows:

$$(\rho C_p)_{eff}\frac{\partial T}{\partial t} + (\rho_w C_p^w \mathbf{v}^w + \rho_g C_p^g \mathbf{v}^g)\cdot \text{grad}\, T - \text{div}(\chi_{eff}\,\text{grad}\, T) = -\dot{m}_{vap}\Delta H_{vap} - \dot{m}_{hydr}\Delta H_{hydr} \tag{3}$$

where χ_{eff} is effective conductivity from experiments while the thermal capacity of the multiphase system and the enthalpy of vaporization and hydration are respectively:

$$(\rho C_p)_{eff} = \rho_s C_p^s + \rho_w C_p^w + \rho_g C_p^g$$
$$\Delta H_{vap} = H^{gw} - H^w$$
$$\Delta H_{hydr} = H^{ws} - H^w \tag{4}$$

with $\beta_{sw} = S_w[(1-n)\beta_s + n\beta_w]$.

– *Linear momentum balance equation* (for the multiphase system). This equation may assume two different forms depending on the problem which is solved: concrete at high temperature (finite form) [2] or concrete leaching, i.e. chemical degradation + creep, (rate form) [6]:

$$\text{div}(\mathbf{t}^{total}) + \rho\mathbf{g} = 0 \tag{5}$$
$$\text{div}(\dot{\mathbf{t}}^{total}) + \dot{\rho}\mathbf{g} = 0 \tag{6}$$

– *Calcium mass balance equation (leaching case)* including the diffusion of ionic species, the advection and the chemical reaction related to the dissolution of the calcium contained in the solid skeleton [4]:

$$(1-n)c_{Ca}S_w\frac{1}{\rho^s}\frac{\partial\rho^s}{\partial\Gamma_{leach}}\frac{\partial\Gamma_{leach}}{\partial t} + nc_{Ca}\frac{\partial S_w}{\partial t} + nS_w\frac{\partial c_{Ca}}{\partial t} + \alpha c_{Ca}S_w\,\text{div}\,\mathbf{v}^s + \frac{1}{\rho^w}div\,\mathbf{J}_d^{Ca} + \frac{1}{\rho^w}\text{div}(c_{Ca}nS_w\rho^w\mathbf{v}^{ls}) = \frac{\dot{m}_{diss}}{\rho^w} - \frac{\dot{m}_{diss}}{\rho^s}c_{Ca}S_w \tag{7}$$

In Eq. (7) c_{Ca} is the calcium concentration in the liquid solution, $\mathbf{v}^{ls}$ is the velocity of the solution with respect to the solid skeleton, and finally $\mathbf{J}_d^{Ca}$ is the diffusion of the Ca^{2+} ions in the water solution.

For the model closure the initial and boundary conditions are needed, for further information see [2, 6, 7].

The governing equations of the model are discretized in space by means of the finite element method [8]. The unknown variables are expressed in terms of their nodal values as

$$p^g(t) \cong \mathbf{N}_p \bar{\mathbf{p}}^g(t), \quad p^c(t) \cong \mathbf{N}_p \bar{\mathbf{p}}^c(t), \quad T(t) \cong \mathbf{N}_t \bar{\mathbf{T}}(t),$$
$$\mathbf{u}(t) \cong \mathbf{N}_u \bar{\mathbf{u}}(t), \quad c_{Ca}(t) = \mathbf{N}_s \bar{\mathbf{c}}_{Ca}(t) \tag{8}$$

The discretized form of the model equations was obtained in [2, 6, 7] by means of Galerkin's method [8] (weighted residuals), and can be written in the following concise matrix form:

$$\mathbf{C}_{ij}(\mathbf{x}) \frac{\partial \mathbf{x}}{\partial t} + \mathbf{K}_{ij}(\mathbf{x})\, \mathbf{x} = \mathbf{f}_i(\mathbf{x}), \tag{9}$$

where $\mathbf{x} = \{\bar{\mathbf{p}}^g, \bar{\mathbf{p}}^c, \bar{\mathbf{T}}, \bar{\mathbf{u}}\}^T$ and the non-linear matrix coefficients $\mathbf{C}_{ij}(\mathbf{x})$, $\mathbf{K}_{ij}(\mathbf{x})$ and $\mathbf{f}_i(\mathbf{x})$ are defined in detail in for concrete at high temperature and in for the case of leaching.

The time discretization is accomplished through a fully implicit finite difference scheme (backward difference),

$$\boldsymbol{\Psi}^i g(\mathbf{x}_{n+1}) = \mathbf{C}_{ij}(\mathbf{x}_{n+1}) \frac{\mathbf{x}_{n+1} - \mathbf{x}_n}{\Delta t} + \mathbf{K}_{ij}(\mathbf{x}_{n+1})\, \mathbf{x}_{n+1} - \mathbf{f}_i(\mathbf{x}_{n+1}) = \mathbf{0}, \tag{10}$$

where superscript i ($i = g, c, T, Ca, \mathbf{u}$) denotes the state variable, n is the time step number and Δt the time step length.

The equation set (10) is solved by means of a monolithic Newton–Raphson type iterative procedure:

$$\boldsymbol{\Psi}^i(\mathbf{x}^k_{n+1}) = -\left.\frac{\partial \boldsymbol{\Psi}^i}{\partial \mathbf{x}}\right|_{\mathbf{x}^k_{n+1}} \Delta \mathbf{x}^k_{n+1}, \quad \mathbf{x}^{k+1}_{n+1} = \mathbf{x}^k_{n+1} + \Delta \mathbf{x}^k_{n+1}, \tag{11}$$

where k is the iteration index and $\partial \boldsymbol{\Psi}^i / \partial \mathbf{x}$ is the Jacobian matrix.

3 Effective Stress Principle

In the approach proposed in this work, cementitius materials are treated as multiphase porous media. In this view, analyzing the stress state and the deformation of the material it is necessary to consider not only the action of an external load, but also the pressure exerted on the skeleton by fluids present in its voids. Hence, the total stress tensor $\mathbf{t}^{total}$ acting in a point of the porous medium may be split into the effective stress $\eta^s \boldsymbol{\tau}^s$, which accounts for stress effects due to changes in porosity, spatial variation of porosity and the deformations of the solid matrix, and a part accounting for the solid phase pressure exerted by the pore fluids [5, 9, 10]

$$\mathbf{t}^{total} = \eta^s \boldsymbol{\tau}^s - \alpha P^s \mathbf{I} \tag{12}$$

where $\mathbf{I}$ is the second order unit tensor, α the Biot coefficient and P^s is some measure of solid pressure acting in the system, also simply called solid pressure. Many different forms of P^s have been proposed in the past decades in Geomechanics, but considering that concrete has a fine microstructure, i.e. the interactions between molecules of water and concrete skeleton on micro-structural level are rather complex, the formulation of Gray and Schrefler [9], which takes into account the degree of contact of each fluid phase with the solid one, has been adopted here. Including the interface in the analysis allows to interpret the Biot coefficient as the ratio of the hydrostatic part of the total stress tensor (p^{total}) to the normal force exerted on the solid surface by the surrounding fluids, i.e. $-\langle \mathbf{n}_s \mathbf{t}_s \mathbf{n}_s \rangle^{ss}$:

$$\alpha = -\frac{p^{total}}{\langle \mathbf{n}_s \mathbf{t}_s \mathbf{n}_s \rangle^{ss}} = 1 - \frac{\tilde{K}_T}{\tilde{K}_S} \tag{13}$$

This relationship accounts for different values of bulk modulus for solid phase (grain) and the skeleton, $\tilde{K}_S$ and $\tilde{K}_T$, respectively. Here $\mathbf{t}_s$ is the stress tensor of the solid phase at microscopic level, $\mathbf{n}_s$ is the unit vector normal to the solid phase in each point, while the Macaulay brackets $\langle\,\rangle^{ss}$ indicate an averaging over the solid surface. With these results, P^s is selected to be the average normal force exerted on the solid surface by the fluids in the pore space:

$$P^s = -\langle \mathbf{n}_s \mathbf{t}_s \mathbf{n}_s \rangle^{ss} \tag{14}$$

By considering the interfaces and by formulating the model from the micro-level, the following form of the so called "standard solid pressure", p^s is obtained:

$$p^s = x_s^{ws} p^w + x_s^{gs} p^g + x_s^{ws} \gamma^{ws} J_{ws}^s + x_s^{gs} \gamma^{gs} J_{gs}^s \tag{15}$$

where x_s^{ws} and x_s^{gs} are the fractions of skeleton area in contact with water and gas, respectively, while J_{ws}^s and J_{gs}^s are the curvature of the water/solid and gas/solid interfaces in that order. γ^{ws} and γ^{gs} are surface tension-like terms. The two forms of the solid pressure are related one each other by means of

$$p^s = \alpha P^s \tag{16}$$

For further details, see [5, 9, 10].

By using the following simplified version of the capillary pressure, valid at thermodynamic equilibrium and neglecting the direct contribution of the fluid-solid interfaces

$$p^c \equiv p^g - p^w = \Pi^f - \gamma^{wg} J_{wg}^w \tag{17}$$

Equation (17) can be transformed into

$$p^s = p^g + x_s^{ws} \gamma^{wg} J_{wg}^w - x_s^{ws} \Pi^f + x_s^{ws} \gamma^{ws} J_{ws}^s + (1 - x_s^{ws})\, \gamma^{gs} J_{gs}^s \tag{18}$$

Equation (17) considers the disjoining pressure Π^f and can be applied in the hygroscopic region (i.e. when the saturation level is lower than the solid saturation point and the water is present only as a thin film on the skeleton surface) as well as in the non-hygroscopic region (i.e. for higher levels of moisture content, for which saturation values exceed the solid saturation point).

In Eq. (18) one can recognize terms corresponding to the main physical phenomena leading to concrete shrinkage: the first term on the rhs describes an effect of gas pressure, the second one of capillary tension, the third one of disjoining pressure, and the last two terms, resulting from action of surface tension of solids on the interfaces with the pore fluids, are negligible. Taking into account such simplifications and relation (16), the so called "effective stress principle", i.e. Eq. (12), can be rewritten in the following manner:

$$\eta^s \boldsymbol{\tau}^s = \mathbf{t}^{total} + (p^g - x_s^{ws} p^c)\mathbf{I} \tag{19}$$

4 Application of the Model to Concrete Structures at Elevated Temperature

The model presented in Sections 2 and 3 can be also applied to the analysis of behavior of concrete structures under severe temperature and pressure conditions. In these conditions concrete structures experience spalling, which results in rapid loss of the surface layers of the concrete at temperature exceeding about 200–300°C. As a result, the core concrete is exposed to these temperatures, thereby increasing the rate of heat transmission to the core part of the element and in particular to the reinforcement, what may pose a risk for the integrity of the concrete structure.

It is commonly believed that the main reasons of the thermal spalling are: build-up of high pore pressure close to the heated concrete surface as a result of rapid evaporation of moisture, and the release of the stored energy due to the thermal stresses resulting from high values of restrained strains caused by temperature gradients. Nevertheless, relative importance of the two mechanisms is not established yet and still needs further studies, both experimental and theoretical. The results of the research performed up to now show, that the fire performance of concrete structures is influenced by several factors, like initial moisture content of the concrete, the rate of temperature increase (fire intensity), porosity (density) and permeability of the concrete, its compressive strength, type of aggregate, dimensions and shape of a structure, its lateral reinforcement and loading conditions. The HSC structures are particularly affected by this phenomenon. In fact, HSC provides better structural performance, especially in terms of strength and durability, compared to traditional, normal-strength concrete (NSC). However, many studies showed that the fire performance of HSC differs from that of NSC which exhibits rather good behavior in these conditions.

For applying the present model to such a peculiar case, some modifications to its basic form are needed. In particular it is necessary to formulate some evolution equations accounting for deterioration processes taking place in concrete under such severe conditions. In the current form of the model the following ones have been considered [2, 11]:

– *Dehydration process evolution law*, considering its irreversibility, has the form (i.e. dehydration degree evolution):

$$\Gamma_{dehydr}(t) = \Gamma_{dehydr}(T_{max}(t)) \tag{20}$$

thus

$$\dot{\Gamma}_{dehydr} = \frac{\partial \Gamma_{dehydr}(T)}{\partial T}\frac{\partial T}{\partial t} \qquad \text{for } T(t) \geq T_{max}(t),$$
$$\dot{\Gamma}_{dehydr} = 0 \qquad \text{for } T(t) < T_{max}(t),$$

where $T_{max}(t)$ is the highest temperature reached by the concrete up to the time instant t. The constitutive relationship $\Gamma_{dehydr}(T)$ can be obtain from the results of thermo-gravimetric (TG or DTA) tests, using the definition of the dehydration degree by means of the mass changes during concrete heating:

$$\Gamma_{dehydr}(T) = \frac{m(T_o) - m(T)}{m(T_o) - m(T_\infty)} \tag{21}$$

where $m(T)$ is mass of concrete specimen measured at temperature T during TG tests, T_o and T_∞ are temperatures when the dehydration process starts and finishes. We assumed here $T_o = 105°\text{C}$ and $T_\infty = 1000°\text{C}$.

– *Thermo-chemical damage evolution equation*, obtained on the basis of the experimental results, takes into account the irreversible character of the material structural changes and may be written as [2, 11]:

$$V(t) = V(T_{max}(t)) \tag{22}$$

– *Mechanical damage evolution equation*, of the following form:

$$d(t) = d(\tilde{\varepsilon}(t)) \tag{23}$$

is expressed in terms of the equivalent strain, $\tilde{\varepsilon}$, given by equations of the classical non-local, isotropic damage theory [2, 11, 12].

The mechanical and thermo-chemical damage parameters, d and V, are defined on basis of the experimentally determined stress-strain profiles at various temperatures, as follows [2, 13]:

$$d = 1 - \frac{E(T)}{E_o(T)}, \qquad V = 1 - \frac{E_o(T)}{E_o(T_a)} \tag{24}$$

where the subscript 'o' refers to the elastic behaviour and $E(T)$ is the Young modulus at temperature$T$ different from the reference (i.e. ambient) temperature T_a.

Taking into account a joint action of mechanical and thermo-chemical degradations, the total damage parameter, D can be defined on the base of the following multiplicative combination [2]:

$$1 - D = \frac{E(T)}{E_o(T_a)} = \frac{E(T)}{E_o(T)}\frac{E_o(T)}{E_o(T_a)} = (1-d)\cdot(1-V) \tag{25}$$

We consider that only the elastic properties of the material are affected by the total damage parameter D, i.e. the material is supposed to behave elastically and to remain isotropic and that the dependence on damage is introduced through the stiffness matrix:

$$\eta^s \boldsymbol{\tau}^s = \boldsymbol{\Lambda}(D) : \mathbf{e}^s \tag{26}$$

In Eq. (20) $\mathbf{e}^s$ is the strain tensor for infinitesimal deformations (i.e the limit of $\mathbf{E}^s$) and $\eta^s \boldsymbol{\tau}^s$ is the effective stress tensor.

The application of the "true stress concept", leads to:

$$\boldsymbol{\tau}^s = (1 - D)\tilde{\boldsymbol{\tau}}^s \tag{27}$$

where $\eta^s \tilde{\boldsymbol{\tau}}^s$ is the "net effective stress" (in the sense of damage mechanics). This results in the following form of the constitutive relationship for the solid phase:

$$\eta^s \tilde{\boldsymbol{\tau}}^s = \boldsymbol{\Lambda}_0 : \mathbf{e}^s \tag{28}$$

where $\boldsymbol{\Lambda}_0$ is the initial stiffness.

Taking into account (24), (23) and (19), the "net" effective stress tensor may be expressed as

$$\eta^s \tilde{\boldsymbol{\tau}}^s - \frac{\mathbf{t}^{total} + p^s}{\mathbf{I}(1 - D)} \tag{29}$$

where p^s is given by (18).

Thus, the linear momentum balance equation has now the following form [2]:

$$\mathrm{div}(\eta^s \boldsymbol{\tau}^s - p^s \mathbf{I}) + \rho \mathbf{g} = 0 \tag{30}$$

where the nature of the effective stress tensor has been discussed in Section 3. By combining Eq. (27) with Eq. (29), $\eta^s \tau^s$ is then given by

$$\eta^s \boldsymbol{\tau}^s = (1 - d)(1 - V)\boldsymbol{\Lambda}_0 : (\boldsymbol{\varepsilon}_{tot} - \boldsymbol{\varepsilon}_{th} - \boldsymbol{\varepsilon}_{tchem} - \boldsymbol{\varepsilon}_{tr}) \tag{31}$$

The evolution equations shown above are not enough for a complete description of the material behavior. Indeed, an unloaded sample of plain concrete or cement stone, exposed for the first time to heating, exhibits considerable changes of its chemical composition, inner structure of porosity and changes of sample dimensions (irreversible in part). The concrete strains during first heating, called load-free thermal strains (LFTS) are usually treated as superposition of thermal and shrinkage components, and often are considered as almost inseparable. LFTS are decomposed in three main contributions [11]:

– *Thermal dilatation strains*

$$d\boldsymbol{\varepsilon}_{th} = \beta_s(T)\,dT \tag{32}$$

– *Capillary shrinkage strains*

$$d\boldsymbol{\varepsilon}_{sh} = \frac{\alpha}{K_T}(dx_s^{ws} p^c + x_s^{ws} dp^c)\mathbf{I} \tag{33}$$

where K_T is the bulk modulus of the porous medium.

– *Thermo-chemical strains*

$$d\boldsymbol{\varepsilon}_{tchem} = \beta_{tchem}(V)\, dV \tag{34}$$

where

$$\beta_{tchem}(V) = \frac{\partial \boldsymbol{\varepsilon}_{tchem}(V)}{\partial V}$$

is obtained from experimental tests (V is the thermo-chemical damage parameter).

As far as the first contribution is concerned, the strains are treated in a manner usual in thermo-mechanics, but considering the thermal expansion coefficient β_s as a function of temperature.

Shrinkage strains are modeled by means of the effective stress principle, in the form derived in Eq. (12) or Eq. (19), Section 3. As already pointed out, for materials with very fine pores and well developed internal pore surface, where water is also present as a thin film (like for example in concrete), the solid pressure relationship contains a coefficient x_s^{ws}, Eqs. (15) or (18), instead of the classical saturation S used in geomechanics [15]. This coefficient is a function of saturation S and takes into account the disjoining pressure, see Eq. (18), which is important in the range of saturation in which only a thin film of water is adsorbed to the wall of the pores, see [11, 16].

In this way the contribution of the term related to the capillary pressure in Eq. (18) or Eq. (19) can be interpreted as a sort of "internal" load for the skeleton of the material. Hence, the associated shrinkage strains are not computed directly in the strain decomposition as it is usual in the classical phenomenological approaches.

In heated concrete, above the temperature of about 105°C, the thermal decomposition of the cement matrix starts, and at higher temperatures also of aggregate (depending on its type and composition). This is a consequence of several complicated, endothermic chemical reactions, called concrete dehydration. As their result a considerable shrinkage of cement matrix (called chemical shrinkage) and usually expansion of aggregate are observed. Due to this contradictory behavior of the material components, cracks of various dimensions are developing when temperature increases, causing an additional change of concrete strains (usually expansion). These strains are modeled as function of thermo-chemical damage which takes into account the thermo-chemical deterioration of the material.

During first heating, mechanically loaded concrete exhibits greater strains as compared to the load-free material at the same temperature. These additional deformations are referred to as load induced thermal strains (LITS) [17]. A part of them originates just from the elastic deformation due to mechanical load, and it increases during heating because of thermo-chemical and mechanical degradation of the material strength properties. The time dependent part of the strains during transient thermal processes due to temperature changes, is generally called thermal creep.

Its physical nature is rather complicated and up to now not fully understood, thus modeling is usually based on the results of special experimental tests. Typically, they are performed at constant heating rate equal to 2 K/min, for various (but constant during a particular test) levels of external load, σ = const, ranging from 0% (load-free measurements) to 60% of the compressive strength of material at room temperature, $f_c(T_a)$.

The formulation employed in the model is due to Thelandersson in its original form, here modified using a coefficient $\bar{\beta}_{tr}$ as a function of thermo-chemical damage V and the effective stresses instead of total stresses, coupling in such a way the thermo-chemo-mechanical damage and capillary shrinkage model with the thermal creep model [11]:

$$d\boldsymbol{\varepsilon}_{tr} = \frac{\bar{\beta}_{tr}(V)}{\mathrm{f_c}(T_a)}\mathbf{Q} : \eta^s \tilde{\boldsymbol{\tau}}^s \, dV \tag{35}$$

In Eq. (35) $\mathbf{Q}$ is a fourth order tensor, $\eta^s \tilde{\boldsymbol{\tau}}^s$ is the "net" effective stress tensor (see Section 3) and finally f_c is the compressive strength of the material at 20°C.

The model in this form has been successfully used for solving several engineering problems, e.g. the case of fire in tunnels [19] or spalling prediction [20].

4.1 Simulation of a Concrete Column under Fire with Fast Cooling

The model presented in Sections 2 and 3 has been applied here to the calculation of some structural elements of high rise buildings, more precisely a concrete column exposed to a fire described by a parametric heating profile.

The latter one can be obtained from the time-temperature parametric curves defined in the Eurocode 1, Part 1–2, formulated on experimental basis. The time-temperature parametric curves defined in the Annex A of the Eurocode 1, Part 1–2 are, on the contrary to the nominal curves, dependent on certain physical parameters such as for example:

- The fire load density (the higher fire load density the longer the fire is),
- The ventilation conditions, depending mainly on the geometry, size and distribution of the compartment openings (big ventilation openings lead to fast fires but less severe),
- The properties of the closing walls of the fire compartment (walls gain energy and limit the fire temperature).

These curves also differ from the nominal curves in the sense that after a heating stage they show a cooling phase (see below), and they are valid for fire sectors with a built surface not larger than 500 m^2, without holes at the roof and with a maximum height of 4 m, assuming a complete combustion of the fire load.

They are defined in the heating stage by

$$\Theta_g = 20 + 1325 \cdot (1 - 0.324 \cdot e^{-0.2 \cdot t^*} - 0.204 \cdot e^{-1.7 \cdot t^*} - 0.472 \cdot e^{-19 \cdot t^*}) \tag{36}$$

where Θ_g (°C) is the air temperature within the fire sector, while t^* is a sort of equivalent time (in hours) that takes into account not only the time, but also the so called "opening coefficient". This parameter can be calculated from the area of total surface of the vertical openings, the total surface closing the fire sector, the averaged height of the windows at all of the walls and the characteristics of the material (specific heat, thermal conductivity and mass density). For further details

Point	Distance from heated surface
A	15.00 cm
B	10.00 cm
C	5.00 cm
D	3.00 cm
E	1.00 cm
F	0.05 cm

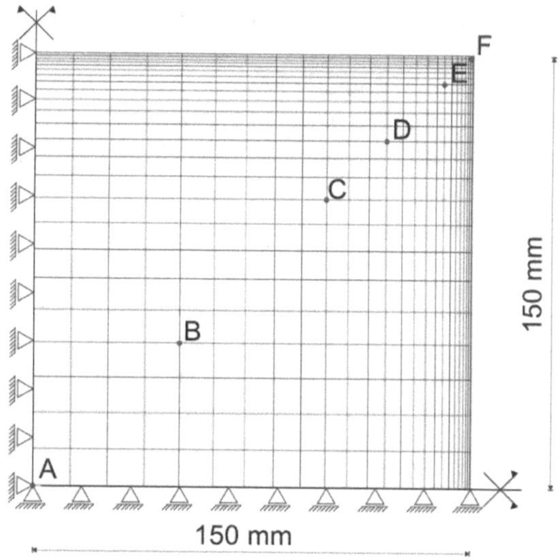

Fig. 1. Cross section of the square column: geometry and FE mesh used in the computations.

about the definition of the possible profiles from the parametric curves, see [21]. From a general viewpoint, when the fire is controlled by the ventilation conditions the increase of the opening coefficient leads to shorter but more severe fires, until fires are controlled by the thermal load. From this point the influence of the opening coefficient is especially observed on the cooling rate of the environment. It is recalled that a fire is defined as "ventilation controlled" if the peak of temperature (calculated on the basis of the fire load density and the opening coefficient) takes place at times longer than a limit value of time t_{lim} (equal to 15, 20 or 25 minutes for slow, medium speed and fast development, respectively). If $t_{max} \leq t_{lim}$ the fire can be classified as "fire load".

By starting from the previous considerations, the heating profile selected for the calculation of the structural element is the following:

$$\Theta_g = 20 + 1325 \cdot (1 - 0.324 \cdot e^{-0.0077 \cdot t} - 0.204 \cdot e^{-0.06545 \cdot t} - 0.472 \cdot e^{-0.7345 \cdot t}) \tag{37}$$

that is shown in Figure 2 (solid line noted as "T_{env}") and that corresponds, for the chosen material, to the minimum value of the opening coefficient allowed in the European rules (i.e. 0.02 $\mathrm{m}^{1/2}$).

The heating stage defined with the profile (37) is followed by a cooling phase defined under the same considerations. It is a fast cooling for which the environmental temperature decreases very quickly (within 20 seconds) from the maximum (i.e. about 709.5 K, reached in 80 minutes) down to the initial value, after a stationary stage at the maximum for 2 minutes. This conditions corresponds to the case of a rapid intervention of the firemen. Hence, the cooling profile is mathematically formulated as follows:

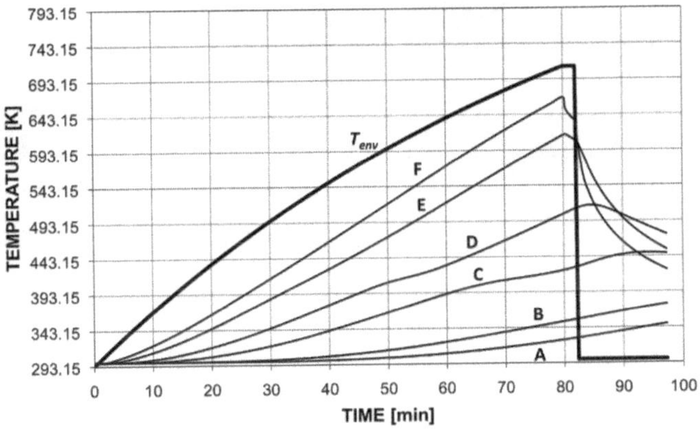

Fig. 2. Temperature histories in the environment (solid blue line), and in points A, B, C, D, E, F at a various depths from the heated sides.

Table 1. Main properties of the materials at 20°C.

Material property	C60
Water/binder ratio, w/b [–]	0.33
MIP porosity, n [%]	8.25
Water intrinsic permeability, k_o [m^2]	$2 \cdot 10^{-18}$
Young modulus, E [GPa]	34.4
Compressive strength, f_c [MPa]	60
Thermal conductivity, λ [W/m·K]	1.92
Specific heat, C_p [J/kg·K]	855

$$\Theta_g = 436.382 - 20.5681 \cdot (t - 4920) \tag{38}$$

As far as the boundary conditions are concerned, a mixed convective-radiative condition has been assumed with a convective exchange coefficient equal to 20 W/m^{2}K and an emissivity equal to 0.9. The mass exchange between the surface of the concrete element and the environment is characterized by a convective flux. The water vapour pressure is kept constant (equal to 1300 Pa). During heating this corresponds to a rapid decrease of the ambient relative humidity down to zero. The mass exchange coefficient is set to 0.025 m/s.

Figure 1 shows the size of the cross section of the square column under consideration (30 × 30 cm). The main properties of the material (C60) assumed in the simulation are listed in Table 1.

Figure 2 shows the temperature histories in six different points located at various depths from the heated surfaces of the column. The sharp decrease of environmental temperature, due to the cooling, leads to a rapid decrease of the temperature in these points, especially those close to the lateral sides of the cross section. The inversion of thermal fluxes has important consequences not only from the thermo-hygral point of view, but also for the mechanical behaviour of the material.

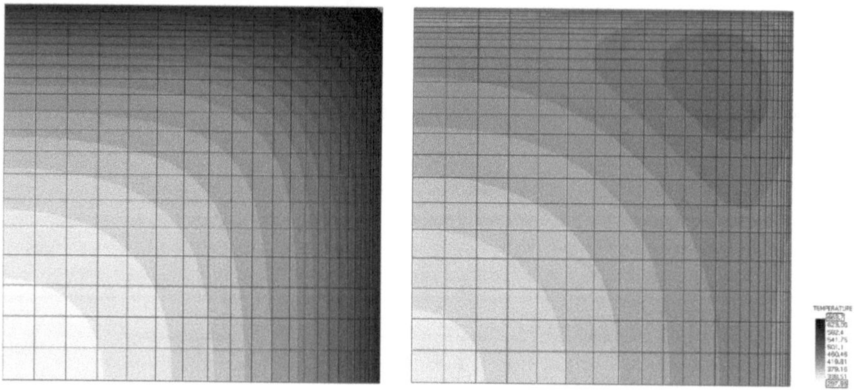

(a) (b)

Fig. 3. Distributions of the temperature [K] at 80 min (a) and at the final time (b).

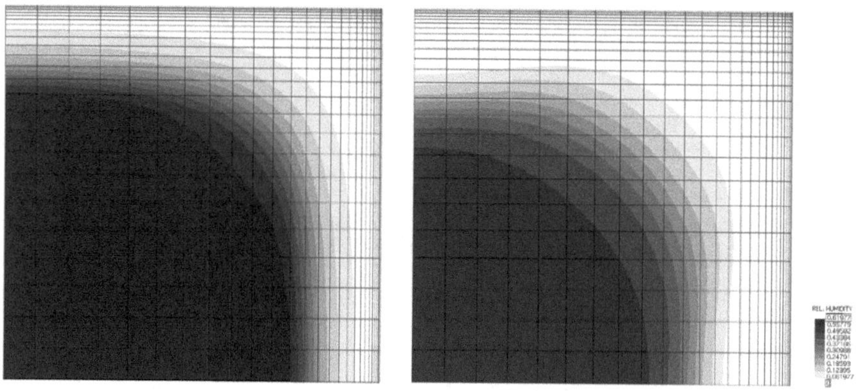

(a) (b)

Fig. 4. Distributions of the relative humidity [–] at 80 min (a) and at the final time (b).

In Figure 3a one can observe the distribution of temperature at $t = 80$ min, corresponding to the time instant of maximum heating. Figure 3b represents the temperature distribution at the end of the computation (approximately 100 minutes). The decrease of temperature is of about 200 K.

Figures 4a and 4b show the distribution of the relative humidity at the two time stations chosen. Again the maximum values of RH are attained at the end of the heating phase in the inner part of the cross section whereas the outer part is almost completely dry. After cooling the relative humidity in the core of the column has values close to the initial ones while the external part is almost fully dry. From these maps one can figure out that drying is progressing also during the cooling phase.

Vapour pressure distribution at 80 min and at the end of calculation is depicted in Figures 5a and 5b. In this case it is worthy to notice the influence of the damage on the

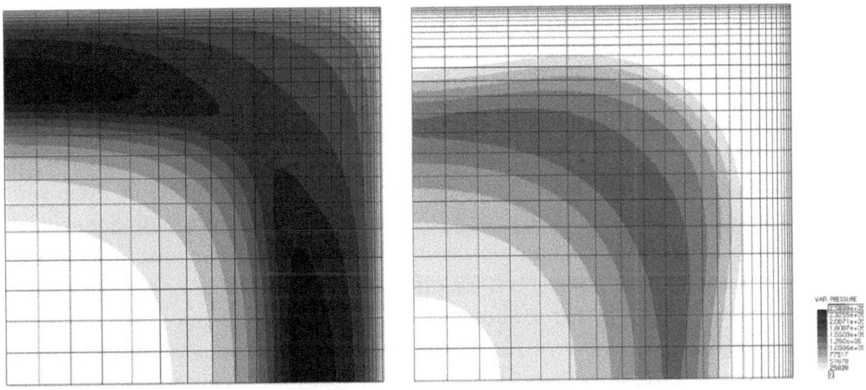

(a) (b)

Fig. 5. Distributions of the vapour pressure [Pa] at 80 min (a) and at the final time (b).

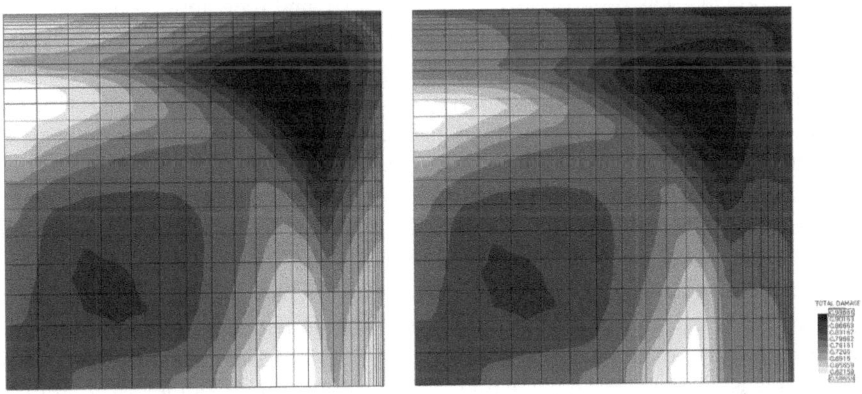

(a) (b)

Fig. 6. Distributions of the damage [–] at 80 min (a) and at the final time (b).

pressure value: the pressure peaks are concentrated in the zones characterized by a lower value of damage parameter, Figure 5a. The pressure is lower where the material is deeply damaged (i.e. close to the corner). At the end of the process the pressure is almost uniformly distributed throughout the cross section of the column and its values are no higher than 1 bar, Figure 5b. Finally, in Figures 6a and 6b one can observe the distribution of damage parameter. Unlike temperature and other thermo-hygral quantities, the damage is continuously increasing because of an additional damaging of the material even during the cooling phase. In Figure 6b it is clearly visible that the zone of damaged concrete is larger than the one in Figure 6a (i.e. before cooling).

5 Application of the Model to Concrete Structures Subject to Leaching Process

Possibly accurate prediction of the service life for structures made of cement-based materials exposed to chemical degradation caused by a long-term contact with the deionized water, when non-isothermal conditions may also occur, is of great practical importance. The most evident example of such a structure is a radioactive waste disposal where concrete is commonly used for containment barriers, liners, waste covers, and encasement of containers. Calcium leaching in the cement-based materials became an important research subject in the recent two decades. For a comprehensive state of the art review of the most important experimental and theoretical works in the field the reader can refer to [22]. The calcium dissolution from cementitious materials was typically modelled by assuming the equilibrium between the calcium concentration in the pore solution and the calcium content in the solid skeleton, see e.g. [23–25]. A general, thermodynamically based model describing kinetics of the process was proposed by Ulm et al. [26], and then applied in numerical simulations in [4]. The calcium ionic transport was usually modelled as a diffusion process, e.g. [23,25,26], sometimes considering also effect of the ions present in the pore solution [27]. Chemical degradation of the strength properties was described in the framework of chemo-elastic damage mechanics, e.g., the chemo-poro-plasticity theory [26] or the theory of reactive porous media (based on the theories of mixture and thermodynamics) [27].

In this section the main modifications to the mathematical model described in the first part of this work for simulating such a complex phenomenon are briefly presented. The advective and thermo-diffusive transport of ions inside the material pores as well as the kinetics of calcium leaching at variable temperature due to thermodynamic imbalance of the calcium in solid and liquid phases is additionally taken into account, as compared to the existing models. Furthermore, boundary conditions of the convective type describing the calcium mass exchange on the surfaces being in direct contact with the deionized water have been formulated and used in the numerical analysis. For the sake of brevity the constitutive relationships describing the transport and strength properties of chemically degraded cement-based materials during evolution of the process as well as the mechanical behaviour of the leached material, are not presented here. The complete formulation of the model, of the constitutive relationships and a full description of the numerical simulations performed, can be found in [4,7,28].

5.1 Modelling Kinetics of Calcium Leaching Process

As already pointed out, in the simulation of leaching process the assumption that there exists local thermodynamic equilibrium between the calcium build up in solid skeleton and the calcium dissolved in pore water is usually applied. Thanks to this assumption, one can use in simulations an equilibrium curve, based on appropriate experimental results, e.g. [23], which describes the values of calcium content in the skeleton being in equilibrium with the pore solution of various calcium ions concentration, $s_{Ca}^{eq}(c_{Ca})$ at a given constant temperature $T = T_{ref}$. The approach used in the present work is based on Thermodynamics.

From Thermodynamics of Irreversible Processes it follows that any progress of calcium leaching, similarly as every chemical reaction, requires thermodynamic imbalance of reactants and products, i.e. calcium compounds within the solid skeleton and calcium ions in the pore water. Thus the kinetics of calcium leaching can be described by means of chemical affinity, A_s, being a driving force of chemical reactions, see e.g. [29]. The leaching rate of material in the actual conditions, described by the set of 3 values, (c_{Ca}, s_{Ca}, T), can be expressed by the following relation [28]:

$$\frac{\partial s_{Ca}}{\partial t} = \frac{1}{\eta} A_s = \frac{1}{RT_{ref}\, \tau_{leach}} A_s \tag{39}$$

where

$$A_s = \int_{s_{Ca}^0}^{s_{Ca}^*} \kappa(\bar{s}, T_{ref}) d\bar{s} - \int_{s_{Ca}^0}^{s_{Ca}} \kappa(\bar{s}, T_{ref}) d\bar{s} \tag{40}$$

with $\eta = RT_{ref}\, \tau_{leach}$ being the coefficient related to the micro-diffusion of the Ca^{2+} ions in the material pores at temperature T and $\kappa(s_{Ca}, T_{ref})$ the equilibrium constant of the leaching process at the reference temperature T_{ref}, and finally $\tau_{leach} = f(s_{Ca}, T)$ the characteristic time of the process, dependent on the actual values of temperature T and solid calcium content s_{Ca} (i.e. the calcium phase being leached from the skeleton). One should underline that the characteristic time τ_{leach} is the only term in Eq. (39) which is dependent on the temperature. Assuming the average diffusion distance of calcium ions from the skeleton surface $l_{leach} \approx 50\ \mu\text{m}$ [4], and considering the temperature dependence of diffusivity of calcium ions, the characteristic time τ_{leach} for different phases of solid calcium can be expressed by the following approximate relationship [28]:

$$\tau_{leach}^{\pi}(T) \cong \tau_{leach}^{\pi}(T_{ref}) \times \exp[A_{\tau}(T - T_{ref})] \tag{41}$$

where $T_{ref} = 298.15$ K, $A_\tau = -0.0239\ \text{K}^{-1}$, and superscript $\pi = por, etr, CSH$ means the portlandite, ettringite and CSH phases, respectively.

One should take into account that calcium leaching is an irreversible process and it proceeds only if the actual values of liquid calcium concentration and solid calcium content, c_{Ca} and s_{Ca}, fulfill the following condition $s_{Ca} > s_{Ca}^{eq}(c_{Ca}, T)$, i.e. $A_s < 0$. On the other hand, the maximum value of leaching rate is limited by thermodynamic constraints, see , assuring that the calcium dissolution process will never lead to the state when the solid calcium content is lower than the equilibrium value corresponding to the actual value of calcium concentration. Hence, for the Finite Difference calculations with time step length of Δt, the following condition must be fulfilled: $s_{Ca}(t) - s_{Ca}^{eq}(c_{Ca}(t)) \leq \dot{s}_{Ca}(t) \cdot \Delta t$.

The values of the equilibrium constant at a given temperature T, $\kappa(s_{Ca}, T)$, can be found from the thermodynamic equilibrium condition in the incremental form,

$$dA_s = RT \frac{dc_{Ca}}{c_{Ca}} - \kappa\, ds_{Ca} = 0$$

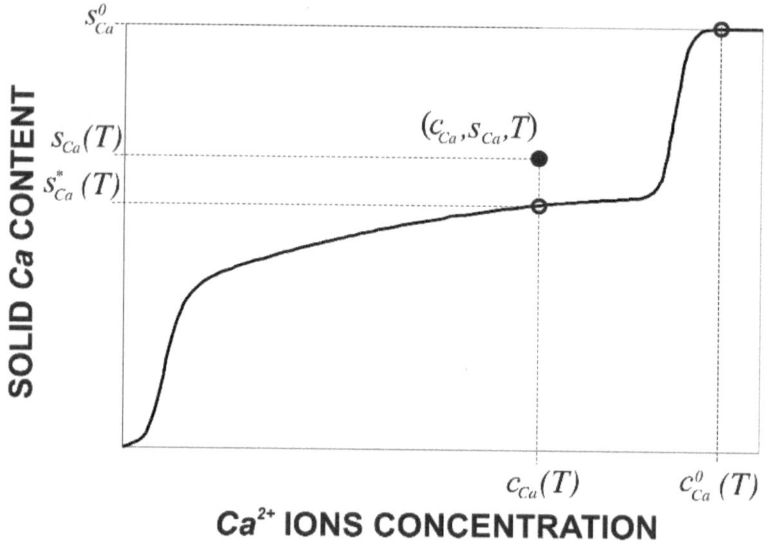

Fig. 7. Explanation of the symbols used in Section 5.

(see [26]), giving the following relation:

$$\kappa(s_{Ca}, T) = \frac{RT}{c_{Ca}} \left(\frac{ds_{Ca}}{dc_{Ca}} \right)^{-1} \tag{42}$$

When considering temperatures different than the standard one (i.e. 25°C) the following Arrhenius-type relationship can be used:

$$s^{eq}_{Ca}(c_{Ca}, T) = s^{eq}_{Ca}[c^{ef}_{Ca}(T),\ T_{ref}] \tag{43}$$

$$c^{ef}_{Ca}(T) = c_{Ca} \times \exp\left[-\frac{E_{leach}}{R} \left(\frac{1}{T} - \frac{1}{T_{ref}} \right) \right] \tag{44}$$

where E_a is the chemical reaction activation energy (see [28]) and c^{ef}_{Ca} is the effective calcium concentration at temperature T.

The value of the integrals on the rhs of Eq. (40) can be calculated in advance for several values of s_{Ca}. See Figure 7 for the explanation of the symbols used.

In this context the leaching degree Γ_{leach} used in the governing equations of the model (eqs. 1–6), and the corresponding leaching degree rate $\dot{\Gamma}_{leach}$ can be defined as follows:

$$\Gamma_{leach}[s_{Ca}(t)] = \frac{s^0_{Ca} - s_{Ca}(t)}{s^0_{Ca}}; \quad \dot{\Gamma}_{leach} = \frac{\partial \Gamma_{leach}}{\partial t} = -\frac{1}{s^0_{Ca}} \frac{\partial s_{Ca}(t)}{\partial t} \tag{45}$$

where $s_{Ca}(t)$ is the current value of calcium concentration in the concrete skeleton. This is equal to the lowest value of $s_{Ca}(t)$ reached at a given point up to the time instant t, $s_{Ca}(t) = s_{Ca,min}(t)$, because the solid dissolution process is irreversible.

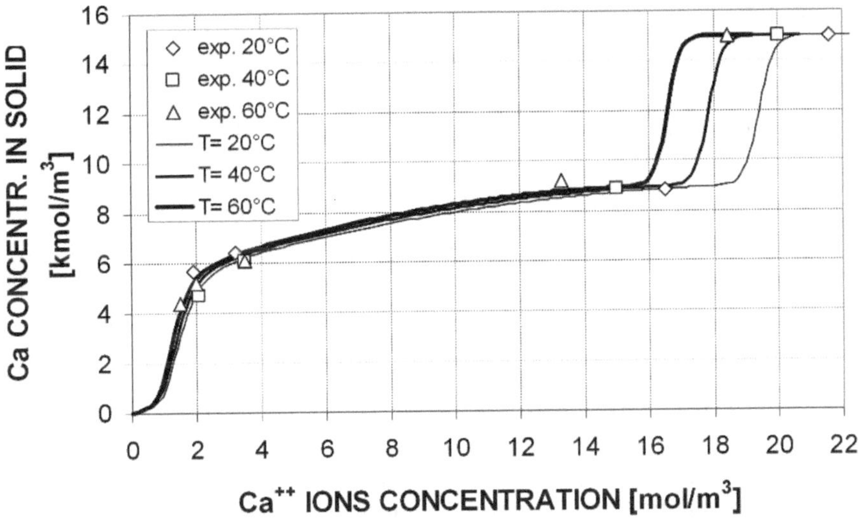

Fig. 8. The equilibrium curves between values of calcium content in skeleton and the pore solution of calcium ions concentration at three different temperatures. The points represent the experimental data of [30].

As far as the mechanical behaviour of the leached material is concerned, the form of the effective stress tensor described in Section 3, has been successfully used in the model to account for the autogenous deformations of concrete structures at early ages or exposed to environments characterized by a lower/higher relative humidity [3, 6, 16]. In the same view such a principle is applied when one considers the long-term mechanical behaviour of the material, which may be important also in case of leaching. Indeed, the aforementioned hydrostatic component of the stress tensor causes a volumetric deformation of the skeleton (shrinkage strain) and it contributes to the creep strains as well. The general constitutive relationship, describing the stress-strain behaviour of matured concrete at non-isothermal conditions, can be written in the following form [6]:

$$\dot{\boldsymbol{\sigma}}_e^s = \mathbf{D}(\dot{\boldsymbol{\varepsilon}}_{tot} - \dot{\boldsymbol{\varepsilon}}_c - \dot{\boldsymbol{\varepsilon}}_{ch} - \dot{\boldsymbol{\varepsilon}}_t) + \dot{\mathbf{D}}(\boldsymbol{\varepsilon}_{tot} - \boldsymbol{\varepsilon}_c - \boldsymbol{\varepsilon}_{ch} - \boldsymbol{\varepsilon}_t) \tag{46}$$

where ϵ_c, ϵ_{ch}, ϵ_t are the creep strain, the chemical strain and the thermal strain respectively. For a detailed description of the constitutive relationships for the definition of such contributions of strain, see [6]. Hence, the linear momentum balance equation has now the following rate form:

$$\operatorname{div}\left[\frac{\partial \eta^s \boldsymbol{\tau}^s}{\partial t} - \left(\frac{\partial p^g}{\partial t} - \frac{\partial x_s^{ws}}{\partial t} p^c - x_s^{ws} \frac{\partial p^c}{\partial t}\right)\right] + \frac{\partial \rho}{\partial t}\mathbf{g} = 0 \tag{47}$$

where $\rho = (1-n)\rho^s + n\, S_w \rho^w + n\, S_g \rho^g$

Table 2. Boundary conditions used in the numerical simulations of leaching process.

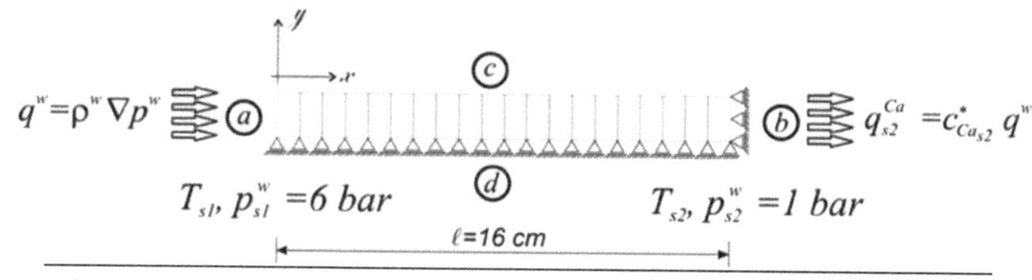

Case	T_0	BC on side *a*		BC on side *b*	
		water	temperature	water	temperature
1	60°C	$p^w_{s1} = 6$ bar	$T_{s1} = 60°C$	$p^w_{s2} = 1$ bar	$T_{s2} = 60°C$
2	60°C	$p^w_{s1} = 6$ bar	$T_{s1} = 25°C$	$p^w_{s2} = 1$ bar	$T_{s2} = 60°C$

5.2 Numerical Simulation of the Non-Isothermal Leaching Process in a Concrete Wall

In this subsection the numerical simulation results obtained by applying the model described briefly in the previous sections are shown. This analysis deals with two concrete slabs having initially the same, uniform temperature of 60°C and exposed to a water pressure gradient of 0.5 MPa. The left side is exposed to an aggressive environment in which the calcium content decreases linearly from an initial value of 17.8 mol/m^3, corresponding to the equilibrium calcium concentration for the temperature of 60°C (Figure 8), to a residual one (1 mol/m^3) in a time span of 1000 days. For the first case the temperature of the left surface remained constant and equal to the initial value of 60°C, while for the second case it decreased linearly in 100 days to the final value of 25°C.

In the simulation the advective-type BCs have been used on the both sides, i.e. the calcium fluxes were directly related to the water fluxes and calcium contents on the surfaces, see Table 2.

The distribution of calcium concentration in the liquid solution and the calcium content in the solid skeleton at various time spans are shown in Figures 9a and 9b respectively. Figure 10 describes the penetration of Portlandite dissolution front as a function of time. One should underline that for the analyzed advection-diffusion-reaction problem, the degradation front penetration is proportional to the time value and not to the square root of time, as in purely diffusive phenomena.

The evolution of leaching process is faster for the case of colder water attack, Figures 9b and 10, which corresponds to a case of a concrete structure used for deep nuclear waste disposal (heated from one side by the radioactive materials) being in contact with cold water. The results clearly show the importance of the temperature for a correct simulation of the real behavior of such a structure with regard to the calcium leaching process.

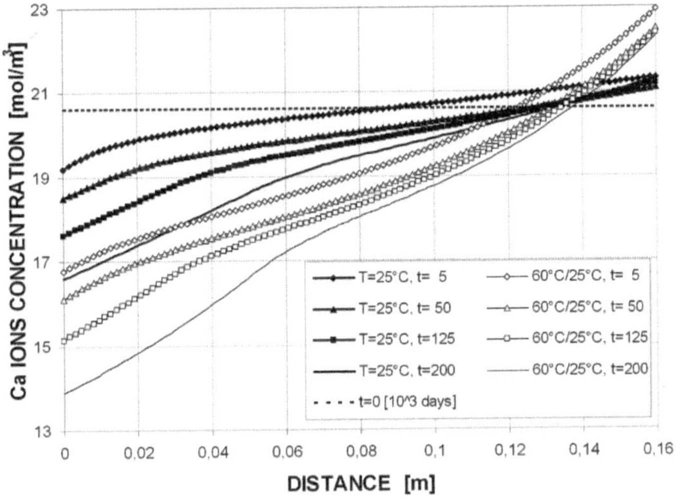

(a)

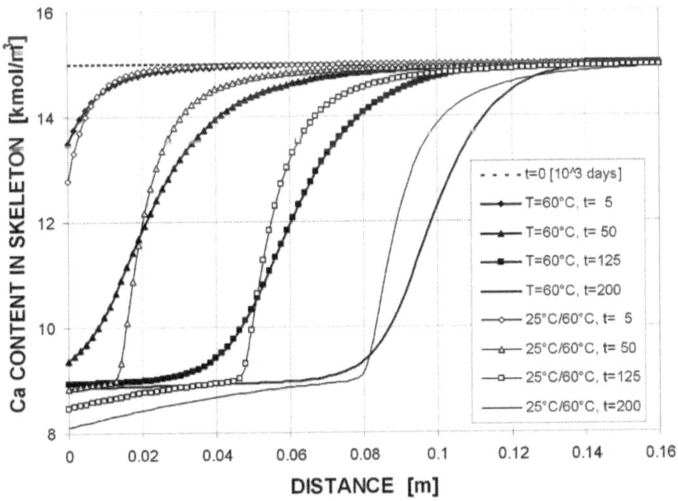

(b)

Fig. 9. Comparison of the space distributions of liquid calcium concentration (a) and solid calcium content (b) at four different time stations, obtained from simulations of the reaction-advection-diffusion problem for the two cases with initial temperature $T_0 = 60°C$: (1) $T_{s1} = T_{s2} = 60°C$ (marked as $T = 60°C$), (2) $T_{s1} = 25°C$ and $T_{s2} = 60°C$ (25°C/60°C). See Table 2 for the description of boundary conditions.

6 Conclusions

A general model for Chemo-Thermo-Hygro-Mechanical behaviour of concrete as a multiphase material has been presented which can be adapted to very different situations such as concrete at early ages and beyond, concrete at high temperatures and

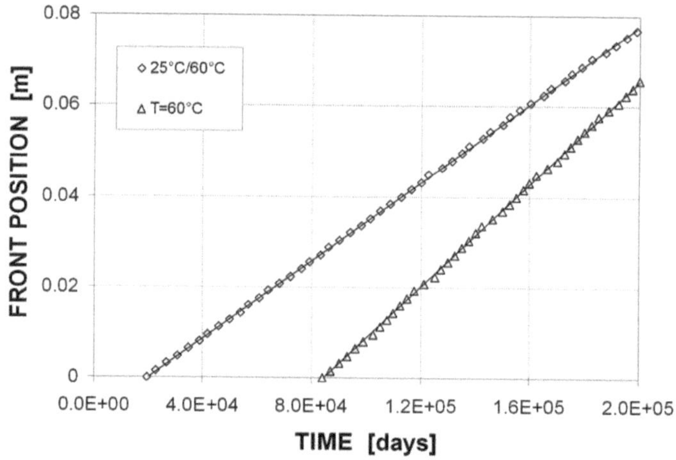

Fig. 10. Comparison of the results concerning progress of the portlandite dissolution front ($s_{Ca} \approx 9$ kmol/m^3) in function of time, obtained for the analyzed cases of the slab leaching process: (1) $T_{s1} = T_{s2} = 60°C$ ($T = 60°C$); (2) $T_{s1} = 25°C$ and $T_{s2} = 60°C$ (25°C/60°C), thin lines show the linear regression lines.

chemical aggressions on concrete, e.g. leaching. The model is thermodynamically consistent and has been obtained from microscopic balance equations by means of Hybrid Mixture Theory. Two applications to concrete at high temperature and leaching show the potentialities of the model.

References

1. Schrefler B.A., Mechanics and thermodynamics of saturated-unsaturated porous materials and quantitative solutions. *Appl. Mech. Rev.*, 55(4):351–388, 2002.
2. Gawin D., Pesavento F., and Schrefler B.A., Modelling of hygro-thermal behaviour of concrete at high temperature with thermo-chemical and mechanical material degradation. *Comput. Methods Appl. Mech. Engrg.*, 192:1731–1771, 2003.
3. Gawin D., Pesavento F., and Schrefler B.A., Hygro-thermo-chemo-mechanical modelling of concrete at early ages and beyond. Part I: Hydration and hygro-thermal phenomena. *Int. J. Numer. Meth. Engng.*, 67(3):299–331, 2006.
4. Gawin D., Pesavento F., and Schrefler B.A., Modeling of cementitious materials exposed to isothermal calcium leaching, with considering process kinetics and advective water flow. Part 1: Theoretical model. *Solids and Structures*, 45:6221–6240, 2008.
5. Pesavento F., Gawin D., and Schrefler B.A., Modeling cementitious materials as multiphase porous media: Theoretical framework and applications. *Acta Mech.*, 201:313–339, 2008.
6. Gawin D., Pesavento F., and Schrefler B.A., Hygro-thermo-chemo-mechanical modelling of concrete at early ages and beyond. Part II: Shrinkage and creep of concrete. *Int. J. Numer. Meth. Engng.,*, 67(3):332–363, 2006.

7. Gawin D., Pesavento F., and Schrefler B.A., Modeling of cementitious materials exposed to isothermal calcium leaching, with considering process kinetics and advective water flow. Part 2: Numerical solution. *Solids and Structures*, 45:6241–6268, 2008.
8. Zienkiewicz O.C. and Taylor R.L., *The Finite Element Method, Volume 1: The Basis*. Butterworth-Heinemann, Oxford, 2000.
9. Gray W.G. and Schrefler B.A., Analysis of the solid phase stress tensor in multiphase porous media. *Int. J. Numer. Anal. Meth. Geomech.*, 31(4):541–581, 2007.
10. Gray W.G., Schrefler B.A., and Pesavento F., The solid phase stress tensor in porous media mechanics and the Hill–Mandel condition. *J. Mech. Phys. Solids*, 57:539–554, 2009.
11. Gawin D., Pesavento, F. and Schrefler B.A., Modelling of de-formations of high strength concrete at elevated temperatures. *Mat. and Struct.*, 37(268):218–236, 2004.
12. Mazars J., Application de la mecanique de l'endommagement au comportament non lineaire et la rupture du beton de structure. Thèse de Doctorat d'Etat, L.M.T., Université de Paris, France, 1984.
13. Nechnech W., Reynouard J.M., and Meftah F., On modelling of thermo-mechanical concrete for the finite element analysis of structures submitted to elevated temperatures. In: de Borst, R., Mazars, J., Pijaudier-Cabot, G., van Mier, J.G.M. (Eds.), *Fracture Mechanics of Concrete Structures*, pp. 271–278, Swets & Zeitlinger, Lisse, 2001.
14. Chaboche J.L., Continuum Damage Mechanics: Part I – General concepts. *J. Appl. Mech.*, 55:59–64, 1988.
15. Lewis R.W. and Schrefler B.A., *The Finite Element Method in the Static and Dynamic Deformation and Consolidation of Porous Media*, 2nd ed. Wiley & Sons, Chichester, 1998.
16. Gawin D., Pesavento F., and Schrefler B.A., Modelling creep and shrinkage of concrete by means of effective stress. *Mat. and Struct.*, 40:579–591, 2007.
17. Khoury G.A., Strain components of nuclear-reactor-type concretes during first heating cycle. *Nucl. Eng. Des.*, 156:313–321, 1995.
18. Thelandersson S., Modeling of combined thermal and mechanical action on concrete. *J. Engrg. Mech. (ASCE)*, 113(6):893–906, 1987.
19. Schrefler B.A., Brunello P., Gawin D., Majorana C.E., and Pesavento F., Concrete at high temperature with application to tunnel fire. *Computation Mechanics* 29:43–51, 2002.
20. Gawin D., Pesavento F., and Schrefler B.A., Towards prediction of the thermal spalling risk through a multi-phase porous media model of concrete. *Comput. Methods Appl. Mech. Engrg.* 195:5707–5729, 2006.
21. UNE-EN 1991-1-2, Eurocodice 1: Azioni sulle strutture. Parte 1-2: Azioni generali. Azioni su strutture esposte al fuoco, May 2004.
22. Kuhl D., Bangert F., and Meschke G., Coupled chemo-mechanical deterioration of cementitious materials. Part I: Modeling. *Int. J. of Solids and Structures*, 41:15–40, 2004.
23. Gerard B., Pijaudier-Cabot G., and Laborderie C., Coupled diffusion-damage modelling and the implications on failure due to strain localization. *Int. J. Solids and Structures*, 35(31-32):4107–4120, 1998.
24. Adenot F. and Buil M., Modelling of the corrosion of the cement paste by deionized water. *Cement and Concrete Research*, 22(2-3): 489–495, 1992.
25. Torrenti J.-M., Mainguy M., Adenot F., and Tognazzi C., Modelling of leaching in concrete. In R. de Borst, N. Bicanic, H. Mang, and G. Meschke (Eds.), *Computational Modelling of Concrete Structures, Proc. of Euro-C 1998*, A.A. Balkema Publishers, Rotterdam, pp. 531–538, 1998.
26. Ulm F.-J., Torrenti J.-M., and Adenot F., Chemoporoplasticity of calcium leaching in concrete. *Journal of Engineering Mechanics*, 15(10):1200–1211, 1999.

27. Kuhl D., Bangert F., and Meschke G., Coupled chemo-mechanical deterioration of cementitious materials. Part I: Modeling. *Int. J. Solids and Structures*, 41:15–40, 2004.
28. Gawin D., Pesavento F., and Schrefler B.A., Modeling deterioration of cementitious materials exposed to calcium leaching in non-isothermal conditions. *Comput. Methods Appl. Mech. Engrg.*, 198:3051–3083, 2009.
29. Atkins P. and De Paula J., *Physical Chemistry*. Oxford University Press, New York, 2002.
30. Yokozekia K, Watanabea K., Sakataa N., and Otsuki N., Modeling of leaching from cementitious materials used in underground environment. *Appl. Clay Science*, 26:293–308, 2004.

Swelling Phenomena in Electro-Chemically Active Hydrated Porous Media

Wolfgang Ehlers, Bernd Markert and Ayhan Acartürk

Institute of Applied Mechanics (CE), University of Stuttgart, Pfaffenwaldring 7, 70569 Stuttgart, Germany; {ehlers, markert, acar}@mechbau.uni-stuttgart.de

Abstract Charged hydrated materials exhibit internal coupling mechanisms stemming from the intrinsic characteristics of the constituents. In this context, the model under consideration consists of a fluid-saturated solid matrix carrying volume-free fixed negative charges, while the pore fluid is given by a mixture of a liquid solvent and the cations and anions of a dissolved salt. Based on the well-founded Theory of Porous Media (TPM), use is made of the assumption of quasi-static processes. The governing equations are given by the volume balance of the fluid mixture governed by the hydraulic pressure, the concentration balance governed by the cation concentration, the overall momentum balance governed by the solid displacement and the electrical continuity equation governed by the electrostatic force. Furthermore, the mechanical solid extra stress is described by an extended neo-Hookean material law, while the viscous fluid flow follows an extended Darcy's law, which includes the gradients of the ion concentrations and the electrical potential. Furthermore, the ion diffusion is described by an extended Nernst–Planck equation. Finally, the model is implemented into the FE tool PANDAS by use of a mixed finite element scheme. The presented examples proceed from boundary conditions depending on internal variables such that certain stabilisation techniques are needed.

Key words: Swelling media, electro-chemomechanical coupling, hydrated porous materials, mixed FEM

1 Introduction

Swelling phenomena of hydrated porous materials occur, for instance, during offshore drilling through active soils like clay or shale, or during physiological loading and unloading of proteoglycan-rich, soft biological tissues such as cartilage. These phenomena are not only very interesting from a scientific point of view, but they are also of great importance for a variety of biomedical and geotechnical applications, for example, tissue engineering or oil and gas exploration to name but a few. From a modelling perspective, ionised porous media can be characterised as multi-component materials consisting of a charged porous solid matrix saturated by a pore fluid, which itself is composed of three components, the liquid solvent (pore water)

R. de Borst and E. Ramm (eds.), *Multiscale Methods in Computational Mechanics*, Lecture Notes in Applied and Computational Mechanics 55, DOI 10.1007/978-90-481-9809-2_20,

and the dissolved anions and cations of some salts in solution such as monovalent Na^+Cl^-.

In fact, these materials have the specific ability to change their volume depending on the chemical characteristics of their environment such as varying ion concentrations in the surrounding solution. It is well known that the swelling behaviour of hydrated media is governed by the charges attached to the solid matrix (fixed charges), which interact with the mobile ions dissolved in the saturating interstitial fluid. Since swelling and shrinking are generally accompanied by large solid deformations, the solid behaviour must be described in the framework of finite elasticity, visco-elasticity or elasto-plasticity depending on the inherent mechanical properties of the matrix material.

Based on the fact that swelling media commonly exhibit a very complex microstructure, one has to decide whether or not this microstructure has to be included into the mechanical description in detail or in a homogenised manner. Apart from a few articles including microstructures in an overall macroscopic framework, the majority of authors rather proceed from a direct macroscopic framework than from micro-to-macro transitions where the microstructure is numerically embedded into the macroscopic procedure. From the macroscopic point of view, bio- and geomaterials are often considered as a single-phasic materials with complex material properties including viscous, plastic and electro-chemical effects in a comprehensive way. However, this procedure will fail to include transport and diffusion processes. If these effects have to be considered, which obviously is true in case of swelling phenomena, one has to switch the modelling procedure from a single-phasic to a multi-phasic approach. In this regard, to achieve an integral description of the electro-chemomechanically coupled problem at hand, it is well accepted to apply multi-phasic mixture theories accounting for the osmotic and electrostatic swelling mechanisms in hydrated porous solids, see [4, 12–15, 19, 20] among others. The advantage of describing multi-phasic materials with multiple components is obvious and can be embedded in the well-founded Theory of Porous Media (TPM).

The TPM, cf. [3, 5, 6], proceeds from the same basis as the Theory of Mixtures (TM), cf. [2], thus describing multi-component media on a homogenised basis by smearing out the individual materials over the control volume and allowing for convenient interaction terms between them. Modelling ionised hydrated materials on a macroscopic level makes use of the basic idea of the TPM describing immiscible phases, such as the solid skeleton and the pore fluid, by smearing out their microstructures. In addition, use is made of the TM while considering the pore fluid as a real mixture of the liquid solvent and the dissolved solutes. When electro-chemical interactions come into play, one has to enrich the TPM by further information describing the electrical and chemical fields, cf. [8, 10, 11].

2 TPM Fundamentals

2.1 Immiscible Components and Volume Fractions

The TPM proceeds from superimposed and interacting but immiscible constituents covering the same amount of space as a result of a real or a virtual averaging process. Following this, an ionised hydrated porous medium φ understood as a multi-component continuum $\mathcal{B}$ generally consists of components φ^α given by the solid part φ^S (solid matrix) and the pore fluid mixture φ^F. As a result, the volume V of $\mathcal{B}$ stems from the sum of the partial volumes of the components φ^α in $\mathcal{B}$:

$$V = \int_{\mathcal{B}} \mathrm{d}v = \bigcup_{\alpha=1}^{k} V^\alpha\,, \qquad \text{where} \qquad V^\alpha = \int_{\mathcal{B}} \mathrm{d}v^\alpha =: \int_{\mathcal{B}} n^\alpha\, \mathrm{d}v\,. \tag{1}$$

Therein, n^α is defined as the local ratio of the volume element $\mathrm{d}v^\alpha$ of a given constituent φ^α with respect to the volume element $\mathrm{d}v$ of the overall medium φ:

$$n^\alpha = \frac{\mathrm{d}v^\alpha}{\mathrm{d}v} \qquad \text{with} \qquad \sum_{\alpha=S,F} n^\alpha = 1\,. \tag{2}$$

By use of the volume fractions n^α, two different density functions can be introduced:

$$\rho^{\alpha R} = \frac{\mathrm{d}m^\alpha}{\mathrm{d}v^\alpha} \qquad \text{and} \qquad \rho^\alpha = \frac{\mathrm{d}m^\alpha}{\mathrm{d}v}\,. \tag{3}$$

The effective density $\rho^{\alpha R}$ is given as the local mass $\mathrm{d}m^\alpha$ over the local constituent volume $\mathrm{d}v^\alpha$, while the partial density defines the same local mass with respect to the local bulk volume $\mathrm{d}v$. As a result,

$$\rho^\alpha = n^\alpha\, \rho^{\alpha R}\,. \tag{4}$$

Based on the above relation, it is immediately evident that the property of material incompressibility ($\rho^{\alpha R} = \text{const.}$) is not equivalent to the property of bulk incompressibility, since the partial density functions ρ^α can still change through changes in the volume fractions n^α.

2.2 Miscible Components and Molar Concentrations

As the pore fluid φ^F has to be treated as a real fluid mixture in the sense of an ionised solution of a solute and dissolved solvents, it is necessary to distinguish differently between the components compared to the introduction of volume fractions. Proceeding from the fact that volume fractions cannot be measured in case of real mixtures, the mixture components φ^β are considered by their partial densities ρ_F^β defined with respect to the compartment of the pore space covered by φ^F. Thus,

$$\rho^\beta =: n^F \rho_F^\beta\,, \qquad \text{where} \qquad \rho^{FR} = \sum_\beta \rho_F^\beta \qquad \text{and} \qquad \rho^F = n^F \rho^{FR}. \tag{5}$$

Comparable to the TM, ρ^{FR} defines the so-called mixture density of the pore fluid given through the sum of the partial densities ρ^{β}_{F} in φ^F. Furthermore, ρ^F relates ρ^{FR} to the volume of the whole aggregate, while ρ^{FR} itself is related to the compartment of the pore space covered by φ^F.

Given ρ^{β}_{F}, the portion of matter is defined through the molar concentration c^{β}_{m} and the molar mass M^{β}_{m} via

$$\rho^{\beta}_{F} = c^{\beta}_{m} M^{\beta}_{m}\,, \qquad \text{where} \qquad c^{\beta}_{F} = \frac{\mathrm{d}n^{\beta}_{m}}{\mathrm{d}v^{F}}. \tag{6}$$

Therein, $\mathrm{d}n^{\beta}_{m}$ is the local number of moles. Since M^{β}_{m} is a constant of the species φ^{β}, the variation of ρ^{β}_{F} is uniquely determined by c^{β}_{m}.

2.3 Constituent Balance Relations

Following the basic ideas and definitions of the TPM, one proceeds from the following balance relations of the components φ^{α} constituting the overall biological aggregate [6]:

$$\begin{aligned}
&\bullet\ \text{mass:} && (\rho^{\alpha})'_{\alpha} + \rho^{\alpha}\,\mathrm{div}\,\overset{\prime}{\mathbf{x}}_{\alpha} = \hat{\rho}^{\alpha},\\
&\bullet\ \text{momentum:} && \rho^{\alpha}\,\overset{\prime\prime}{\mathbf{x}}_{\alpha} = \mathrm{div}\,\mathbf{T}^{\alpha} + \rho^{\alpha}\,\mathbf{b}^{\alpha} + \hat{\mathbf{p}}^{\alpha}\,,\\
&\bullet\ \text{m. o. m.:} && \mathbf{0} = \mathbf{I} \times \mathbf{T}^{\alpha} + \hat{\mathbf{m}}^{\alpha},\\
&\bullet\ \text{energy:} && \rho^{\alpha}(\varepsilon^{\alpha})'_{\alpha} = \mathbf{T}^{\alpha}\cdot\mathbf{L}_{\alpha} - \mathrm{div}\,\mathbf{q}^{\alpha} + \rho^{\alpha}\,r^{\alpha} + \hat{\varepsilon}^{\alpha},\\
&\bullet\ \text{entropy:} && \rho^{\alpha}\,(\eta^{\alpha})'_{\alpha} = \mathrm{div}\left(-\frac{1}{\theta^{\alpha}}\,\mathbf{q}^{\alpha}\right) + \frac{1}{\theta^{\alpha}}\rho^{\alpha}\,r^{\alpha} + \hat{\zeta}^{\alpha}.
\end{aligned} \tag{7}$$

Therein, $\overset{\prime}{\mathbf{x}}_{\alpha}$ and $\overset{\prime\prime}{\mathbf{x}}_{\alpha}$ are the velocity and acceleration functions of φ^{α}. Furthermore, $\mathbf{T}^{\alpha}$ stands for the partial Cauchy stress, $\mathbf{b}^{\alpha}$ for the body force density, ε^{α} for the mass-specific internal energy , $\mathbf{L}_{\alpha}$ for the spatial velocity gradient, $\mathbf{q}^{\alpha}$ for the heat influx vector, r^{α} for the radiation term, η^{α} for the mass-specific entropy and θ^{α} for the Kelvin's temperature. Moreover, $\mathrm{div}(\,\cdot\,)$ is the divergence operator corresponding to the gradient operator $\mathrm{grad}(\,\cdot\,) = \partial(\,\cdot\,)/\partial\mathbf{x}$, while $(\,\cdot\,)'_{\alpha}$ characterises the material time derivative following the individual motion of φ^{α}.

The direct production terms of mass, momentum, moment of momentum, energy and entropy given through $\hat{\rho}^{\alpha}$, $\hat{\mathbf{p}}^{\alpha}$, $\hat{\mathbf{m}}^{\alpha}$, $\hat{\varepsilon}^{\alpha}$ and $\hat{\zeta}^{\alpha}$ are related to the respective total productions via

$$\begin{aligned}
\hat{\mathbf{s}}^{\alpha} &= \hat{\mathbf{p}}^{\alpha} + \hat{\rho}^{\alpha}\,\overset{\prime}{\mathbf{x}}_{\alpha}\,,\\
\hat{\mathbf{h}}^{\alpha} &= \hat{\mathbf{m}}^{\alpha} + \mathbf{x} \times (\hat{\mathbf{p}}^{\alpha} + \hat{\rho}^{\alpha}\,\overset{\prime}{\mathbf{x}}_{\alpha})\,,\\
\hat{e}^{\alpha} &= \hat{\varepsilon}^{\alpha} + \hat{\mathbf{p}}^{\alpha}\cdot\overset{\prime}{\mathbf{x}}_{\alpha} + \hat{\rho}^{\alpha}\left(\varepsilon^{\alpha} + \tfrac{1}{2}\,\overset{\prime}{\mathbf{x}}_{\alpha}\cdot\overset{\prime}{\mathbf{x}}_{\alpha}\right),\\
\hat{\eta}^{\alpha} &= \hat{\zeta}^{\alpha} + \hat{\rho}^{\alpha}\eta^{\alpha}.
\end{aligned} \tag{8}$$

By summing up the relations (7) over all constituents φ^α, one obtains the following constraints of the production terms:

$$\sum_\alpha \hat{\rho}^\alpha = 0\,, \quad \sum_\alpha \hat{\mathbf{s}}^\alpha = \mathbf{0}\,, \quad \sum_\alpha \hat{\mathbf{h}}^\alpha = \mathbf{0}\,, \quad \sum_\alpha \hat{e}^\alpha = 0\,, \quad \sum_\alpha \hat{\eta}^\alpha \geq 0\,. \tag{9}$$

For details concerning the above setting of the balance laws of the TPM, the interested reader is again referred to [6] and the quotations therein.

Given the relations (7)–(9), the entropy principle for the overall medium reads

$$\sum_{\alpha=1}^{k} \left[\rho^\alpha (\eta^\alpha)'_\alpha + \hat{\rho}^\alpha \eta^\alpha + \operatorname{div}\left(\frac{1}{\theta^\alpha}\mathbf{q}^\alpha\right) - \frac{1}{\theta^\alpha}\rho^\alpha r^\alpha \right] \geq 0\,. \tag{10}$$

Introducing mass-specific constituent free energy functions ψ^α via Legendre transformations

$$\psi^\alpha := \varepsilon^\alpha - \theta^\alpha \eta^\alpha\,, \tag{11}$$

one obtains with the aid of the energy balance relations $(7)_3$ the mostly used form of the entropy principle for multi-component materials, viz.:

$$\begin{aligned} \sum_{\alpha=1}^{k} \frac{1}{\theta^\alpha} \Big[& \mathbf{T}^\alpha \cdot \mathbf{L}_\alpha - \rho^\alpha \left[(\psi^\alpha)'_\alpha + (\theta^\alpha)'_\alpha \eta^\alpha \right] - \hat{\mathbf{p}}^\alpha \cdot \overset{\prime}{\mathbf{x}}_\alpha - \\ & - \hat{\rho}^\alpha \left(\psi^\alpha + \tfrac{1}{2}\, \overset{\prime}{\mathbf{x}}_\alpha \cdot \overset{\prime}{\mathbf{x}}_\alpha \right) - \frac{1}{\theta^\alpha}\, \mathbf{q}^\alpha \cdot \operatorname{grad} \theta^\alpha + \hat{e}^\alpha \Big] \geq 0\,. \end{aligned} \tag{12}$$

3 Swelling Media as Biphasic, Four-Component Aggregates

In the mechanics of charged hydrated porous media such as cartilagineous tissue or active soil as well as in various other fields, swelling problems occur and have to be described on a computational basis. The phenomenon of swelling is triggered by the fact that negative electrical charges are fixed to the skeleton material such that the chemical equilibrium can change as a result of the composition of an external solution. To give an example for this behaviour, consider a soot-coloured hydrogel disc, cf. Figure 3 of the numerical example section, that is taken from a given bathing solution, where it was in chemical equilibrium, and put into another one with a lower concentration of both cations and anions. As a result, the chemical equilibrium in the disc is violated and can only be gathered again, when the disc absorbs a certain amount of liquid such that the internal and external solutions are again in a state of chemical equilibrium. It is furthermore seen from Figure 3 that swelling phenomena are often accompanied by large deformations of the solid skeleton.

To describe swelling processes of ionised hydrated materials, one has to consider a biphasic medium of a solid skeleton φ^S with adhering fixed charges φ^{fc} and a pore fluid φ^F understood as a mixture of a liquid solvent φ^L, e.g. water, and dissolved solutes, e.g. positive and negative ions φ^+ and φ^- of a salt such as the monovalent

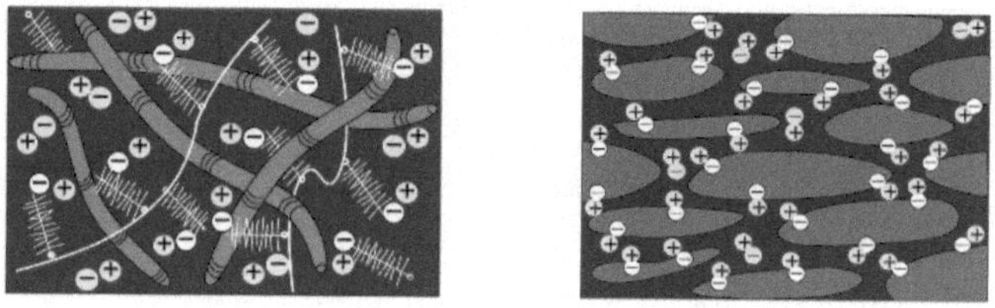

Fig. 1. Sketch of the charged hydrated microstructures of biological tissues (e.g. cartilage, left) and geomaterials (e.g. clay, right).

Na^+Cl^-, see Figure 1. Based on the balance equations $(7)_{1,2}$, the biphasic, four-component material under study is governed by the following mass and momentum balances:

$$\left.\begin{aligned} (\rho^\alpha)'_\alpha + \rho^\alpha \operatorname{div} \overset{\prime}{\mathbf{x}}_\alpha &= 0\,, \\ \mathbf{0} &= \operatorname{div} \mathbf{T}^\alpha + \rho^\alpha \mathbf{b}^\alpha + \hat{\mathbf{p}}^\alpha \end{aligned}\right\} \quad \alpha = \{S,\ L,\ +,\ -\} \quad \text{and} \quad \varphi^F = \bigcup_{\beta = L,\ +,\ -} \varphi^\beta . \tag{13}$$

To obtain (13), quasi-static processes ($\overset{\prime\prime}{\mathbf{x}}_\alpha = \mathbf{0}$) at constant temperature ($\theta^\alpha = \theta =$ const.) and a general absence of phase transitions ($\hat{\rho}^\alpha = 0$) have been assumed. Based on the concept of material incompressibility ($\rho^{\alpha R} =$ const.) of the porous solid including the fixed charges and the fluid components, liquid, cations and anions, the mass balances reduce to volume balances

$$(n^\alpha)'_\alpha + n^\alpha \operatorname{div} \overset{\prime}{\mathbf{x}}_\alpha = 0\,. \tag{14}$$

Although the component densities $\rho^{\alpha R}$ are constant, the partial densities ρ^β_F of the fluid components as well as the effective pore-fluid density ρ^{FR} given by (5) and (6) can vary as a result of varying molar concentrations c^β_m in the pore-fluid mixture:

$$\rho^{FR} = \sum_{\beta = L,\,\gamma} \rho^\beta_F \quad \text{with} \quad \rho^\beta_F = c^\beta_m M^\beta_m \quad \text{and} \quad \gamma = \{+,\ -\}\,. \tag{15}$$

Since the molar masses M^β_m of the fluid components are constant, $(13)_1$ is rewritten to yield

$$\begin{aligned} &(n^F c^\beta_m)'_\beta + n^F c^\beta_m \operatorname{div} \overset{\prime}{\mathbf{x}}_\beta = 0 \qquad \text{or} \\ &n^F (c^\beta_m)'_\beta + c^\beta_m \operatorname{div} (\mathbf{u}_S)'_S + n^F c^\beta_m \operatorname{div} \mathbf{w}_\beta + c^\beta_m \operatorname{grad} n^F \cdot \mathbf{w}_\beta = 0\,, \end{aligned} \tag{16}$$

respectively. Proceeding from the fact that the fixed charges adhere to the skeleton material, the fixed charges volume balance reads

$$(\rho^{fc})'_S + \rho^{fc} \operatorname{div} \overset{\prime}{\mathbf{x}}_S = 0\,, \qquad \text{where} \qquad \rho^{fc} =: n^F c^{fc}_m M^{fc}_m\,. \tag{17}$$

Therein, c^{fc}_m and M^{fc}_m are the molar concentration and the molar mass of φ^{fc}.

Proceeding from an isothermal four-component formulation of a biphasic material, quasi-static computational procedures are based on a basic set of seven primary variables given by the solid displacement $\mathbf{u}_S$, the pore-fluid seepage velocity $\mathbf{w}_F$, the relative ion velocities $\mathbf{w}_+$ and $\mathbf{w}_-$, the hydraulic excess pore-fluid pressure $\wp$ and the cation and anion concentrations c_m^+ and c_m^-. Under quasi-static conditions, one obtains a coupling between $\mathbf{w}_F$ and $\wp$ as well as between $\mathbf{w}_\gamma$ and c_m^γ resulting from the individual fluid momentum balances and the constitutive setting yielding Darcy-like and Nernst–Planck-like relations. Following this reduces the set of primary variables from seven to four: the solid displacement $\mathbf{u}_S$, the pore-fluid pressure $\wp$ and the ion concentrations c_m^γ. Thus, the corresponding set of governing equations is given by the overall momentum balance corresponding to $\mathbf{u}_S$, the fluid volume balance corresponding to $\wp$ and the ion concentration balances corresponding to c_m^δ. Following this, one obtains

$$\begin{aligned} \mathbf{0} &= \operatorname{div}\mathbf{T} + \rho\,\mathbf{g}, \\ 0 &= (n^F)'_S + n^F \operatorname{div}(\mathbf{u}_S)'_S + \operatorname{div}(n^F\,\mathbf{w}_F) + \frac{n^F}{\rho^{FR}}(\rho^{FR})'_F, \\ 0 &= n^F(c_m^\gamma)'_S + c_m^\gamma \operatorname{div}(\mathbf{u}_S)'_S + \operatorname{div}(n^F c_m^\gamma \mathbf{w}_\gamma). \end{aligned} \tag{18}$$

Therein, $(18)_1$ was obtained by summing up $(13)_2$, where use was made of $\rho := \sum_\alpha \rho^\alpha$, $\mathbf{T} := \sum_\alpha \mathbf{T}^\alpha$ and of constant body force densities equivalent to the overall gravitation $\mathbf{g}$ through $\mathbf{b}^\alpha = \mathbf{b} =: \mathbf{g}$. Furthermore, the constraint $(9)_2$ was considered with the aid of $(8)_1$ and $\hat{\rho}^\alpha = 0$. Describing the pore-fluid mass balance $(18)_2$ and the ion concentration balances $(18)_3$ with respect to the moving solid skeleton, the relative velocities $\mathbf{w}_F = \overset{\prime}{\mathbf{x}}_F - (\mathbf{u}_S)'_S$ and $\mathbf{w}_\gamma = \overset{\prime}{\mathbf{x}}_\gamma - (\mathbf{u}_S)'_S$ come into play, where $(\mathbf{u}_S)'_S = \overset{\prime}{\mathbf{x}}_S$. To obtain $(18)_2$ and $(18)_3$, additional use was made of (5) and (6).

3.1 Restrictions Obtained from the Entropy Inequality

To close the model under consideration, constitutive equations are required for the partial Cauchy stresses $\mathbf{T}^\alpha$ and the linear momentum productions $\hat{\mathbf{p}}^\beta$ of the fluid components. Note in passing that defining $\mathbf{T}^\alpha$ will include the constitutive equations for the chemical potentials and the osmotic pressure.

An evaluation of the overall entropy inequality

$$\mathbf{T}^S \cdot \mathbf{L}_S - \rho^S(\psi^S)'_S + \sum_\beta [\,\mathbf{T}^\beta \cdot \mathbf{L}_\beta - \rho^\beta(\psi^\beta)'_\beta - \hat{\mathbf{p}}^\beta \cdot \mathbf{w}_\beta\,] \geq 0 \tag{19}$$

obtained from (12) under consideration of constant temperatures and vanishing density productions imposes the necessary thermodynamical restrictions on the process. However, since the model includes a pore-fluid mixture instead of an immiscible combination of pore liquids and pore gases, the free energy functions of the fluid components will be formulated per fluid volume [2] instead of per constituent mass as in (19). Thus,

$$\rho^\beta\,\psi^\beta = n^F(\rho_F^\beta\psi^\beta) =: n^F\,\Psi_F^\beta, \qquad \text{where} \qquad \Psi_F^\beta := \rho_F^\beta\psi^\beta. \tag{20}$$

In addition, the pore-fluid free energy yields

$$\Psi^F_F := \sum_\beta \Psi^\beta_F\,. \tag{21}$$

Inserting (20) and (21) into (19), the entropy inequality transforms towards

$$\begin{aligned}&(\mathbf{T}^S - n^S\Psi^F_F\,\mathbf{I})\cdot\mathbf{L}_S - \rho^S\,(\psi^S)'_S + \sum_\beta[\,(\mathbf{T}^\beta - n^F\Psi^\beta_F\,\mathbf{I})\cdot\mathbf{L}_\beta - n^F\,(\Psi^F_F)'_F -\\ &-\operatorname{grad}\Psi^\beta_F\cdot(n^F\mathbf{d}_{\beta F}) - (\hat{\mathbf{p}}^\beta + \Psi^\beta_F\operatorname{grad}n^F)\cdot\mathbf{w}_\beta\,] \geq 0\,,\end{aligned} \tag{22}$$

where the diffusion velocities $\mathbf{d}_{\beta F}$ are given through $\mathbf{d}_{\beta F} = \mathbf{w}_\beta - \mathbf{w}_F$. Furthermore, the process is affected by two constraints, the saturation constraint and the electroneutrality constraint. These constraints read:

$$\begin{aligned}\Big(n^S + \sum_\beta n^\beta\Big)'_S &= -\,n^S\operatorname{div}\overset{\prime}{\mathbf{x}}_S - \sum_\beta\Big(n^\beta\operatorname{div}\overset{\prime}{\mathbf{x}}_\beta + \operatorname{grad}n^\beta\cdot\mathbf{w}_\beta\Big) = 0\,,\\ \Big(\rho^{fc}_e + \sum_\gamma \rho^\gamma_e\Big)'_S &= -\,\rho^{fc}_e\operatorname{div}\overset{\prime}{\mathbf{x}}_S - \sum_\gamma\Big(\rho^\gamma_e\operatorname{div}\overset{\prime}{\mathbf{x}}_\gamma + \operatorname{grad}\rho^\gamma_e\cdot\mathbf{w}_\gamma\Big) = 0\,.\end{aligned} \tag{23}$$

To obtain the first constraint, the solid material time derivative of the saturation condition $(2)_2$ has been taken, while to obtain the second constraint, the solid material time derivative has been applied to the electroneutrality condition

$$\rho_e = \rho^{fc}_e + \sum_\gamma \rho^\gamma_e = 0\,. \tag{24}$$

Therein, the densities of the volumetric electrical charges of the free moving ions φ^γ and of the ions φ^{fc} adhering at the skeleton are given by

$$\rho^\gamma_e := n^F z^\gamma c^\gamma_m F \quad\text{and}\quad \rho^{fc}_e := n^F z^{fc} c^{fc}_m F\,, \tag{25}$$

where z^γ and z^{fc} are the valances of the ions, and F is the Faraday constant.

Multiplying the saturation constraint $(23)_1$ by the hydraulic pressure $\mathcal{P}$ and the electroneutrality constraint $(23)_2$ by the electrical potential $\mathcal{E}$, both acting as Lagrangean multipliers, the entropy inequality (22) yields after having been combined with these equations

$$\begin{aligned}&(\mathbf{T}^S_E - n^S\Psi^F_F\,\mathbf{I})\cdot\mathbf{L}_S - \rho^S\,(\psi^S)'_S + \sum_\beta\Big[(\mathbf{T}^\beta_E - n^F\Psi^\beta_F\,\mathbf{I})\cdot\mathbf{L}_\beta - n^F\,(\Psi^F_F)'_F -\\ &-\operatorname{grad}\Psi^\beta_F\cdot(n^F\mathbf{d}_{\beta F}) - (\hat{\mathbf{p}}^\beta_E + \Psi^\beta_F\operatorname{grad}n^F)\cdot\mathbf{w}_\beta\Big] \geq 0\,.\end{aligned} \tag{26}$$

Therein, both the hydraulic pressure $\mathcal{P}$ and the electrical potential $\mathcal{E}$ occur as new fields via the extra quantities indicated by $(\,\cdot\,)_E$ which are not matched by governing equations yet. In particular, in (26), the extra stresses $\mathbf{T}^S_E$ and $\mathbf{T}^\beta_E$ as well as the extra momentum productions $\hat{\mathbf{p}}^\beta_E$ with $\beta = \{L,\ \gamma\}$ are defined as

$$\begin{aligned}&\mathbf{T}^S_E = \mathbf{T}^S + n^S\mathcal{P}\,\mathbf{I} + \rho^{fc}_e\,\mathcal{E}\,\mathbf{I}\,,\\ &\mathbf{T}^\gamma_E = \mathbf{T}^\gamma + n^\gamma\,\mathcal{P}\,\mathbf{I} + \rho^\gamma_e\;\mathcal{E}\,\mathbf{I}\,, &&\hat{\mathbf{p}}^\gamma_E = \hat{\mathbf{p}}^\gamma - \mathcal{P}\operatorname{grad}n^\gamma - \mathcal{E}\operatorname{grad}\rho^\gamma_e\,,\\ &\mathbf{T}^L_E = \mathbf{T}^L + n^L\mathcal{P}\,\mathbf{I}\,, &&\hat{\mathbf{p}}^L_E = \hat{\mathbf{p}}^L - \mathcal{P}\operatorname{grad}n^L\,.\end{aligned} \tag{27}$$

From the above relations, it is seen that the extra quantities and the total quantities are related to each other by the action of the incompressibility constraint $(23)_1$ multiplied by $\mathcal{P}$ and by the electroneutrality constraint $(23)_2$ multiplied by $\mathcal{E}$.

Combining (26) with the constitutive assumptions

$$\psi^S = \psi^S(\mathbf{F}_S) \quad \text{and} \quad \Psi_F^\beta = \Psi_F^\beta(c_m^\beta) \tag{28}$$

yields with the aid of the standard evaluation procedure:

$$\begin{aligned} \mathbf{T}^S_{E\,mech.} &:= \mathbf{T}^S_E + n^S\Big(\sum_\beta c_m^\beta \frac{\partial \Psi_F^\beta}{\partial c_m^\beta} - \Psi_F^F\Big)\,\mathbf{I} = \rho^S \frac{\partial \psi^S}{\mathbf{F}_S}\mathbf{F}_S^T\,, \\ \mathbf{T}^\beta_{E\,dis.} &:= \mathbf{T}^\beta_E + n^F\Big(c_m^\beta \frac{\partial \Psi_F^\beta}{\partial c_m^\beta} - \Psi_F^\beta\Big)\,\mathbf{I} \quad \approx \mathbf{0}\,, \\ \hat{\mathbf{p}}^\beta_{E\,dis.} &:= \hat{\mathbf{p}}^\beta_E - \Big(c_m^\beta \frac{\partial \Psi_F^\beta}{\partial c_m^\beta} - \Psi_F^\beta\Big)\,\mathrm{grad}\, n^F\,. \end{aligned} \tag{29}$$

Therein, $\mathbf{T}^S_{E\,mech.}$ is the purely mechanical part of the solid extra stress, while $\mathbf{T}^\beta_{E\,dis.}$ and $\hat{\mathbf{p}}^\beta_{E\,dis.}$ are the frictional stresses and the drag and diffusion forces of φ^β. Note that a dimensional analysis reveals that the frictional stresses $\mathbf{T}^\beta_{E\,dis.}$ can be neglected in comparison with the momentum productions, cf. [17]. In addition to the results obtained with (29), it remains the dissipation inequality

$$\mathcal{D} = -\sum_\beta \hat{\mathbf{p}}^\beta_{E\,dis.} \cdot \mathbf{w}_\beta \geq 0\,, \tag{30}$$

which is fulfilled with

$$\begin{aligned} \hat{\mathbf{p}}^\gamma_{E\,dis.} &= -\mathbf{S}^{\gamma S}\,\mathbf{w}_\gamma - \mathbf{S}^{L\gamma}\,(\overset{\prime}{\mathbf{x}}_\gamma - \overset{\prime}{\mathbf{x}}_L)\,, \\ \hat{\mathbf{p}}^L_{E\,dis.} &= -\mathbf{S}^{LS}\,\mathbf{w}_L - \textstyle\sum_\gamma \mathbf{S}^{L\gamma}\,(\overset{\prime}{\mathbf{x}}_L - \overset{\prime}{\mathbf{x}}_\gamma)\,. \end{aligned} \tag{31}$$

Therein,

$$\mathbf{S}^{\beta S} = n^F\,\frac{n^\beta \mu^{\beta R}}{K^S}\,\mathbf{I} \quad \text{and} \quad \mathbf{S}^{L\gamma} = n^F\,\frac{n^L R\,\theta\, c_m^\gamma}{D^\gamma}\,\mathbf{I} \tag{32}$$

are the positive definite permeability and diffusion tensors under isotropic conditions, where D^γ is the ion diffusion coefficient of φ^γ, $\mu^{\beta R}$ is the effective shear viscosity of φ^β, and K^S is the intrinsic permeability of φ^S.

3.2 The Fluid Components

Based on the the assumption of vanishing frictional stresses $\mathbf{T}^\beta_{E\,dis.}$, the extra stresses of the fluid components φ^β read

$$\mathbf{T}^\beta_E = -n^F\Big(c_m^\beta \frac{\partial \Psi_F^\beta}{\partial c_m^\beta} - \Psi_F^\beta\Big)\,\mathbf{I} \quad \text{yielding} \quad \mathbf{T}^\beta_E - n^F\,\Psi_F^\beta\,\mathbf{I} = n^F c_m^\beta \frac{\partial \Psi_F^\beta}{\partial c_m^\beta}\,\mathbf{I}\,. \tag{33}$$

From these relations, it is easily concluded that $(33)_2$ contains the molar chemical potential μ^β_m of φ^β, while $(33)_1$, as the "extra pressure" of φ^β, is the contribution π^β of φ^β to the overall osmotic pressure π. Thus,

$$\mu^\beta_m = \frac{\partial \Psi^\beta_F}{\partial c^\beta_m}, \qquad \pi^\beta = c^\beta_m \frac{\partial \Psi^\beta_F}{\partial c^\beta_m} - \Psi^\beta_F = c^\beta_m \mu^\beta_m - \Psi^\beta_F \quad \text{and} \quad \pi = \sum_\beta \pi^\beta . \tag{34}$$

Combining $(27)_{(1-3)}$, (29) and (33), one obtains the constituent partial stresses as

$$\begin{aligned} \mathbf{T}^S &= -n^S(\wp + \pi)\,\mathbf{I} \qquad - \rho_e^{fc}\, \mathcal{E}\, \mathbf{I} + \mathbf{T}^S_{E\,mech.}, \\ \mathbf{T}^\gamma &= -n^\gamma\, \wp\, \mathbf{I} - n^F \pi^\gamma\, \mathbf{I} - \rho^\gamma_e\, \mathcal{E}\, \mathbf{I}, \\ \mathbf{T}^L &= -n^L\, \wp\, \mathbf{I} - n^F \pi^L\, \mathbf{I}, \end{aligned} \tag{35}$$

where (21) has been used. Furthermore, one recovers the total momentum productions $(27)_{4,5}$ to yield

$$\begin{aligned} \hat{\mathbf{p}}^\gamma &= \wp \operatorname{grad} n^\gamma + \pi^\gamma \operatorname{grad} n^F + \mathcal{E} \operatorname{grad} \rho^\gamma_e + \hat{\mathbf{p}}^\gamma_{dis.}, \\ \hat{\mathbf{p}}^L &= \wp \operatorname{grad} n^L + \pi^L \operatorname{grad} n^F + \hat{\mathbf{p}}^L_{dis.}. \end{aligned} \tag{36}$$

Finally, summing over the components of the pore-fluid mixture yields the overall fluid stress and the overall fluid momentum production, viz.:

$$\begin{aligned} \mathbf{T}^F &= -n^F(\wp + \pi)\,\mathbf{I} - \sum_\gamma \rho^\gamma_e\ \mathcal{E}\, \mathbf{I}, \\ \hat{\mathbf{p}}^F &= (\wp + \pi) \operatorname{grad} n^F + \mathcal{E} \sum_\gamma \operatorname{grad} \rho^\gamma_e + \sum_\beta \hat{\mathbf{p}}^\beta_{E\,dis.}. \end{aligned} \tag{37}$$

3.3 Ion Diffusion and Fluid Flow

Proceeding from the standard assumption that the free energy of the fluid components can be given as

$$\Psi^\beta_F = c^\beta_m \mu^\beta_{0m} + c^\beta_m\, R\theta\, (\ln c^\beta_m - 1)\,, \tag{38}$$

the chemical potentials and the osmotic pressure read

$$\mu^\beta_m = \mu^\beta_{0m} + R\theta \ln c^\beta_m \quad \text{and} \quad \pi = R\theta \sum_\beta c^\beta_m . \tag{39}$$

Therein, μ^β_{0m} is known as the constant standard potential.

Inserting the partial ion stress $\mathbf{T}^\gamma$ from $(35)_2$ and the ion momentum production $\hat{\mathbf{p}}^\gamma$ from $(36)_1$ into the ion momentum balance $(13)_2$ yields

$$\mathbf{0} = -n^\gamma\,(\operatorname{grad} \wp - \rho^{\gamma R}\mathbf{g}) - n^F \operatorname{grad} \pi^\gamma - \rho^\gamma_e \operatorname{grad} \mathcal{E} + \hat{\mathbf{p}}^\gamma_{E\,dis.}. \tag{40}$$

Since partial osmotic pressures cannot be measured, π^γ must be substituted by μ^γ_m with the aid of $(34)_2$. Thus, with the aid of $(28)_2$ and $(34)_1$, the ion momentum balance (40) yields

$$\mathbf{0} = -n^\gamma(\operatorname{grad} \wp - \rho^{\gamma R}\mathbf{g}) - n^F c_m^\gamma \operatorname{grad} \mu_m^\gamma - \rho_e^\gamma \operatorname{grad} \mathcal{E} + \hat{\mathbf{p}}_{E\,dis.}^\gamma . \tag{41}$$

Proceeding from the assumption that the ion volume fraction n^γ is negligible compared to the liquid volume fraction n^L including

$$n^\gamma \ll n^L \quad \text{and} \quad n^L \approx n^F , \tag{42}$$

one obtains with the aid of (5)

$$\rho^{FR} \approx \rho_F^L \quad \text{with} \quad \sum_\gamma \rho_F^\gamma \approx 0 \quad \text{and} \quad \overset{\prime}{\mathbf{x}}_F \approx \overset{\prime}{\mathbf{x}}_L , \tag{43}$$

and (41) transforms towards the extended Nernst–Planck equation

$$n^F \mathbf{d}_{\gamma F} = -\frac{D^\gamma}{R\,\theta\, c_m^\gamma}\,(c_m^\gamma \operatorname{grad} \mu_m^\gamma + \rho_{eF}^\gamma \operatorname{grad} \mathcal{E}) , \tag{44}$$

where $\mathbf{d}_{\gamma F} = \overset{\prime}{\mathbf{x}}_\gamma - \overset{\prime}{\mathbf{x}}_F \approx \overset{\prime}{\mathbf{x}}_\gamma - \overset{\prime}{\mathbf{x}}_L$ has been used. Furthermore, in analogy to $(5)_1$, $\rho_e^\gamma = n^F \rho_{eF}^\gamma$. In order to obtain (44) describing the ion diffusion process from (41), the first term on the rhs of (41), the drag term, has been neglected with respect to $(43)_1$. Furthermore, $\hat{\mathbf{p}}_{E\,dis.}^\gamma$ from (31) and (32) has been simplified with the same argument as before by dropping $\mathbf{S}^{\gamma S}$, cf. $(32)_1$. Expressing the chemical potentials μ_m^γ by the concentrations c_m^γ yields the final version of the diffusion equation (44), viz.:

$$n^F \mathbf{d}_{\gamma F} = -\frac{D^\gamma}{R\,\theta\, c_m^\gamma}\,(R\,\theta \operatorname{grad} c_m^\gamma + \rho_{eF}^\gamma \operatorname{grad} \mathcal{E}) . \tag{45}$$

Therein, (39) together with the assumption of isothermal processes has been used.

Inserting the fluid stress and the fluid momentum production given by (37) into the fluid momentum balance obtained by summing $(13)_2$ over $\beta = \{L, +, -\}$ yields

$$\mathbf{0} = -n^F \Big[\operatorname{grad}(\wp + \pi) - \rho^{FR}\mathbf{g} - \sum_\gamma \rho_{eF}^\gamma \operatorname{grad} \mathcal{E} \Big] + \hat{\mathbf{p}}_{E\,dis.}^F . \tag{46}$$

Therein, $\hat{\mathbf{p}}_{E\,dis.}^F$ is obtained from (31), (32) and $(43)_1$ as

$$\hat{\mathbf{p}}_{E\,dis.}^F = \hat{\mathbf{p}}_{E\,dis.}^L + \sum_\gamma \hat{\mathbf{p}}_{E\,dis.}^\gamma = -(n^F)^2 \frac{\mu^{FR}}{K^S}\,\mathbf{w}_F . \tag{47}$$

Combing (46) and (47), one recovers the extended Darcy equation

$$n^F \mathbf{w}_F = -\frac{K^S}{\mu^{FR}} \Big[\operatorname{grad}(\wp + \pi) - \rho^{FR}\mathbf{g} - \sum_\gamma \rho_{eF}^\gamma \operatorname{grad} \mathcal{E} \Big] \tag{48}$$

governing the fluid flow.

3.4 The Electrical Potential

The electrical potential $\mathcal{E}$ entering the process through the entropy inequality (26) together with the electroneutrality constraint $(23)_2$ reveals that there is no matching governing equation. Thus, this additional equation must be found and added to the process. To overcome this situation, use is made of the Poisson equation of the electrical potential of electrostatics yielding

$$\operatorname{div}\operatorname{grad}\mathcal{E} = -\frac{1}{\epsilon^F}\,\rho_e\,, \tag{49}$$

where $\operatorname{div}\operatorname{grad}(\,\cdot\,) =: \Delta(\,\cdot\,)$ is the Laplace operator, ϵ^F the constant electric fluid permittivity, while the density ρ_e of the electrical charges is given through (24) and (25). In contrast to the electroneutrality condition (24) stating $\rho_e = 0$, ρ_e is usually obtained from the Gaussian law

$$\operatorname{div}\mathbf{d}_e = \rho_e\,, \tag{50}$$

where $\mathbf{d}_e$ is the electric displacement obtained from the electrical field $\mathbf{e}$ via

$$\mathbf{d}_e = \epsilon^F\,\mathbf{e} \quad \text{with} \quad \mathbf{e} = -\operatorname{grad}\mathcal{E}\,. \tag{51}$$

Combining (49) with (25) finally yields

$$\operatorname{div}\operatorname{grad}\mathcal{E} = -\frac{n^F\,F}{\epsilon^F}\,\big(\sum_\gamma z^\gamma c_m^\beta + z^{fc} c_m^{fc}\big)\,. \tag{52}$$

This equation will be used in the numerical setting to match $\mathcal{E}$.

3.5 The Solid Skeleton

As was mentioned at the beginning of this section, swelling phenomena are usually combined with large solid deformations. Following this results in the fact that a non-linear elasticity law is needed to describe the elastic solid deformations. Following [9], a combination of the Neo-Hookean law with a non-linear volumetric deformation yields by use of $(29)_1$

$$\begin{aligned}\mathbf{T}^S_{E\,mech.} &= \rho^S\frac{\partial\psi^S}{\mathbf{F}_S}\,\mathbf{F}_S^T\\ &= (\det\mathbf{F}_S)^{-1}\Big[\,2\,\mu^S\,\mathbf{K}_S + \lambda^S\,(1-n^S_{0S})^2\Big(\frac{J_S}{1-n^S_{0S}} - \frac{J_S}{J_S - n^S_{0S}}\Big)\,\mathbf{I}\,\Big],\end{aligned} \tag{53}$$

where μ^S and λ^S are the Lamé constants. Concerning the description of finite volumetric deformations of a materially incompressible solid skeleton, recall that the property of material incompressibility ($\rho^{SR} = \text{const.}$) does not include total incompressibility ($\rho^S = \text{const.}$), since ρ^S can still change through n^S. As a result, the volumetric deformation included in (53) is only due to variations of the pore space, such as those initiated by swelling processes, and is not due to volumetric changes of the matrix material itself. Possible extensions to viscoelastic solid properties can be carried out as described in [18].

4 Weak Forms and Basic Numerical Setting

As was discussed before, the numerical treatment of initial boundary-value problems is based on the weak formulations of the governing field equations together with discretisation methods in the space and time domains. Concerning swelling media described as a biphasic, four-component material of a swellable solid skeleton φ^S including the fixed charges φ^{fc} and the pore fluid φ^F consisting of the liquid solvent φ^L and the cations φ^+ and anions φ^-, the isothermal problem under consideration is basically governed by eight independent fields: the solid displacement $\mathbf{u}_S$, the seepage and diffusion velocities $\mathbf{w}_F$, $\mathbf{d}_{+F}$ and $\mathbf{d}_{-F}$, the hydraulic pressure $\wp$ as well as the ion concentrations c_m^+ and c_m^-. In addition, the electrical potential $\mathcal{E}$ has to be considered. However, under quasi-static conditions, Nernst–Planck-like and Darcy-like relations, cf. (45) and (48), have been found to eliminate the seepage and diffusion velocities by the effective hydraulic pressure $\wp$ and the ion concentrations c_m^γ. Thus, the remaining set of five independent fields must be matched by a corresponding set of five governing equations given through (18) by the overall momentum balance corresponding to $\mathbf{u}_S$, the fluid volume balance corresponding to $\wp$ and the ion concentration balances corresponding to c_m^γ. In addition, the Poisson equation (52) is considered corresponding to $\mathcal{E}$.

Here, we obtain the weak forms of the governing equations in the framework of the standard Galerkin procedure. In particular, the overall momentum balance $(18)_1$ is multiplied by a test function $\delta\mathbf{u}_S$ and integrated over the domain Ω. Integration by parts together with the Gaussian integral theorem yields the following result:

$$\int_\Omega [\,\mathbf{T}^S_{E\,mech.} - (\wp + \pi)\,\mathbf{I}\,] \cdot \operatorname{grad}\delta\mathbf{u}_S\,\mathrm{d}v = \int_\Omega \rho\,\mathbf{g}\cdot\delta\mathbf{u}_S\,\mathrm{d}v + \int_{\Gamma_t} \bar{\mathbf{t}}\cdot\delta\mathbf{u}_S\,\mathrm{d}a\,. \tag{54}$$

Therein, $\bar{\mathbf{t}} = [\,\mathbf{T}^S_{E\,mech.} - (\wp + \pi)\,\mathbf{I}\,]\,\mathbf{n}$ is the external load vector acting on the Neumann boundary Γ_t of the overall aggregate.

Proceeding from the assumption $\rho^{FR} \approx \rho^L_F$ resulting from $n^\gamma \ll n^L$, cf. (43), $(\rho^{FR})'_F$ vanishes and the fluid volume balance $(18)_2$ reduces to

$$\operatorname{div}[\,(\mathbf{u}_S)'_S + n^F\,\mathbf{w}_F\,] = 0\,, \tag{55}$$

where (14) with $\alpha = S$ has been used. Multiplying (55) by $\delta\wp$ and using the same procedure as before results in

$$\int_\Omega n^F\,\mathbf{w}_F\cdot\operatorname{grad}\delta\wp\,\mathrm{d}v - \int_\Omega \operatorname{div}(\mathbf{u}_S)'_S\,\delta\wp\,\mathrm{d}v = \int_{\Gamma_v} \bar{v}^F\,\delta\wp\,\mathrm{d}a\,. \tag{56}$$

Therein, $\bar{v}^F = n^F\,\mathbf{w}_F\cdot\mathbf{n}$ is the efflux of fluid volume through the Neumann boundary Γ_v.

In a next step, the ion concentration balances $(18)_3$ have to be considered. After having multiplied these equations by δc_m^γ, one obtains in analogy to the above procedure

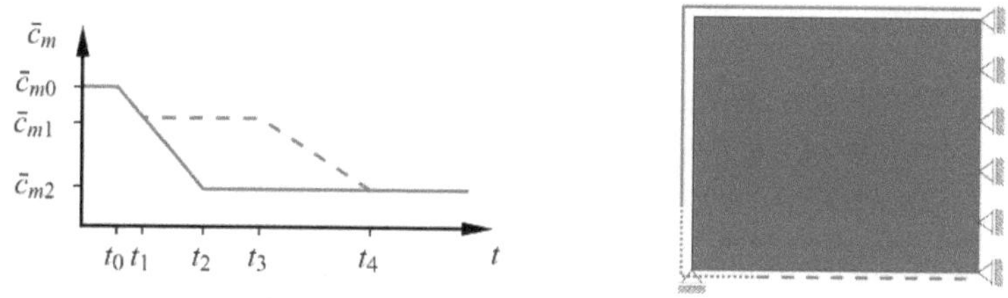

Fig. 2. Chemical loading and geometry of a 2-d free swelling simulation. The solid and the dashed concentration curves on the left side belong to the respective boundaries on the right side. The dotted line at the bottom's left corner of the right-hand picture denotes the interval with the transition zone from the value of the solid concentration curve to the dashed one.

$$\int_\Omega n^F\, c_m^\gamma\, (\mathbf{d}_{\gamma F} + \mathbf{w}_F) \cdot \operatorname{grad} \delta c_m^\gamma \, dv - \\ - \int_\Omega [\, n^F\, (c_m^\gamma)'_S + c_m^\gamma \operatorname{div} (\mathbf{u}_S)'_S \,]\, \delta c_m^\gamma \, dv = \int_{\Gamma_{\bar{J}^\gamma}} \bar{J}^\gamma\, \delta c_m^\gamma \, da\,. \tag{57}$$

Therein, $\bar{J}^\gamma = n^F\, c_m^\gamma\, \mathbf{w}_\gamma \cdot \mathbf{n}$ is the ionic efflux through the Neumann boundary $\Gamma_{\bar{J}^\gamma}$.

Finally, multiplying the Poisson equation (52) by $\delta\mathcal{E}$ and applying the same procedure as before results in

$$\int_\Omega \operatorname{grad} \mathcal{E} \cdot \operatorname{grad} \delta\mathcal{E} \, dv - \int_\Omega \frac{n^F\, F}{\epsilon^F} \big(\sum_\gamma z^\gamma c_m^\gamma + z^{fc} c_m^{fc}\big)\, \delta\mathcal{E}\, dv = - \int_{\Gamma_e} \bar{e}\, \delta\mathcal{E}\, da\,, \tag{58}$$

where $\bar{e} = \mathbf{e} \cdot \mathbf{n}$ is the electrical field across the Neumann boundary Γ_e.

The constitutive setting of the problem and the weak forms given through (54)–(58) are sufficient to solve initial boundary-value problems in the framework of swelling media.

5 Numerical Examples

5.1 Free Swelling Hydrogel

Based on the weak forms of the general swelling model, cf. [1, 7, 8], the numerical scheme governing swelling media is based on extended Taylor-Hood elements with quadratic shape functions for the solid displacement $\mathbf{u}_S$ and linear shape functions for hydraulic fluid pressure $\wp$, the ion concentrations c_m^+ and c_m^- and the electrical potential $\mathcal{E}$.

The following numerical example concerns a 2-d plane-strain simulation of the swelling experiment carried out by the group of Jacques Huyghe in Eindhoven (the Netherlands), cf. Figure 3 (left). To simulate this experiment, a rectangular cross section is discretised under symmetry conditions. During the computation, the chemically uncharged hydrogel is firstly equilibrated in a bathing solution with the initial external concentration $\bar{c}_{m0}$. This is necessary in order to transform the hydrogel from a fictitious non-loaded reference configuration in the sense of a natural state at $t = t_0$ towards a pre-swollen initial configuration. Note in passing that there is no uncharged configuration in any real experiment, since swelling materials are always somehow in a swollen state. In the fictitious non-loaded reference configuration, there is neither a solid stress ($\mathbf{T}^S_{E\,mech.\,0S} = \mathbf{0}$) nor an osmotic pressure ($\pi_0 = 0$). To avoid numerical problems during initial loading, the initial osmotic pressure is applied slowly by increasing the referential concentration of the fixed charges from 0 to the prescribed value c^{fc}_{m0S} over time.

After a certain time of computation, an equilibrium state is reached defined as the "initial configuration" at which the initial osmotic pressure π^*_0, the initial concentration c^*_{m0} and the initial purely mechanical solid extra stress $\mathbf{T}^{S\,*}_{E\,mech.\,0S}$ fits the volumetric deformation $(\det \mathbf{F}_S)^*_{0S}$ through the balance equations and the constitutive setting.

Concerning the numerical simulation prior to the actual swelling process, the specimen is equilibrated for a period of $t_0 = 7\,000\,\mathrm{s}$ in a $\bar{c}_{m0} = 4.1$ molar NaCl solution, where c^{fc}_{m0S} is increased from 0 to 0.8 meq/l within the first 4 000 s. After an equilibrium state is obtained, the initial configuration is reached, and the hydrogel is placed in another NaCl solution. There, the external salt concentration is decreased at all boundaries from $\bar{c}_{m0} = 4.1$ mol/l to $\bar{c}_{m1} = 2.7$ mol/l until $t_1 = 7\,080\,\mathrm{s}$. Thereafter, only the boundaries denoted by the solid line are exposed to the solution. Here, the concentration is further decreased to $\bar{c}_{m2} = 0.3$ mol/l until $t_2 = 7\,220\,\mathrm{s}$, while the concentration at the bottom is kept constant. At $t_3 = 8\,000\,\mathrm{s}$, the bottom of the hydrogel gets in contact with the bathing solution as well. Therefore, $\bar{c}_{m1}$ is decreased here to $\bar{c}_{m2} = 0.3$ mol/l until $t_4 = 10\,000\,\mathrm{s}$. In Figure 3, the simulation results on the left-hand side are compared to the experiment on the right-hand side. Besides the comparison of the deformation, the contour plots show the development of the overall fluid pressure within the material. The top left picture shows the computed initial configuration ($t_0 = 7\,000\,\mathrm{s}$) after having increased c^{fc}_{0S} from 0 to its full amount and after having reached chemical equilibrium. This is a swollen state, i. e., there is an osmotic pressure and a deformation in comparison to the artificial stress-free reference configuration at $t = 0\,\mathrm{s}$ (not shown). The picture below shows the phase of the negative osmosis ($t = 7\,107\,\mathrm{s}$), during which the hydrogel shrinks initially in its inner region. Note in passing that this effect results from the state of chemical non-equilibrium at the domain boundary yielding an influx of liquid both from the bathing solution through the domain boundary and from the inner part of the specimen to outer parts. Thereafter, the state with the greatest bending is shown ($t = 7\,224\,\mathrm{s}$). The subsequent pictures present how the final equilibrium state is reached ($t = 7\,650\,\mathrm{s}$ and $t = 11\,000\,\mathrm{s}$). At the end of the simulation process, the cross-sectional area of the specimen is approximately 2.2 times larger than the area of the computed initial state.

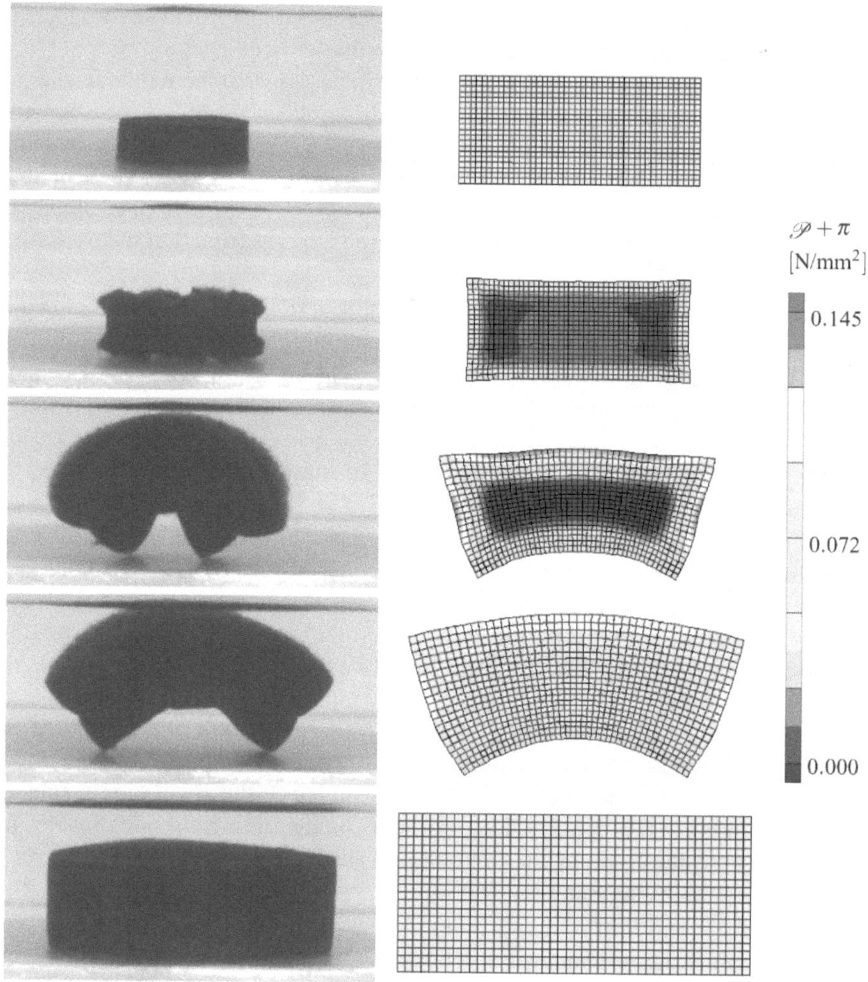

Fig. 3. Qualitative comparison between the free swelling experiment of a soot-coloured hydrogel disc carried out by the group of Jacques Huyghe at the TU Eindhoven and our numerical simulation. The swelling simulation starts from a pre-swollen initial state (top, right). The following contour plots show the deformation and the overall pressure (hydraulic and osmotic) up to the final equilibrium state.

5.2 Electro-Active Polymers

In order to simulate the response of an ionic electro-active polymer (EAP) to an applied electrical field, the second example concerns a 3-d hydrogel cantilever (1 mm by 1 mm by 10 mm) which is exposed to a voltage of 0.8 V. Following this, the displacements at the bottom surface are disabled such that the applied voltage leads to a bending of the ionic EAP. In particular, the electrical potential $\mathcal{E}$ included in

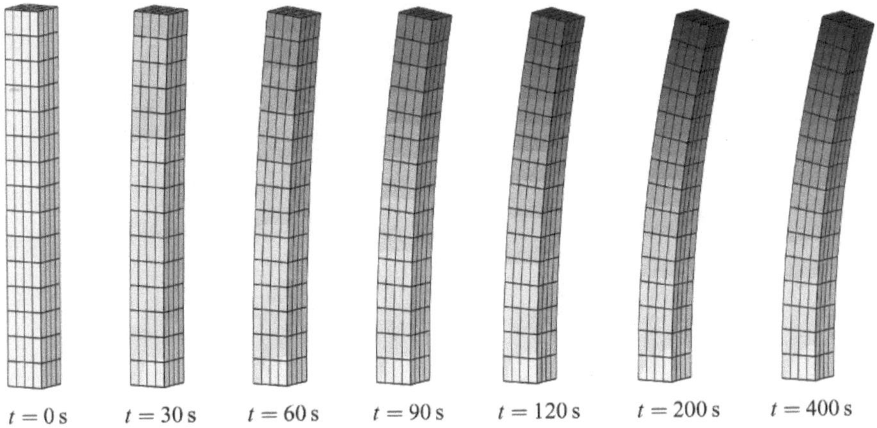

Fig. 4. Deformation behaviour of the cantilever made of an ionic EAP. The largest deflection of 1.3 mm is displayed in black, while white represents a deflection of 0 mm.

the boundary term $\bar{e} = -\operatorname{grad}\mathcal{E}\cdot\mathbf{n}$ with outward-oriented unit surface normal $\mathbf{n}$ is linearly applied within 4 s, thereby increasing the left side from 0 V to 0.4 V, while decreasing the voltage from 0 V to −0.4 V on the opposite (right) side. Moreover, since the EAP is placed in a bathing solution, the molar concentration $\bar{c}_m^{\gamma} = 1.0$ mol/l of both ions is kept constant at all boundaries. Figure 4 illustrates the deformation behaviour of the cantilever at different time steps, where the shading indicates the deflection of the cantilever.

5.3 Borehole Instability in Active Soil

This example is concerned with a geotechnical and petroleum engineering problem which can occur during borehole drilling. Consider a drilling well through a layer of chemically active soil such as clay or shale where to some reason (e.g. water inrush) the salt concentration of the stabilising drilling mud is reduced. What happens is that the soil starts to swell causing the borehole cross section to shrink and possibly the drill to get stuck. In the qualitative simulation, the symmetry of the problem is exploited and only a quarter of the borehole is computed. The active soil layer is restrained by rigid rock strata, cf. Figure 5. Swelling is initiated by decreasing the ion concentration in the well from 0.20 mol/l to 0.10 mol/l within 10 seconds. The computed swelling deformation and the contours of the cation concentration (dark-coloured high, light-coloured low) are depicted in Figure 5.

5.4 Swelling of an Intervertebral Disc

Applying the theoretical framework presented before to swelling phenomena of biological tissue such as the intervertebral disc, one is able to compute the disc-bulging observed in an in-vitro experiment by the group of Gerhard Holzapfel in

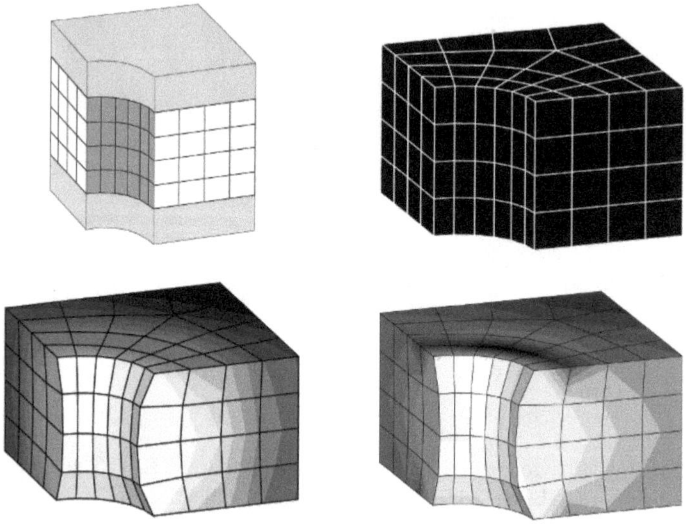

Fig. 5. Discretised borehole with upper and lower rock stratum (top left) and computed cation concentration at different times.

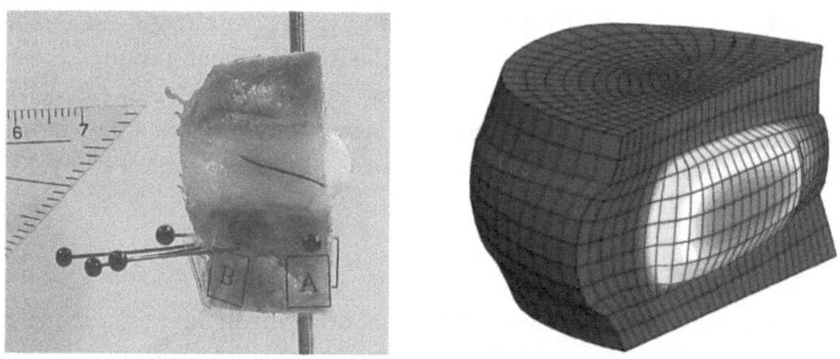

Fig. 6. Swelling of an intervertebral disc. Experiment by Gerhard Holzapfel (left) and numerical simulation (right).

Graz (Austria), cf. Figure 6. The computations are carried out by the additional use of Lanir's assumption [16] of an immediate chemical equilibrium not only at the domain boundary but also in the overall domain. This assumption proceeds from the fact that the ion diffusion is much faster than the solid deformation coming along with the corresponding swelling or shrinking processes. Furthermore, one proceeds from the idea that biological tissue, especially under in-vivo conditions, where it is generally in an equilibrium state, cannot be exposed suddenly to a rapid decrease or increase of the molar ion concentration of the surrounding interstitial fluid. As a result, the general model can be simplified by only using the solid deformation and the fluid pressure as primary variables solving the week forms of the overall momentum and volume balances. The chemical information is included by using the external concentration as material parameters in combination with van't Hoff's law $\Delta\pi = R\theta[(c^+ + c^-) - (\bar{c}^+ + \bar{c}^-)]$, where the concentrations $\bar{c}^\delta$ represent the boundary values of the bathing solution.

6 Conclusion

Based on porous media and mixture theories a convenient model for the finite swelling and shrinking of electro-chemically active porous media has been presented. Proceeding from the established Theory of Porous Media (TPM), the governing balance relations and a thermodynamically consistent constitutive setting for the solid and fluid components accounting for electrostatic and osmotic effects have been derived. Recast in variational form, the coupled multi-field problem is efficiently treated within the mixed Finite Element Method (FEM). Finally, four demonstrative numerical examples reveal the principle applicability of the TPM model to real engineering problems. Especially modern bio-materials with multifunctional and adaptive properties such as hydrogels used as tissue replacement or electro-active polymers applied as artificial muscles may guide future improvements of the presented approach.

References

1. Acartürk, A., Simulation of charged hydrated porous materials. Dissertation, Report No. II-18 of the Institute of Applied Mechanics (CE), University of Stuttgart, 2009.
2. Bowen, R.M., Theory of mixtures. In: A.C. Eringen (Ed.), *Continuum Physics*, vol. III, pp. 1–127. Academic Press, New York, 1976.
3. Bowen, R.M., Incompressible porous media models by use of the theory of mixtures. *International Journal of Engineering Sciences*, 18:1129–1148, 1980.
4. Dormieux, L., Lemarchand, E., and Coussy, O.: Macroscopic and micromechanical appproaches to modelling of the osmotic swelling in clays. *Transport in Porous Media*, 50:75–91, 2003.
5. Ehlers, W., Constitutive equations for granular materials in geomechanical context. In: K. Hutter (Ed.), *Continuum Mechanics in Environmental Sciences and Geophysics*, CISM Courses and Lectures No. 337, pp. 313–402. Springer-Verlag, Wien, 1993.
6. Ehlers, W.: Foundations of multiphasic and porous materials. In: W. Ehlers and J. Bluhm (Eds.), *Porous Media: Theory, Experiments and Numerical Applications*, pp. 3–86. Springer-Verlag, Berlin, 2002.
7. Ehlers, W., Challenges of porous media models in geo- and biomechanical engineering including electro-chemically active polymers and gels. *International Journal for Advances in Engineering Sciences and Applied Mathematics*, 1:1–24, 2009.
8. Ehlers, W. and Acartürk, A., The role of weakly imposed Dirichlet boundary conditions for numerically stable computations of swelling phenomena. *Computational Mechanics*, 43:545–557, 2009.
9. Ehlers, W. and Eipper, G.: Finite elastic deformations in liquid-saturated and empty porous solids. *Transport in Porous Media*, 34:179–191, 1999.

10. Ehlers, W., Karajan, N., and Markert, B., An extended biphasic model for charged hydrated tissues with application to the intervertebral disc. *Biomechanics and Modeling in Mechanobiology*, 8:233–251, 2009.
11. Ehlers, W., Markert, B., and Acartürk, A., Swelling phenomena of hydrated porous materials. In: Y.N. Abousleiman, A.H.D. Cheng, and F.J. Ulm (Eds.), *Poromechanics III, Proceedings of the 3rd Biot Conference on Poromechanics*, pp. 781–786. A.A. Balkema Publishers, Leiden, 2005.
12. Heidug, W.K. and Wong, S.: Hydration swelling of water-absorbing rocks: A constitutive model. *International Journal for Numerical and Analytical Methods in Geomechanics*, 20:403–430, 1996.
13. Huyghe, J.M. and Janssen, J.D., Quadriphasic mechanics of swelling incompressible porous media. *International Journal of Engineering Sciences*, 35:793–802, 1997.
14. Huyghe, J.M. and Janssen, J.D., Thermo-chemo-electro-mechanical formulation of saturated charged porous solids. *Transport in Porous Media*, 34:129–141, 1999.
15. Lai, W.M., Hou, J.S., and Mow, V.C., A triphasic theory for the swelling and deformation behaviours of articular cartilage. *ASME Journal of Biomechanical Engineering*, 113:245–258, 1991.
16. Lanir, Y., Biorheology and fluidflux in swelling tissues. I. Biocomponent theory for small deformations, including concentration effects. *Biorheology*, 24:173–187, 1987.
17. Markert, B., A constitutive approach to 3-d nonlinear fluid flow through finite deformable porous continua. *Transport in Porous Media*, 70:427–450, 2007.
18. Markert, B., A biphasic continuum approach for viscoelastic high-porosity foams: Comprehensive theory, numerics, and application. *Archive for Computational Methods in Engineering*, 15:371–446, 2008.
19. Mow, V.C., Ateshian, G.A., Lai, W.M., and Gu, W.Y., Effects on fixed charges on the stress-relaxation behavior of hydrated soft tissues in a confined compression problem. *International Journal of Solids and Structures*, 35:4945–4962, 1998.
20. Mow, V.C., Sun, D.D., Guo, X.E., Likhitpanichkul, M., and Lai, W.M., Fixed negative charges modulate mechanical behaviours and electrical signals in articular cartilage under confined compression. In: W. Ehlers and J. Bluhm (Eds.), *Porous Media: Theory, Experiments and Numerical Applications*, pp. 227–247. Springer-Verlag, Berlin, 2002.

Propagating Cracks in Saturated Ionized Porous Media

F. Kraaijeveld and J.M. Huyghe

Materials Technology, Eindhoven University of Technology, P.O. Box 513, 5600 MB Eindhoven, The Netherlands; famkekraaijeveld@gmail.com, j.m.r.huyghe@tue.nl

Abstract Ionized porous media swell or shrink under changing osmotic conditions. Examples of such materials are shales, clays, hydrogel and biological tissues. The presence of the fixed charges causes an osmotic pressure difference between the material and surrounding fluid and concomitantly prestressing of the material. Understanding the mechanisms for fracture and failure of these materials are important for design of oil recovery, medical treatment and materials. The aim has therefore been to study with the Finite Element Method the effect of osmotic conditions on propagating discontinuities.

The work uses the partition of unity modeling of a crack in a swelling medium. The modeling of the fluid flow around the crack is essentially different for mode-I compared to mode-II. In mode-I, the pressure is assumed continuous in the crack area, while in mode-II the pressure is assumed discontinuous across the crack. Step-wise crack propagation through the medium is observed both for mode-II as for mode-I. Furthermore, propagation is shown to depend on the osmotic prestressing of the medium. In mode-II the prestressing has an influence on the angle of growth. In mode-I, the prestressing is found to enhance crack propagation or protect against failure depending on the load and material properties.

Key words: Porous media, crack propagation, partition of unity, cohesive zone, step-wise propagation

1 Introduction

Shales, clays, hydrogel and biological tissues swell and shrink under changing osmotic conditions. These materials are described as multi-phase materials consisting of a solid part with ions embedded in the porous solid matrix (fixed charges) and a fluid part containing mobile counter charges. That is why they are characterized as ionized porous media. The presence of the fixed charges causes an osmotic pressure difference between the material and surrounding fluid resulting in prestressing of the material. In addition, the response of the material to mechanical and chemical load is highly influenced by the presence of prestress, fluid and cracks. Ionized porous media can undergo phenomena like instability, localization and even fracturing under changing osmotic conditions [26, 33]. Understanding of the mechanisms for fracture

R. de Borst and E. Ramm (eds.), *Multiscale Methods in Computational Mechanics*, Lecture Notes in Applied and Computational Mechanics 55, DOI 10.1007/978-90-481-9809-2_21,

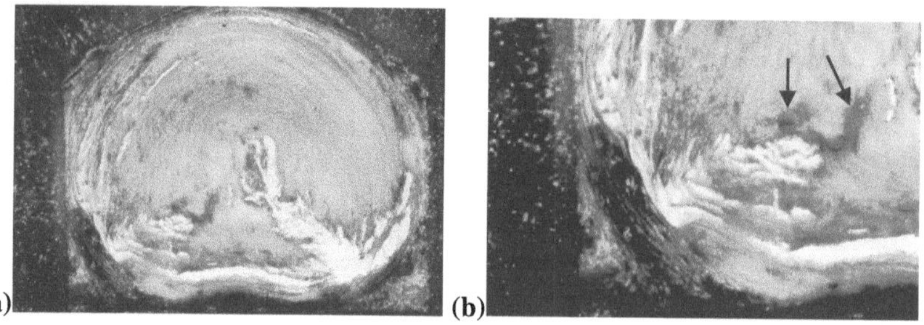

(a) (b)

Fig. 1. (a) Herniation of a full disc. (b) Close-up from a herniation with micro-damage at the crack-tip pointed out by the arrows. (Courtesy of T. Videman)

and failure of these materials together with the fluid redistribution are important for material and construction design.
Experiments in these type of materials can be expensive and difficult. Therefore experiments are usually complemented by computational studies. There is extensive literature on fracture in solids, but the relationship between the presence of cracks and fluid flow has had little attention and the relation between failure and osmotic conditions has had even less attention. The aim has therefore been to study the effect of osmotic conditions on propagating discontinuities under different types of loads for saturated ionized porous media using the Finite Element Method (FEM).

Fractures of interest in this case are discrete cracks, for instance herniation (Figure 1). A small damage zone, the so-called process zone, ahead of the growth of the macro-crack is seen in which micro-separations and crack-bridging (fibers running from one crack-surface to the other) takes place (Figure 1b). This paragraph will treat the modeling of fractures of these kinds.

Models based on linear elastic fracture mechanics (LEFM) cannot capture this nonlinear behavior in the process zone. Since a prominent crack is present, the discontinuity kinematics is preferably not spread over a finite volume and a discrete approach is required. Several methods have been developed in solid mechanics, such as embedded discontinuity model and the use of geometrical entities: interface elements. Unfortunately, these methods seem not to overcome mesh-dependencies [8].

An alternative is adaptive remeshing, but this method requires extensive procedures. Alternatively, partition of unity or extended finite element methods [3, 5] allow for the inclusion of discontinuous functions to the standard shape functions of the FE. The use of these enhanced functions makes it possible to preserve continuity along the crack surface. The discontinuity is projected on the element by additional degrees of freedom and not nodes. It is not necessary to align elements with the crack path, no dummy stiffness is needed nor remeshing is required. Furthermore, it allows for standard discretization. The partition of unity is a powerful property which can also be used for other scale separation than crack growth, for instance material separation [12, 13]. The benefit of this method is also that relatively coarse meshes (and with that relative large time stepping) is used. Unfortunately, the method is not easy to implement in commercial codes.

The quasi-brittle property of the crack means that before growth micro-damage occurs ahead of the macro crack. A widely accepted method to include the damage is the cohesive zone model. Barenblatt [4], Dugdale [9] and Hillerborg [14], were the pioneers to cohesive fracture in solids. The cohesive zone model lumps all damage preceding crack propagation to a model acting on a virtual extension of the crack. This model relates the material integrity (traction forces) to deformation of the crack. Where in previous research the cohesive zone model has been introduced either at all element boundaries [34] or in the path of interest [25, 27], the application of partition of unity on discontinuity kinematics allowed for the cohesive zone to cross continuum elements [20, 21, 32].

An increasing amount of papers is published on applying these tools in biomechanics [11], in which fluid contribution is unfortunately usually neglected. The presence of fluid highly affects the mechanical response of the structure, fracture initiation and fracture growth. Most computational poromechanical models involving strong discontinuities consider shear banding (mode-II) in porous media where fluid exchange from one crack surface to the other is assumed continuous [2] or decreased [22]. The partition of unity method has been combined with Delta-dirac function [19] for simulating an undrained localization band.

In tensile mode (mode-I), hydraulic fracturing has been studied where benchmarks are available when assuming little exchange of fluid between the fracture and surrounding fluid [6, 7, 10]. With efficient mesh generators and transfer schemes, fracturing of porous media has been successfully modeled by adaptive meshing without making assumptions on the shape of the pressure field or displacement [28–30]. Mesh refinement plays a large role in these models.

The partition of unity method allows for relatively coarse meshes in solid mechanics. Capturing the pressure field corresponding to a displacement discontinuity is not straightforward. Attempts for combining partition of unity model with porous media in tensile mode are a double porosity model for the fluid flow without enhancing the fluid phase [1], a moisturizing model without enhancing the fluid phase [24] and a weak discontinuity model for the fluid phase with Couette flow in the crack [23]. Still, numerical simulations of a fracturing process in a porous medium are limited and it is difficult to verify and validate model performance with a benchmark. In any case no osmotic effects were taken into account.

In this chapter, both shearing and tensile mode are addressed as separate cases, since the different deformations cause different fluid flows.

2 Bulk Material

The governing equations consist of equations for the bulk and for the discontinuity, dominated by mass balance, momentum balance and constitutive behavior. Figure 2 shows a body Ω with external boundary Γ with a traction force on Γ_t and fluid supply on Γ_f, with $\mathbf{n}$ the normal unit vector on the boundary Γ directed outwards. The body is cut by a discontinuity Γ_d in two domains, Ω^+ and Ω^-. The normal of the discontinuity $\mathbf{n}_\Gamma$ is directed towards Ω^+.

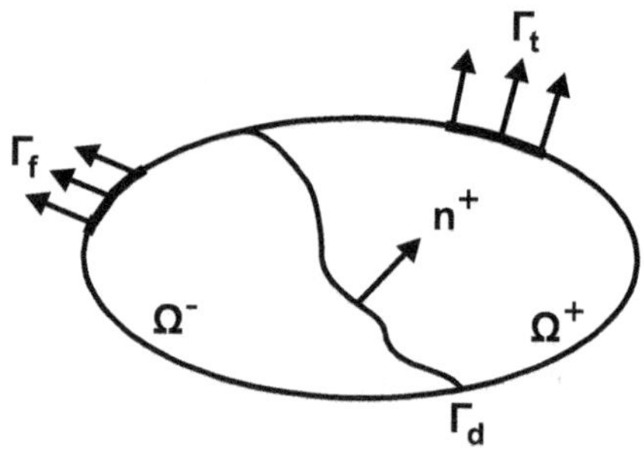

Fig. 2. Schematic representation of the body Ω: the two parts are Ω^+ and Ω^-, separated by crack Γ_{d}. A traction force holds on Γ_{t} and fluid supply at Γ_{f}.

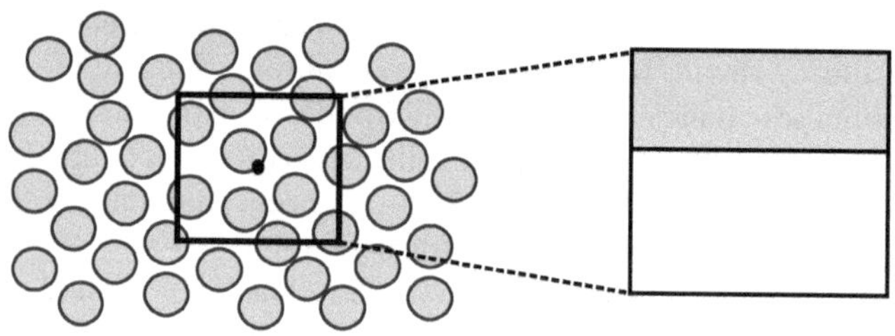

Fig. 3. Terzaghi: quantities are averaged over a volume RVU, and represented by volume fractions of the components.

To simulate osmotically swelling porous media Lanir's model is used for the case of plane strain [18]. This model assumes that free ions are always in equilibrium with the external salt concentration. Ion contribution is therefore neglected and the medium is described by two constituents only: the solid (s) and the fluid (f). Instead of describing each particle in the medium, the properties of the constituents are spread over a representative volume unit (RVU), represented in Figure 3. The constituents are assumed to be incompressible. The material is assumed linear-elastic, isothermal, isotropic, homogeneous and fully saturated.

In addition, Terzaghi's rule of effective stress σ_e [31] is used. This rule means that the stress is split in two parts, the effective stress (σ_{e}) and hydraulic pressure (p). The effective stress contains all deformation dependent effects. If inertia and body forces are neglected, the momentum balance for fully saturated porous media reads

$$\nabla \cdot \sigma_{\mathrm{e}} - \nabla p = \mathbf{0}. \tag{1}$$

Here, effective stress is assumed to follow Hooke's law. The other relation driving load (re-)distribution in porous media is the mass balance. This balance relates seepage flux $\mathbf{q}$ to the permeability of the matrix and gradients according to Darcy's law.

The presence of the fixed charge causes a deformation dependent pressure difference between the sample and surrounding fluid. The chemical potential of the fluid is assumed to be the driving force for fluid flow. This chemical potential is a measure for the free energy of the fluid. The chemical potential of the fluid μ^{f} is defined

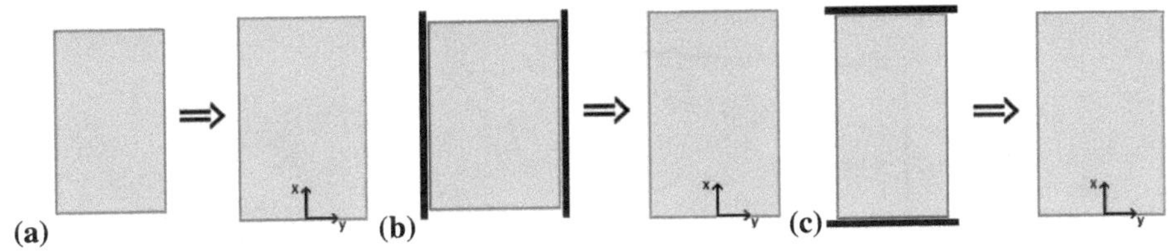

Fig. 4. Representation of the three different types of prestress, namely as result of free swelling (a) in both directions, (b) in x-direction and (c) in y-direction.

per unit volume fluid and relates to hydrostatic pressure by (gravitational forces are neglected)

$$\mu^{\mathrm{f}} = p - \Delta\pi. \tag{2}$$

The osmotic pressure π is determined by the empirical Van 't Hoff equation. This relation defines the osmotic pressure in terms of concentrations of free cations, c^{+}, anions c^{-}, gas constant R and temperature T and depends indirectly on the fixed charge density c^{fc} and the external salt concentration c^{ex}. Van 't Hoff equation is given by

$$\Delta\pi = \pi - \pi^{ex} = RT\sqrt{(c^{\mathrm{fc}})^2 + 4(c^{\mathrm{ex}})^2} - 2RTc^{\mathrm{ex}}. \tag{3}$$

In the external salt solution, this relation holds also, but with the fixed charge density c^{fc} zero and the osmotic coefficient may be different. Electro-neutrality holds, therefore the amount of negative charges are equal to the amount of positive charges: $c^{-} + c^{\mathrm{fc}} = c^{+}$.

The presence of ions fixed to the solid matrix results in prestress of the solid matrix at initial condition. This initial stress is associated with an initial straining of the solid matrix. This strain does not have to be isotropic, while it could be limited by the surroundings. In this chapter three types of prestress are considered (Figure 4), namely the initial condition is the result of swelling in both directions (free swelling), in only horizontal direction or in only vertical direction.

In general, the constitutive relations hold relative to stress free configuration, i.e. without any prestrain in the solid matrix. Therefore, the strain tensor ϵ is separated in an initial strain ϵ_{i} and the deformation from initial to current state, $\epsilon_{\Delta} = \epsilon - \epsilon_{\mathrm{i}} = \nabla^{s}\mathbf{u} = 1/2\{\nabla\mathbf{u} + (\nabla\mathbf{u})^{T}\}$. Similarly, the fixed charge density is calculated from the fixed charge density at stress-free configuration c_0^{fc} and initial volume fraction $\varphi_{\mathrm{i}}^{\mathrm{f}}$. Equations (4) summarize the governing equations for the bulk material:

$$\begin{aligned} \frac{\partial tr(\epsilon)}{\partial t} - \nabla^{s}\cdot(\mathsf{K}\cdot\nabla^{s}\mu^{\mathrm{f}}) &= 0, \\ 2\mu\nabla^{s}\cdot\epsilon + \lambda\nabla^{s}tr(\epsilon) - \nabla^{s}(\mu^{\mathrm{f}} + \Delta\pi(\epsilon)) &= \mathbf{0}, \end{aligned} \tag{4}$$

with λ and μ the Lamé constants and $\mathsf{K} = K\mathsf{I}$ isotropic permeability tensor. Whilst the system is time dependent, Euler implicit is used to calculate subsequent time increments.

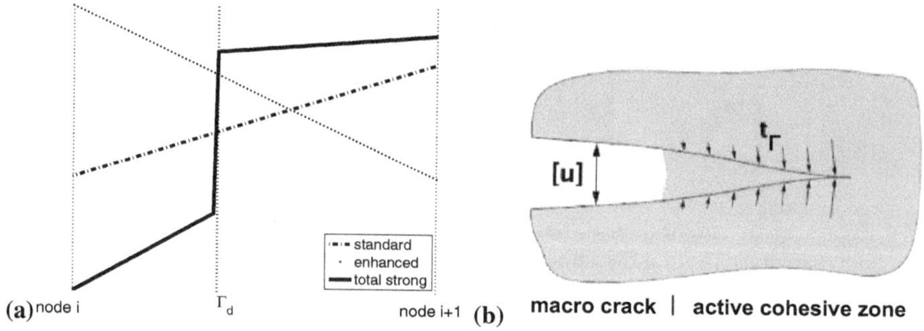

Fig. 5. (a) A 1-D representation of the effect of the Heaviside function on a linear shape function. The total field built from the standard field and the enhanced field around the crack Γ_d. (b) Schematic representation of the cohesive zone.

3 Discontinuity in the Solid Part

Cracks are mathematically represented by discontinuities in the solid matrix. The treatment of the solid part is similar for shear loading and for tensile loading.

The opening or sliding of a crack is incorporated as a jump in the displacement. These jumps are included by employing the partition of unity property [3]. A jump in displacement is mathematically represented by a function operating on a smooth continuous function: the Heaviside function. Then the displacement field $\mathbf{u}$ of the body is additively decomposed into the continuous part $\hat{\mathbf{u}}$ and the enhanced part $\tilde{\mathbf{u}}$ [21]:

$$\mathbf{u}(\mathbf{x}) = \hat{\mathbf{u}}(\mathbf{x}, t) + \mathcal{H}_{\Gamma_d}(\mathbf{x})\tilde{\mathbf{u}}(\mathbf{x}, t). \quad (5)$$

The Heaviside function $\mathcal{H}_{\Gamma_d}$ is defined by (6) with the jump at the discontinuity (with h the unit jump size):

$$\mathcal{H}_{\Gamma_d} = \begin{cases} -h/2 \; \mathbf{x} \in \Omega^+ \\ +h/2 \; \mathbf{x} \in \Omega^- \end{cases} \quad (6)$$

The jump at the discontinuity Γ_d is given by $[\mathbf{u}]$ and represents the opening of the crack:

$$[[\mathbf{u}(\mathbf{x}, t)]] = h\tilde{\mathbf{u}}(\mathbf{x}, t), \quad \mathbf{x} \in \Gamma_d. \quad (7)$$

This is represented in 1D by Figure 5a. A local coordinate system (n,s) is introduced where component n is in the direction of the normal vector $\mathbf{n}^+$ to Ω^+ and component t is orthogonal in the direction of propagation $\mathbf{t}^+$.

3.1 Cohesive Zone

A discrete crack is preceded by local damage. This micro-damage is lumped into one constitutive relation (cohesive zone) and projected on the crack [21], Figure 5b. The model relates the decohesion, i.e. softening of the traction forces, as result of opening of the crack. This behavior is typically nonlinear and therefore Newton–Rhapson procedure is used to calculate the solution for each time increment.

For the traction forces at each surface holds $\sigma \cdot \mathbf{n}^+ = \mathbf{t}_\Gamma^+ = -\mathbf{t}_\Gamma^- = -\sigma \cdot \mathbf{n}^-$ and

$$(\sigma_e - (\mu^f + \Delta\pi)\mathrm{I})^\pm \cdot \mathbf{n}^\pm = \mathbf{t}_\Gamma^\pm(\mathbf{x}, t), \ \mathbf{x} \in \Gamma^\pm \tag{8}$$

The cohesive law holds locally and is described in the local coordinate system of the discontinuity. The cohesive zone model is usually nonlinear and acts on the effective stress. The system is linearized in order to use Newton–Rhapson. The following equation holds for the change in traction forces w.r.t. previous step ($\Delta\mathbf{t}_d$)

$$\Delta\mathbf{t}_d = (\mathbf{n}^+ \ \mathbf{t}^+)\underset{\sim}{t}_d, \ \Delta\underset{\sim}{t}_d = h\underline{T}_d\underset{\sim}{u}_d, \tag{9}$$

with $\underset{\sim}{t}_d$ column vector of the traction forces and $\underset{\sim}{u}_d$ column vector of displacement in the local coordinates. Matrix $\underline{T}_d$ is the tangent stiffness of the cohesive zone model. A transition of local to global coordinates is made to express the traction forces in terms of nodal displacements. The change in traction forces $\Delta\mathbf{t}_\Gamma^\pm$ in global coordinates is given by

$$\Delta\mathbf{t}_d^\pm = (\mathbf{e}_1 \ \mathbf{e}_2)\Delta\underset{\sim}{t}_d^\pm. \tag{10}$$

Then the transition from local coordinates to global coordinates occurs through a mapping $\underline{Q}$

$$(\mathbf{e}_1 \ \mathbf{e}_2) = (\mathbf{n}^+ \ \mathbf{t}^+)\underline{Q}^T. \tag{11}$$

The change in traction forces $\Delta\mathbf{t}_\Gamma^\pm$ in the local coordinates is expressed in terms of enhanced nodal displacements (with $\Delta\pi^+ = 0$ at the crack surface and $T_{e,d}$ the stiffness tangent of the cohesive law in local coordinates):

$$\Delta\underset{\sim}{t}_\Gamma^+ = \underline{Q}\underset{\sim}{t}_d^+ - (\mu^f + \Delta\pi)^+\underset{\sim}{n}^+ = \underline{Q}\underline{T}_{e,d}\underline{Q}^T\underset{\sim}{\mathbf{u}} - \mu^{f+}\underset{\sim}{n}^+ \tag{12}$$

Then the tangent stiffness in element coordinates is given by $\underline{T}_e = \underline{Q}\underline{T}_{e,d}\underline{Q}^T$. In general the cohesive zone parameters, such as the fracture energy $\mathcal{G}_c$ and the ultimate strength t_{ult}, are determined from experiments.

3.2 Nonlocal Stress

Crack growth is determined by damage in the solid matrix. Therefore the yield criterion, next to the cohesive zone, is related to the effective stress. The effective stress at the crack-tip varies locally, therefore the critical effective stress state is calculated nonlocally using Gaussian functions following Wells et al. [32].

$$\begin{aligned}
\sigma_{tip} &= \sum_{i=1}^{n_{tot}} \frac{w_i}{w_{tot}}\sigma_i \\
w_{tot} &= \sum_{j=1}^{n_{tot}} w_j \\
w_i &= \frac{(2\pi)^{2/3}}{l_a^3} e^{-\frac{r_i^2}{2l_a^2}}.
\end{aligned} \tag{13}$$

Herein, r_i is the distance between integration point i and the crack-tip and l_a is a length scale parameter which determines the influence of a sample point.

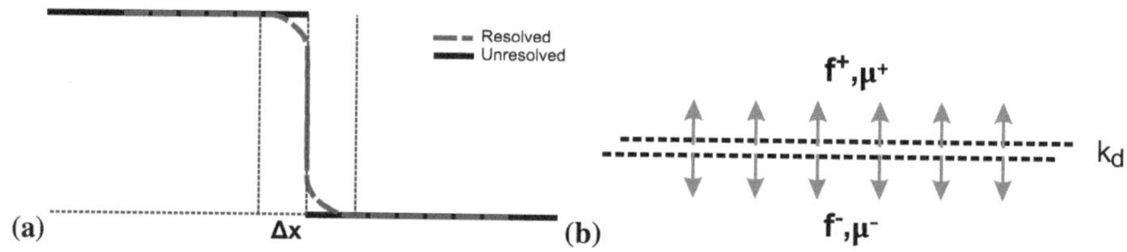

Fig. 6. (a) Representation of the gradient in the chemical potential near a discontinuity approximated by a jump. (b) Schematic representation of the fluid flow at the crack surface.

4 Shearing Mode

Shear failure is a common failure mode in porous media. In this section, shear loading of one prominent quasi-brittle crack is considered. The cohesive zone method is applied together with the partition of unity method for simulating crack propagation. Compression of the material results in fluid flow. The modeling of the fluid flow in shearing mode is essential different from tensile failure. In this case, the crack is closed and no additional relations are needed to model the fluid flow in the crack such as in tensile mode.

4.1 Discontinuity in the Fluid Part

On initial growth a small opening arises, i.e. small displacement jump, and a steep gradient in the chemical potential occurs over a small transition zone (Figure 6a). While the opening grows, the gradient in the chemical potential decreases in magnitude and spreads over a wider transition zone. High gradients as result of shear loading are approximated by a jump $\tilde{\mu}^{\mathrm{f}}(\mathbf{x}, t)$ over the crack surfaces, Figure 6.

In this case the decomposition becomes

$$\mu^{\mathrm{f}}(\mathbf{x}) = \hat{\mu}^{\mathrm{f}}(\mathbf{x}, t) + \mathscr{H}_{\Gamma_{\mathrm{d}}}(\mathbf{x})\tilde{\mu}^{\mathrm{f}}(\mathbf{x}, t) \tag{14}$$

and the difference in chemical potential at the discontinuity is given by $[[\mu^{\mathrm{f}}]]$:

$$[[\mu^{\mathrm{f}}(\mathbf{x}, t)]] = h\tilde{\mu}^{\mathrm{f}}(\mathbf{x}, t), \quad \mathbf{x} \in \Gamma_{\mathrm{d}}. \tag{15}$$

Figure 6 shows a schematic overview of the local mass balance. The flow across each of the crack surfaces ($f_{\Gamma}^{\pm}$) are equal and depend on the local pressure gradient and material properties. When the flow is resolved, it follows locally

$$\mathbf{q} \cdot \mathbf{n}^{+} = f_{\Gamma}^{+}(\mathbf{x}, t) = -f_{\Gamma}^{-}(\mathbf{x}, t) = -K\frac{\mu^{\mathrm{f}+} - \mu^{\mathrm{f}-}}{2\Delta x} \tag{16}$$

where Δx is the distance the fluid flow has overcome and depends on the time Δt since discontinuity developed at that place, the bulk permeability K and aggregate modulus of the material ($c = 2\mu + \lambda - \frac{\partial\pi}{\partial tr(\epsilon)}$), namely $2\Delta x^2 \approx \Delta t K c$. Then for the flow it holds that

Table 1. Material properties for compression test.

R	= 8.3145	[N mm/mmol K]	T	= 298	[K]
E	= 90.0	[MPa]	ν	= 0.20	[-]
$\varphi_{\mathrm{i}}^{\mathrm{f}}$	= 0.80	[-]	K	= 0.28e-3	[mm^4/Ns]
c^{ex}	= 0.15e-3	[mmol/mm^3]	$c_{\mathrm{i}}^{\mathrm{fc}}$	= −0.2e-3	[mmoleq/mm^3]
K_{d}	= 0.28e-3	[mm^3/Ns]			
$\mathcal{G}_c$	= 2.0	[J/m^2]	τ_{ult}	= 0.4	[MPa]
l_a	= 0.2	[mm]	v	= 1.5e-3 mm/s	

$$f_{\Gamma}^{+} = -k_{\mathrm{d}}(\mu^{\mathrm{f}+} - \mu^{\mathrm{f}-}) = -k_{\mathrm{d}}[[\mu^{\mathrm{f}}]] = -hk_{\mathrm{d}}\tilde{\mu}^{\mathrm{f}}. \tag{17}$$

Equation (17) is Darcy's equation on the unresolved scale and should correspond to the same equation on the resolved scale Eq. (16). This is true if – unlike the paper [22] where a diaphragm in the crack is assumed – we choose the unresolved hydraulic permeability as

$$k_{\mathrm{d}} = \sqrt{\frac{K}{c\Delta t}}. \tag{18}$$

The unresolved hydraulic permeability is therefore derived from intrinsic material properties. Furthermore, while local deformation is halted, the fluid is still redistributed. This model has been tested by comparison with an analytical solution for a singularity [15, 17].

4.2 Numerical Example

To show that in shearing the osmotic prestress plays an important role a compression test is considered. For the compression test the sample is boxed except for the contact area with the piston. On the right the sample is in contact with a filter, which causes an equilibrium with an external salt solution ($\mu_{\mathrm{in}}^{\mathrm{f}} = \mu_{\mathrm{ex}}^{\mathrm{f}}$). An initial crack is imposed away from element interfaces (Figure 7).

Damage is initiated when maximum shear stress (τ_{max}) is larger than the critical value τ_{ult} (Tresca criterion). For the cohesive damage model an exponential law is used. This law denotes the softening behavior after reaching the critical stress state. Defined is critical length $\delta_s = \mathcal{G}_c/\tau_{ult}$, with τ_{ult} the ultimate traction forces and $\mathcal{G}_c$ fracture toughness. The cohesive law is then

$$t_s = \tau_{ult}\frac{[[u]]_s}{\delta_s}e^{-(\frac{[[u]]_s}{\delta_s})^2} \tag{19}$$

External load is applied through the piston. The piston is moved with constant speed $v = 0.15\text{e-}3$ mm/s. Three types of prestress are considered (Figure 4).

In Figure 8b the resulting distribution of chemical potential is given at different time points. The movement of the piston results in initially straight crack growth which after a while deflects. With prestress in y direction, the crack propagates relatively straight longer than with prestress in x-direction only. The osmotic conditions

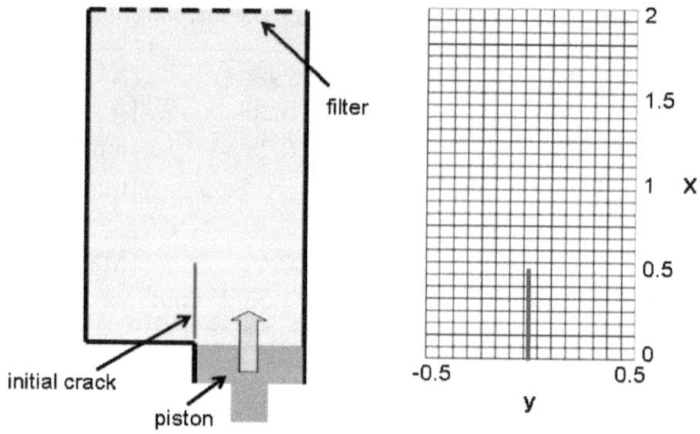

Fig. 7. (a) Schematic representation of the compression test. (b) Representation of the mesh for the compression test.

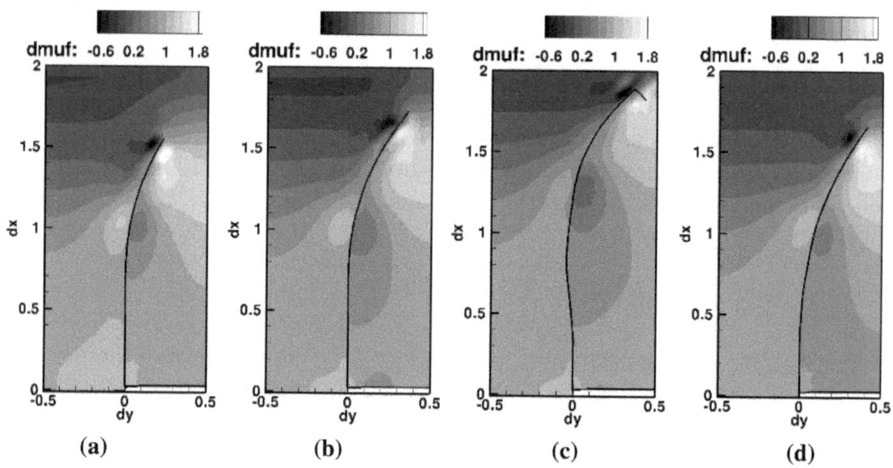

Fig. 8. Distribution of chemical potential (a) for no fixed charges ($\epsilon_{\rm i} = 0$, $dt = 12.5$ s), and for prestress in (b) both directions ($\epsilon_{\rm i}$ <> 0, $dt = 16.6$ s) (c) x-direction only ($\epsilon_{\rm ix}$ <> 0, $dt = 23.3$ s) (d) y-direction only ($\epsilon_{\rm iy}$ <> 0, $dt = 18.9$ s). Crack-path is independent of mesh. Results depend on prestress.

alter the angle of crack propagation when nearing boundaries. In general, a higher prestress perpendicular to the direction of loading causes faster growth and more deflection.

The influence of crack growth on flow is seen when considering fluid flow evolution in Figure 9 in case of free swelling across the crack $\mathbf{q} \cdot \mathbf{n}^+$ for two points, in the initial crack ($dx = 0.28$ mm) and in the crack ($dx = 0.48$ mm). The flow in the initial crack is nonzero. Every time the crack propagates a peak in flow takes place which is also felt in the already existing crack. The crack grows for several elements

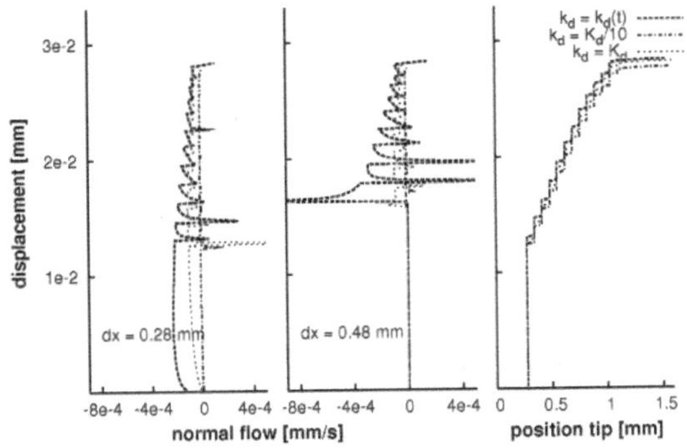

Fig. 9. Normal flow for free swelling along the crack at points $dx = 0.28$ mm and $dx = 0.48$ mm. Crack growth causes a peak in flow after which relaxation takes place. The use of the unresolved hydraulic permeability ($k_d = k_d(t)$) results in a nonzero initial flow and high peaks in case of growth followed by relaxation.

after which stress is built up again. crack propagation alternates with fluid flow. This is the result of consolidation.

When the local permeability is taken constant and similar to bulk permeability ($k_d - K_d$), the overal crack propagation is not changed much, but fluid exchange is lower and more slowly initiated. When the permeability is taken much lower, there is hardly any fluid flow. The use of the unresolved hydraulic permeability enables flow over the crack surface, even in the existing crack.

5 Tensile Mode

Deformation around the discontinuity is strongly linked to fluid flow. Fluid flow takes place at the surface of the discontinuity from the medium into the crack and a flow along the crack when opening of the crack increases.

5.1 Discontinuity in the Fluid Part

Opening of a crack causes a high gradient in the chemical potential at both crack surfaces. In opening mode there is no pressure difference between the crack surfaces but a pressure difference between the crack surface and the middle of the crack, as shown in Figure 10. Using the partition of unity property, a distance function $\mathcal{D}_{\Gamma_d}$ similar to Réthoré et al. [23] (Figure 11a) is introduced [16]. This function allows for the natural introduction of steep gradients without enforcing it. The effect on the basis functions is shown in Figure 11. In this case the decomposition becomes

$$\mu^{\mathrm{f}}(\mathbf{x}) = \hat{\mu}^{\mathrm{f}}(\mathbf{x}, t) + \mathcal{D}_{\Gamma_d}(\mathbf{x})\tilde{\mu}^{\mathrm{f}}(\mathbf{x}, t) \tag{20}$$

The distance function is defined with respect to the coordinates of the crack $\mathbf{x}_{\Gamma}$, namely by

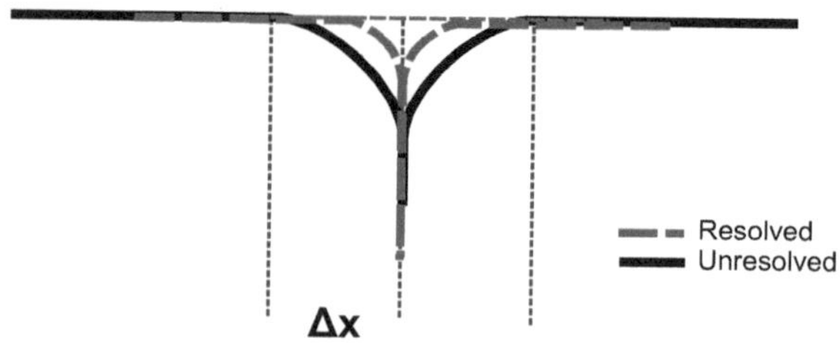

Fig. 10. Distribution of chemical potential as result of crack growth. Crack growth causes an instant peak in flow at the location of the crack (dark grey). This is approximated by a weak discontinuity (lighter grey).

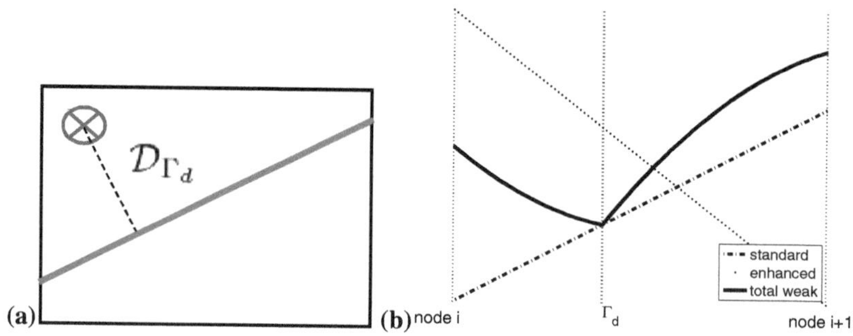

Fig. 11. (a) Distance function for simulating a weak discontinuity. (b) Schematic representation of the enhancement on a linear shape function.

$$\mathcal{D}_{\Gamma_d} = \frac{h}{2}|(\mathbf{x} - \mathbf{x}_\Gamma) \cdot \mathbf{n}^+|, \quad \mathbf{x} \in \Omega. \tag{21}$$

The gradient of the distance function $\partial \boldsymbol{\mathcal{D}}$ is given by

$$\partial \boldsymbol{\mathcal{D}} = \nabla^s \mathcal{D}_{\Gamma_d} = \begin{cases} \frac{h}{2}\mathbf{n}^+ & \mathbf{x} \in \Omega^+ \\ \frac{h}{2}\mathbf{n}^- & \mathbf{x} \in \Omega^- \end{cases} = \mathcal{H}_{\Gamma_d}\mathbf{n}^+ \tag{22}$$

Deformation around the discontinuity is strongly linked to fluid flow. Fluid flow takes place at the surface of the discontinuity from the medium into the crack and a flow along the crack when opening of the crack increases. When the crack is closed, the normal fluid flow $f_c^{\pm}$ over surface $\Gamma_d^{\pm}$ determines the amount of fluid exchange. When the crack opens, additional terms are included. Figure 12 shows a schematic overview of the local mass balance. In case of tensile mode, there is a balance between tangential flow and normal flow.

$$\begin{aligned} \mathbf{q}_\Gamma \cdot \mathbf{n}^+ + \mathbf{q}_\Gamma \cdot \mathbf{n}^- &= f_\Gamma^+ + f_\Gamma^- = -\frac{\partial \mathbf{q}_\Gamma \cdot \mathbf{t}^+}{\partial s} - [[\dot{u}]]_n, \\ \mathbf{q}_\Gamma \cdot \mathbf{t}^+ &= -|[[u]]_n|^k K_d \frac{\partial \mu_\Gamma^f}{\partial s}. \end{aligned} \tag{23}$$

where s represents the distance along the crack, with $s = 0$ the crack-tip and s positive in direction of $\mathbf{t}^+$. Tangential flow assumes Couette flow.

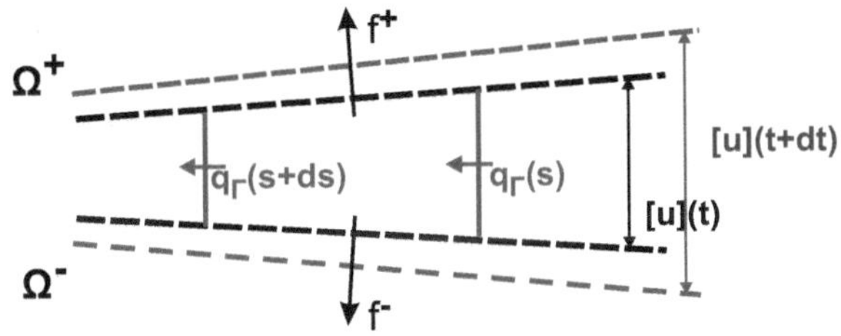

Fig. 12. Schematic representation of the fluid flow at the crack surface with parameter s the distance along the crack.

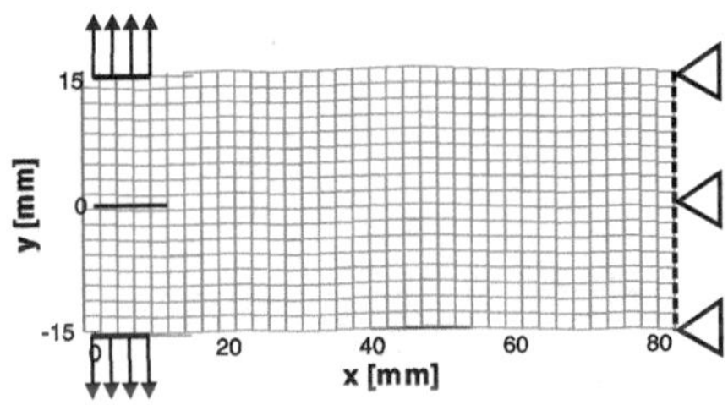

Fig. 13. The mesh and boundary conditions for delamination consisting of 575 elements. Material is pulled at the top and bottom on the left and is on the right in contact with a filter.

5.2 Numerical Example

A delamination test is performed with a predefined angle of 0° (weak interface). An initial defect of length 13 mm is inserted on the lefthandside, Figure 13. The sample is fixed on the righthandside and is in contact with a filter ($\mu^{\mathrm{f}} = 0$). Crack propagation is initiated by pulling the sample on the top and bottom over approximate 8 mm with a fixed velocity of 1.0e-3 mm/s. In the example a cohesive law is used for the damage evolution, i.e. softening behavior after reaching the critical stress state. Defined is critical length $\delta_n = \mathcal{G}_c/\tau_{ult}$, with τ_{ult} the ultimate traction forces and $\mathcal{G}_c$ fracture toughness. The cohesive law is then for positive opening

$$t_n = \tau_{ult} e^{-(\frac{[[u]]_n}{\delta_n})}, \; [[u]]_n \geq 0. \tag{24}$$

The local fluid distribution is determined by Eq. (23).

The exact material properties are given in Table 2. This means that $2RT\Gamma c^{\mathrm{ex}} = 4.96$ N/mm^2 holds. A time step of 0.1 s is used.

Considering the chemical potential distribution, Figure 14, the figures show localization at the crack-tip with a negative chemical potential. This low chemical potential is relaxed by fluid redistribution towards the crack-tip. The chemical potential is largest at the left due to largest opening of the crack.

Crack growth occurs slightly faster in the case of prestress. With further opening of the crack, the chemical potential decreases and the tangential flow increases. Numerical oscillations seem to be present, but the oscillations are actual changes in chemical potential due to crack growth and redistribution of load.

Table 2. Material properties for delamination test.

R	= 8.3145	[N mm mmol^{-1} K^{-1}]	T	= 298	[K]
E	= 1.4e+4	[N/mm^2]	ν	= 0.33	[-]
φ_i^f	= 0.10	[-]	K	= 2.0e-1	[mm^{4}N^{-1}s^{-1}]
c^{ex}	= 1.0e-3	[mmol/mm^3]	c_i^{fc}	= −1.0e-3	[mmoleq/mm^3]
K_d	= 2.0e-1	[mm^{4}N^{-1}s^{-1}]	k	= 2	[-]
$\mathcal{G}_c$	= 0.020	[N/mm]	τ_{ult}	= 1.1	[N/mm^2]
l_a	= 7.8	[mm]	v	= 1.0e-3	[mm/s]

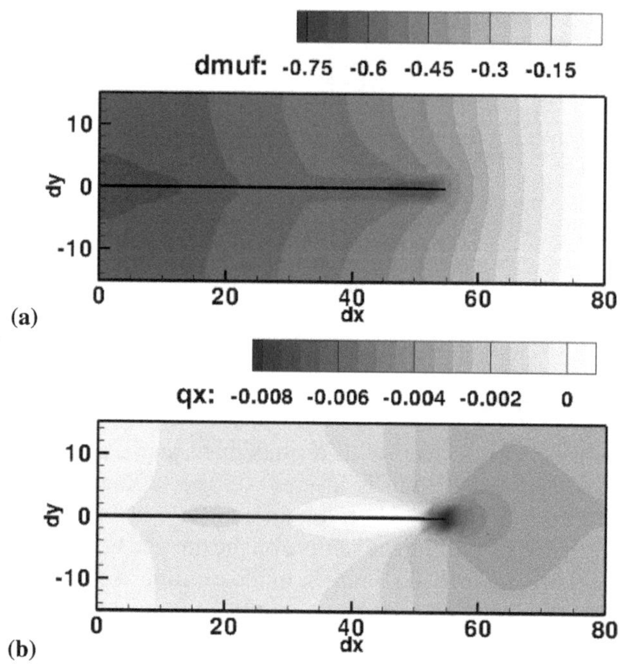

Fig. 14. Delamination in case of prestress in both directions after 950 time increments (i.e. 9.5e-2 mm displacement of top boundary). Distribution of (a) chemical potential in [N/mm^2] and (b) flow in x-direction in [mm/s].

Different loading simulations on less stiff materials has shown that crack propagation causes flow, which is preferable from the crack, since the permeability in the crack is much higher. This can cause the crack to close, hence slowing down crack propagation. But also due to osmotic difference, the crack can temporarily close. In addition, we have seen that for less stiff/more porous materials, the assumption of a weak discontinuity is insufficient to resolve the flow locally.

When a sample is considered without external mechanical load and with an initial crack, the initial crack can open and grow due to a global decrease in osmotic pressure. This phenomenon is similar to the drying of clay, but in fully saturated conditions.

6 Concluding Remarks

In this chapter a short insight in the interchange between fluid flow, osmotic conditions and crack propagation is given. A mesh-independent 2D finite element model of crack propagation in osmotic swelling porous media has been developed. The model has the capability of describing both mode-II and mode-I crack propagation. The numerical results have been compared to a specially designed analytical solution of a dislocation of a non-propagating crack in swelling porous media. The model has been applied to mode-II failure under confined compression and to a delamination test.

The results show step-wise crack propagation through the medium both in mode-I and in mode-II. This is because the propagation of the crack alternates with pauses in which the crack-tip area consolidates. The consolidation results in a progressive transfer of the load from the fluid to the solid. As the load on the solid increases, the failure load is reached and the crack propagates again.

Furthermore, propagation is shown to depend on the osmotic prestressing of the medium. The dependence is present for mode-II and mode-I. In mode-II the prestressing has an influence on the angle of growth. In mode-I, the prestressing is found to enhance crack propagation or protect against failure depending on the load and material properties. It is found that osmotic prestressing in itself can propagate fractures without external mechanical load, such as the spontaneous formation of cracks during the drying of clay, but then in fully saturated conditions. This mechanism may explain the tears observed in intervertebral discs as degeneration progresses.

List of Symbols

$[[*]]$	[-]	Jump over crack surface
c	[MPa]	Deformation constant, $c = 2\mu + \lambda + \frac{\partial \Delta\pi}{\partial tr(\epsilon)}$
c^{ex}	[mmol/mm^3]	External salt concentration
c^{fc}	[mmoleq/mm^3]	Fixed charge concentration:
E	[MPa]	Young's modulus
ϵ	[-]	Strain tensor
$f_\Gamma^\pm$	[mm/s]	Normal flow over crack surface $\Gamma^\pm$
φ^f	[-]	Volume fraction of the fluid
Γ	[-]	External boundary of body Ω
$\Gamma_d^\pm$	[-]	Discontinuity boundaries: $\Gamma^\pm = \Gamma_d \cap \Omega^\pm$
$\mathcal{H}_{\Gamma_d}$	[-]	Heaviside function at discontinuity Γ_d
h	[-]	Size of the Heaviside jump
K	[mm^4/Ns]	Bulk permeability

K_d	[mm^5/Ns]	Crack permeability
k_d	[mm^5/Ns]	Unresolved hydraulic permeability
l_a	[mm]	Nonlocal length, measure for stress singularity
μ^f	[MPa]	Chemical potential
$\mathbf{n}$	[-]	Normal unit vector on Γ, directed outward Ω
$\mathbf{n}^{\pm}$	[-]	Normal unit vector on Γ_d, directed to $\Omega^{\pm}$
∇^s	[mm^{-1}]	Symmetric divergence, $\nabla^s * = 1/2\{\nabla * + (\nabla *)^T\}$
$\Omega^{\pm}$	[-]	Body of interest in direction $\mathbf{n}^{\pm}$ of Γ_d
$\Delta\pi$	[MPa]	Osmotic pressure difference $\pi - \pi_{ex}$
$\mathbf{q}$	[mm/s]	Seepage flux
R	[N mm/mmol K]	Gas constant
$\sigma,\ \sigma_e$	[MPa]	Total and effective stress
T	[K]	Temperature
$\underline{T}_d$	[MPa/mm]	Tangent stiffmess of cohesive law in local coordinates
t	[s]	Time parameter
$\mathbf{t}^{+}$	[-]	Tangential vector unit directed towards propagation
$\mathbf{t}_{\Gamma}^{\pm}$	[-]	Traction force at discontinuity surface $\Gamma^{\pm}$
$\mathbf{u}$	[mm]	Displacement

Acknowledgement

This research is performed as part of the collaboration "Multiscale Methods in Computational Mechanics" funded by the German and Dutch organizations for scientific research, DFG and NWO. The research itself was supported by the Technology Foundation STW, applied science division of NWO and the technology program of the Ministry of Economic Affairs. Further financial support was kindly provided by the Royal Dutch Shell.

References

1. R. Al-Khoury and L.J. Sluys. A computational model for fracturing porous media. *International Journal for Numerical Methods in Engineering*, 70(4):423–444, 2007.
2. F. Armero and C. Callari. An analysis of strong discontinuities in a saturated poro-plastic solid. *International Journal for Numerical Methods in Engineering*, 46(10):1673–1698, 1999.
3. I. Babuska and J.M. Melenk. The partition of unity method. *International Journal for Numerical Methods in Engineering*, 40(4):727–758, 1997.
4. G.I. Barenblatt. The mathematical theory of equilibrium cracks in brittle fracture. *Advances in Applied Mechanics*, (7):55–129, 1962.
5. T. Belytschko and T. Black. Elastic crack growth in finite elements with minimal remeshing. *International Journal for Numerical Methods in Engineering*, 45(5):601–620, 1999.
6. T.J. Boone and A.R. Ingraffea. A numerical procedure for simulation of hydraulically-driven fracture propagation in poroelastic media. *International Journal for Numerical and Analytical Methods in Geomechanics*, 14(1):27–47, 1990.

7. T.J. Boone, A.R. Ingraffea, and J.C. Roegiers. Simulation of hydraulic fracture propagation in poroelastic rock with application to stress measurement techniques. *International Journal of Rock Mechanics and Mining Sciences & Geomechanics Abstracts*, 28(1):1–14, 1991.
8. R. de Borst, J.J.C. Remmers, A. Needleman, and M.A. Abellan. Discrete vs smeared crack models for concrete fracture: Bridging the gap. *International Journal for Numerical and Analytical Methods in Geomechanics*, 28(7/8):583–607, 2004.
9. D.S. Dugdale. Yielding of steel sheets containing slits. *Journal of the Mechanics and Physics of Solids*, 8(2):100–104, 1960.
10. S.H. Emerman, D.L. Turcotte, and D.A. Spence. Transport of magma and hydrothermal solutions by laminar and turbulent fluid fracture. *Physics of the Earth and Planetary Interiors*, 41(4):249–259, 1986.
11. T.C. Gasser and G.A. Holzapfel. Modeling plaque fissuring and dissection during balloon angioplasty intervention. *Annals of Biomedical Engineering*, 35(5):711–723, 2007.
12. T. Hettich, A. Hund, and E. Ramm. Modeling of failure in composites by x-fem and level sets within a multiscale framework. *Computer Methods in Applied Mechanics and Engineering*, 197(5):414–424, 2008.
13. T. Hettich and E. Ramm. Interface material failure modeled by the extended finite-element method and level sets. *Computer Methods in Applied Mechanics and Engineering*, 195(37-40):4753–4767, 2006.
14. A. Hillerborg, M. Modeér, and P.E. Petersson. Analysis of crack formation and crack growth in concrete by means of fracture mechanics and finite elements. *Cement and Concrete Research*, (6):773–782, 1976.
15. F. Kraaijeveld, J.M. Huyghe, and F.P.T. Baaijens. Singularity solution of Lanir's osmoelasticity: Verification of discontinuity simulations in soft tissues. *Journal of Biomechenical Engineering*, 2009.
16. F. Kraaijeveld, J.M. Huyghe, J.J.C. Remmers, R. de Borst, and F.P.T.Baaijens. A mesh-independent model for mode-*i* fracture in osmoelastic saturated porous media. *International Journal for Numerical Methods in Engineering*, 2009.
17. F. Kraaijeveld, J.M. Huyghe, J.J.C. Remmers, R. de Borst, and F.P.T. Baaijens. Shear fracture in osmoelastic saturated porous media: A mesh-independent model. *Engineering Fracture Mechanics*, 2009.
18. Y. Lanir. Biorheology and fluid flux in swelling tissues. 1. Bicomponent theory for small deformations, including concentration effects. *Biorheology*, 24(2):173–187, 1987.
19. J. Larsson and R. Larsson. Localization analysis of a fluid-saturated elastoplastic porous medium using regularized discontinuities. *Mechanics of Cohesive-Frictional Materials*, 5(7):565–582, 2000.
20. N. Moes and T. Belytschko. Extended finite element method for cohesive crack growth. *Engineering Fracture Mechanics*, 69(7):813–833, 2002.
21. J.J.C. Remmers, R. de Borst, and A. Needleman. A cohesive segments method for the simulation of crack growth. *Computational Mechanics*, 31(1-2):69–77, 2003.
22. J. Rethore, R. de Borst, and M.A. Abellan. A discrete model for the dynamic propagation of shear bands in a fluid-saturated medium. *International Journal for Numerical and Analytical Methods in Geomechanics*, 31(2):347–370, 2007.
23. J. Rethore, R. de Borst, and M.A. Abellan. A two-scale approach for fluid flow in fractured porous media. *International Journal for Numerical Methods in Engineering*, 71(7):780–800, 2007.
24. S. Roels, P. Moonen, K. De Proft, and J. Carmeliet. A coupled discrete-continuum approach to simulate moisture effects on damage processes in porous materials. *Computer Methods in Applied Mechanics and Engineering*, 195(52):7139–7153, 2006.

25. J.G. Rots. Smeared and discrete representation of localized fracture. *International Journal of Fracture*, 51:45–59, 1991.
26. F.J. Santarelli, D. Dahen, H. Baroudi, and K.B. Sliman. Mechanisms of borehole instability in heavily fractured rock media. *International Journal of Rock Mechanics and Mining Sciences & Geomechanics Abstracts*, 29(5):457–467, 1992.
27. J.C.J. Schelleken and R. de Borst. Free edge delamination in carbon-epoxy laminates: a novel numerical/experimental approach. *Composite Structures*, 28:357–373, 1994.
28. B.A. Schrefler, S. Secchi, and L. Simoni. On adaptive refinement techniques in multi-field problems including cohesive fracture. *Computer Methods in Applied Mechanics and Engineering*, 195(4-6):444–461, 2006.
29. S. Secchi, L. Simoni, and B.A. Schrefler. Mesh adaptation and transfer schemes for discrete fracture propagation in porous materials. *International Journal for Numerical and Analytical Methods in Geomechanics*, 31(2):331–345, 2007.
30. L. Simoni and S. Secchi. Cohesive fracture mechanics for a multi-phase porous medium. *Engineering Computations*, 20(5/6):675–698, 2003.
31. K. Terzaghi. *Theoretical Soil Mechanics*. John Wiley and Sons, New York, 1943.
32. G.N. Wells and L.J. Sluys. Discontinuous analysis of softening solids under impact loading. *International Journal for Numerical and Analytical Methods in Geomechanics*, 25(7):691–709, 2001.
33. S. Wognum, J.M. Huyghe, and F.P.T. Baaijens. Influence of osmotic pressure changes on the opening of existing cracks in 2 intervertebral disc models. *Spine*, 31(16):1783–1788, 2006.
34. X.P. Xu and A. Needleman. Void nucleation by inclusion debonding in a crystal matrix. *Modelling and Simulation in Materials Science and Engineering*, 1(2):111–132, 1993.

Author Index

Subject Index

Kolibri 360 GmbH
Ferdinand-Jühlke-Straße 7
99095 Erfurt, Deutschland
produktsicherheit@kolibri360.de